Non-Invasive Health Systems based on Advanced Biomedical Signal and Image Processing

This book contains up-to-date noninvasive monitoring and diagnosing systems closely developed by a set of scientists, engineers, and physicians. The chapters are the results of different biomedical projects and theoretical studies that were coupled by simulations and real-world data.

Non-Invasive Health Systems based on Advanced Biomedical Signal and Image Processing provides a multifaceted view of various biomedical and clinical approaches to health monitoring systems. The authors introduce advanced signal- and image-processing techniques as well as other noninvasive monitoring and diagnostic systems such as inertial sensors in wearable devices and novel algorithm-based hybrid learning systems for biosignal processing. The book includes a discussion of designing electronic circuits and systems for biomedical applications and analyzes several issues related to real-world data and how they relate to health technology including ECG signal monitoring and processing in the operating room. The authors also include detailed discussions of different systems for monitoring various conditions and diseases including sleep apnea, skin cancer, deep vein thrombosis, and prosthesis controls.

This book is intended for a wide range of readers including scientists, researchers, physicians, and electronics and biomedical engineers. It will cover the gap between theory and real life applications.

Prospects in Biomedical Engineering and Applications

Series Editors:
Mohammad M. Banat, Jordan University of Science and Technology, Jordan
Adel Al-Jumaily, Faculty of Engineering, University of Technology Brunei

5G Impact on Biomedical Engineering
Jacques Bou Abdo, Jacques Demerjian, and Abdallah Makhou

Non-Invasive Health Systems based on Advanced Biomedical Signal and Image Processing
Edited by Adel Al-Jumaily, Paolo Crippa, Ali Mansour, and Claudio Turchetti

For more information about this series, please visit: https://www.crcpress.com/ Prospects-in-Biomedical-Engineering-and-Applications/book-series/PBIOMED

Non-Invasive Health Systems based on Advanced Biomedical Signal and Image Processing

Edited by
Adel Al-Jumaily, Paolo Crippa,
Ali Mansour, and Claudio Turchetti

CRC CRC Press
Taylor & Francis Group
Boca Raton London New York

CRC Press is an imprint of the
Taylor & Francis Group, an **informa** business

Designed cover image: Shutterstock

MATLAB® is a trademark of The MathWorks, Inc. and is used with permission. The MathWorks does not warrant the accuracy of the text or exercises in this book. This book's use or discussion of MATLAB® software or related products does not constitute endorsement or sponsorship by The MathWorks of a particular pedagogical approach or particular use of the MATLAB® software.

First edition published 2024
by CRC Press
2385 NW Executive Center Drive, Suite 320, Boca Raton FL 33431

and by CRC Press
4 Park Square, Milton Park, Abingdon, Oxon, OX14 4RN

CRC Press is an imprint of Taylor & Francis Group, LLC

© 2024 selection and editorial matter, Adel Al-Jumaily, Paolo Crippa, Ali Mansour, and Claudio Turchetti; individual chapters, the contributors

Library of Congress Cataloging-in-Publication Data
Names: Al-Jumaily, Adel, editor. | Crippa, Paolo (Associate professor of electronics), editor. | Mansour, Ali, 1969- editor. | Turchetti, Claudio, editor.
Title: Non-invasive health systems based on advanced biomedical signal and image processing / edited by Adel Al-Jumaily, Paolo Crippa, Ali Mansour, and Claudio Turchetti.
Description: First edition. | Boca Raton, FL : CRC Press, 2024. | Includes bibliographical references and index.
Identifiers: LCCN 2023038133 (print) | LCCN 2023038134 (ebook) | ISBN 9781032386942 (hardback) | ISBN 9781032387697 (paperback) | ISBN 9781003346678 (ebook)
Subjects: MESH: Medical Informatics Computing | Monitoring, Physiologic–instrumentation | Artificial Intelligence | Image Processing, Computer-Assisted–methods | Signal Processing, Computer-Assisted–instrumentation | Biomedical Engineering–instrumentation
Classification: LCC R857.M3 (print) | LCC R857.M3 (ebook) | NLM W 26.5 |
DDC 610.28/4–dc23/eng/20231108
LC record available at https://lccn.loc.gov/2023038133
LC ebook record available at https://lccn.loc.gov/2023038134

ISBN: 978-1-032-38694-2 (hbk)
ISBN: 978-1-032-38769-7 (pbk)
ISBN: 978-1-003-34667-8 (ebk)

DOI: 10.1201/9781003346678

Typeset in Times
by codeMantra

Contents

Editors

Adel Al-Jumaily is a researcher and academic leader with more than two decades of experience. He is a Professor at the University of Technology Brunei, a professor research fellow at ENSTA-Bretagne, and adjunct Professor at the University of Western Australia. His research area is Computational Intelligence and Humanized Computational Intelligence. He has published more than 250 peer-reviewed papers. He has 13 patents, 12 of which are sponsored by industry. Adel has supervised more than 40 PhD and master's students and received two supervision awards.

Paolo Crippa is an Associate Professor of electronics at the Department of Information Engineering of the Università Politecnica delle Marche, Ancona, Italy. His research interests include micro and nanoelectronics, statistical device modeling, mixed-signal and RF integrated circuit design, biomedical circuits, systems and signal processing, neural networks, and non-linear system identification. He has published more than 120 papers in international journals, edited books, and conference proceedings. He is a member of the editorial boards and technical program committees of several international scientific journals and conferences. He is an IEEE senior member and a member of the Italian AEIT.

Ali Mansour has held many positions: Postdoctoral at LTIRF-INPG (Grenoble-France), Researcher at BMC–RIKEN (Nagoya-Japan), Teacher-Researcher at ENSIETA (Brest-France), Senior-Lecturer at Curtin-University (Perth-Australia), Invited-Professor at ULCO (Calais-France), Professor at Tabuk University (KSA), and recently Professor at ENSTA-Bretagne (Brest-France). He has published numerous refereed publications, several books, and book chapters, and has supervised many Post-Docs, PhDs, and MScs. He is interested in statistics, signal processing, robotics, telecom, biomedical engineering, electronic warfare, and cognitive radio.

Claudio Turchetti received the Laurea degree in electronics engineering from the University of Ancona, Italy, in 1979. He joined the Università Politecnica delle Marche, Ancona, in 1980, where he is currently a Full Professor of Embedded Systems design. He has published more than 160 papers, the most relevant are in *IEEE Journal of Solid-State Circuits*, *IEEE Transactions on Electron Devices*, *IEEE Transactions on Neural Networks and Learning Systems*, *IEEE Transaction on Signal Processing*, *IEEE Transaction on Cybernetics*, *IEEE Journal of Biomedical and Health Informatics*, *IEEE Transaction on Consumer Electronics*, *IEEE Journal on Emerging and Selected Topics in Circuits and Systems*, *IEEE Access*, and *IEEE Open Journal on Circuits and Systems*.

Contributors

Paolo Crippa
Department of Information Engineering
Università Politecnica delle Marche
Ancona, Italy

Claudio Turchetti
Department of Information Engineering
Università Politecnica delle Marche
Ancona, Italy

Abdullah Al Hamid
Saudi Ministry of Health
Najran, Kingdom of Saudi Arabia

Adel Al-Jumaily
Faculty of Engineering
University of Technology Brunei
Bandar Seri Begawan, Brunei
Department of Computer Science &
    Software Engineering
University of Western Australia
Perth, Australia
ENSTA Bretagne
Brest, France
and
School of Science
Edith Cowan University
Joondalup, Australia

Ahmed A. Al Taee
School of Computing, Mathematics and
    Engineering
Charles Sturt University
Wagga Wagga, Australia

Adnan Al-Anbuky
School of Engineering, Computer &
    Mathematical Sciences
Auckland University of Technology
Auckland, New Zealand

Dhiya Al-Jumeily OBE
School of Computer Science and
    Mathematics
Liverpool John Moores University
Liverpool, United Kingdom

Khalid Al-Naime
School of Engineering, Computer &
    Mathematical Sciences
Auckland University of Technology
Auckland, New Zealand

Fady Alnajjar
College of Information Technology
United Arab Emirates University
Al Ain, United Arab Emirates
and
Intelligent Behavior Control Unit
CBS-TOYOTA Collaboration Center
Nagoya, Japan

Philippe Aries
Department of Reanimation & Intensive
    Therapy
Military Teaching Hospital "Clermont
    Tonnerre"
Brest, France

Jolnar Assi
Traders Island Ltd
London, United Kingdom

Sulaf Assi
School of Pharmacy and Biomolecular
    Sciences
Liverpool John Moores University
Liverpool, United Kingdom

Mohamed A. Bahloul
Electrical Engineering Department
College of Engineering
Alfaisal University
Riyadh, Saudi Arabia

Zehor Belkhatir
School of Electronics and Computer
    Science
University of Southampton
Southampton, United Kingdom

Mohammed Bennamoun
School of Physics, Maths and
    Computing, Computer Science and
    Software Engineering
University of Western Australia
Perth, Australia

Kahina Bensafia
Departement of Electronics
Laboratory LAMPA
University of Mouloud Mammeri
Tizi Ouzou, Algeria
and
LABSTICC (UMR 6285), CNRS
ENSTA-Bretagne
Brest, France

Thibaud Berthomier
LABSTICC (UMR 6285), CNRS
ENSTA-Bretagne
Brest, France
and
EA 3878 (GETBO), CIC Inserm 1412
CHU de la Cavale Blanche
Brest, France

Giorgio Biagetti
Department of Information Engineering
Università Politecnica delle Marche
Ancona, Italy

Abdel Ouahab Boudraa
Ecole navale
Lanvéoc-Poulmic
Lanvéoc, France

Luc Bressollette
GETBO UMR 13-04
CHRU Cavale Blanche
Brest, France

Benoit Clement
LABSTICC (UMR 6285), CNRS
ENSTA-Bretagne
Brest, France

Istvàn Defrançais
CHRU de Brest, Blvd T. Prigent
Brest, France

Girish Dwivedi
Department of Advanced Clinical
    and Translational Cardiovascular
    Imaging
Harry Perkins Institute of Medical
    Research
University of Western Australia
Perth, Australia
and
Department of Cardiology and Exercise
    Physiology
Fiona Stanley Hospital
Murdoch, Australia

Laura Falaschetti
Department of Information Engineering
Università Politecnica delle Marche
Ancona, Italy

Najmeh Fayyazifar
Curtin School of Allied Health
Faculty of Health Sciences
Curtin University
Perth, Australia
and
Department of Advanced Clinical
    and Translational Cardiovascular
    Imaging
Harry Perkins Institute of Medical
    Research
University of Western Australia
Perth, Australia
and
Department of Cardiology and Exercise
    Physiology
Fiona Stanley Hospital
Murdoch, Australia

Matthew Fynn
Faculty of Science and Engineering
Curtin University
Perth, Australia

Daniel Vélez Gutiérrez
Ubiquitous Computing Laboratory,
    Department of Computer Science
HTWG Konstanz
Konstanz, Germany

Sallah Haddab
Departement of Electronics
Laboratory LAMPA
University of Mouloud Mammeri
Tizi Ouzou, Algeria

Abdullah Al Hamid
College of Clinical Pharmacy,
    Department of Pharmacy Practice
King Faisal University
AlAhsa, Saudi Arabia

Nur Amirah Abd Hamid
Malaysia-Japan International Institute
    of Technology
University Technology Malaysia
Skudai, Malaysia

Mohammad Afiq Hassan
Electrical and Electronic Engineering
University Teknologi Brunei
Bandar Seri Begawan, Brunei

Wit Heartlè
Angiology Department
Centre Hospitalier Hôtel Dieu
Pont l'Abbé, France

Clément Hoffman
GETBO UMR 13-04
CHRU Cavale BlancheBrest, France

Azadeh Noori Hoshyar
Institute of Innovation, Science and
    Sustainability
Federation University Australia
Ballarat, Australia

Md Jakir Hossen
Department of Electrical Engineering,
    Faculty of Engineering
University of Malaya
Kuala Lumpur, Malaysia

Mohammad Aminul Islam
Department of Electrical Engineering,
    Faculty of Engineering
University of Malaya
Kuala Lumpur, Malaysia

Syed Mohammed Shamsul Islam
School of Science
Edith Cowan University
Joondalup, Australia

Manoj Jayabalan
School of Computer Science and
    Mathematics
Liverpool John Moores University
Liverpool, United Kingdom

Sandrine Jousse-Joulin
EA 3878 (GETBO), CIC Inserm 1412
CHU de la Cavale Blanche
Brest, France

Rami N. Khushaba
School of Aerospace, Mechanical and
    Mechatronic Engineering
The University of Sydney
Sydney, Australia

Daphne Teck Ching Lai
School of Digital Science
University Brunei Darussalam
Bandar Seri Begawan, Brunei

Dina Shona Laila
Electrical and Electronic Engineering
University Teknologi Brunei
Bandar Seri Begawan, Brunei

Taous-Meriem Laleg-Kirati
Computer, Electrical, and Mathematical
    Sciences and Engineering
King Abdullah University of Science
    and Technology
Makkah, Saudi Arabia
and
National Institute for Research in
    Digital Science and Technology
    (INRIA)
Saclay, France

Foo Wah Low
Department of Electrical & Electronic
    Engineering, Faculty of Engineering
    & Science
Universiti Tunku Abdul Rahman
Kajang, Malaysia

Natividad Martínez Madrid
Internet of Things Laboratory, School
    of Informatics
Reutlingen University
Reutlingen, Germany

Andrew Maiorana
Curtin School of Allied Health
Faculty of Health Sciences
Curtin University
Perth, Australia
and
Department of Cardiology and Exercise
    Physiology
Fiona Stanley Hospital
Murdoch, Australia

Ali Mansour
LABSTICC (UMR 6285), CNRS
ENSTA-Bretagne
Brest, France

Hamidreza Mohafez
Department of Electrical Engineering,
    Faculty of Engineering
University of Malaya
Kuala Lumpur, Malaysia

Kazi Zehad Mostofa
Department of Electrical Engineering,
    Faculty of Engineering
University of Malaya
Kuala Lumpur, Malaysia

Dominique Mottier
GETBO UMR 13-04
CHRU Cavale Blanche
Brest, France

Ghada Muneer Bani Musa
Melbourne Institute of Technology
Sydney, Australia
and
University of Technology Sydney
Sydney, Australia

Sven Nordholm
Electrical and Computer Engineering
Curtin University
Perth, Australia

Mohammad Nur-E-Alam
School of Science
Edith Cowan University
Joondalup, Australia

Vaibhav Parakh
Liverpool John Moores University
Liverpool, United Kingdom

Yue Rong
School of Electrical Engineering,
    Computing and Mathematical
    Sciences
Curtin University
Perth, Australia

Najmeh Samadiani
CSIRO Manufacturing
Clayton, Australia

Ralf Seepold
Ubiquitous Computing Laboratory,
    Department of Computer Science
HTWG Konstanz
Konstanz, Germany

Mohd Ibrahim Shapiai@Abd Razak
Malaysia-Japan International Institute
    of Technology
University Technology Malaysia
Skudai, Malaysia

Shingo Shimoda
Intelligent Behavior Control Unit
CBS-TOYOTA Collaboration Center
Nagoya, Japan

Juan M. Vargas
Computer, Electrical, and Mathematical
    Sciences and Engineering
King Abdullah University of Science
    and Technology
Makkah, Saudi Arabia

Mikhail Vasiliev
ClearVue PV technologies
Perth, Australia

Rahimi Zahari
Electrical and Electronic Engineering
University Teknologi Brunei
Bandar Seri Begawan, Brunei

Tanveer Zia
School of Computing, Mathematics and
    Engineering
Charles Sturt University
Wagga Wagga, Australia
and
Centre of Excellence in Cybercrimes
    and Digital Forensics
Naif Arab University for Security
    Sciences
Riyadh, Saudi Arabia

1 Upper Limb Recovery Prediction Based on Multilevel Mixed Effect EMG Synergy and Biomarker Values

Ghada Muneer Bani Musa
Melbourne Institute of Technology
University of Technology Sydney

Fady Alnajjar
United Arab Emirates University
CBS-TOYOTA Collaboration Center

Shingo Shimoda
CBS-TOYOTA Collaboration Center

Adel Al-Jumaily
University of Technology Brunei
University of Western Australia
ENSTA Bretagne
Edith Cowan University

1.1 INTRODUCTION

IMMEDIATELY after a stroke, the primary concern of the patient, relatives, and care-givers is the prospect of recovery and the possibility of daily rehabilitation. To address this concern and to aid in clinical management, improved patient recovery predictions are needed. Information based on observation of average recovery patterns may have little relevance to an individual patient or clinician. Analytical studies on outcomes after stroke have tended to concentrate on predicting the outcome at a specific time point (e.g. 3 months after stroke). This type of prediction does not aid clinical decisions about whether or not to continue an intervention, such as a rehabilitation program, nor does it identify the causes of recovery failure [1,2].

DOI: 10.1201/9781003346678-1

The patient's rate of recovery also has important implications for the costs of care, especially the length of hospital stay. Such probabilities for recovery may vary according to treatment regimen or patient characteristics. It has been emphasized that standard patterns of recovery from a stroke should be established as a guide to monitoring the future recovery of patients. However, there is a lack of carefully designed statistical models for the development of such recovery patterns [3,4].

This chapter proposes a system that helps stroke patients achieve specific benchmarks by predicting their ability to recover. The proposed system also assists physiotherapists in deciding whether or not to change the exercise regimen for patients. Such exercise aids recovery and independence in activities of daily living by a given time point after stroke, allowing the patient to recover part or all past rehabilitation as well as helping to reduce hospital stay costs. Recent research has focused on rehabilitation using different types of robots, as our work in Ref. [5] required a certain type of control [6]. In contrast, electromyography (EMG) muscle signals have been used as one of the control signals as well as rehabilitation performance measurement indications and developed many algorithms to process for such applications as our published work in Refs. [7,8]. Research on the prediction and accuracy of rehabilitation performance is limited and could be improved by an appropriate selection of inputs [9,10]. To achieve more accurate prediction results and to better predict rehabilitation performance recovery, we developed a multilevel mixed effect method (MME) to create a more flexible and accurate framework to predict recovery [11,12].

The purpose of this chapter is to describe the use of robots and clinical biomarkers associated with upper limb function. This is quantified in the first length of time post-stroke, which could be days or months, depending on the willingness of the patient to allow data to be collected from them. With data collected from patients, it is possible to estimate the ability of the patients to perform daily activities and to predict their skills at 3 months post-stroke. In particular, recent developments in robot-based behavioural tasks provide a rich set of biomarkers on sensory and motor function, including the performance of both the affected and unaffected upper limbs. In this study, we observe possible rehabilitation recovery and compare the results with the Functional Independence Measure (FIM) and the Stroke Impairment Assessment Set (SIAS) scores prior to and following rehabilitation.

To start the observation, we train the model using the recorded EMG signals from the upper limb muscles of the patients during their initial rehabilitation sessions. We target such a research direction and results that could be essential to motivate the patient to complete the designed rehabilitation program. The results can then be used by the therapist to tailor an appropriate rehabilitation program for each patient.

1.2   STROKE ASSESSMENT

Generally, stroke causes long-term disability, with the patient suffering from physical weakness or paralysis of the limbs, usually on one side of the body. The patient will also experience difficulty gripping or holding objects and will have reduced joint mobility, which makes biomarker assessment problematic [13]. The assessment procedure can be time-consuming, which can increase patient recovery time and cost. FIM and SIAS are two types of biomarker assessments. These assessments generally describe the patient's ability to recover or improve after stroke to perform daily activities.

1.2.1 Functional Independence Measure

FIM is widely accepted as a body functional assessment of the patient's ability to perform daily rehabilitation tasks [14,15]. FIM contains a total of 18 tasks, rated on up to seven scales [16]. Table 1.1 summarizes the FIM measurement features. These are divided into two categories: 13 motor items (assessing self-care, sphincter control, mobility, and locomotion) and 5 cognitive items (assessing communication, psychosocial, and cognitive) [17,18].

Generally, FIM reliability and validity have a good interrater reliability score. FIM evaluates the recovery rehabilitation depending on the independence score from one to seven to complete the 18 items. The maximum score for FIM is $18 \times 7 = 126$ (for motor items and cognitive items), which indicates complete independence ability. The minimum score is $18 \times 1 = 18$, which means the end of the task dependence [17]. The maximum score for motor and cognitive items is $13 \times 7 = 91$ and $5 \times 7 = 35$, respectively. The minimum score for motor and cognitive items is $13 \times 1 = 13$ and $5 \times 1 = 5$, respectively.

1.2.2 Stroke Impairment Assessment Set

The SIAS is a standardized measure of stroke damage and is used clinically to evaluate upper limb and lower body movement ineptness [19]. Generally, the range of SIAS scores, according to symposium recommendations, is between one and five for each item [20]. The SIAS score is divided into sub-categories of motor function, tone, and sensory function, including a range of motion, pain, feeling, visualization function, speech, and sound function, among others, as illustrated in Table 1.2. The scores are rated from 0 (severe) to 3 (normal) for some categories or 0 (severe) to 5 (normal) for the motor function category [21–23].

TABLE 1.1
FIM Features of Measurement

Motor Items	Cognitive Items
Self-Care Items	**Communication Items**
1. Feeding	1. Comprehension
2. Grooming	2. Expression
3. Bathing	
4. Upper body dressing	**Psychosocial Adjustment**
5. Lower body dressing	1. Social interactions
6. Toileting	
7. Bladder control	**Cognitive**
8. Bowel control	1. Problem-solving
	Memory
9. Bed, chair, wheelchair	
10. Toilet	
11. Tub or shower	
12. Walking, wheelchair locomotion	
13. Stairs locomotion	
Motor Score/91	Cognitive Score/35

TABLE 1.2
Stroke Impairment Assessment Set

Item	Upper Limb and Lower Limb Scaling
Categories	U/E L/E
Motor Function	
1. Proximal	0–5 (0–5) hip
2. Distal	(0–5) knee
	0–5 (0–5)
Muscle Tone	
1. Deep tendon reflexes (DTR)	0–3 (0–3)
2. Tone	0–3 (0–3)
Sensory Function	
1. Touch	
2. Position	0–3 (0–3)
	0–3 (0–3)
Range of Motion (ROM)	
Pain	0–3 (0–3)
Trunk balance	0–3
Visuospatial	0–3
Speech	0–3
Grip	0–3
Unaffected side	0–3

The total score for the upper limb side and the lower side is 76 (normal function).

The scores in Table 1.2 are used to plot a chart on a radar chart program to quickly indicate the stroke patient's status [22,23]. This chart is considered another SIAS feature; it can show the patients' scores in a different elegant way. Figures 1.1 and 1.2 show two examples of a SIAS radar chart.

Figure 1.1 shows the example of a 73-year-old male. This patient was first tested 26 days after the stroke; his SIAS assessment score was severe for the upper and lower limbs. After 251 days, the test was readministered, and the SIAS score was recorded as moderate. Thus, the patient's status improved from severe to moderate, and he fully recovered from some severe stroke, like motor simulation, and the pain reduced over time.

Figure 1.2 shows the SIAS radar chart for a 59-year-old male. This patient was first tested 52 days after the stroke and again after 187 days. His SIAS assessment score shows some recovery in his lower and upper limbs [23].

Healthcare providers such as physiatrists, nurses, or therapists score the FIM scale or SIAS scale based on direct input from the patient [21]. Generally, for FIM and SIAS, standard errors (SE) are taken into account [21]. Numerous studies have also been conducted to predict the clinical assessment (FIM and SIAS) value and predict the patient's length of stay [24]. In this study, we use either FIM or SIAS clinical assessment or both as a biomarker to compare the biomarker value with our prediction results. In our study, we have both values, so we want to use both for more accurate results. If the results towards the biomarker of FIM, or SIAS results, then

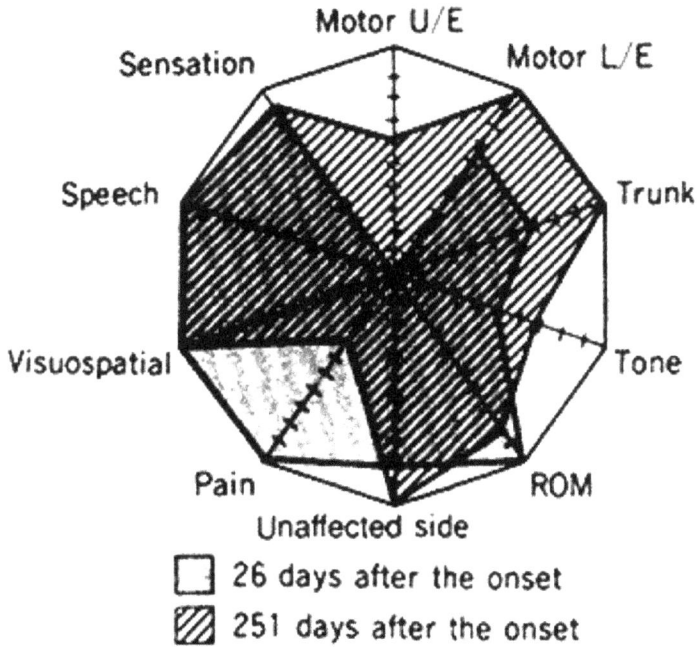

FIGURE 1.1 SIAS radar chart for a 73-year-old male patient.

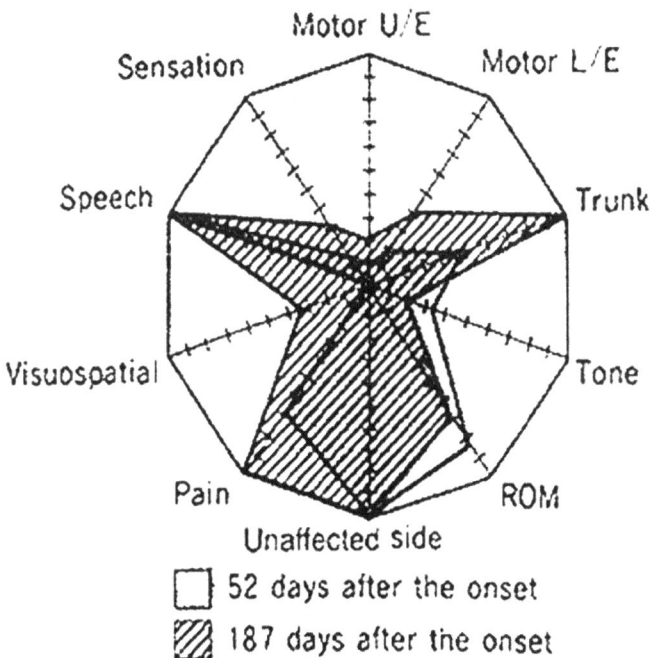

FIGURE 1.2 SIAS radar chart for a 59-year-old male patient.

the proposed method successfully predicts future rehabilitation recovery. This study represents a development insofar as it uses MME to predict the patient's rehabilitation performance as well as evaluate that prediction.

The following section explains the use of MME as a prediction method, followed by a detailed description of the materials and methodology used in this study.

1.3 MULTILEVEL MIXED EFFECTS PREDICTION

The MME model is generally used to identify how to process a variety of nonlinear predictions. Thus, it helps to process single EMG channels. In this study, we extended the MME to predict variability in the human body for multiple EMG channels. This method provides an accurate result when based on FIM/SIAS biomarkers by adding regression coefficient parameters (discussed in greater depth in the materials and methodology section). This step reduces the final prediction error in predicting the rehabilitation recovery for the affected side of the body by using the unaffected side as a reference. The MME for each EMG signal has different intercepts and slopes across time. MMEs have become important tools. They allow the analyst to treat the effects of the collected data through the prediction process. EMG data vary from patient to patient and muscle to muscle; there are even variations within one muscle depending on movement type, movement force, movement angle, and surface electrodes. Thus, such a data structure lends itself to analysis via mixed effects models. The most important element in rehabilitation application is the modelling and especially, the prediction of rehabilitation [25]. In this study, we describe how we improve the MME model to fit multiple EMG channels.

1.4 MATERIALS AND METHODOLOGY

This section specifies how we modify and predict rehabilitation movement using the EMG data. It also describes how we predict movement using multivariate EMG data.

1.4.1 PARTICIPANTS

Three stroke patients were recruited for this experiment. The data were collected from each patient at different times, as shown in Table 1.3. In the table, ID refers to the three patients identified as P1 (Patient 1), and so on. G/A indicates gender

TABLE 1.3
Patients' Data Collection Dates

ID	G/A	A-Side	2016 (Year)	2017 (Year)					
			12	1	2	3	4	5	6
P1	F/93	R	1 7 14 16	30 6 13 20	27 6 13 27				
P2	F/71	L		30 6 13	27 6 13 27 3 10 17		15 22 5 12		
P3	M/52	L		13	27 6 13 27 3 10 17 8				

TABLE 1.4

Patient Information

Ref. ID	G/A	Stroke Date	Stroke Type	SIAS Before Training		SIAS After Training		FIM Before Training		FIM After Training	
				Date	UE	Date	UE	Date	Motor	Date	Motor
1 P1	F/93	07.11.16	Moderate	01.12.16	1	02.04.17	3	01.12.16	20	02.04.17	71
2 P2	F/71	25.12.16	Severe	30.01.17	0	12.06.17	2	30.01.17	20	12.06.17	62
3 P3	M/527	23.01.17	Light	13.02.17	3	23.06.17	4	13.02.17	58	23.06.17	87

and age, respectively. A-side indicates the affected side. For Patient 1, the date on which the stroke occurred was 7 November 2016 (see Table 1.4). The first data collection for this patient occurred on 1 December 2016. Overall, data was collected from this patient 12 times over 4 months. For Patient 2, data were collected 14 times over 7 months, whereas for Patient 3, data were collected 9 times over 4 months. The age of Patient 1 was 93 years, Patient 2 was 72 years, and Patient 3 was 52 years. The first two patients were female; the third patient was male. The protocols for this experiment were approved by the RIKEN ethics committee.

Table 1.4 shows the patients' stroke strength based on SIAS/FIM biomarkers rated as severe, moderate, or mild. The table also shows the stroke date for each patient, the SIAS biomarker and the FIM biomarker reading date, and results before and after treatment, respectively. The table shows SIAS biomarker results before treatment and after treatment. These reflected an improvement in the patient's status, meaning that the patient had recovered. The same data are presented for the FIM biomarkers, with the motor items (see Table 1.1) also showing improvement.

1.4.2 REHABILITATION DEVICE

The dual steering system is used to perform symmetrical tasks at the upper limb level, as shown in Figure 1.3a. Figure 1.3b illustrates the graphical interface, which signals the patient to synchronize subject movement with the desired cycling frequency. The system allows three working modes depending on the interaction between the steering axes: accessible mode (FREE), in which the axes rotate independently; asynchronous mode (ASYM); and synchronous locked mode (SYM), in which both axes are connected, experiencing the same angle of rotation in an asymmetric or asymmetric direction (Figure 1.3c). The system also permits the use of different rotating elements, such as steering wheels and cranks, as shown in Figure 1.4. First, the patient makes a voluntary movement before using the dual steering system. This movement trains the central nervous system (CNS), indicating that there is a new movement that the body will make. The CNS will assign new neurons to carry out this exercise. This movement was discussed in detail in one of our previous studies [26]. Second, the patient is seated in front of the dual steering system. For accessible EMG recordings, the data were collected with a sampling frequency of 1,000 Hz by using surface electrodes positioned according to the surface electromyography (sEMG) guidelines

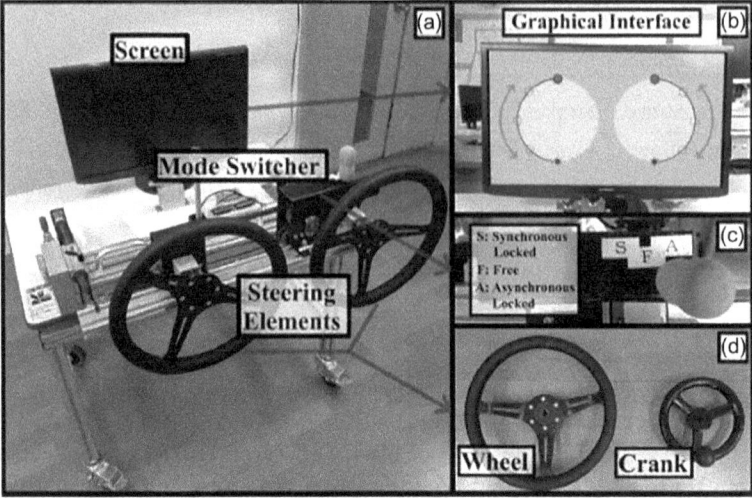

FIGURE 1.3 Experimental environment. (a) Dual-steering system. (b) Graphical interface to synchronize subject movement with the desired cycling frequency. (c) Switcher to select working mode. (d) Steering elements.

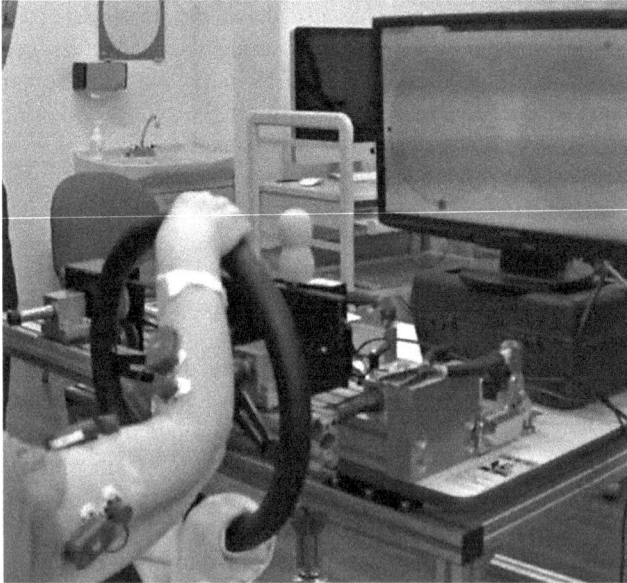

FIGURE 1.4 Position of the participant while holding the dual steering wheel.

[26,27], as shown in Figure 1.4. Several independent sessions were conducted with rest in between to minimize potential fatigue. Some of these sessions took place on the same day, some took place on different days, and some were conducted after a few months (see Table 1.3).

1.4.3 MME PREDICTION AND TIME SERIES PREDICTION

MME is commonly used in economics, biostatistics, and sociology. It is usually used for single prediction items, such as the performance of the economy to predict the growth of a country in the coming year [11,28].

We developed and calculated the MME prediction based on the FIM/SIAS biomarker by combining groups of EMG data for the affected muscles to predict rehabilitation recovery. We took the unaffected side as a reference to determine the prediction recovery for the affected side. EMG data were collected from the affected and unaffected sides using the dual system with a sampling frequency of 1,000 Hz.

EMG data were obtained using 18 EMG channels, which recorded data from the upper limbs and back and chest muscles of the patients while performing the driving simulation tasks [27,29]. Nine surfaces of EMG were recorded from the muscles on the affected side, and nine sEMG data were recorded from the unaffected side according to the sEMG guidelines [27,30] with the following distribution: brachio radialis (BP), protanor teres (PT), biceps (B), triceps (T), anterior deltoid (AD), posterior deltoid (PD), pectoralis major (PM), infra spinatus (IS), and elector spinae (ES). We use the unaffected EMG signals as an indication of the prediction procedure to determine whether the affected side prediction results are going towards the unaffected side EMG signal value. If the affected prediction result values were close to the unaffected side EMG signal value, this indicated that the affected side had recovered.

1.4.4 METHODOLOGY AND RESULTS

As shown in Figure 1.5, the MME prediction process started with raw EMG data that were collected using the dual steering system. The data were then sampled at 1,000 Hz and filtered between 20 and 450 Hz using the Butter filter. Then, the signal was smoothed using a moving average filter to obtain better results, as in Figure 1.6.

To calculate the average signal values, as in Equation (1.1)

$$Y(n) = \frac{X(n-1) + X(n) + X(n+1)}{3} \tag{1.1}$$

where $X(n)$ is the noise-effected signal and $Y(n)$ stands for the averaged signal.

Note: To validate the real meaning of $(n-1)$ and $(n+1)$, the average loop starts from $2:n-1$.

Time series prediction is a useful tool for predicting future behaviour. Time series is usually modelled through a random probability distribution process and uses the data based on time [31,32]. We built a new time series model to predict future rehabilitation based on extracted synergy.

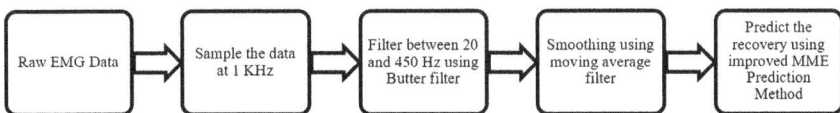

FIGURE 1.5 MME prediction for EMG signal.

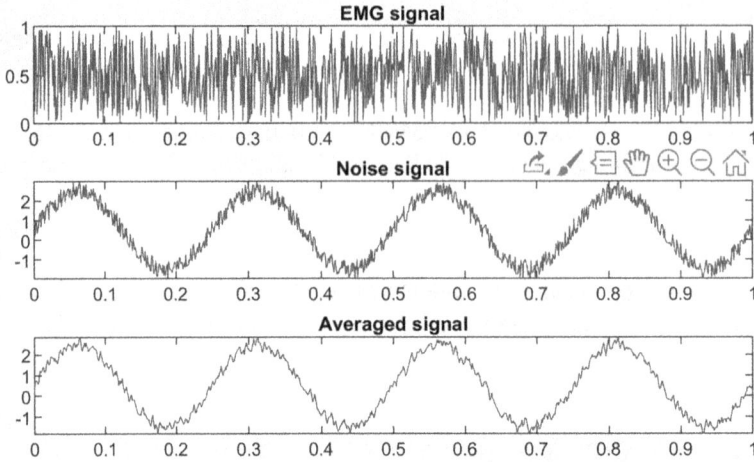

FIGURE 1.6   MME prediction for EMG signal.

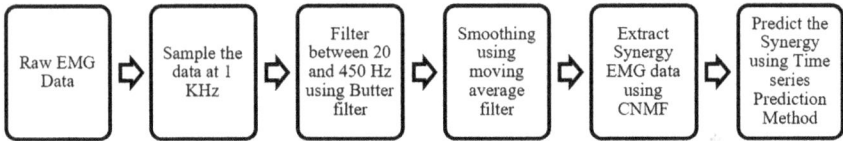

FIGURE 1.7   Time series prediction for synergy EMG.

Figure 1.7 illustrates our time series prediction methodology model. The model starts with raw EMG data, but after smoothing the signal, we extracted the synergy EMG using concatenated non-negative matrix factorization (CNMF) for the unaffected and affected sides. In particular, we used the CNMF extracting method rather than the NMF method to improve the efficiency of the CNMF, as in one of our previous studies [26].

We calculated the final prediction error (FPE) for the MME as in Equation (1.2) and the mean square error (MSE) as in Equation (1.3) between the filtered EMG data and the prediction EMG data. This is defined in the equation below:

$$\text{FPE} = \left(\frac{1 + \dfrac{d}{N}}{1 - \dfrac{d}{N}} \right) * V \tag{1.2}$$

Where d is the regression coefficient, which is the estimated value that describes the prediction relationship value between the unaffected side and the affected side. The coefficient sign (+/−) indicates the direction of the relationship between the unaffected and affected sides. The positive sign means that the muscle will recover, whereas the negative sign means that the muscle will not recover [33]. N is the length of the data record (observation data). V is the loss function for the structure [34].

$$\text{MSE} = \frac{1}{n} \sum_{i=1}^{n} \left| \theta_i - \widehat{\theta}_i \right|^2 \tag{1.3}$$

Where n is the number of observation values in each EMG channel (number of the collected data), θ is the true value of the estimate of interest, and $\widehat{\theta}_i$ is the estimate of interest obtained from the ith simulation [35].

The simple regression model explains the relation between y (the regression output) and one independent variable, x, as in Equation (1.4). We used this equation to predict the EMG single channel.

$$y = \beta_0 + \beta_1 x + \varepsilon \tag{1.4}$$

For the MME regression model, we developed (1.4) to fit the verity EMG data. Equation (1.5) explains the relationship between y and more than one independent variable:

$$y = \beta_0 + \beta_1 x_1 + \beta_2 x_2 + \cdots + \beta_n x_n + \varepsilon \tag{1.5}$$

where $\beta_0, \beta_1, \beta_2, \ldots, \beta_n$ are the parameters, and ε is the error. We developed (1.5) to fit our EMG data. We used a sample to estimate the multiple prediction equation based on the least square method. Since the calculation for the parameters β_0, β_1, and β_2, is not known, it was necessary to use sample statistics to estimate these parameters. We calculated the coefficient for each EMG channel to calculate the recovery prediction. We used the sample statistics to develop the parameters in Equation (1.5), as in Equation (1.6) to obtain the multiple prediction equation. This allowed us to find the best line that fitted our data by adding the coefficient parameter to Equation (1.5).

$$\hat{y} = b_0 + b_1 c_1 + b_2 c_2 + \cdots + b_n c_n + \varepsilon \tag{1.6}$$

where $b_0, b_1, \ldots, b_n$ are the EMG values for each observation, c is the coefficient for each muscle, and n is the signal value, which is 9 in our study (9 EMG channel).

Equation (1.7) is the least square equation; it calculates the least square error between the observed EMG data and the prediction value:

$$\min \sum \left(y_n - \hat{y}_n \right)^2 \tag{1.7}$$

where y_n is the observed value of the y for the nth observation and $\hat{y}_n$ is the predicted value of y for the nth observations. From Equation (1.7), we obtained two values: the observed value (the unaffected side reference value) and the predicted value (which is the prediction recovery for the affected side). The best regression line that affects the data can be found by minimizing the mean square error between the observed value $\left(y_n \right)$ and the predictive value $\left(\hat{y}_n \right)$.

By testing the significance of the individual parameters, we suppose the following

$$
\begin{cases}
H_0 : \beta_i = 0 & p\text{-value} \leq \alpha, \alpha = 0.025. \quad \text{The value is sygnificant, possibility of recovery} \\
H_n : \beta_n \neq 0 & \text{otherwise, No recovery}
\end{cases}
\tag{1.8}
$$

In Equation (1.8), we predicted H_0 if the p-value $\leq \alpha$ or by using the critical value approach, which is: if the p-value $\leq \alpha$, that means $\beta_i = 0$. In turn, this means that there is a relationship between the affected and unaffected side prediction values. Therefore, it is possible to predict the recovery of the patient.

By calculating the p-value in each session for each EMG channel, we can predict whether the muscle will recover or not. Table 1.5 presents the results for each muscle to determine an indication parallel with the SIAS biomarker to enable recovery prediction. The table shows the sample values for the three patients for multiple sessions with moderate, severe, and mild stroke, respectively. We calculate the intercept, multiple R, R squared and the standard error to find the signal prediction output.

Table 1.5 shows that the values for the multiple R and the R squared are towards 1. This means that the patients recovered, but they needed more practice. When these results are compared with those in Tables 1.1 and 1.2, we find that after training, the first patient showed a moderate recovery (which is 3 SIAS and 71 FIM), the second patient showed minimal recovery (which is 2 SIAS and 62 FIM) and the third patient showed the greatest recovery (which is 4 SIAS and 87 FIM).

After calculating the previous parameters for each EMG for all sessions for each patient, the results are presented in Figures 1.8–1.10. As in Equation (1.8), after calculating $(\alpha - (p\text{-value}))$ for each single EMG channel, we placed zeros in the negative value, as shown in Table 1.6, so that just to simplify and understand the prediction results.

Figures 1.11–1.13 indicate which muscles will help recover during the single sessions for each patient.

As shown in Table 1.6 and Figure 1.11, Patient 1 had 12 data collection sessions. With each session, some muscles recovered, whereas others did not. Therefore, we can conclude that this patient's recovery was moderate. As shown in Table 1.6 and Figure 1.12, for Patient 2, only a few muscles recovered with each session. Therefore, we can conclude that this patient had a severe stroke. As shown in Table 1.6 and Figure 1.13, for Patient 3, many muscles recovered with each session. Therefore, we can conclude that this patient recovered after rehabilitation. When we compare our results and the previous conclusion with the SIAS results in Table 1.4. We can see that the results match because the high value of five for the SIAS assessment indicates full recovery, whereas the lowest value of 0 indicates no recovery. Our prediction method and results match the SIAS results in Table 1.4. This means that the MME prediction for multi-EMG successfully predicts stroke patients' rehabilitation.

We demonstrate the time series prediction method for synergy EMG signal for stroke patients to predict rehabilitation recovery of the upper limbs for the near future results for the first, second and third patients as shown in Figures 1.14–1.16, respectively.

TABLE 1.5
MME Prediction for the sEMG Channel for Unaffected Side

	Sessions	Intercept	Es (EMG 9)	Multiple R	R Square	Standard Error
First Patient	1	−3.8E−11	−0.021995	0.67	0.45	2.61433E−05
	2	7.65E−11	8.61E−02	0.76	0.57	1.10175E−05
	3	−8.3E−10	−2.33E−02	0.90	0.81	3.51754E−05
	4	−2.9E−09	1.21E−03	0.85	0.73	1.82951E−05
	5	−1.1E−10	−3.02E−03	0.28	0.08	8.62527E−06
	6	1.74E−11	−1.21E−03	0.69	0.48	1.02802E−05
	7	−2.1E−11	−1.70E−02	0.74	0.55	1.9213E−05
	8	−6.2E−11	2.40E−02	0.56	0.31	1.85957E−05
	9	−4.6E−10	−1.30E−01	0.86	0.74	1.42564E−05
	10	−1.8E−10	−6.26E−01	0.92	0.85	1.51882E−05
	11	−5.8E−11	1.52E−01	0.49	0.24	1.23125E−05
	12	−2.8E−11	−1.13E+00	0.94	0.87	1.20597E−05
Second Patient	1	2.78E−11	3.29E−01	0.85	0.73	1.25563E−05
	2	1.83E−10	5.97E−03	0.31	0.10	1.07049E−05
	3	−3.9E−11	8.38E−02	0.91	0.82	1.67023E−05
	4	1.94E−10	−8.17E−03	0.86	0.74	9.71386E−06
	5	7.18E−12	1.93E−01	0.34	0.11	1.02255E−05
	6	1.19E−10	2.59E−02	0.84	0.70	1.03589E−05
	7	−2.7E−10	4.77E−02	0.92	0.85	1.46256E−05
	8	4.96E−10	−9.07E−02	0.82	0.67	5.25837E−05
	9	7.75E−11	1.70E−03	0.05	0.00	9.84463E−06
	10	−3.4E−11	−2.27E−02	0.18	0.03	9.41904E−06
	11	−9.7E−12	−1.96E−04	0.35	0.12	8.95578E−06
	12	−3.1E−10	−2.44E−03	0.04	0.00	1.51845E−05
Third Patient	1	8.36E−12	4.48E−04	0.89	0.79	7.6842E−06
	2	1.23E−11	5.71E−03	0.57	0.32	1.70211E−05
	3	2.12E−12	1.24E−01	0.66	0.43	1.73212E−05
	4	7.07E−11	1.15E−01	0.74	0.54	1.1864E−05
	5	1.37E−10	−6.00E−02	0.59	0.35	3.68439E−05
	6	−2.1E−11	1.65E−02	0.83	0.69	7.70275E−06
	7	−2.3E−10	2.52E−01	0.82	0.68	1.27693E−05
	8	9.51E−12	6.85E−01	0.55	0.31	1.61569E−05
	9	2.97E−10	3.26E−01	0.60	0.36	9.80951E−06
	10	4.81E−11	−1.11E−02	0.69	0.48	1.48361E−05
	11	5.08E−10	1.72E−02	0.73	0.53	1.0308E−05
	12	−1.1E−10	−4.86E−02	0.92	0.85	7.30509E−06

Multiple R for The First Patient

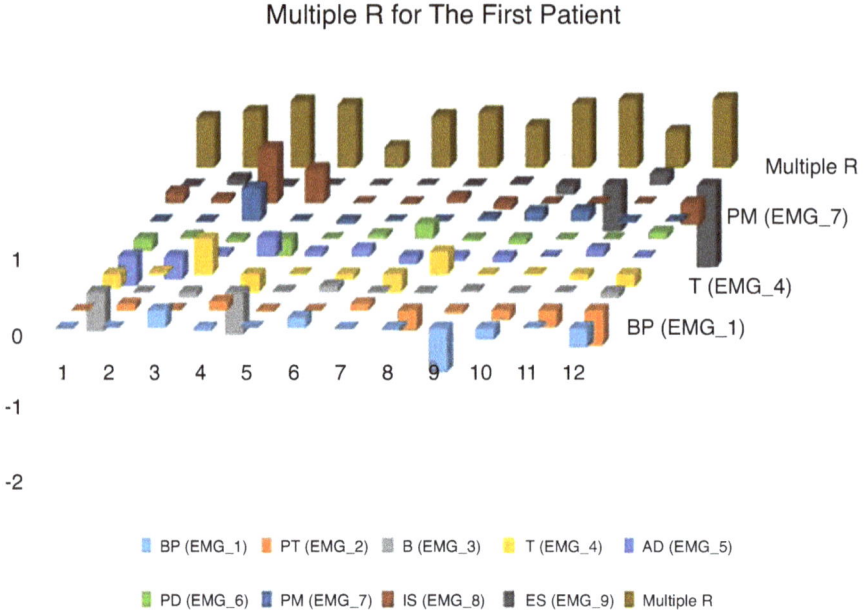

FIGURE 1.8   Multiple R for Patient 1.

Multiple R for The Second Patient

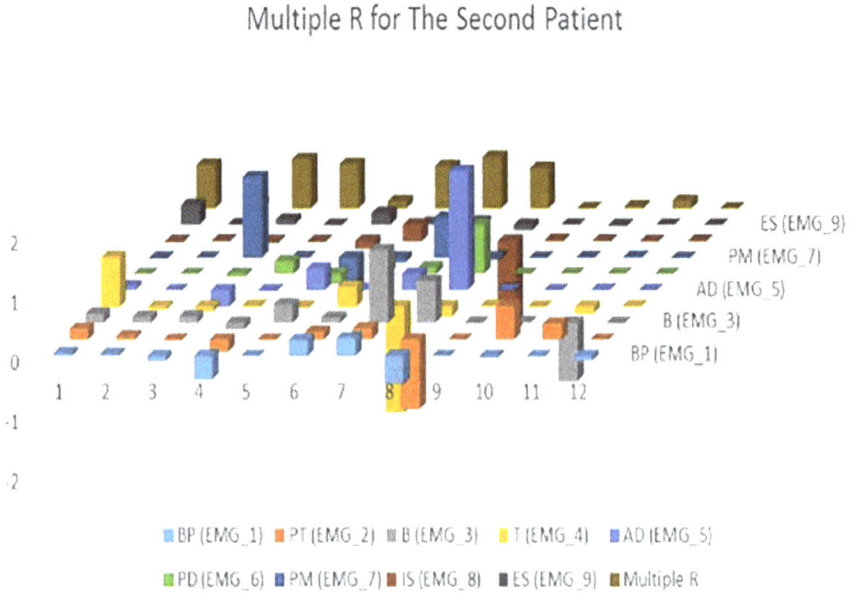

FIGURE 1.9   Multiple R for Patient 2.

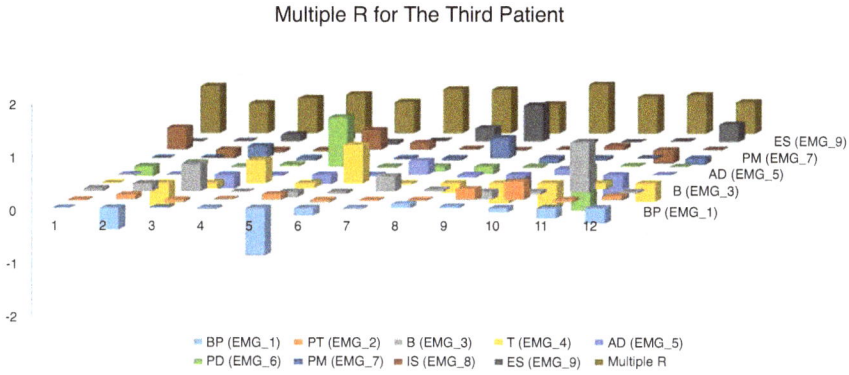

FIGURE 1.10   Multiple R for Patient 3.

In Figures 1.14–1.16, the first simulation shows the synergy prediction for a short time. The second simulation shows the time series prediction after 3 months. The first patient has recovered moderately, taking into account the amplitude range. The second patient has recovered with a small amplitude. This means that the stroke was severe, not all muscles will recover, and recovery time will be longer. The third patient also has an increased amplitude range, which means that his recovery is different from the first and second patients'. These results are compatible with the SIAS biomarkers to predict future rehabilitation recovery.

1.5   CONCLUSION

In this study, we predicted rehabilitation for post-stroke patients by developing the MME and predicting rehabilitation performance ahead of 3 months based on time series prediction. It also presented the synergy of the affected and non-affected sides of the body for post-stroke patients, which can assist in predicting their ability to recover daily activity. This can indicate the extent of a patients' recovery and improve the potential of post-stroke patients and help guide rehabilitation strategies. This chapter presents the prediction recovery of severe, moderate, and mild stroke patients during their rehabilitation program. We successfully anticipated their rehabilitation possibility using the MME method and predicted their future recovery performance based on muscle synergy using time prediction series. In this chapter, we modelled and developed the MME for the collected EMG data, demonstrating that the MME provides a significantly superior method to predict the rehabilitation recovery performance of stroke patients for multi-EMG channels. This improved prediction method can help researchers identify the possibility of stroke patients' recovery and ability to perform daily rehabilitation tasks. This was proved by comparing our results with the clinical FIM or SIAS biomarkers, which are collected by clinical experts. This study also demonstrated predicted future rehabilitation movement recovery for stroke patients based on synergy data, which facilitates the efficient selection of essential muscles.

TABLE 1.6

(α-(p_value)) for All EMG Signals and Replacing 0 in the Negative Values

		BP	PT	B	T	AD	PD	PM	IS	ES
	Sessions	(EMG 1)	(EMG 2)	(EMG 3)	(EMG 4)	(EMG 5)	(EMG 6)	(EMG 7)	(EMG 8)	(EMG 9)
First Patient	1	2.500E−05	2.611E−01		0	0	0	2.227E−14	0	8.073E−51
	2	6.907E−04	0	2.625E−02	2.146E−136	0	3.848E−77	2.297E−07	0	0
	3	0	1.976E−01	1.706E−211	0	3.097E−18	3.366E−76	0	0	1.955E−19
	4	5.338E−23	9.467E−36	0	0	1.404E−163	5.424E−285	1.188E−02	0	7.173E−01
	5	6.997E−60	4.711E−23	3.520E−200	1.242E−63	0	6.381E−06	3.435E−133	1.458E−07	1.764E−03
	6	0	5.349E−02	2.241E−219	0	0	0	2.202E−02	4.892E−56	5.641E−05
	7	4.833E−50	0	3.673E−12	0	0	0	3.292E−20	0	1.245E−101
	8	7.701E−66	0	4.266E−27	0	0	1.088E−62	2.821E−57	0	1.648E−32
	9	0	2.774E−166	1.818E−108	2.334E−23	1.256E−258	0	0	1.731E−03	0
	10	0	0	6.782E−07	1.448E−31	1.433E−01	3.607E−02	0	1.113E−03	0
	11	1.501E−08	0	1.766E−10	3.833E−189	0	1.579E−02	8.616E−14	1.360E−01	0
	12	0	0	0	0	3.135E−29	4.292E−212	5.263E−02	0	0
	1	1.085E−01	6.169E−04	2.511E−02	8.587E−70	0	2.928E−13	1.379E−288	6.971E−01	7.584E−03
	2	0	1.319E−220	4.281E−161	0	6.994E−14	7.273E−01	1.610E−22	6.031E−17	0
	3	2.364E−92	1.920E−26	0	7.233E−85	4.892E−01	6.703E−01	4.476E−01	1.576E−02	1.067E−13
	4	0	3.024E−03	4.628E−01	7.397E−68	0	0	1.418E−164	1.459E−248	0
	5	0	6.101E−37	0	0	0	3.313E−214	0	2.018E−58	1.655E−75
	6	1.959E−13	7.226E−31	3.264E−01	8.350E−58	6.061E−04	8.671E−88	7.440E−01	4.219E−028	5.447E−34
	7	0	0	0	9.166E−242	9.672E−25	0	7.422E−53	5.726E−02	1.391E−27
	8	4.659E−80	2.489E−115	0	2.108E−149	1.149E−63	3.450E−02	2.911E−17	4.670E−40	6.361E−54
	9	5.325E−254	0	0	7.605E−16	0	0	1.074E−37	0	1.710E−55
	10	9.611E−01	4.148E−04	0	2.278E−02	9.018E−157	6.047E−105	0	6.224E−08	2.607E−187

(Continued)

TABLE 1.6 (Continued)

(α-(p_value)) for All EMG Signals and Replacing 0 in the Negative Values

Sessions	BP (EMG 1)	PT (EMG 2)	B (EMG 3)	T (EMG 4)	AD (EMG 5)	PD (EMG 6)	PM (EMG 7)	IS (EMG 8)	ES (EMG 9)
11	0	1.529E-100	6.515E-176	1.275E-27	1.113E-15	1.340E-01	3.656E-22	7.776E-08	3.157E-08
12	0	0	0.00E+00	0	5.748E-05	0	1.161E-135	0	1.559E-08
13	0	2.045E-82	0.00E+00	0	0	6.242E-202	0	2.874E-110	1.026E-61
14	0	2.932E-230	1.918E-129	0	0	8.501E-120	1.185E-13	4.527E-100	1.723E-90
15	1.282E-128	0	0	1.075E-241	0	0	8.468E-21	0	4.551E-158
16	3.221E-01	5.953E-02	7.588E-01	4.301E-01	2.490E-11	3.970E-01	6.958E-08	3.766E-02	0
17	5.370E-01	5.582E-01	1.358E-01	8.076E-06	3.118E-03	9.578E-03	4.495E-01	3.129E-54	4.296E-01
18	1.230E-06	0	1.648E-22	2.688E-01	6.069E-01	2.342E-02	4.152E-03	2.577E-02	2.207E-10
19	1.971E-01	1.083E-210	0	3.301E-262	2.454E-12	1.671E-05	1.184E-14	1.293E-01	9.186E-01
20	2.006E-01	3.382E-01	1.292E-01	1.317E-02	8.644E-01	7.437E-02	5.889E-01	5.474E-01	4.369E-02
21	3.388E-03	6.418E-01	3.512E-01	1.408E-04	1.201E-01	4.432E-01	3.946E-01	3.623E-02	5.664E-01
22	5.210E-01	2.433E-08	5.327E-09	1.176E-135	0	8.191E-32	0	0	1.003E-160
Second Patient 1	0	0	0	5.302E-01	0	0	1.100E-10	0	6.188E-01
2	0	0	0	0	6.899E-01	1.203E-33	1.244E-01	0	1.513E-02
3	0	0	1.555E-215	0	2.020E-41	3.526E-01	2.152E-03	6.649E-19	1.624E-09
4	1.29918E-01	5.135E-106	0	2.957E-117	0	1.860E-18	2.088E-242	2.319E-28	0
5	3.040E-19	1.305E-22	9.040E-04	1.716E-177	1.378E-64	2.190E-115	9.965E-234	6.028E-24	0
6	1.617E-13	0	7.954E-74	0	5.295E-35	4.217E-42	0	0	0
7	0	0	4.531E-11	1.174E-27	0	0	0.000E+00	0	0
8	0	6.820E-12	3.266E-88	5.899E-20	9.621E-52	0	0	2.823E-185	3.699E-22
9	5.631E-218	8.204E-31	2.988E-116	8.697E-24	9.621E-52	0	1.425E-11	0	3.699E-22

(Continued)

TABLE 1.6 (Continued)
(α-(p_value)) for All EMG Signals and Replacing 0 in the Negative Values

	Sessions	BP (EMG 1)	PT (EMG 2)	B (EMG 3)	T (EMG 4)	AD (EMG 5)	PD (EMG 6)	PM (EMG 7)	IS (EMG 8)	ES (EMG 9)
Third Patient	10	0	6.559E–232	0	0	8.800E–110	9.819E–77	1.486E–126	0	8.075E–40
	11	1.135E–11	3.608E–14	0	3.955E–28	0	3.615E–154	4.510E–100	1.866E–15	0
	12	6.240E–172	4.530E–17	1.069E–04	1.131E–128	2.402E–37	1.215E–281	0	3.476E–10	0
	13	0	2.395E–11	8.557E–01	1.078E–07	6.405E–67	0	0	4.992E–164	2.256E–82
	14	8.293E–225	3.292E–217	0	7.145E–32	0	1.419E–114	1.586E–275	6.884E–106	0
	15	0	3.384E–12	9.868E–27	1.560E–05	1.015E–30	1.288E–09	2.775E–08	2.015E–34	4.583E–19
	16	2.672E–05	2.099E–286	0	5.848E–153	0	1.317E–81	6.909E–277	1.738E–10	1.043E–51
	17	4.063E–166	0	5.782E–160	0	0	0	1.525E–147	4.108E–23	1.146E–02
	18	7.223E–17	0	1.047E–59	0	7.894E–175	0	4.470E–12	4.271E–164	6.737E–02
	19	0	5.499E–138	0	1.879E–134	0	2.200E–25	7.746E–16	0	8.373E–31
	20	1.124E–97	3.699E–03	0	2.341E–13	1.299E–12	0	8.770E–02	0	1.151E–03
	21	7.561E–274	6.555E–127	5.724E–120	0	4.532E–11	5.395E–02	6.489E–107	7.803E–13	0

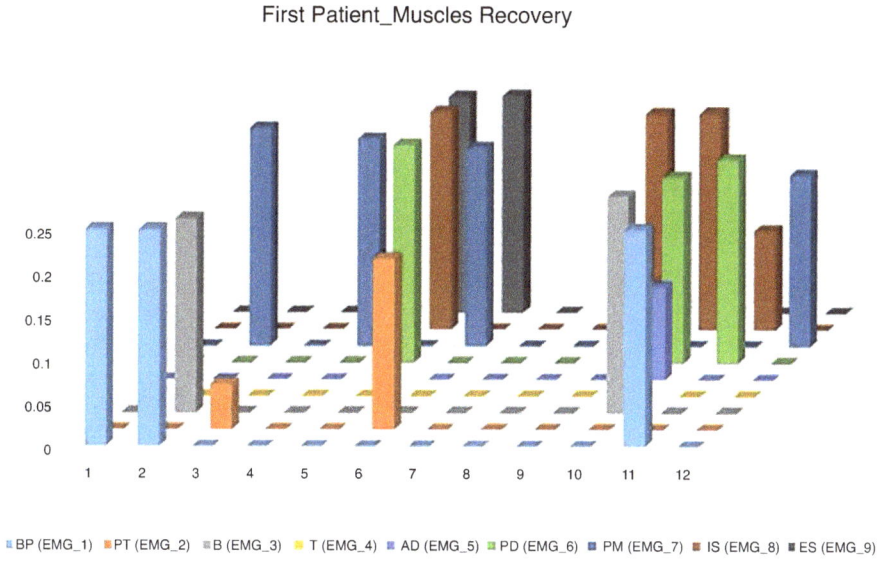

FIGURE 1.11 Predicted muscle recovery for Patient 1.

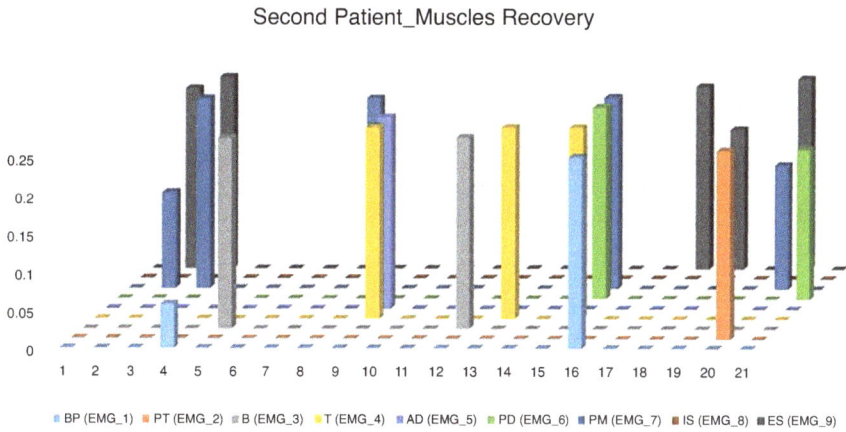

FIGURE 1.12 Predicted muscle recovery for Patient 2.

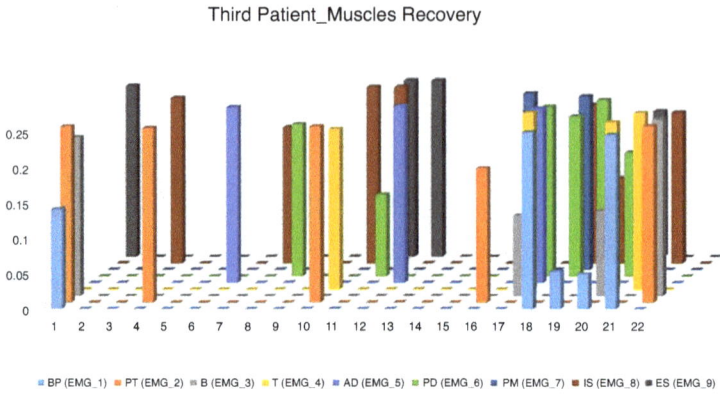

FIGURE 1.13 Predicted muscle recovery for Patient 3.

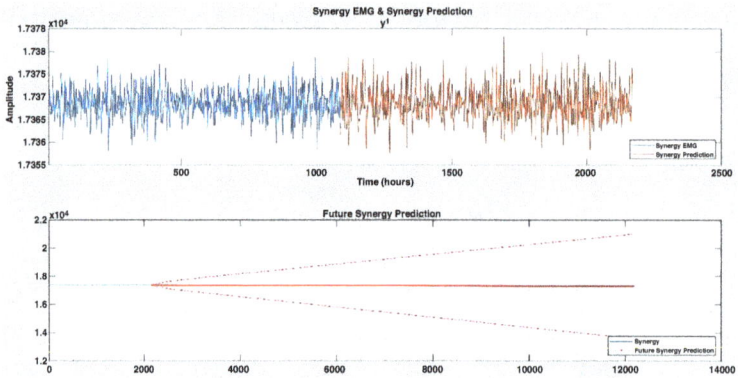

FIGURE 1.14 Patient 1 synergy prediction.

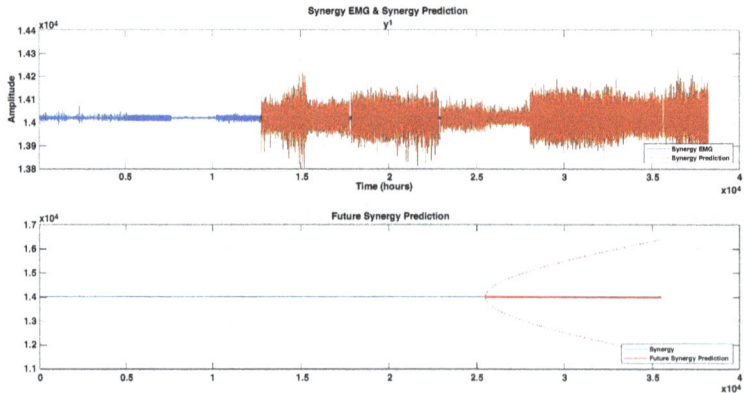

FIGURE 1.15 Patient 2 synergy prediction.

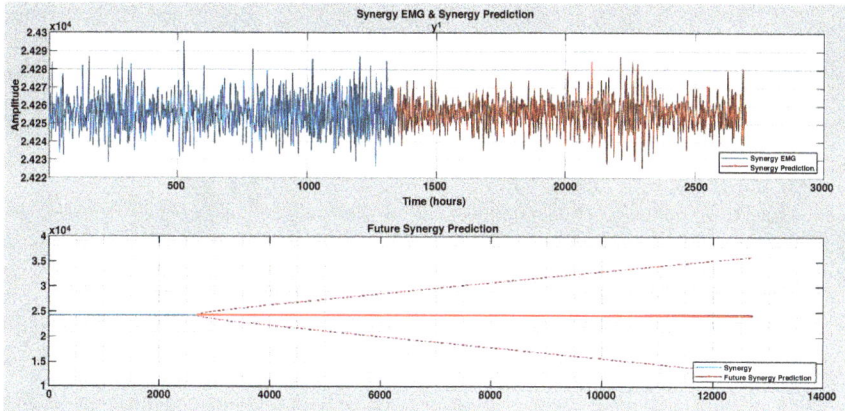

FIGURE 1.16   Patient 3 synergy prediction.

REFERENCES

1. G. Chen, P. A. Taylor, Y.-W. Shin, R. C. Reynolds, and R. W. Cox, "Untangling the relatedness among correlations, Part II: Inter-subject correlation group analysis through linear mixed-effects modeling," *YNIMG Neuroimage,* vol. 147, pp. 825–840, 2017.
2. J. C. Pinheiro and D. M. Bates, "Non-linear mixed-effects models: Basic concepts and motivating examples," In J. Pinheiro and D. Bates (eds.), *Mixed-Effects Models in S and S-PLUS,* pp. 273–304. Springer Science & Business Media: Berlin, Germany, 2000.
3. S. M. Mostafavi, J. I. Glasgow, S. P. Dukelow, S. H. Scott, and P. Mousavi, "Prediction of stroke-related diagnostic and prognostic measures using robot-based evaluation," *IEEE International Conference on Rehabilitation Robotics (ICORR),* Seattle, WA, vol. 2013, pp. 1–6, 2013.
4. P. M. Pilarski, T. B. Dick, and R. S. Sutton, "Real-time prediction learning for the simultaneous actuation of multiple prosthetic joints," *IEEE 13th International Conference on Rehabilitation Robotics (ICORR),* Seattle, WA, pp. 1–8, 2013. doi: 10.1109/ICORR.2013.6650435
5. A. Rahman and A. Al-Jumaily, "Design and development of a bilateral therapeutic hand device for stroke rehabilitation," *International Journal of Advanced Robotic Systems,* vol. 10, no. 12, p. 405, 2013.
6. K. Anam and A. A. Al-Jumaily, "Active exoskeleton control systems: State of the art," *Procedia Engineering,* vol. 41, pp. 988–994, 2012.
7. R. N. Khushaba, A. Al-Ani, and A. Al-Jumaily, "Feature subset selection using differential evolution and a statistical repair mechanism," *Expert Systems with Applications,* vol. 38, no. 9, pp. 11515–11526, 2011.
8. K. Anam and A. Al-Jumaily, "Evaluation of extreme learning machine for classification of individual and combined finger movements using electromyography on amputees and non-amputees," *Neural Networks,* vol. 85, pp. 51–68, 2017.
9. J. C. Sanchez, J. M. Carmena, M. A. Lebedev, M. A. L. Nicolelis, J. G. Harris, and J. C. Principe, "Ascertaining the importance of neurons to develop better brain-machine interfaces," *IEEE Transactions on Biomedical Engineering,* vol. 51, no. 6, pp. 943–953, 2004.

10. D. T. Westwick, E. A. Pohlmeyer, S. A. Solla, L. E. Miller, and E. J. Perreault, "Identification of multiple-input systems with highly coupled inputs: Application to EMG prediction from multiple intracortical electrodes," *Neco Neural Computation,* vol. 18, no. 2, pp. 329–355, 2006.

11. G. Chen, Z. S. Saad, J. C. Britton, D. S. Pine, and R. W. Cox, "Linear mixed-effects modeling approach to FMRI group analysis," *YNIMG NeuroImage,* vol. 73, pp. 176–190, 2013.

12. D. Lu, Y. Tripodis, L. C. Gerstenfeld, and S. Demissie, "Clustering of temporal gene expression data with mixtures of mixed effects models with a penalized likelihood," *Bioinformatics,* vol. 35, no. 5, pp. 778–786, 2019.

13. G. Le Sant, A. Nordez, F. Hug, R. Andrade, T. Lecharte, P. J. McNair, and R. Gross, "Effects of stroke injury on the shear modulus of the lower leg muscle during passive dorsiflexion," *Journal of Applied Physiology Journal of Applied Physiology,* vol. 126, no. 1, pp. 11–22, 2019.

14. K. J. Ottenbacher, Y. Hsu, C. V. Granger, and R. C. Fiedler, "The reliability of the functional independence measure: A quantitative review," *Archives of Physical Medicine and Rehabilitation,* vol. 77, no. 12, pp. 1226–1232, 1996.

15. B. B. Hamilton, J. A. Laughlin, R. C. Fiedler, and C. V. Granger, "Interrater reliability of the 7-level functional independence measure (FIM)," *Scandinavian Journal of Rehabilitation Medicine,* vol. 26, no. 3, pp. 115–119, 1994.

16. V. J. Chehata, M. Shatzer, and A. Cristian, "Inpatient rehabilitation outcome measures in persons with brain and spinal cord Cancer," In A. Cristian (ed.), *Central Nervous System Cancer Rehabilitation,* pp. 19–25. Elsevier: Amsterdam, Netherlands, 2019.

17. G. Gillen, "Overview of cognitive and perceptual rehabilitation," In G. Gillen (ed.), *Cognitive and Perceptual Rehabilitation,* pp. 1–31. Mosby, St. Louis, 2009.

18. D. Cech and S. T. Martin, "Evaluation of function, activity, and participation," In D. J. Cech and S. T Martin (eds.), *Functional Movement Development Across the Life Span (Third Edition),* pp. 88–104. W.B. Saunders: St. Louis, 2012.

19. R. Osu, K. Ota, T. Fujiwara, Y. Otaka, M. Kawato, and M. Liu, "Quantifying the quality of hand movement in stroke patients through three-dimensional curvature," *Journal of Neuroengineering and Rehabilitation,* vol. 8, p. 62, 2011.

20. M. Liu, N. Chino, T. Tuji, Y. Masakado, K. Hase, and A. Kimura, "Psychometric properties of the Stroke Impairment Assessment Set (SIAS)," *Neurorehabilitation & Neural Repair,* vol. 16, no. 4, pp. 339–351, 2002.

21. T. Tsuji, M. Liu, S. Sonoda, K. Domen, and N. Chino, "The stroke impairment assessment set: Its internal consistency and predictive validity," *Archives of Physical Medicine and Rehabilitation,* vol. 81, pp. 863–868, 2000.

22. N. Chino, S. Sonoda, K. Domen, E. Saitoh, and A. Kimura, "Stroke Impairment Assessment Set (SIAS)," In N. Chino and J. L. Melvin (eds.), *Functional Evaluation of Stroke Patients,* pp. 19–31. Springer: Tokyo, 1996.

23. N. Chino and J. L. Melvin, *Functional Evaluation of Stroke Patients.* Springer: Tokyo 1996.

24. G. Sprint, D. J. Cook, D. L. Weeks, and V. Borisov, "Predicting functional independence measure scores during rehabilitation with wearable inertial sensors," *IEEE Access,* vol. 3, pp. 1350–1366, 2015.

25. D. B. Hall and M. Clutter, "Multivariate multilevel non-linear mixed effects models for timber yield predictions," *Biometrics,* vol. 60, no. 1, pp. 16–24, 2004.

26. G. M. Bani Musa, A. Al-Jumaily, F. Alnajjar, and S. Shimoda, "Analyze the human movements to help CNS to shape the synergy using CNMF and pattern recognition," *Procedia Computer Science,* vol. 105, pp. 170–176, 2017.

27. A. Costa, M. Itkonen, H. Yamasaki, F. S. Alnajjar, and S. Shimoda, "Importance of muscle selection for EMG signal analysis during upper limb rehabilitation of stroke patients," *2017 39th Annual International Conference of the IEEE Engineering in Medicine and Biology Society (EMBC),* Jeju, Korea (South), pp. 2510–2513, 2017.

28. C. M. Stinear, W. D. Byblow, S. J. Ackerley, P. A. Barber, and M.-C. Smith, "Predicting recovery potential for individual stroke patients increases rehabilitation efficiency," *Stroke,* vol. 48, no. 4, pp. 1011–1019, 2017.

29. G. M. Bani Musa, F. Alnajjar, A. Al-Jumaily, and S. Shimoda, "Upper limb recovery prediction after stroke rehabilitation based on regression method," In *Converging Clinical and Engineering Research on Neurorehabilitation III: Proceedings of the 4th International Conference on NeuroRehabilitation (ICNR2018),* October 16–20, 2018, Pisa, Italy, vol. 21, Springer, pp. 380–384, 2019.

30. H. J. Hermens, B. Freriks, R. Merletti, D. Stegeman, J. Blok, G. Rau, C. Disselhorst-Klug, and G. Hägg, "European recommendations for surface electromyography," *Roessingh Research and Development,* vol. 8, no. 2, pp. 13–54, 1999.

31. U. Thissen, R. Van Brakel, A. De Weijer, W. Melssen, and L. Buydens, "Using support vector machines for time series prediction," *Chemometrics and Intelligent Laboratory Systems,* vol. 69, no. 1–2, pp. 35–49, 2003.

32. N. Sapankevych and R. Sankar, "Time series prediction using support vector machines: A survey," *IEEE Computational Intelligence Magazine,* vol. 4, no. 2, pp. 24–38, 2009.

33. S. Wang and H. Cui, "Test for high dimensional regression coefficients of partially linear models," *Communications in Statistics-Theory and Methods,* vol. 49, no. 17, pp. 4091–4116, 2020.

34. G. J. Fryer, M. E. Odegard, and G. H. Sutton, "Deconvolution and spectral estimation using final prediction error," *Geophysics,* vol. 40, no. 3, pp. 411–425, 1975.

35. I. Rombach, C. Jenkinson, A. M. Gray, D. W. Murray, and O. Rivero-Arias, "Comparison of statistical approaches for analyzing incomplete longitudinal patient-reported outcome data in randomized controlled trials," *Patient Related Outcome Measures,* vol. 9, p. 197, 2018.

2 EMG Feature Extraction Based on Cardinality with Deep Learning Concepts

Ahmed A. Al Taee
Charles Sturt University

Rami N. Khushaba
The University of Sydney

Tanveer Zia
Charles Sturt University
Naif Arab University for Security Sciences

Adel Al-Jumaily
University of Western Australia
University of Technology Brunei
Edith Cowan University
ENSTA Bretagne

2.1 INTRODUCTION

The primary characteristic of myoelectric signals is that the relevant information regularly varies in time, frequency domain, or both. EMG signals acquired from the upper limb involve multiple issues, including changes in the signal's characteristics with time, electrode location change, muscle fatigue, and limb position variations [1]. Therefore, analysing such signals needs a flexible method for extracting relevant information or features. Simultaneously, this method must be reliable in terms of simplicity, computing time, and precision. EMG pattern recognition algorithms have utilised various myoelectric feature extraction strategies for decades. The output of these algorithms can be utilised to control rehabilitation and prosthetic limb devices. Choosing a reliable method for feature extraction is one of the crucial components. Recent research has identified the quality of the traditional EMG features extraction method as a factor that may hinder the transition from laboratory to clinical applications. To address this limitation, researchers have changed their focus from traditional feature extraction methods to deep learning models, which can identify the optimal feature representations. While deep learning models trained on raw

DOI: 10.1201/9781003346678-2

EMG data produce promising outcomes, their clinical use is occasionally limited by their high computational costs (because of the significant volumes of data needed for training and the significant number of model parameters). Meanwhile, time-domain features typically incur less computational cost than frequency and time-frequency feature extraction methods [2,3]. This approach combines the simplicity and low processing costs of traditional feature extraction with concepts adapted from deep learning models to effectively extract the spatial-temporal properties of EMG signals, making it suitable for real-time applications. To accomplish this, a novel contribution to cardinality-based feature extraction is presented and compared to well-known time-domain features. Very limited research has explored cardinality as a feature extraction approach for biomedical signal analysis. Cardinality was reported to have outperformed the other traditional features in terms of simplicity, not being affected by sliding window length, sampling frequency and the number of classes [4]. Unlike earlier studies, this study also proposes using cardinality to evaluate the connection between the different muscles utilised to accomplish a movement. This technique analyses spatial information by getting the correlation between the available channels for the LD-EMG signals. In addition, new cardinality features have been developed and evaluated in this work. The proposed approaches are validated using two known and published EMG datasets: The first is force [5], and the second is the BioPatRec dataset [6].

The following is a brief description of this chapter's contribution:

- Examine cardinality as a framework for EMG feature extraction compared to other simple and cost-effective traditional TD feature extraction approaches.
- New cardinality feature sets were introduced and evaluated.
- The idea of cardinality is examined for generating innovative logical combinations of EMG channels to extract spatial features.
- Adopting the memory principle from deep learning structures in the feature extraction method to capture EMG signals' temporal dynamics over short and long-time scales. Usually, the features retrieved from each analysis window may not be able to consider how the features are extracted from previous windows (an interplay between controller delay and window sizes and overlaps). Simply put, the standard method is cross-sectional, which means it ignores any temporal information between consecutive windows. Therefore, one of the most important aspects of the deep learning network is the possibility that prior information could be related to the current one. This is the exact concept that we bring to cardinality feature extraction to form an interplay between spatial and temporal resolution.

2.2 CARDINALITY

The field of data profiling involves a broad range of approaches and procedures for analysing a dataset and determining its characteristics. The results provide a variety of statistics regarding the columns, their linkages, and their dependencies. The number of distinct values (cardinality), minimum/maximum values, and the number of null values are significant column dependency statistics. Alon, Marias, and Szegedy, who introduced the frequency moments of a subset, described the number of unique values in a subset, also known as the zeroth-frequency moment, as one of the most significant sorts of relationships [7,8]. Cardinality is a widely contested subject in academia due to its large number of applications, particularly in computer science domains.

Furthermore, cardinality is a core part of database query processing and optimisation, making it an extremely important function, and it is considered one of the most researched and valuable methods in network security monitoring, data flow and data mining. Cardinality is also used as one of the best techniques to optimise memory size because determining it is a straightforward process. Even so, such memory requirements are excessive for some applications. However, many methods have been developed to efficiently approximate the cardinality of a dataset while minimising the amount of storage space and computing power required [9,10].

The fundamental idea of cardinality can be shown as follows: Suppose we have a collection denoted by the letter A. This is denoted by the symbol Card (A) or $|A|$, defined as the number of unique values contained within set A. Cardinality is defined as the number of distinct values that exist inside two different data sets and is represented by the notation $|A \cap B|$. If we consider dataset A equal to the numbers (4, 5, 6, 7) and dataset B equal to the numbers (4, 4, 5, 5, 6, 7), then the cardinality is four items. The following is a mathematical definition of what is meant by the term "cardinality" in reference to a subset:

Consider a subset $X = (x_1, x_2, ..., x_n)$ of n items where x_i represents a universe member with K potential values, where many items might be assigned similar values. Assume $m_i = |\{j : x_j = i\}|$ represents the number of events of item i in the subset X. For each $k \geq 0$, the k-the frequency moment F_k is evaluated as below [11].

$$F_k = \sum_{i=1}^{n} m_i^k \tag{2.1}$$

where the 0-th frequency moment F_0 is the number of distinct elements appearing in the subset X. F_1 is the number of elements in the subset X. F_2 is a uniformity measure. In most cases, NULL values are ignored.

One of the primary reasons for signal processing's use of cardinality as a feature is that it is unaffected by DC offsets, which are often generated by mismatched electrode impedance. Cardinality is unlike other amplitude-dependent features such as RMS, ZC and MAV, as they are affected by DC offsets of the electrodes [6].

Soft cardinality was proposed by Jimenez et al. [12] and presented as a method for grouping identical and similar components. Its basic idea extrapolates from the traditional cardinality concept by incorporating the similarities between the items in a set, enabling a more obvious measurement of the number of those items. This technique is used for text applications, where texts are modelled as sets of words, and a similarity function is used to identify the similarity between the two words. Soft cardinality can reflect syntactic similarity, semantic relatedness and other concepts. Soft cardinality has already been demonstrated to be a reasonable threshold for various applications such as information extraction, object matching and plagiarism detection.

A similarity function in soft cardinality provides the idea of element similarity by comparing two elements like k_i and k_j, and returning a score between 0 and 1. For example, the following expression expresses soft cardinality, which is denoted as $|K|_{\text{sim}}$:

$$|K|_{\text{sim}} = \sum_{i=1}^{|K|} \omega_{k_i} \left(\sum_{j=1}^{|K|} \text{sim}\left(k_i, k_j\right)^p \right)^{-1} \qquad (2.2)$$

where K is a collection with elements $k_1, k_2, \ldots, k_{|K|}$. It is obvious that $|K| = |K|_{\text{sim}}$ if p is very large or sim is a fixed value, then it returns 1 for identical items; otherwise, it returns 0. Therefore, p defines the degree of 'softness'; this characteristic demonstrates how soft cardinality extends from classical cardinality. The weight ω_{k_i} is a value associated with every element k_i; the default value of this weight can have a value of 1. For example, suppose we have two datasets, A and B. In that case, the cardinality between them could be empty using $|A \cap B|$. However, by using the similarity function, it is possible to represent the intersection of two datasets, A and B, by employing soft cardinality in the process as $|A \cap B|_{\text{sim}} = |A|_{\text{sim}} + |B|_{\text{sim}} - |A \cup B|_{\text{sim}}$ instead of calculating it directly from $|A \cap B|$. As a result, soft cardinality has proven to be superior to classic cardinality in terms of textual similarity. Since Jaccard created the cardinality set in 1901, the similarity can be derived from any one of three algebraic combinations of $|A|$, $|B|$ and $|A \cup B|$ or $|A \cap B|$ or $(|A\Delta B|)$. When comparing two sets of data, these three cardinalities unambiguously reflect all of the areas in the Venn diagram. In this particular instance, the fundamental cardinality can be expanded in order to obtain an additional four relationships, as demonstrated in Table 2.1 [13].

Later, the basic combination of cardinality was expanded to represent more features and employed as a feature in a machine learning regression approach. The extension features based on soft cardinality are shown in Table 2.2, which is an extension of the basic one [13].

These features have been utilised in the past to determine semantic textual similarity (STS). During this research, these features were analysed to determine how effectively they might be extracted from an EMG signal. In addition to that, in this

TABLE 2.1

The Extension of Cardinality Sets

#	Extension of Basic Cardinality Sets
1	$|A \Delta B| = |A \cup B| - |A \cap B|$
2	$|A \cap B| = |A| + |B| - |A \cup B|$
3	$|B/A| = |B| - |A \cap B|$ $|A/B| = |A| - |A \cap B|$

TABLE 2.2
Extensive Set of Logical Features

#	Feature
1	$|A|/|A \cup B|$
2	$|B|/|A \cup B|$
3	$|A| - |A \cap B|/|A|$
4	$|A| - |A \cap B|/|A \cup B|$
5	$|B| - |A \cap B|/|B|$
6	$|B| - |A \cap B|/|A \cup B|$
7	$|A \cap B|/|A|$
8	$|A \cap B|/|B|$
9	$|A \cap B|/|A \cup B|$
10	$|A \cup B| - |A \cap B|/|A \cap B|$
11	$|A \cup B| - |A \cap B|$

body of work, a different innovative cardinality features combination has been researched, tested, and evaluated to express its efficacy and compare it to the already existing cardinality features.

2.3   COSINE SIMILARITY

Cosine similarity is a method that is used to determine the degree of similarity between two vectors that are located in the same space. Calculating it requires first identifying whether or not two vectors point in the same general direction and then taking the cosine of the angle that exists between the two vectors. Even though the Euclidean distance puts some distance between two comparable vectors, the cosine similarity can help determine whether or not they can be oriented so that they are closer together. The greater the cosine similarity between two vectors, the smaller the angle that separates them. If (a) and (b) are two vectors that are to be compared with the cosine measure as a similarity function, then the result that we get is:

$$\text{sim}(a,b) = \frac{a \cdot b}{\|a\| \cdot \|b\|} \tag{2.3}$$

where $\|a\|$ is the Euclidean norm of vector a, which is equal to $\sqrt{a_1^2 + a_2^2 + \cdots + a_n^2}$ and $\|b\|$ is the same for vector b.

Equation (2.3) determines the value of the cosine of the angle formed by vectors a and b. A cosine value of 0 shows that the two vectors are perpendicular to one another, meaning there is no possible match between them. When the cosine value is closer to one, the angle is lower, and the vector match is better [14].

The Otsuka-Ochiai definition applies to this biological notion. When sets are represented as bit vectors, the cosine similarity and the Otsuka-Ochiai coefficient are equivalent to one another. The Otsuka-Ochiai coefficient can be expressed as the following formula:

$$\frac{|a \cap b|}{\sqrt{|a| \cdot |b|}} \tag{2.4}$$

where a and b are considered to be subsets, and $|a|$ and $|b|$ represent the total number of components that are contained within a and b, respectively.

2.4   METHODOLOGY

Pattern recognition (PR) relies heavily on feature extraction, particularly when the pattern in question is encoded in the form of a temporal or spatial signal (like an EMG signal). Before being transferred to the classifier, this signal is first broken into small segments, sampled, and then transformed into a digital representation. All of these steps are performed as part of the PR system. In recent years, a large number of researchers have concentrated on the temporal aspect while ignoring the possibility of benefiting from spatial correlations and muscle crosstalk [15].

Finding the dependencies of temporal characteristics through the concept of recurrent deep learning method. Finding the dependencies of spatial features through the concepts of the spatial filter with cardinality logics. The following is a synopsis of these research methodologies:

2.4.1   SPATIAL DEPENDENCIES

Usually, important signal features are retrieved from each channel independently, without taking into account the interplay between the numerous muscles utilised to perform a movement. This study proposed a unique (U) and Union (Un) correlation that can be employed to solve this issue. Unique correlations used to extract all the values of channel A are numerically different from the values of channel B. In contrast, the Union combines the values of channels A and B into a distinct single set.

By integrating each pair of channels, unique correlation effectively enhances the number of EMG channels for feature extraction. As depicted in Figure 2.1, this process can be characterised using NC as the total number of EMG channels. Given every possible combination of channels (denoted by k), the first stage of the proposed method is to determine the unique value between these channels' values. Here, for the sake of simplicity, we shall consider the combination of every two channels; therefore, $k=2$. This expression can be written as follows:

$$\left(\frac{NC}{k} \right) = \frac{NC(NC-1)\ldots(NC-k+1)}{k(k-1)\ldots1} \tag{2.5}$$

In factorials format, it will be $\dfrac{NC!}{k!(NC-k)!}$ where $k \le NC$

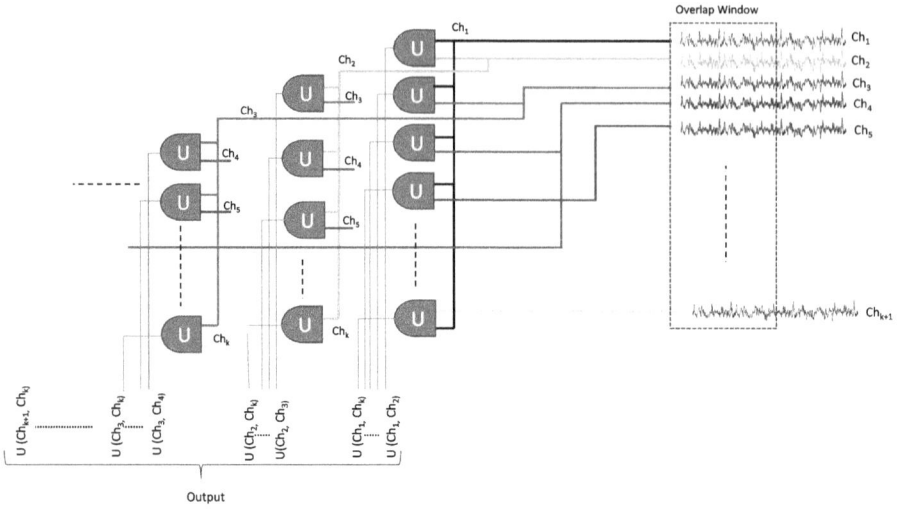

FIGURE 2.1   Block diagram for the unique correlation.

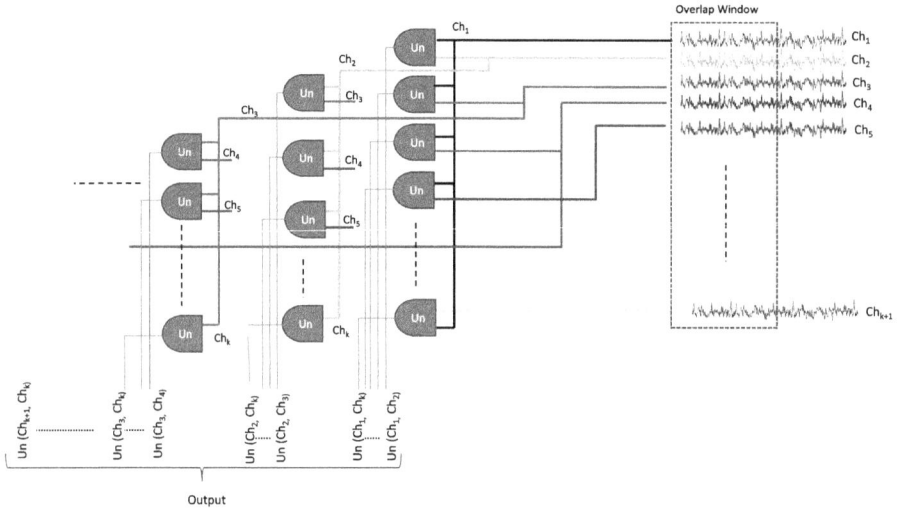

FIGURE 2.2   Block diagram for the **Union** correlation.

This will end up with $\dfrac{NC(NC-1)}{2}$ logical combinations of two channels. Similarly, we determined the union values between channels, as illustrated in Figure 2.2. Obtaining the unique and union values is the first step in determining the output channel cardinality, which is applied in the subsequent section.

2.4.2 Features Extraction

The novel feature extraction approach provided is based on the cardinality principle. This work proposes and verifies a novel set and combination of cardinality features. Figure 2.3 depicts a block diagram of the proposed feature set.

All the features in Table 2.2 have been analysed and investigated to develop new feature sets consisting of the combinations listed below; this set confirms its efficacy by having a lower error rate than other cardinality features. Additionally, the sequential forward selection (SFS) method has been utilised to obtain the best cardinality feature sets. To limit the influence of noise, the extracted feature values must be normalised after getting the features from each channel. The logarithmically transformed amplitudes are utilised for power normalisation.

The first and second features represent the cardinality of the first and second derivatives for the unique correlation between EMG channels. This concept is justified by the time differentiation characteristic of the FT, which states that the nth derivative of a TD function equals the FT multiplied by a factor $j\omega^n$ in the FD. This, in turn, is equivalent to the description of the nth moment of the power spectrum. Hence, instead of transforming the generated logical signals combination into the FD and extracting their moments, we directly do that from the TD to save processing time. Therefore, the EMG signal's first and second derivatives may improve EMG classification accuracy, driven by applying the power spectrum moments. The first feature is represented below [16]:

$$F_1 = \left| \Delta^1 \ U\big(\text{Ch}[k], \text{Ch}[k+1]\big) \right| \tag{2.6}$$

where $i = 1, 2, 3, \ldots, N$, where N is the total number of channels.

The second feature represents the cardinality of the second derivative of the Unique correlation channel.

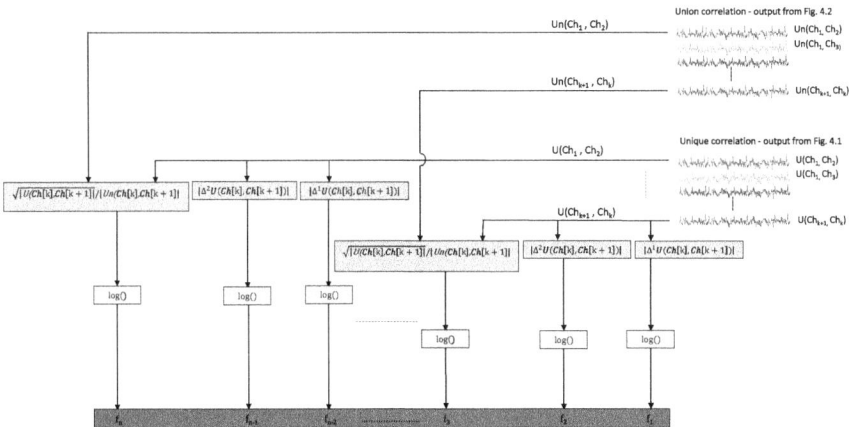

FIGURE 2.3   Block diagram for the feature extraction method.

$$F_2 = \left| \Delta^2 U \left(\mathrm{Ch}[k], \mathrm{Ch}[k+1] \right) \right| \tag{2.7}$$

The third feature is the square root of the cardinality for the unique correlation divided by the cardinality of the union correlations between each pair of EMG channels.

$$F_3 = \sqrt{\left| \left(\mathrm{Ch}_k \cup \mathrm{Ch}_{k+1} \right) \right|} \Big/ \left| \left(\mathrm{Ch}_k \cup_n \mathrm{Ch}_{k+1} \right) \right| \tag{2.8}$$

where $k = 1, 2, 3, \ldots Nc$.

The same method is then used to extract features from a nonlinear version of the original signal. Equation (2.4) was then used to determine the cosine similarity between these vectors. The block diagram of the proposed feature extraction approach for each sliding window is depicted in Figure 2.4.

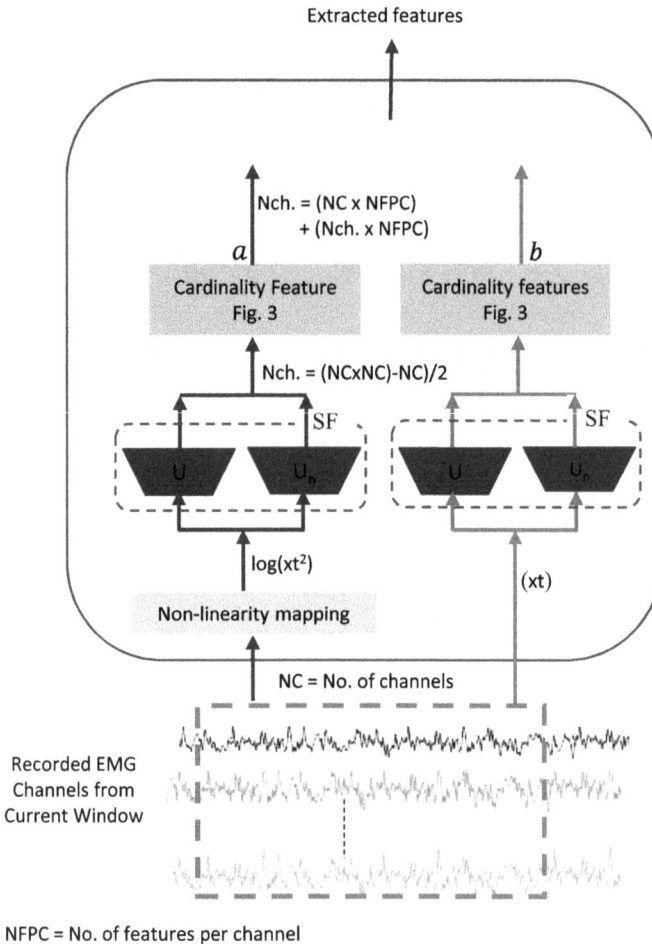

FIGURE 2.4 Block diagram for feature extraction approach.

2.4.3 TEMPORAL DEPENDENCIES

To successfully extract the temporal dynamics of EMG signals, this research intends to combine the simplicity and low computational cost of traditional feature extraction with memory concepts from long short-term memory models (LSTM). The long short-term memory works by processing the prior outputs with the current inputs, i.e., it permits the processing of earlier information with each neural network chunk [17,18]. Theoretically, recurrent networks can employ backpropagation to store activation representations of prior input events in "short-term memory" or "long-term memory." This study examines the impact of both short- and long-term memory on classification accuracy, as outlined below:

2.4.3.1 Short-Term Memory

For short-term memory, we multiplied (fused) the present window features by the nth prior window features (which could be a near or far window), and the resulting features were sigmoid to maintain a tolerable range. In this fusion, correlation values increase when features in the current and preceding windows have the same class and decrease when the classes are distinct. It can be additionally correlated with the second, third and other windows in such a process. Figure 2.5 represents a block schematic of the whole feature extraction procedure using the short-term concept.

2.4.3.2 Long-Term Memory

We integrated, for short-term memory, the features derived from the current analysis window with those taken from a previous window for the same window size (as in Figure 2.5). It should be emphasised that such studies can be conducted by fusing features from any previous window; however, we select the first previous window for convenience. Combining short-term and long-term memory components is vital to

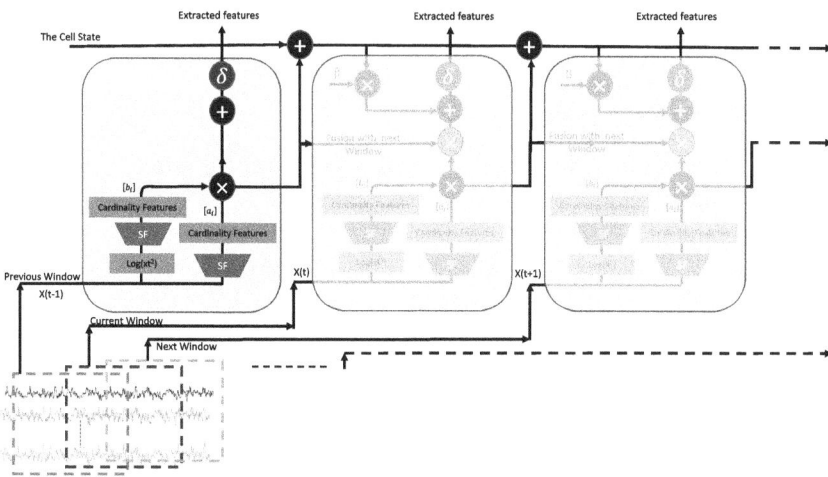

FIGURE 2.5 Block diagram of short-term feature extraction approach using.

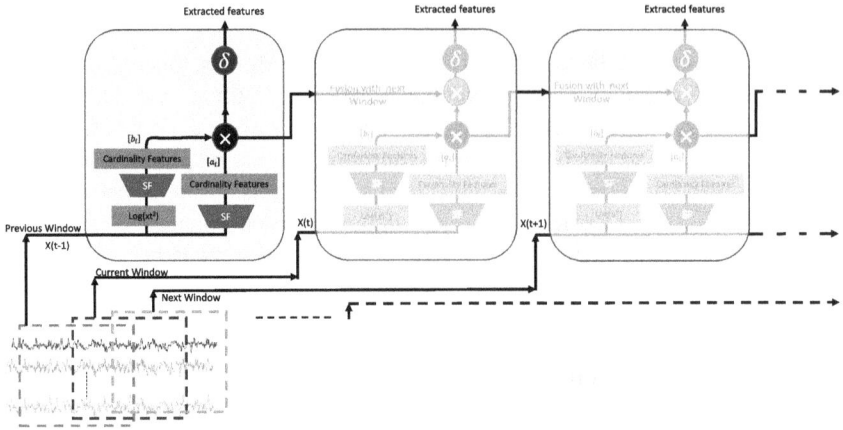

FIGURE 2.6　Block diagram of long-term feature extraction approach.

capture the dynamics of any time-series data. As shown in Figure 2.6, this prompted us to add a long-term memory component to the proposed short-term memory with fixed convolutions and construct it while gradually extracting features from the recorded EMG signals. As with LSTM models, the cell state component must be scaled appropriately to prevent overflow. This is a result of our model's continuous input of data to the cell state, which LSTM models control via gates that attenuate particular dimensions based on the inputs. A weighting factor (β) was given to the incoming information and the information going to the cell state in order to account for the long memory required during and between gestures. Thus, the (β) can be empirically changed to balance the influence of the long-term component while extracting features with larger or smaller values, as increases or decreases will have an effect on the historical components.

2.5　EXPERIMENTS AND RESULTS

Here, we illustrate the effectiveness of the cardinality-based feature extraction method by applying it to the BioPatRec and Force datasets. All the datasets used in this work are summarised briefly in Table 2.3. In this research, we evaluate the effectiveness of the suggested feature extraction approach and evaluate it against several well-known features.

Following feature extraction, dimensionality reduction was performed using the spectral regression feature projection technique (SR). Finally, three distinct kinds of classifiers were tested: the traditional linear discriminant analysis (LDA), support vector machines (SVM) and k-nearest neighbours (KNN). The efficacy of the proposed method was further investigated using the Wilcoxon signed rank test; this statistical tool can assist us in determining if the variations in error rates attained by various procedures can be related to the advantages or drawbacks of each method. The significance threshold was chosen at a p-value < 0.05 being significant. Additionally,

TABLE 2.3

A Summary of the Electromyogram Datasets Used to Evaluate the Suggested Algorithms

	Subjects #	Channels #	Classes #	Freq. (Hz)	Intact/ Amputees	Train/Test
Dataset 1 - force Dataset	9	8	6	2,000	Transradial amputees	Train: 1, 2, 3 Test: Remaining trials
Dataset 2 - BipPatrec dataset	17	8	26	2,000	Healthy participants	Train: 80% of the data Test: 20% of the data

Cohen's effect size, denoted by the letter d, was applied as a metric to evaluate the significance of the differences that were found.

2.5.1   EXPERIMENT 1: A COMPARISON OF THE ORIGINAL CARDINALITY FEATURES WITH WELL-KNOWN FEATURES AND FEATURE SET

As a first step in analysing the data, we may examine how well the original cardinality features from Table 2.2 perform on the BioPatRec dataset. In this method, the Support Vector Machine classifier (SVM) is utilised. Figure 2.7 depicts the percentage of misclassifications for each of these features; it is clear to see that the error rates for $|A \cup B| - |A \cap B|/|A \cap B|$ and $|A \cup B| - |A \cap B|$ are quite similar. However, they significantly outperformed other methods ($p < 0.01$ and $d > 2$). Figure 2.8 also illustrates how the original cardinality features performed in comparison to the widely used feature sets (Table 2.4) when analysed with the SVM classifier. Cardinality features are clearly outperformed by AR, Horth Hargrove's set, Hudgins' set, Englehart's set, fTDD, and TSD, as evidenced by their p-value < 0.001 and $d > 2$ for all tests.

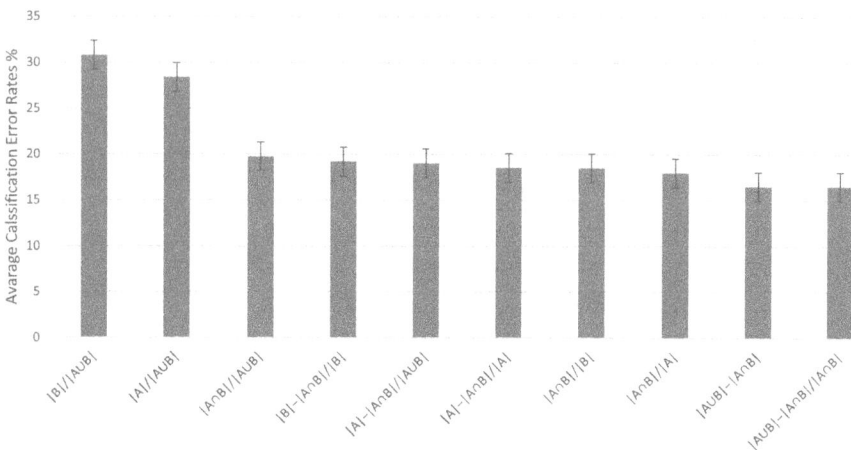

FIGURE 2.7   Average classification error using BioPatRec data set for the original cardinality feature set.

TABLE 2.4

State-of-the-Art Feature Sets to Compare with the Proposed Algorithms

Feature Set Name	No. of Features	Features
Hudgins [19]	5	MAV, MAVS, ZC, SSC and WL
Englehart [20]	4	MAV, ZC, SSC and WL
Hargrove [21]	12	MAV, MAVS, SSC, ZC, WL, RMS and R6
fTDD [16]	6	An irregularity factor, a sparsity measure, a WL ratio represent and the first three even power spectrum moments
TSD [22]	6	The temporal-spatial descriptors feature extraction approach

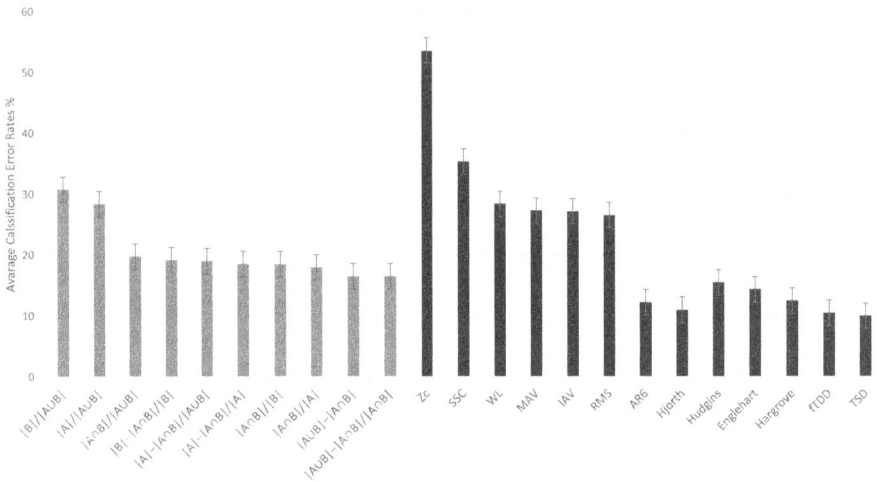

FIGURE 2.8   Average classification error results across 17 subjects in the BioPatRec data set to compare the original cardinality features with well-known features.

Furthermore, the results reveal that fTDD and TSD are similar in terms of performance and outperform the others with $p < 0.01$ and $d = 1.652$.

2.5.2   EXPERIMENT 2: COMPARE THE PROPOSED CARDINALITY FEATURE SET WITH THE ORIGINAL CARDINALITY FEATURES WITHOUT USING SHORT AND LONG-TERM MEMORY CONCEPT

Results from this experiment compare the suggested feature set to those of original cardinality features (Table 2.2) and demonstrate its effectiveness in lowering the classification error rate. In Figure 2.9, we see that the SVM classifier was applied to the BioPatRec dataset without taking into account the concepts of short-term and long-term memory; however, a spatial filter was still used. There was a significant

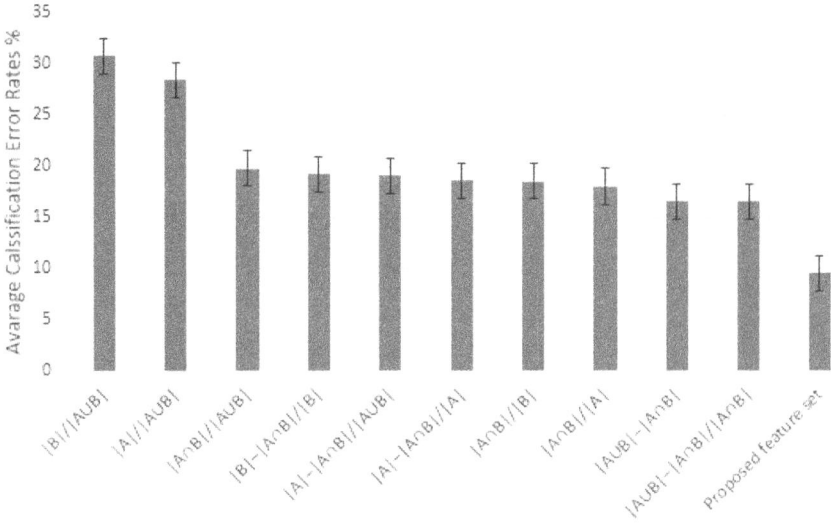

FIGURE 2.9 Average classification error rate for the proposed feature set compared with other cardinality features using SVM classifier for BioPatRec dataset.

improvement in performance using the proposed feature set compared to competing approaches ($p < 0.01$ and $d > 2$).

2.5.3   EXPERIMENT 3: THE EFFECTIVENESS OF SPATIAL FILTER WITH THE PROPOSED FEATURE SET USING DIFFERENT CLASSIFIERS TYPE

Figure 2.10 was generated to illustrate the effectiveness of the suggested spatial filter in identifying the relationship between the EMG channels. The error rate is lowered, as seen in the figure, when the recommended cardinality feature set is used in conjunction with the spatial filter (SF). In this study, we tested three distinct classifiers: KNN, LDA, and SVM. Given that the regularisation parameter C and the kernel function parameter γ significantly impact SVM's performance, the SVM parameters were customised for each dataset. The value of $K = 5$ was settled upon for the KNN classifier. The BioPatRec dataset and the force dataset were used with these classifiers. The SVM classifier significantly outperformed the KNN and LDA ($p < 0.01$ and $d = 2.853$), while LDA also outperformed KNN ($p < 0.01$ and $d = 1.7144$).

2.5.4   EXPERIMENT 4: USING THE PROPOSED METHOD WITH THE SHORT-TERM MEMORY CONCEPT

Following feature extraction, as shown in Figure 2.4, this approach multiplies the features from the current window by those from the previous window and then sigmoid maps the result to keep the range of the features within acceptable limits.

BioPatrec Dataset

Force dataset

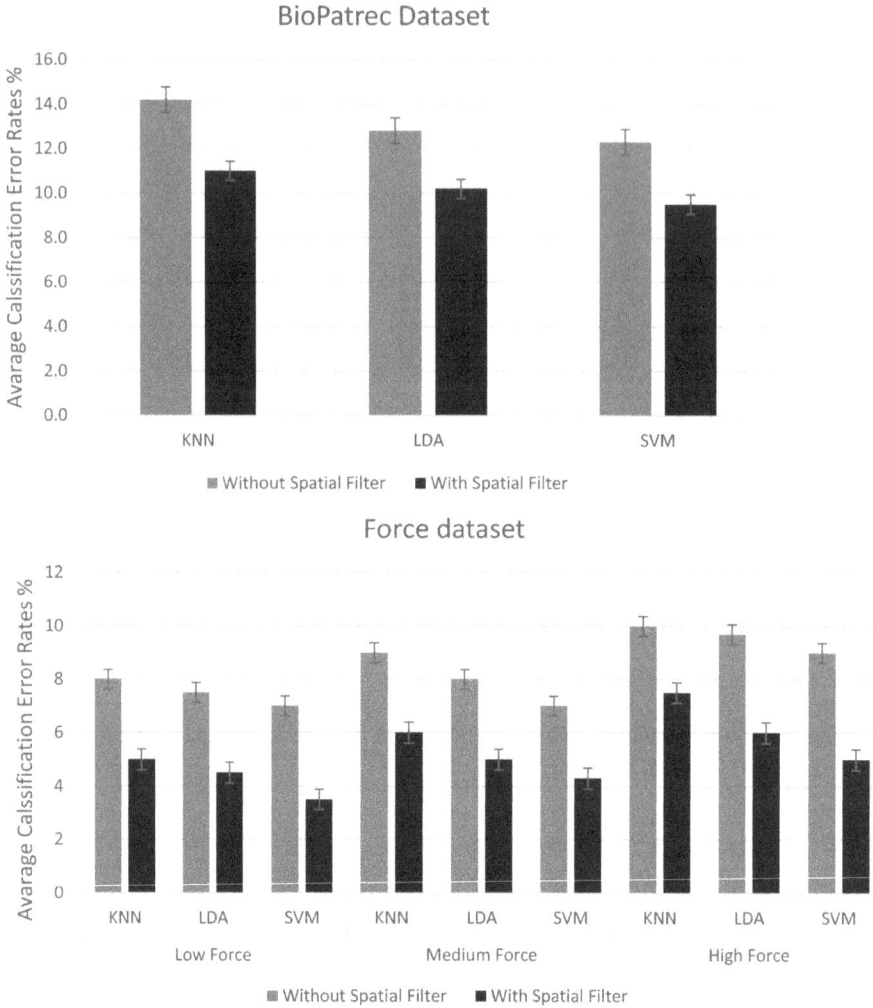

FIGURE 2.10   Average classification error rate with and without using spatial filter using different classifiers for BioPatRec and force dataset.

2.5.4.1   Results of Dataset 1 (Force Dataset)

Figure 2.11 displays the classification results for each force level using the LDA classifier, showing that cardinality performed better than commonly used features and feature sets for the nine transradial amputee force datasets. LDA was chosen because it outperformed SVM and KNN in the third experiment. When compared to other approaches from the literature, the proposed method's performance greatly decreases the classification error, as shown by the Wilcoxon test's findings ($p < 0.001$ and $d > 2$). The proposed method's near-zero error performance at low force is a major improvement over the prior approach. The proposed method also reduces the

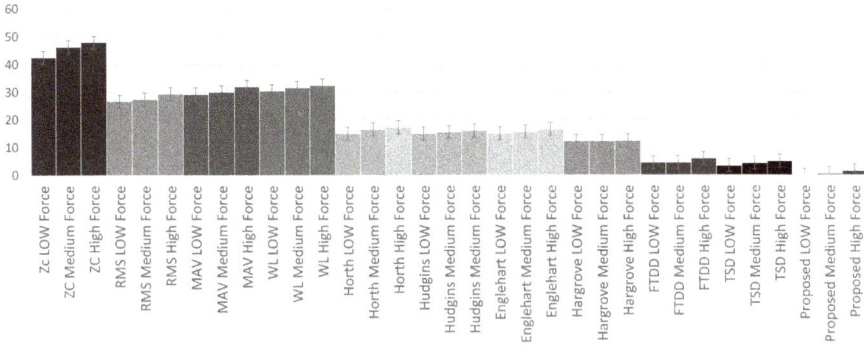

FIGURE 2.11 Shows the effectiveness of the proposed method with different TD features and feature sets using the LDA classifier for the forcing dataset.

error rate for medium and high force, making it the best solution. This experiment demonstrates that the error rate can be greatly reduced with this procedure.

2.5.4.2 Results of Dataset 2 (BioPatRec Data set)

The Bio-Patrec Dataset was used to evaluate the proposed method and compare it to other features and feature sets. The average classification errors across the 17 participants are shown in Figure 2.12 for different feature extractions and the three classifiers of LDA, SVM and KNN. The results demonstrate many important facts. First, the suggested cardinality feature extraction methodology resulted in significantly

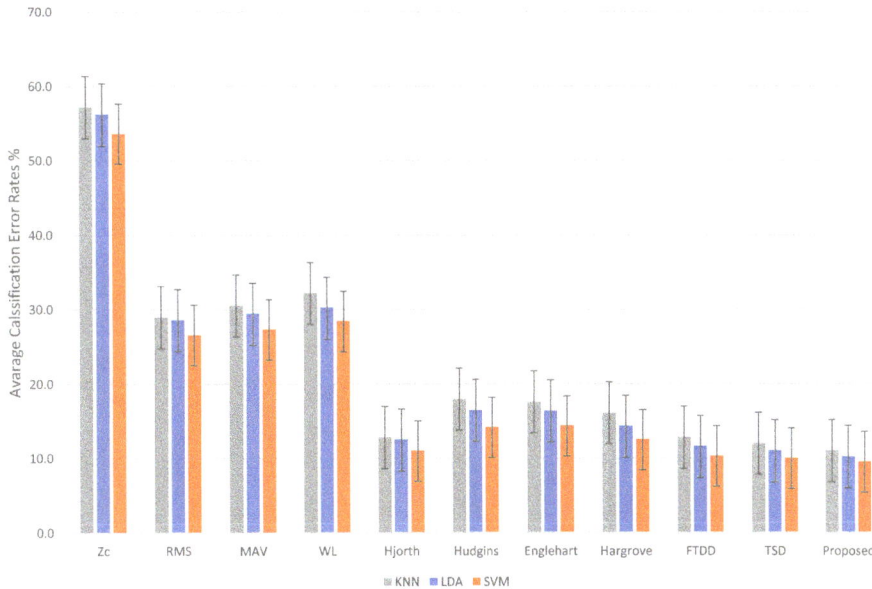

FIGURE 2.12 Average classification error results across 17 subjects using EMG features and feature sets. The LDA, SVM and KNN classifiers were employed.

lower error rates than any other method across all classifiers (p-value < 0.001 and $d > 2$ for all tests except for the comparison against TSD, giving $d = 0.751$). Also, the proposed feature extraction's performance did not differ significantly amongst the three different classifiers. This essential attribute illustrates that the extracted features were resistant to classifier changes.

2.5.5 EXPERIMENT 5: USING THE PROPOSED METHOD WITH THE LONG-TERM MEMORY CONCEPT

This experiment uses the BioPatRec dataset to evaluate the relative efficacy of the long-term memory concept vs. the short-term idea. Figure 2.13 shows that long-term results were significantly better than short-term results ($p < 0.01$). The significance of choosing an appropriate value for the parameter β is illustrated in Figure 2.13 as well. Larger values of β did not significantly improve the performance of the suggested approach.

2.5.6 EXPERIMENT 6: USING THE PROPOSED METHOD WITH DIFFERENT WINDOW SIZE

This experiment aims to determine the best window length for classification by gradually altering the window length from 50 to 500 ms with a constant window increment of 50 ms. Although the experiment with the longest window size achieves the best precision level, the longer window length is not necessarily the best. Researchers claim between 150 and 250 ms is the perfect balance for the window's duration [23]. Figure 2.14 illustrates that the average accuracy increases as the window length increases. The recognition system processing time, window length, and window

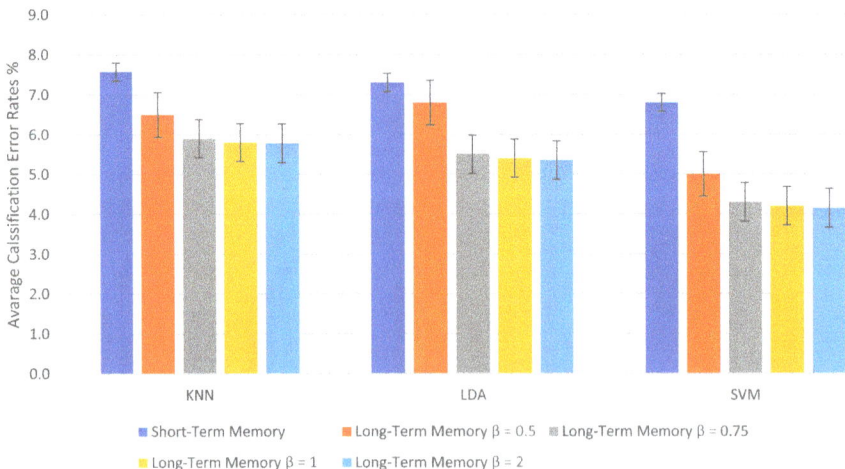

FIGURE 2.13   Average error rate between long-term and short-term memory concept.

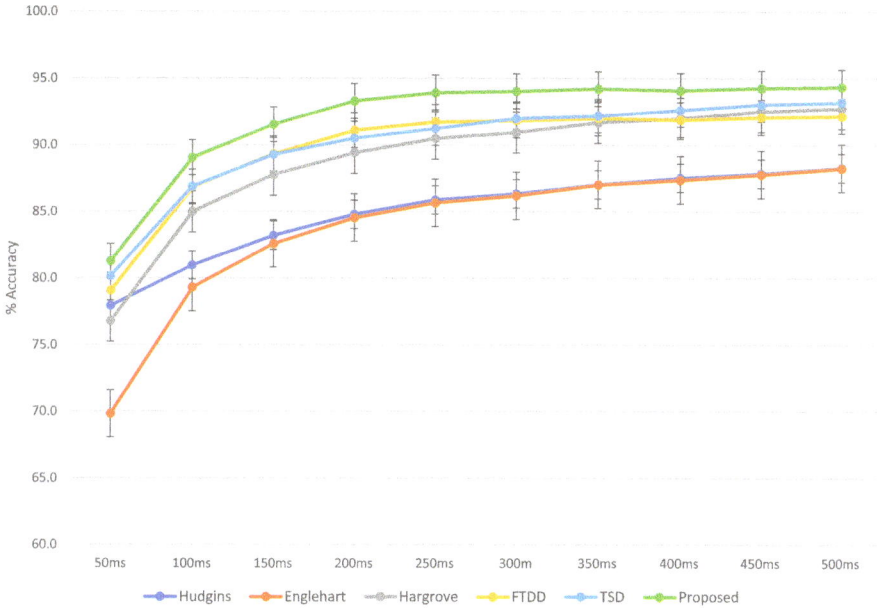

FIGURE 2.14   Classification accuracy averaged across ten different window lengths using the BioPatRec dataset.

increment affect the myoelectric delay time. Ideally, the delay period should be anywhere between 100 and 125 ms, as suggested by T. R. Farrell and Weir [24].

Six feature sets were employed in this work to evaluate the suggested method: Hudgins, Englehart, Hargrove, Hjorth Time Domain (HTD) parameters, Fusion of Time Domain Descriptors (fTDD), and Temporal-Spatial Descriptors (TSD). Figure 2.14 displays the reliability of different feature sets across all participants for windows of 50–500 ms in time. There is a constant increase in accuracy with increasing window length across all feature sets. The proposed feature outperforms other ways in some contexts when compared to combined feature sets, but it does not differ significantly from two other methods (fTTD and STD) when the window size is changed.

2.5.7   Experiment 7: Computing Time

This portion of the study compares the computational time of well-known feature set extraction methods to that of the proposed method. A dataset containing 150 randomly produced samples over ten dimensions was used for this test, which is equivalent to 150 ms with ten channels sampled at 1,000 Hz. The time required to extract the various features was then calculated in MATLAB by running the analysis 1,000 times for each feature type and averaging the results, as shown in Table 2.5.

TABLE 2.5
Computation Time for Feature Extraction Methods

Feature Set	Computing Time (ms)
Hudgins's set	0.0942
Englehart's set	0.0885
Hargrove's set	0.1193
HTD set	0.2636
fTDD	0.1351
TSD	0.8147
Cardinality (proposed)	0.1225

2.6 SUMMARY

The accuracy of myoelectric pattern identification is highly dependent on the signal features employed. Signal processing has a tremendous effect on the values of such features and, consequently, the accuracy of the classification. Various myoelectric features have been utilised as inputs to pattern recognition algorithms for decoding motions. In this study, we propose and assess cardinality in connection to well-known classical time-domain feature sets and additional recently disclosed myoelectric feature sets (like fTTD and TSD).

In this study, we suggested a novel method to use soft cardinality for feature extraction and to obtain spatial information between channels and muscles, combining this with the LSTM-inspired concept of memory. Cardinality is also used to determine the vectors' cosine similarity. The advantages of the suggested method included the ease with which it is implemented based on time-domain features without any complicated processes, and the low levels of classification errors attained based on testing with EMG signals acquired from various databases. This research bridges the gap between academics and industry/clinical practice. In addition, cardinality consistently outperformed other methods despite variations in signal processing, movement sets, and classifiers.

The results indicate that the suggested feature sets of cardinalities with a spatial filter improve classification accuracy by around 40% when compared to other cardinality features. Classification accuracy increased significantly when the concept of short memory was applied to this collection by considering the preceding window. The error was practically eradicated for the Force data set, a feat never before accomplished by any published approach. When utilising long-term memory, the classification accuracy can reach as high as 50%, which substantially improves over more contemporary feature sets such as fTTD and TSD. This method is also shown to be resilient regardless of the classifier employed.

Cardinality features demonstrate good outcomes when compared to other well-known individual features using EMG signals. We believe this method has the potential to significantly improve the performance of feature extraction by modifying certain parameters, such as the selected n^{th} previous window. We further enhance this approach and conduct real-time performance tests to generalise the results.

REFERENCES

1. Roberts, T.J. and A.M. Gabaldón, Interpreting muscle function from EMG: Lessons learned from direct measurements of muscle force. *Integrative and Comparative Biology*, 2008, **48**(2): pp. 312–320.
2. Rahimian, E., et al., FS-HGR: Few-shot learning for hand gesture recognition via electromyography. *IEEE Transactions on Neural Systems and Rehabilitation Engineering*, 2021, 29: pp. 1004–1015.
3. Li, Z., et al. Intelligent classification of multi-gesture EMG signals based on LSTM. In *2020* International Conference on Artificial Intelligence and Electromechanical Automation (AIEA), Tianjin, China, IEEE, 2020,
4. Ortiz-Catalan, M., Cardinality as a highly descriptive feature in myoelectric pattern recognition for decoding motor volition. *Frontiers in Neuroscience*, 2015, **9**: p. 416.
5. Al-Timemy, A.H., et al., Improving the performance against force variation of EMG controlled multifunctional upper-limb prostheses for transradial amputees. *IEEE Transactions on Neural Systems and Rehabilitation Engineering*, 2015, **24**(6): pp. 650–661.
6. Ortiz-Catalan, M., R. Brånemark, and B. Håkansson, BioPatRec: A modular research platform for the control of artificial limbs based on pattern recognition algorithms. *Source Code for Biology and Medicine*, 2013, **8**(1): pp. 1–18.
7. Alon, N., Y. Matias, and M. Szegedy, The space complexity of approximating the frequency moments. *Journal of Computer and System Sciences*, 1999, **58**(1): pp. 137–147.
8. Abedjan, Z., L. Golab, and F. Naumann, Profiling relational data: A survey. *The VLDB Journal*, 2015, **24**(4): pp. 557–581.
9. Heule, S., M. Nunkesser, and A. Hall. Hyperloglog in practice: Algorithmic engineering of a state of the art cardinality estimation algorithm. In *Proceedings of the 16th International Conference on Extending Database Technology*, Genoa, Italy, 2013.
10. Metwally, A., D. Agrawal, and A.E. Abbadi. Why go logarithmic if we can go linear? Towards effective distinct counting of search traffic. In *Proceedings of the 11th International Conference on Extending Database Technology: Advances in Database Technology*, Nantes, France, 2008.
11. Alon, N., Y. Matias, and M. Szegedy, The space complexity of approximating the frequency moments. *Journal of Computer and System Sciences*, 1999, **58**(1): pp. 137–147.
12. Jimenez, S., F. Gonzalez, and A. Gelbukh, Text comparison using soft cardinality. In: E. Chavez and S. Lonardi, S. (eds.), *String Processing and Information Retrieval. SPIRE 2010*. Lecture Notes in Computer Science, vol. 6393. Springer, Berlin, Heidelberg, 2010. https://doi.org/10.1007/978-3-642-16321-0_31
13. Harmouch, H. and F. Naumann, Cardinality estimation: An experimental survey. *Proceedings of the VLDB Endowment*, 2017, **11**(4): pp. 499–512.
14. Rahutomo, F., T. Kitasuka, and M. Aritsugi. Semantic cosine similarity. In *The 7th International Student Conference on Advanced Science and Technology ICAST*, University of Seoul, South Korea, 2012.
15. Geng, W., et al., Gesture recognition by instantaneous surface EMG images. *Scientific Reports*, 2016, **6**(1): pp. 36571.
16. Al-Timemy, A.H., et al., Improving the performance against force variation of EMG controlled multifunctional upper-limb prostheses for transradial amputees. *IEEE Transactions on Neural Systems and Rehabilitation Engineering*, 2016, **24**(6): pp. 650–661.
17. Hochreiter, S. and J. Schmidhuber, Long short-term memory. *Neural Computation*, 1997, **9**(8): pp. 1735–1780.

18. Al Taee, A.A., et al. Cardinality and short-term memory concepts based novel feature extraction for myoelectric pattern recognition. In *2021 43rd Annual International Conference of the IEEE Engineering in Medicine & Biology Society (EMBC)*, Mexico, IEEE, 2021.

19. Hudgins, B., P. Parker, and R.N. Scott, A new strategy for multifunction myoelectric control. *IEEE Transactions on Biomedical Engineering*, 1993, **40**(1): pp. 82–94.

20. Englehart, K., B. Hudgins, and A.D. Chan, Continuous multifunction myoelectric control using pattern recognition. *Technology and Disability*, 2003, **15**(2): pp. 95–103.

21. Hargrove, L.J., K. Englehart, and B. Hudgins, A comparison of surface and intramuscular myoelectric signal classification. *IEEE Transactions on Biomedical Engineering*, 2007, **54**(5): pp. 847–853.

22. Hushaba, R.N., et al., A framework of temporal-spatial descriptors-based feature extraction for improved myoelectric pattern recognition. *IEEE Transactions on Neural Systems and Rehabilitation Engineering*, 2017, **25**(10): pp. 1821–1831.

23. Smith, L.H., et al., Determining the optimal window length for pattern recognition-based myoelectric control: Balancing the competing effects of classification error and controller delay. *IEEE Transactions on Neural Systems and Rehabilitation Engineering*, 2010, **19**(2): pp. 186–192.

24. Farrell, T.R. and R.F. Weir, The optimal controller delay for myoelectric prostheses. *IEEE Transactions on Neural Systems and Rehabilitation Engineering*, 2007, **15**(1): pp. 111–118.

3 Surface Electromyography Sensors for Human Activity Recognition
Recent Advancements and Perspectives

Paolo Crippa, Giorgio Biagetti,
Laura Falaschetti, and Claudio Turchetti
Università Politecnica delle Marche

3.1 INTRODUCTION

According to the *2022 Revision of World Population Prospects*, the estimated total world population aged 65 years and above has reached 761 million in 2020, and is expected to more than double by 2050 reaching nearly 1.6 billion. Globally, the number of people aged 80 years or over is growing even faster than the number aged 65 or over. By 2050, the world is expected to have around 459 million persons aged 80 or more, almost tripling the number of around 155 million in 2021 [1].

The impact on family, community and society caused by these people's aging is becoming more and more relevant in a large part of the world. Thus, more and more efforts are needed to be spent to help and assist elderly people's daily lives, resulting in increasing interest in research and development of methodologies and technologies for healthcare and ambient assisted living (AAL).

Among the age-related diseases, the neurodegenerative ones, such as Alzheimer's disease (AD), Parkinson's disease (PD) and dementia, are often characterized by a progressive decline in functional and cognitive abilities [2]. Thus, the management of people affected by these diseases is very hard and stressful, especially due to the behavioral and psychiatric progressive symptoms that are often present in this kind of patients. Therefore identifying systems and technologies able to help and support caregivers in patient daily management will be of great value. In this manner, the quality of life of patients and their caregivers will be improved, and the social costs linked to the diseases will be reduced. In particular, caregivers in a hospital environment as well as caregivers at home can classify certain activities as unsafe, dangerous

DOI: 10.1201/9781003346678-3

or potentially harmful, and it is of paramount importance to detect such activities in order to provide alarm functions to these patients. At the same time, the detection of other activities performed by these patients could give some insights about their health status [3].

With these considerations in mind, the human activity recognition (HAR) represents one of the most important and promising assistive technologies. It aims to recognize specific activities or a series of specific activities or movements or discover long-term patterns from a set of signals acquired from a large variety of sensors in real life scenarios. On the one hand, monitoring daily activities using HAR technologies can provide a significant contribution to the evaluation of the patient's health status. For example, the distribution of sleep during the 24 hours (e.g., not getting up from bed at regular hours, sleeping during the day, not sleeping during the night) can be related to medical and psychiatric problems, to insomnia or simply to fatigue or tiredness; similarly, changes in routine activities or performing long-term sedentary activities can be related to suffer from certain cognitive problems, or have early symptoms of dementia as well as of other psychomotor pathologies such as multiple sclerosis, and so on. On the other hand, recognizing and monitoring different human activities such as some simple activities performed at home e.g. climbing or descending stairs, opening or closing doors/windows, walking, lying, along with recognizing complex activities e.g. driving a car, cycling, cooking, sleeping, can greatly benefit the surveillance, assistance, and caregiving of elderly people or of people affected by age-related and neurodegenerative diseases. Thus, for quite similar reasons, HAR technologies and methods are of paramount importance in a number of research fields such as ambient assisted living, sport, healthcare, and fitness technologies.

Among different HAR technologies, the ones based on lightweight tiny sensors that can be worn by subjects performing activities are the most promising due to their low invasivity. More specifically, wearable-sensors-based HAR (WSHAR) systems are able to identify personal activities by means of data mining techniques based on data acquired from sensing nodes placed across the human body. Due to their excellent characteristics, such as high recognition performance, user-friendliness and wearability, increased flexibility, wireless capabilities, and increased lifespan, the WSHAR systems have been attracting more and more interest in recent years.

The advancements of the state of the art in micro and nanoelectronics, flexible electronics, integrated circuits, mobile electronics systems, sensor technologies [4], signal processing, and communication network protocols have created a new generation of low-power, low-cost, tiny and lightweight sensors that can easily be embedded into clothes, belts, shoes, sunglasses, watches and smartphones, or positioned directly on the body in order to collect a large amount of data such as body acceleration, orientation and movement, heart rate, muscle fatigue, skin temperature, and so on [5–8]. Data-driven wireless WSHAR systems exploiting all these technologies follow similar procedures that can be schematically described as shown in Figure 3.1.

In the last decade, a number of studies have demonstrated that a quite large set of physical activities can be efficiently recognized by acquiring acceleration signals from sensors placed across the human body. Indeed, the accelerometer signal acquisition is an easy and noninvasive method to evaluate the human activity as it allows to monitor variations in speed and orientation and easily detects body movements

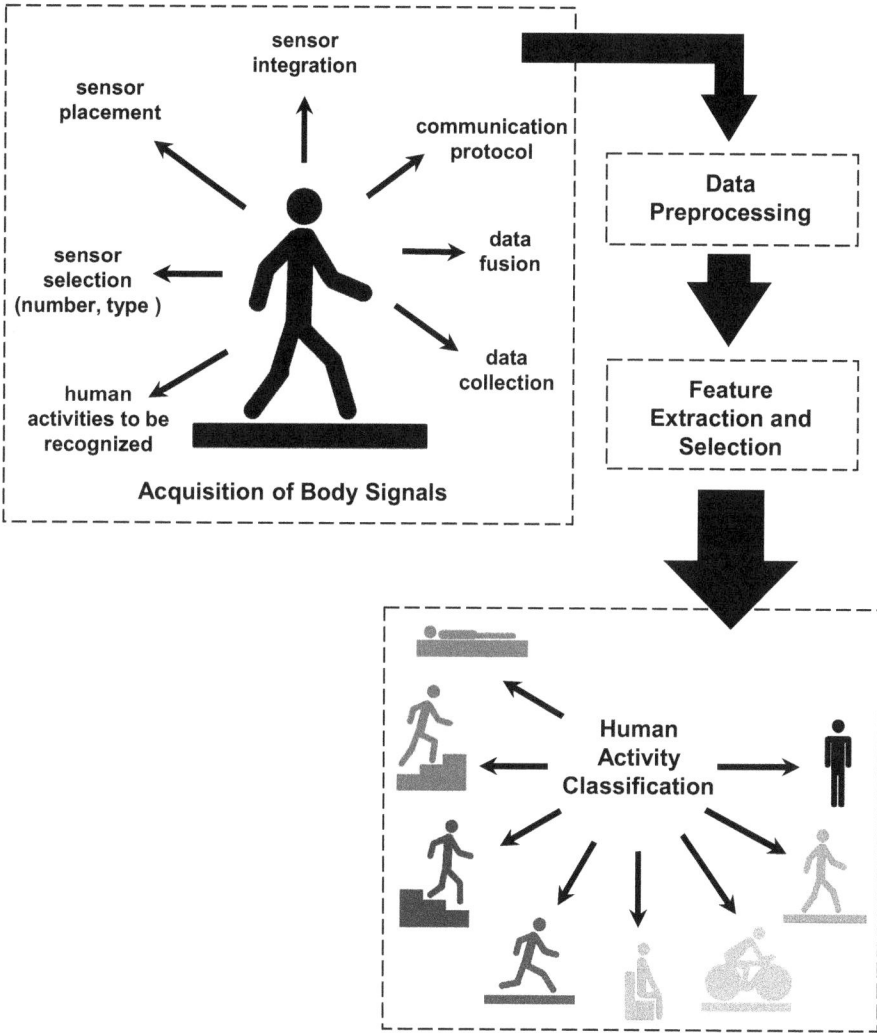

FIGURE 3.1   Flowchart of a typical data-driven wireless WSHAR system.

using a reduced number of sensors. In selected cases, and for a particular set of activities, a single accelerometer sensor is enough to monitor human activity in AAL [9–11]. In particular, Naranjo-Hernández et al. [9] presented both the hardware and the software design of a wearable and low-cost smart accelerometer sensor for the monitoring of physical activity, to be applied in the subject homes or in residential centers for seniors, for real-time remote monitoring or as a local application to be executed offline.

The characterization of some basic activities using a single triaxial accelerometer located at the waist has been performed in Ref. [10]. Here a novel postural detection algorithm based on support vector machines (SVM) method for detection and identification has been optimized to be easily implemented in real-time systems for online monitoring applications.

Similarly, in Ref. [11], the use of a single accelerometer, placed alternatively at the wrist or ankle, has been proposed for activity recognition. Besides, an algorithm to process sensor raw data and to classify behavior into four broad activity classes: sedentary, cycling, ambulation, and other activities, has been presented.

Nowadays, lightweight wireless sensor devices can be comfortably worn during activities of daily living (ADL), e.g. walking, cooking, driving, and sleeping. Thus, many accelerometry-based wearable systems for the pervasive monitoring of ADL have been developed that are simply based on just one inertial sensor. Again, in recent years, the huge development of the smartwatch and smartphone market worldwide has made wearable accelerometer sensors available everywhere to a large number of people. Thus, many WSHAR systems have been developed exploiting techniques based on single sensor data acquired from these personal devices [12–14].

However, they are effective in monitoring and classifying all those activities that involve repetitive body motions, such as running, walking, climbing stairs, cycling, or lifting weights [15]. Thus, to improve the activity detection and classification performance, the placement of more accelerometer sensors across the body has been recently taken into consideration and several systems and methods based on them have been developed [16,17]. For elderly people, these systems could be used for the detection of alarm conditions generated by unusual behaviors of the person (not getting up from bed, no activity during a defined time interval) or changes in routine activities related to psychomotor pathologies or neurodegenerative diseases such as AD. For healthy people, they could be used during sports activities, for evaluating the training level, for counting exercise repetitions in order to track a workout routine as well as determine the energy expenditure of individual movements [18]. Several WSHAR methods using a number of tri-axis acceleration sensors to digitalize patient behavior during their daily activities are therefore available [19]; however, they are very inaccurate giving out multiple readings at once, so they are unreliable in the case of an emergency.

Besides accelerometer signals, other inertial signals such as magnetometric and gyroscope have become easily available. At the same time, a large number of biological signals have been becoming easy to capture from the human body without invasive procedures for obtaining additional accurate and reliable information about body movements and positions, heart rate, muscle fatigue, and human activity and behavior. With the progress of the biosignal processing, more and more information has been derived from biosignals [20–25]. A more accurate detection of activities can be easily reached by acquiring information both on acceleration and other biosignals at the same time. In particular, the surface electromyography (sEMG) signal, which is directly related to muscle's activity, has been demonstrated to be very useful in monitoring a person's body posture and physiological activity [26,27].

With the speech, the most commonly captured biosignals are the electrocardiography (ECG), the electromyography (EMG), the electroencephalography (EEG), and the photoplethysmography (PPG) signals. The recent progress in microelectronics and novel nanotechnologies, in wireless wearable communication systems, as well as in machine learning and big data analysis [28,29] are leading to a variety of innovative health monitoring systems [30] and interfaces for easy interaction with the real world. Tiny, very-low-power wireless multimodal sensors have been embedded into garments, smartwatches, and smartphones, or bends placed straight to the skin for body data collection [31–34]. Thus, systems based on these wearable sensors are becoming very common in many application fields such as healthcare, sport, fitness, ambient assisted living, entertainment, autonomous driving systems, and surveillance-based security [35,36]. Among this plethora of biosignals, recent works have demonstrated how the sEMG signals are very promising in monitoring the body posture, physical performance, and fitness level of a person [7,18,37–40]. This is due to the fact that they can be obtained using intrinsically noninvasive measurement devices, they are relatively easy to acquire and can give a consistent amount of information about body movements.

In this chapter, we will briefly illustrate the recent advancements in wearable sensors and systems for surface electromyography signal acquisition and processing. Then an investigation on human activity recognition systems that are based on two of the most common types of data collected by wearable devices, namely the sEMG and the accelerometer signals, will be presented. Finally, we throw light on recently proposed algorithms and techniques that take advantage of the multimodal signal processing paradigm to improve the accuracy of activity detection and classification.

3.2 RECENT ADVANCEMENTS IN SURFACE ELECTROMYOGRAPHY SENSORS AND SYSTEMS

In the last few years, a new generation of healthcare systems has been proposed due to the recent advances in nanotechnologies, flexible circuits, mobile electronics systems, signal processing, and communication technologies that make it possible to efficiently obtain more and more information from biosignals [21–25]. Telehealthcare and fitness monitoring systems are two examples of areas where research and development in wearable and portable technology have mainly progressed [30].

Surface electromyography has received great attention in the last decade because a large number of research works have demonstrated how this kind of signal is relatively easy to acquire, and thus, it can be considered an important means by which a person's health status [41], body posture [37], fitness level [18], and physical performance can be assessed [38–40]. As an application example, in the automotive scenario, EMG signals are recently being used for developing driving interfaces that allow drivers to accelerate, brake, and steer cars [42] or for monitoring the driver's health and cognitive status in traffic safety. This is due to intrinsically noninvasive sEMG measurement devices. Besides, the electromyographic signal originates from the electrical potentials generated by the contraction of skeletal muscles and can be simply collected with electrodes contacting the surface of the skin [21,43]. The relatively low amplitude of this signal requires carefully designed sensor circuits characterized by high-input-impedance low-noise amplifiers for its recording and processing [44–46]. In some cases, devices capable of acquiring EMG signals could be easily obtained or adapted from those used for measuring the ECG signals [44,47]. However, in general, sensors employed to record bioelectrical potentials generated by muscles are different from the ones used to detect heart activity [48]. The main reason for this differentiation is related to the fact that the ECG signal bandwidth is typically below 100 Hz, whereas EMG signals have significant spectral components extending at least up to 500 Hz, are more likely to be acquired during movements, and are hence affected by artifacts induced by body and cable movements at frequencies typically below 5 Hz. To avoid saturation, these low-frequency motion artifacts must be rejected by the high-input-impedance, low-noise amplifier that should be carefully designed to capture the relatively low amplitude of the EMG signal. Additionally, data derived from a triaxial accelerometer have been demonstrated to be very useful in the reduction of motion artifacts corrupting the surface EMG (sEMG) signals recorded from the wrist, arm, or anywhere else on the body of the subjects that are performing various activities [49]. Therefore, the coexistence of both EMG and accelerometer (inertial) acquisition capabilities in the same sensor can definitively improve its performance when applied in activity detection and classification purposes taking relevant advantage of the multimodal signal processing paradigms.

In the following section, a detailed investigation of the state of the art of multifunctional EMG body sensors capable of acquiring the electromyography signal along with other important vital signals and/or inertial data is proposed.

One of the main problems in building wearable sEMG sensing nodes was their ability to acquire, amplify, digitize, and transmit the biological signals to one or more base stations in a wireless manner satisfying low-cost and low-power constraints for the body area networks based on them. Several of these networks use transmission methods based on a 2.4 GHz radio link that is based on the IEEE 802.15.4 standard protocol [50]. For networks having short range (usually <30 m), operating at low rate speeds (LR-WPAN), the IEEE 802.15.4 standard defines the physical (PHY) and media access control (MAC) layers. In Ref. [51], the maximum throughput and the minimum delay for both the beaconed and the unbeaconed versions of the IEEE 802.15.4 standard have been investigated. In particular, the exact capacity formulae for transmissions between one sender and one receiver for the unbeaconed version are provided. In the unbeaconed case, it is shown that the maximum throughput is not higher than 163 kbit/s when no addresses are used, and that the maximum throughput drops when other address schemes are used. The low cost of this kind of connectivity, related to the low complexity of the wireless link, along with the low power consumption make the IEEE 802.15.4 protocol suitable for vehicular, residential, and industrial sensor networks with relaxed throughput [52]. However, many other attempts were made to convey streaming data over IEEE 802.15.4 [47,53–57]. The applications go from streaming audio over 802.15.4 compliant radios [55], to streaming data in medical applications [47,56,57]. As an example, a dual radio ZigBee homecare gateway to support remote patient monitoring has been presented in Ref. [56]. The ZigBee wireless technology for data/command communications is a 2.4 GHz RF digital communication technology standard by ZigBee Alliance (www.zigbeee.com) based on IEEE 802.15.4 standard and is mainly targeted for short range, less complex and slower applications. Additionally, ZigBee appears to be practically more "open" for researchers, scientists, or noncommercial developers than other technologies such as Bluetooth. Indeed many low-cost evaluation kits and development boards are available in the consumer market. Here the proposed dual radio implementation supports seven polling service sensors (but only one streaming service sensor), increases the ZigBee transmission data rate, and meets the general latency standard of 2 s for telemedicine services as stipulated by the U.S. NIST. In addition, its zero-configuration design makes it capable of guaranteeing a user-friendly, low latency, and highly accurate telehealth service to elderly people and long-term patients at home.

In the same perspective, with the goal to exploit existing low-power and low-cost transceivers, most of body area networks have been designed to use transmission methods based on a 2.4 GHz radio link that is based on the IEEE 802.15.4 protocol.

A wireless biosignal acquisition system, employing ZigBee wireless technology has been designed and implemented by Kobayashi [47,53,54]. The proposed system consists of two basic components: an intelligent electrode device that transmits biosignals and a data acquisition host that receives them. The core of this system is the active electrode called Variable-Adjustable Multi-Purpose Intelligent Remote-controlled Electrode for Biosignal Amplification and Transmission (VAMPIRE-BAT) that performs amplification, filtering, A/D conversion, binary encoding, and wireless transmission of biosignals such as EMG or ECG. It employs the Cypress Semiconductor's Programmable System on Chip CY8C29666 as a micro controller unit (MCU). It streams the data at up to 2 kSps (kilo samples per second) and its parameters can be changed online. However, if many electrodes are connected to the same data acquisition host they share the same bandwidth, and so the capacity of the channel is divided between the electrodes. To achieve full speed transmission when using more than one electrode each one must be connected to a different acquisition host.

Exploiting the IEEE 802.15.4 technology, a wireless surface EMG measurement system, comprised of a preamplifier including an electrode for the measurement of the EMG signal, a main amplifier for signal processing, a digital signal processor (DSP) for A/D conversion and digital signal processing, and a Bluetooth module for wireless communication has been designed in Ref. [57]. Band reject filtering, implemented both in analog (in the amplifier) and digitally (in the DSP), was used to attenuate the 60 Hz power line frequency. As a result, a signal-to-noise ratio of 59 dB has been obtained.

In Ref. [44], an inexpensive and flexible wireless system, called WiSE, that is able to acquire the sEMG signals across the human body using ultralight wearable sensor nodes has been proposed (see Figure 3.2 as a reference). As an additional feature, the bandwidth of the sensor nodes is software-selectable: thanks to this property the system can be easily configured to capture and process ECG signals [58] as well as motion-related signals. The EMG/ECG sensing nodes are able to acquire, amplify, digitize, and transmit the biological signals to one or more base stations through a 2.4 GHz radio link using a custom-made communication protocol designed on top of the IEEE 802.15.4 physical layer, in order to exploit existing low-cost and low-power transceivers but also to enable the possibility of higher throughput and better synchronization than the standard would have allowed.

In Figure 3.3, the WiSE system block diagram is reported. The WiSE sensor comprises an AC-coupled instrumentation amplifier, cascaded with a programmable gain

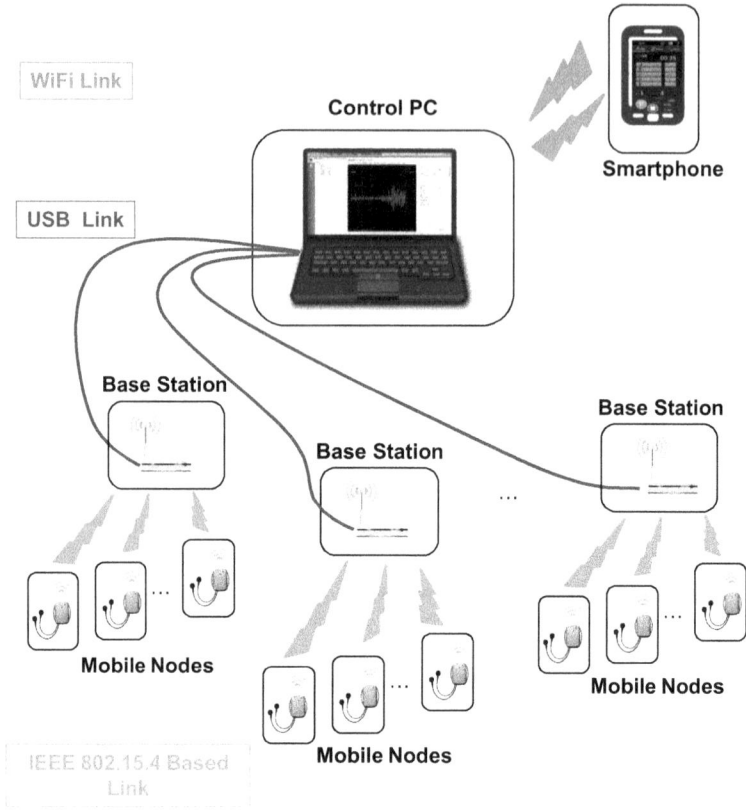

FIGURE 3.2   Flowchart of WiSE system.

amplifier, a low pass filter, and a circuit for the common mode biasing of the instrumentation amplifier and for measuring the electrode contact resistance (magenta
box), a three-axis accelerometer (green box), a circuit for the power supply and
recharging of the battery (red box), and an 8 bit microcontroller (cyan box). The
microcontroller acquires the analog signal from the PGA and transmits it to a base
station using the standard IEEE 802.15.4 physical layer, with a custom protocol. It
communicates through an SPI with an integrated RF transceiver operating in the
2.4 GHz ISM band. It also controls the battery charger and the circuit for measuring
contact resistance. The base station can be powered either by an external power supply or by its USB interface and contains an RF transceiver for wireless connectivity
to mobile nodes, a system for simultaneous charging of up to six mobile nodes, and
a 32 bit microcontroller that manages all the different components. Each base station
is connected through a USB link to a control PC running a user interface software
for displaying, analyzing, and storing the raw data. On the basis of the WiSE system
for acquiring signals from wearable nodes, an activity tracker has been developed
by combining the WiSE system with a smartphone application. The application has
been designed to monitor some fitness metrics extracted from the signals. To this

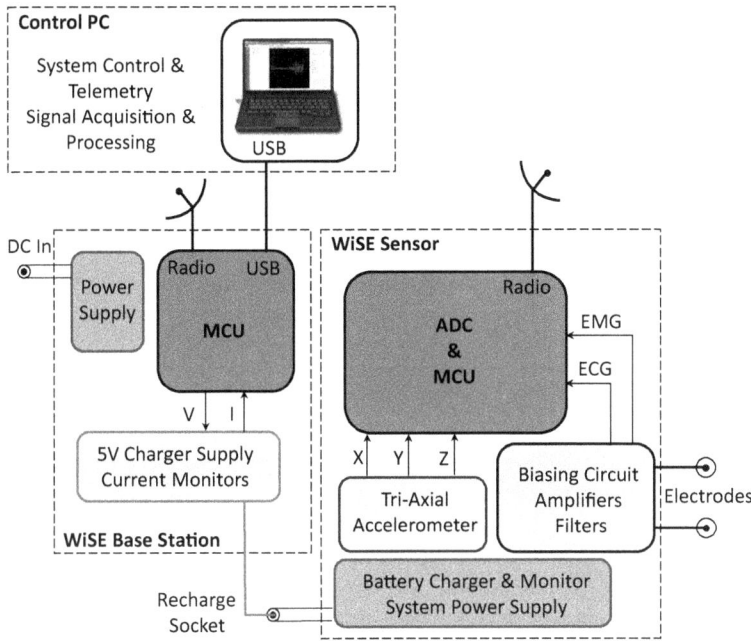

FIGURE 3.3   WiSE system block diagram: view of the three main WiSE system components highlighting the major internal functions.

end, specific algorithms have been implemented at the server side in order to derive the desired metrics.

In Figure 3.4, it is reported that a photo of the WiSE system shows one base station connected via USB to the control PC and four wearable sensor nodes.

The key features of the WiSE system can be summarized as follows: (i) high input impedance: 5 and $20 \, M\Omega$ in common and differential mode, respectively; (ii) 5–500 Hz (for sEMG) and 160 mHz to 500 Hz (for ECG) selectable bandwidth; (iii) over 46 dB of regulation from a, eight-step scope-scaled adjustable gain; (iv) CMRR > 100 dB, (v) 10-bit, 2 kHz (analog to digital converter) ADC sampling; (vi) IEEE-802.15.4 compliant physical layer, 2.4 GHz band for the radio link; (vii) latencies as low as 40 ms thanks to highly-specialized radio protocol; (viii) low energy consumption: more than 8 hours of continuous acquisition; (ix) on-board battery charger and monitor with Coulomb counter for accurate residual run time prediction; (x) reduced PCB sensor size: 41 mm × 26 mm, 12 mm height.

Modifying an off-the-shelf commercial product, Pinto et al. [59] developed a system for acquiring sEMG, ECG, and acceleration signals during swimming and other physical exercises in water environments. In more detail, in order to allow the use in water-related bio-signal research scenario, the BITalino board kit, including EMG, ECG and accelerometer, is placed inside the patient's swimming cap, and the sensors are connected to the prototype as follows: a ground electrode was placed behind the ear over the edge of the skull, three ECG leads were placed in the horizontal plane precordial position, the triaxial accelerometer sensor was placed on the *lumbodorsal*

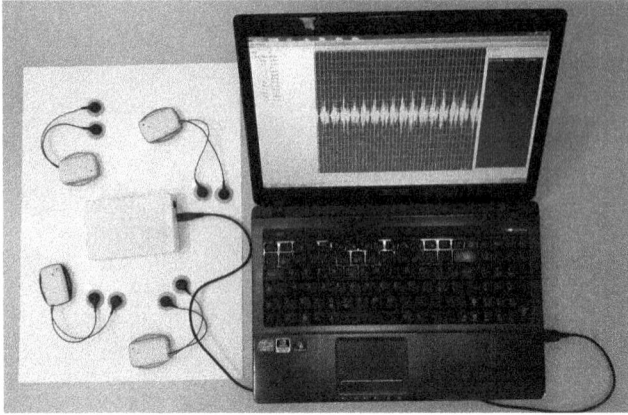

FIGURE 3.4 A WiSE system photo showing one base station connected via USB to the control PC, and four wearable sensor nodes.

fascia and four EMG sensors were placed on *triceps brachii*, *caput laterale* and on *rectus femoris*. It can reliably stream up to 24 kbit/s over BLE.

In Ref. [60], a wearable bracelet for automotive applications was developed to monitor the driver's health status collecting live data through embedded multiple bio-sensors such as EMG, ECG, accelerometer, electrodermal activity, and temperature. Multiple testing scenarios created to simulate real time events have demonstrated the ability of the system to detect health-threatening and emergency situations that can be exploited in semi-self-driving vehicle technologies.

A low-cost, low-power EMG and ECG sensor with high sensitivity, capable of capturing and identifying even a small muscle movement quite distinctly has been proposed in Ref. [61] for wearable biometric and medical applications. The size of the sensor readout PCB is 59 mm × 15 mm and weight <10 g. The input stage is realized with two single-ended passive filters for also removing the dc-offset which enables the design to achieve a CMRR >100 dB. The sensor uses three op-amp instrumenta-tion amplifiers for a total gain ranging from 100 to 15,000. With a gain of 11,400, the noise of the sensor was measured to be 1.68 mVpp, with an estimated power consumption of 1.65 mW. This enables the sensor to work on 3.3 V coin-cell battery for ~700 hours.

Research has also been done in developing an on-silicon system-on-chip (SoC). At chip level, a number of integrated solutions that implement biosignal circuit sen-sor and signal processing have been manufactured using state-of-the-art silicon technologies.

In Ref. [29], a highly integrated batteryless sensor node was implemented in TSMC 0.18 μm CMOS technology with a size of 1.7 mm × 2.5 mm. The SoC sensor node is capable of acquiring EMG and ECG signals with wireless powering and compressed sensing functionalities. The overall implementation contains an analog front-end for biosignal amplification, a high-resolution low-power ADC, a digital signal process-ing stage for data compression, a power management unit, and an RF-to-DC rectifier for wireless energy acquisition.

In Ref. [45], a low-voltage, ultra-low power sensor interface for EMG signal acquisition was fabricated in 65 nm CMOS process. The chip is 0.22 mm² and consumes 3.8 nW. It consists of an amplifier and a successive approximation register (SAR) ADC that works from a 0.3 V supply. The low-voltage amplifier, which is a pseudo-differential topology without a tail current source, provides a noise level of 26 μV_{rms}, 40 dB gain and a state-of-the-art power efficiency factor (PEF) of 2.2 for a 20–425 Hz bandwidth.

In Ref. [62] a wearable platform tied on the chest of the test subject to measure both the EMG and the ECG signals, was developed. The system also presents an additional inertial measurement unit (IMU) sensor to detect the motion artifacts to be removed. Bringing all the sensors on single platform allows us to address the sensor fusion issues. The proposed platform successfully detected the motion artifacts during all the events and the necessary features of ECG and EMG waveforms are very well extracted by using signal processing techniques. The work is a contribution towards the development of integrated wearable biomedical sensors for long-term monitoring.

In Ref. [63], a compact bio-potential measuring system using a custom-designed ultra-low noise instrumentation amplifier (INA) as analog front-end (AFE), was proposed. Both the gain and the bandwidth of the bio-signal measuring system are programmable. Therefore, several bio-potential signals in addition to the EMG one, such as ECG, electrooculography (EOG) and EEG, can be successfully recorded using disposable, off-the-shelf wet Ag/AgCl electrodes. Two versions of the system were proposed: (i) with one INA and measures all the ExG signals in time-multiplexed manner, and (ii) with four INAs that can measure up to four parallel channel ExG simultaneously. In addition, an efficient signal processing algorithm has been developed and integrated within a PC/smartphone user interface.

In Ref. [64], a five-channel EMG and three-axis accelerometer sensor system were implemented for recognizing hand gestures. The accelerometer was placed on the back of the forearm near the wrist to capture information about hand movements whereas the five surface EMG sensors were placed over five forearm muscles *extensor digiti minimi*, *palmaris longus*, *extensor carpi ulnaris*, *extensor carpi radialis*, and *brachioradialis*, respectively. For the EMG sensors, the signals in each channel pass through a two-stage amplifier, with a total gain of 60 dB, over a band from 20 to 1,000 Hz. For each EMG sensor, there are two silver barshaped electrodes with a 10 mm × 1 mm contact dimension and a 10 mm electrode-to-electrode spacing.

In a similar manner, in Ref. [65], a four-channel EMG sensor and a nine-axis motion sensor implemented a framework for real-time American sign language recognition. The motion sensor is based on the InvenSense MPU9150, a combination of three-axis accelerometer, three-axis gyroscope, and three-axis magnetometer. A 32-bit MCU controls the sensor. The board also includes a microSD storage unit and a dual-mode Bluetooth module. The system can be used for real-time data streaming or can store data for later analysis at the sampling rates (for accelerometer and gyroscope) of 100 Hz. It also has an IEEE 802.15.4 wireless module for low-power proximity measurement or ZigBee communication. The four channel sEMG acquisition system uses a TI low-power analog front end to capture the sEMG signals and a TIMSP430 microcontroller to forward data to a PC via Bluetooth. A resolution of 0.4

µV is achieved with a gain of 1 and a sampling rate of 1 kHz is chosen as in several sEMG-based pattern recognition systems.

Again, in Ref. [66], a simple low-cost wireless four-channel EMG sensor network for hand gesture recognition was designed and implemented by having wearable electrode units with multi-channels, portable wireless pocket EMG container. The signals captured from the muscle sensors are transmitted to the 16-bit ADC and then passed into Arduino Nano device. Muscle data were wirelessly transmitted via the Bluetooth protocol.

Following a similar methodology, a low-cost wireless surface EMG sensor network which consists of four surface EMG sensors and a computer software was designed for hand gesture recognition in Ref. [67]. A single-chip microcomputer and a WiFi module are employed in the EMG sensor. Signals acquired by the electrodes are amplified, filtered, level transferred by the analog front end and are digitalized by the C8051F320 built-in 10-bit ADC. The EMG signal sampling rate can be set to 200–1,000 Hz. The digitalized EMG signals are transmitted to the USR-C322 through the UART communication unit under the control of the MCU. The sensor is powered by a 3.7 V, 300 mAh rechargeable lithium battery. Data transmission over computer networks is realized based on the TCP/IP protocol, which uses the WSASsyncSelect socket model for server computer to receive synchronized data from the EMG sensors.

In Ref. [31], a sock-type wearable five-channel sEMG sensor was designed and manufactured for estimating lower leg muscle activity from the distal EMG signals around the ankles. Ten conductive fabric electrodes are placed around the subject ankle to acquire the five EMG signals and one conductive fabric electrode is placed on the medial malleolus for a ground signal. The signals are obtained through 11 snaps from the outer sock surface. Each snap is connected to the commercial data logger BioLog DL-2000 (S&ME Corp.) by wires. The data logger stores all measured signals for further analysis.

New sEMG-based biosensors with the multimodal signal acquisition, based on near-infrared spectroscopy (NIRS), mechanomyography (MMG), forcemyography (FMG) and vibration detection, have been recently proposed.

In Ref. [68] a hybrid sEMG, NIRS, and MMG sensor system that is able to monitor the muscle motion from the modalities of electrophysiology, optics, and acoustics was proposed. Using the hybrid sensor system, an incremental grip force experiment is carried out to explore the relationship between the three signals, blood oxygen metabolism and grip force. The electrodes for collecting the EMG signals include two differential electrodes and one reference electrode. The detection depth of near-infrared light is determined by the distance between the near-infrared light source (LED) and the photodetector. Finally, a high-sensitivity microphone with an electret condenser is used to detect the MMG signal.

In Ref. [69] a modular multimodal EMG–FMG acquisition system with floating electrodes was presented. The overall system consists of eight identical hybrid EMG–FMG sensor modules and a base station. The hybrid EMG–FMG sensor modules implement the signal conditioning and amplification, while the base station is used for analog-to-digital conversion and wireless data transmission. The EMG signal extraction is done by two electrode input buffers, an instrumentation ampli-fier, a fourth-order Butterworth band-pass filter (10–500 Hz), a twin-T notch filter for power line interference removal, and a variable gain circuit. The FMG is measured by a commonly used force-sensing resistor (FSR400, Interlink Electronics). An inno-vative floating electrode design for the mechanical structure of the sensor modules guarantees the time and space synchronization of the EMG and FMG signals.

In Ref. [30], a multimodal EMG, ECG, vibration, and temperature sensing system for elderly care and precision sport study, was proposed. The system contains an AFE for biosignal sensing, implemented in 180-nm TSMC technology. It includes an EMG/ECG signals acquisition chain, a piezo wave sensing circuit for vibration sensing, a low-power ADC with pulse-width modulation digital output, and a band-gap reference circuit for temperature tracking. A digital signal controller for AFE control and calibration, an ARM-like MCU for compression/communication, and a phase-lock loop circuit for system clock generation have been used. Analog circuit was integrated and tested with Xilinx Zynq System-on-Module (SoM).

However, most of the research that has been devoted to these sEMG-based body sensors has been focused on data acquisition and signal conditioning for data com-pression and activity recognition, while no particular attention has been devoted to the data transmission through the wireless channel and to the overall system optimi-zation. Standard protocols are often used such as BLE, resulting in sub-optimal per-formance, like low throughput [59], or 100 ms latency and only 6 hours of operation as in the work just cited above [30].

In Ref. [70] a three-channel wireless sensor for either EMG or ECG signal acqui-sition with an inertial platform for simultaneously capturing movement information was proposed. In Figure 3.5, a photo of the SoM of the sensor is displayed. It uses a BLE radio for real-time data transmission so as to minimize energy consumption, and it is characterized by a carefully designed transmission scheduling for maximiz-ing the data throughput and maintaining latency reasonably low, i.e. latencies lower than and net data rates (not counting protocol overhead) in excess of 80 kbits/s, as displayed in Figures 3.6 and 3.7.

FIGURE 3.5 Photograph of the sensor prototype built for testing and evaluation purposes. The EMG/ECG sensing chip is U2, the inertial sensor is U5.

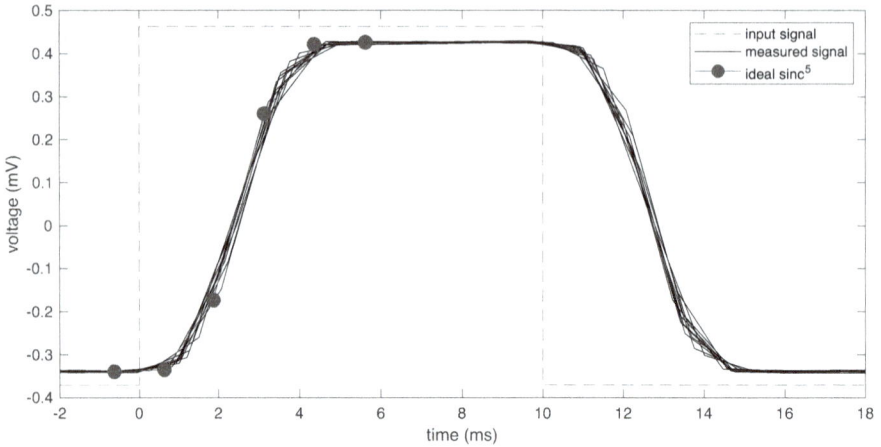

FIGURE 3.6 Measured output from the EMG signal chain corresponding to a rectangular impulse digitally provided (and hence shown not to scale). An ensemble of ten different measures across a two-minute time span is reported to show clock synchronization effectiveness. For reference, the ideal step response of a $sinc^5$ digital filter like those used in the analog front-end is reported as obtained by a VHDL simulation. The ODR was set to 800 Hz and so samples are apart. (Reprinted, under the terms and conditions of the Creative Commons Attribution (CC BY) license, from Ref. [70].)

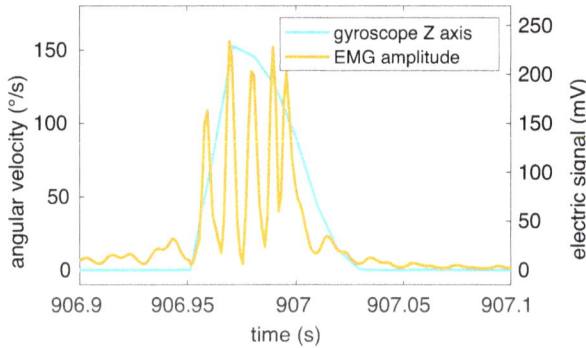

FIGURE 3.7 Example of the synchronization between the EMG and the gyroscope sensors. A stimulus causing simultaneous electrical and mechanical excitations of the device was applied after about 15 minutes of operation. Since the two sensors have independent clocks, without synchronization a drift of several seconds would have accumulated after 15 minutes. The reclocked signals are instead synchronized to within sampling resolution. (Reprinted, under the terms and conditions of the Creative Commons Attribution (CC BY) license, from Ref. [70].)

An adaptation layer on top of BLE allows tight time synchronization and provides the reliability of the data transmission by ensuring lossless communication with up to one-third of the radio channel capacity compromised, making this sensor suitable for performing research or real-time monitoring. Moreover, lossless compression

TABLE 3.1

Key Feature Comparison of Some of the Most Relevant Commercial and Non Commercial sEMG Sensors Available in Literature

References	Acquired Signals	Power (mW)	Size (mm²)	Weight (g)	#Ch	S.	Freq (kHz)
[70]	sEMG/ECG/acc.	18.4	1,377	40	3		3.2
[24]	sEMG	96	N.A.	N.A.	1–32		N.A.
[72]	sEMG	N.A.	N.A.	13	20		1
[61]	sEMG/ECG	1.65	885	10	1		N.A.
[47]	sEMG/ECG	220	N.A.	12	≤4		≤2
[36]	sEMG/ECG	169.3	N.A.	N.A.	1		0.256
[73]	sEMG/ECG/temp./acc.	50	978	23	4		2
[74]	sEMG	16.2	N.A.	N.A.	1–2		2

algorithms tuned for EMG signals, like a dedicated variable-length differential encoding, can also be implemented inside the MCU that controls the sensor [71], further reducing power consumption and radio channel occupancy without increasing latency. The overall system is characterized by energy efficiency and ease of use: indeed, the motion-triggered wake-up capabilities provided by the accelerometer helped achieve both goals simultaneously, as the system just activates itself when worn, without needing any operator intervention. Additionally, using BLE allows to use commodity hardware such as smartphones or laptops as data receivers, further simplifying deployment.

Finally, to have a complete scenario of the sEMG wearable technology, Table 3.1 reports the comparison of some of the most relevant commercial and non-commercial sEMG sensors available in the literature.

3.3 SURFACE ELECTROMYOGRAPHY FOR HUMAN ACTIVITY RECOGNITION: PAST, PRESENT, AND FUTURE PERSPECTIVES

As it has been discussed in the previous section, the sEMG signal is well suited for monitoring person's body posture, physical performance, and fitness level because it can be obtained using intrinsically noninvasive measurement devices and is relatively easy to acquire [7,18,37–40]. Indeed, the sEMG signals are generated from the electrical potentials produced during muscle contractions [21,43] and they can be collected simply by placing electrodes on the skin surface. However, these electrical signals have relatively low amplitudes, and to be processed, they need carefully designed high-input-impedance, low-noise amplifiers [44].

In Refs. [27,73] a human activity detection system based on sEMG signals and inertial data has been proposed. The measurements were acquired using the low-cost flexible wireless sEMG WiSE system displayed in Figure 3.4 with three sEMG/ acceleration wearable wireless sensor nodes. Here some simple daily activities have been considered such as opening a window (OW) and closing a window (CW), opening a drawer (OD) and closing a drawer (CD), rotating a steering wheel left (SL) and right (SR), opening a sliding door (OS) and closing a sliding door (CS). First of all, the placement and number of sensors on the upper limb have been accurately optimized. Four healthy subjects, three males and one female, were asked to perform these simple daily activities while wearing three sensing nodes on their right arm. The first node, called "A", acquired EMG and acceleration signals from the forearm, specifically from the *flexor carpi radialis* muscle, the other nodes acquired the same set of signals from the upper arm, namely, "B" acquired the EMG signal from the *biceps brachii*, and "C" from the *triceps brachii*.

Each activity was repeated multiple times. Table 3.2 summarizes the number of recordings for each of them. Data from two subjects (one male and one female) was used as training material, and the other two subjects for testing purposes.

The accelerations and the EMG signal captured by the three sensors are reported in Figures 3.8 and 3.9 where two samples for each of the eight selected activities are displayed. Triaxial accelerations (represented by blue, red and green lines) are sampled at 125 Hz and the sEMG signals (represented by black lines) at 2 kHz. Raw data are shown in thin, grayish lines. Post-processed data are shown in thick lines. Data processing for the acceleration signals consisted of a low-pass filter with a 2.5 Hz cut-off frequency. The EMG signal was first rectified, then low-pass filtered with the same 2.5 Hz cut-off frequency, and finally resampled to match the 125 Hz sampling rate of the acceleration signal. The black thick lines in Figures 3.8 and 3.9 are multiplied by $\pi/2$ only for display purposes (to visually compensate for the crest factor of the signals) and this amplification has not been used in the classification algorithm. The start and the end points of the eight activities have been marked by manually annotating the recorded data.

The classification has been performed using the k-nearest neighbors (kNN) algorithm with $k = 3$ and the *dynamic time warping* distance was used as the classifier metric. No further processing has been made on the acquired data. In order to identify which sensor locations are more useful to discriminate this set of activities, the activity classification was performed either including or excluding data from a particular sensor. Starting from the three EMG signals and three (three-dimensional) acceleration signals recorded from the three EMG/acceleration sensor nodes, a total of $2^3 \times 2^3 = 64$ combinations have been considered either including or excluding data from a particular sensor. A classification trial on the basis of these combinations has allowed us to investigate which sensors (in terms of kind and location) are more useful to identify this set of activities. As a result of this investigation, there is

TABLE 3.2

Database Consistency

Split	OW	CW	OD	CD	SR	SL	OS	CS	All
Train	20	19	20	20	19	18	20	20	156
Test	12	12	28	28	27	27	32	33	199
Total	32	31	48	48	46	45	52	53	355

Number of recorded samples of each activity, split between training and testing material.

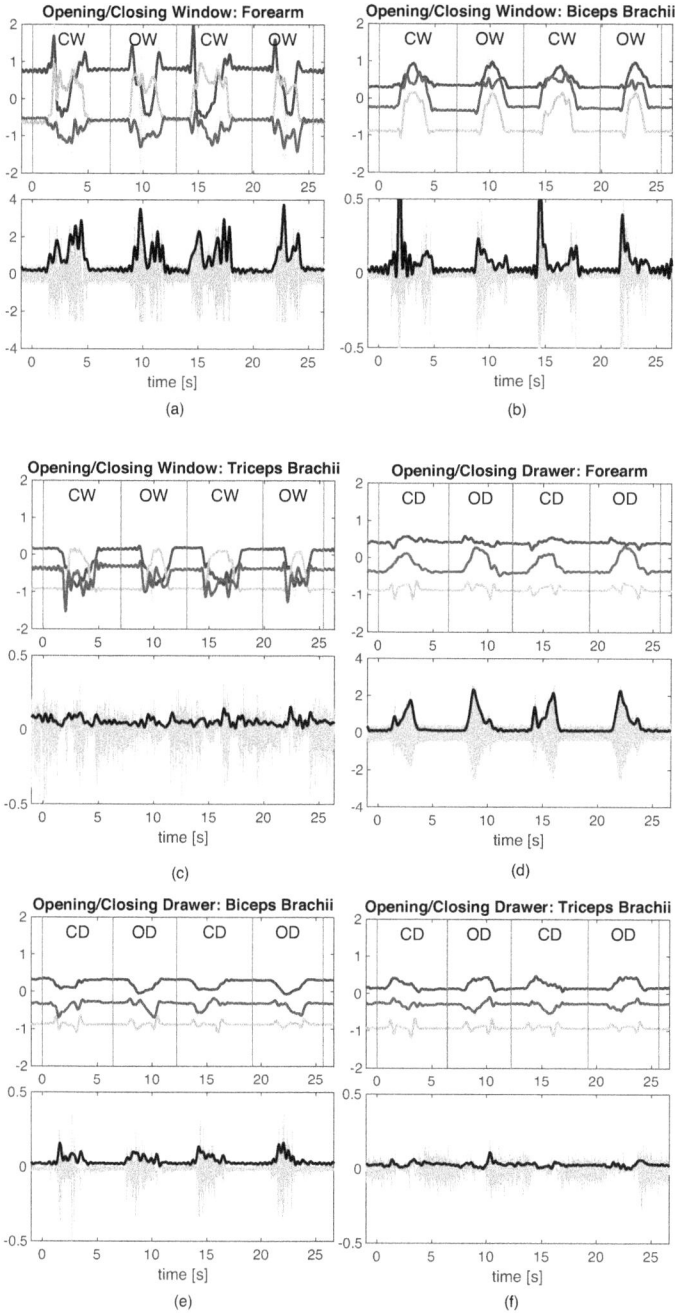

FIGURE 3.8 In the left column: the accelerations and the sEMG signals recorded from the right arm while opening (OW) and closing (CW) a window. In the right column: the accelerations and the sEMG signals recorded from the right arm while opening (OD) and closing (CD) a drawer.

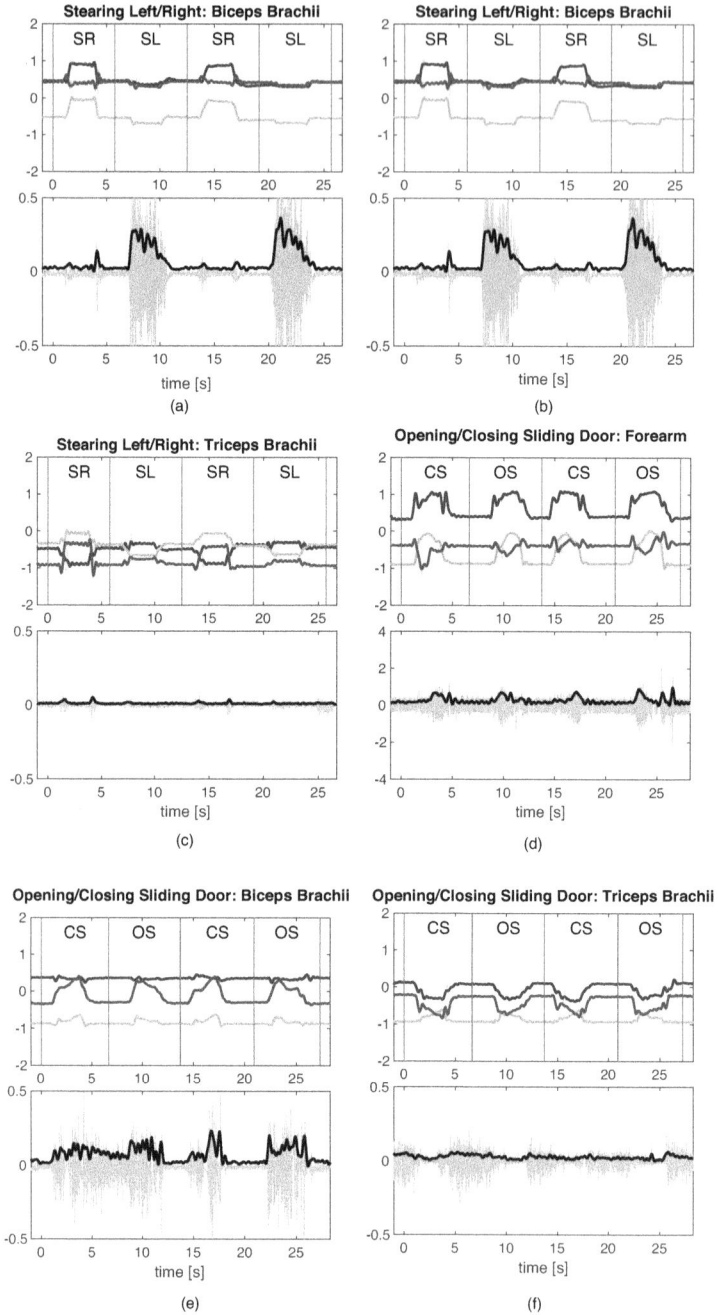

FIGURE 3.9 In the left column: the accelerations and the sEMG signals recorded from the right arm while rotating a steering wheel left (SL) and right (SR). In the right column: the accelerations and the sEMG signals recorded from the right arm while opening (OS) and closing (CS) a sliding door.

basically only some confusion distinguishing between opening and closing a drawer. Indeed these two activities are quite similar from an acceleration point of view and could have benefited from using the EMG signals recorded from the sensor placed on biceps brachii. Indeed, the combination, acceleration from the forearm and triceps brachii and EMG from biceps brachii, has been demonstrated to be the best option.

In its best case configuration, the confusion matrix derived from the classifier is displayed in Table 3.3, and an analysis of the classification performance is shown in Table 3.4, which reports for each of eight classes the sensitivity (or recall), the precision along with their harmonic mean, i.e. the F_1-score, where these metrics are defined as:

$$\text{Sensitivity} = \text{TP}/(\text{TP} + \text{FN}) \tag{3.1}$$

$$\text{Precision} = \text{TP}/(\text{TP} + \text{FP}) \tag{3.2}$$

$$F_1\text{-score} = 2 \cdot \text{sensitivity} \cdot \text{precision}/(\text{sensitivity} + \text{precision}) \tag{3.3}$$

where TP are the true positives (the diagonal elements of the confusion matrix), FN are the false negatives (the sum of the other elements on the same row of the confusion matrix), and FP are the false positives (the sum of the other elements on the same column of the confusion matrix).

Finally, in this case of a single sEMG sensor, an overall recognition rate of 92.5% was obtained using only EMG data from the sensor placed on biceps brachii and the acceleration data from the remaining two sensors placed on the forearm and triceps brachii.

Moreover, in the area of construction safety and health, the human activity recognition systems based on wearable sensors have gained increased interest due to low cost, ease of use, and non-invasivity. Indeed, classifying workers' activity allows to easily monitor and manage the safety, job quality, and efficiency of workers. In

TABLE 3.3

Confusion Matrix of the Classifier, Using Accelerometer Data Collected from Nodes A and C Only, Placed on the Forearm and Upper Arm, and EMG Data Collected from the Biceps Brachii, Resulting in a 92.46% Overall Accuracy

	OW	CW	OD	CD	SR	SL	OS	CS
OW	12	0	0	0	0	0	0	0
CW	0	12	0	0	0	0	0	0
OD	0	0	27	1	0	0	0	0
CD	0	0	10	18	0	0	0	0
SR	0	0	0	0	27	0	0	0
SL	0	0	0	0	0	27	0	0
OS	0	0	0	0	0	0	31	1
CS	0	0	0	0	0	0	3	30

TABLE 3.4
Classification Performance List for
the Eight Activities (%)

Activity	Sensitivity	Precision	F_1-Score
OW	100.00	100.00	100.00
CW	100.00	100.00	100.00
OD	96.43	72.97	83.08
CD	64.29	94.74	76.60
SR	100.00	100.00	100.00
SL	100.00	100.00	100.00
OS	96.88	91.18	93.94
CS	90.91	96.77	93.75

the construction domain, Bangaru et al. [75] reported a study that evaluates the data quality and reliability of forearm EMG and IMU of armband sensors for human activity classification. To achieve the proposed investigation, the forearm EMG and IMU data were collected from eight workers while performing construction activities such as screwing, wrenching, lifting, and carrying and different classification experiments were performed. The results conclude that the combined EMG and IMU signals classify activities with higher accuracies compared to individual sensor data giving similar findings of Refs. [27,73].

Additionally, it has been demonstrated that the EMG signal can be usefully used for simultaneously obtaining the repetition frequency and evaluating the muscular fatigue during the accomplishment of exercises, repetitive activities and cyclic movements such as walking, cycle riding, weight lifting, and so on. In Ref. [7] a methodology, based on multicomponent AM-FM modeling uses only the EMG signal acquired by a lightweight wireless electromyograph applied close to the involved muscles or evaluating fatigue and repetition rate. To do this, the mean frequency of the amplitude spectrum of the EMG signal has been considered as a function of time and has directly been related to the dynamics of the movement performed and to the fatigue of the muscles. This allows to conclude that wearing tiny lightweight sEMG sensors with IMU on the body, or embedding them on smart clothing, can easily monitor the type of activity we are doing, and at which level of intensity.

3.4 CONCLUSIONS

This chapter presented recent advances in wearable systems based on sEMG wireless sensors for recognizing, monitoring, and evaluating body movements. In particular, the state of the art of circuits and systems for acquiring sEMG signals together with easily detectable inertial data available in consumer electronic devices has been reported and discussed here. Also, recent advances on algorithms and techniques based on multimodal signal processing for the recognition of human activities from sEMG-based sensors have been investigated. Finally, some application examples, taken from recent literature, have been presented in order to shed light on the potential of these wearable systems for monitoring and managing the health and safety of patients and sporting persons as well as for assessing the quality of work, the level of safety, and productivity of the workers. Moreover, with the continuing advances of the technology related to wearable sensors, and with the foreseeable perspective of embedding them into regular clothing or similarly commonly worn items, the usage of sEMG can be predicted to become as common as the inertial sensing, opening the road to a widespread adoption and diffusion of advanced, personalized healthcare and wellness.

REFERENCES

1. United Nations, "World Population Prospects – Population Division," https://population.un.org/wpp/, accessed: 06-March-2023.
2. G. M. McKhann, D. S. Knopman, H. Chertkow, B. T. Hyman, C. R. Jack, C. H. Kawas, W. E. Klunk, W. J. Koroshetz, J. J. Manly, R. Mayeux, R. C. Mohs, J. C. Morris, M. N. Rossor, P. Scheltens, M. C. Carrillo, B. Thies, S. Weintraub, and C. H. Phelps, "The diagnosis of dementia due to Alzheimer's disease: Recommendations from the National Institute on Aging-Alzheimer's Association workgroups on diagnostic guidelines for Alzheimer's disease," *Alzheimer's & Dementia*, vol. 7, no. 3, pp. 263–269, 2011.
3. N. K. Vuong, S. Chan, C. T. Lau, S. Y. W. Chan, P. L. K. Yap, and A. S. H. Chen, "Pre- liminary results of using inertial sensors to detect dementia-related wandering patterns," in *37th Annual Int. Conf. IEEE Engineering in Medicine and Biology Society (EMBC)*, Milan, Italy, Aug. 2015, pp. 3703–3706.
4. A. De Vita, G. D. Licciardo, L. D. Benedetto, D. Pau, E. Plebani, and A. Bosco, "Low-power design of a gravity rotation module for HAR systems based on inertial sensors," in *IEEE 29th Int. Conf. Application-specific Systems, Architectures and Processors*, July 2018, pp. 1–4.
5. H. Yu, S. Cang, and Y. Wang, "A review of sensor selection, sensor devices and sensor deployment for wearable sensor-based human activity recognition systems," in *10th Int. Conf. Software, Knowledge, Information Management Applications*, Dec. 2016, pp. 250–257.
6. A. Bacà, G. Biagetti, M. Camilletti, P. Crippa, L. Falaschetti, S. Orcioni, L. Rossini, D. Tonelli, and C. Turchetti, "CARMA: A robust motion artifact reduction algorithm for heart rate monitoring from PPG signals," in *23rd European Signal Processing Conference*, Sept. 2015, pp. 2696–2700.
7. G. Biagetti, P. Crippa, A. Curzi, S. Orcioni, and C. Turchetti, "Analysis of the EMG signal during cyclic movements using multicomponent AM-FM decomposition," *IEEE Journal of Biomedical and Health Informatics*, vol. 19, no. 5, pp. 1672–1681, Sept. 2015.
8. J. Taborri, J. Keogh, A. Kos, A. Santuz, A. Umek, C. Urbanczyk, E. van der Kruk, and S. Rossi, "Sport biomechanics applications using inertial, force, and EMG sensors: A literature overview," *Applied Bionics and Biomechanics*, vol. 2020, 2020.
9. D. Naranjo-Hernández, L. M. Roa, J. Reina-Tosina, and M. A. Estudillo-Valderrama, "SoM: A smart sensor for human activity monitoring and assisted healthy ageing," *IEEE Trans. Biomed. Eng.*, vol. 59, no. 11, pp. 3177–3184, Nov. 2012.
10. D. Rodriguez-Martin, A. Samà, C. Perez-Lopez, A. Català, J. Cabestany, and A. Rodriguez-Molinero, "SVM-based posture identification with a single waist-located tri-axial accelerometer," *Expert Systems with Applications*, vol. 40, no. 18, pp. 7203–7211, 2013.

11. A. Mannini, S. S. Intille, M. Rosenberger, A. M. Sabatini, and W. Haskell, "Activity recognition using a single accelerometer placed at the wrist or ankle," *Medicine and Science in Sports and Exercise*, vol. 45, no. 11, pp. 2193–2203, 2013.

12. F. Miao, Y. He, J. Liu, Y. Li, and I. Ayoola, "Identifying typical physical activity on smartphone with varying positions and orientations," *BioMedical Engineering Online*, vol. 14, no. 1, 2015.

13. S. Dernbach, B. Das, N. C. Krishnan, B. L. Thomas, and D. J. Cook, "Simple and complex activity recognition through smart phones," in *8th International Conference on Intelligent Environments*, June 2012, pp. 214–221.

14. A. M. Khan, Y.-K. Lee, S. Y. Lee, and T.-S. Kim, "Human activity recognition via an accelerometer-enabled-smartphone using kernel discriminant analysis," in *2010 5th International Conference on Future Information Technology*, May 2010, pp. 1–6.

15. G. Biagetti, P. Crippa, L. Falaschetti, S. Orcioni, and C. Turchetti, "An efficient technique for real-time human activity classification using accelerometer data," in *Intelligent Decision Technologies*. Cham, Switzerland: Springer, 2016, pp. 425–434.

16. A. Mannini and A. M. Sabatini, "Machine learning methods for classifying human physical activity from on-body accelerometers," *Sensors*, vol. 10, no. 2, pp. 1154–1175, 2010.

17. C. Catal, S. Tufekci, E. Pirmit, and G. Kocabag, "On the use of ensemble of classifiers for accelerometer-based activity recognition," *Applied Soft Computing*, vol. 37, pp. 1018– 1022, 2015.

18. G. Biagetti, P. Crippa, L. Falaschetti, S. Orcioni, and C. Turchetti, "A rule based framework for smart training using sEMG signal," in *Intelligent Decision Technologies*, ser. Smart Innovation, Systems and Technologies. Cham, Switzerland: Springer, 2015, vol. 39, pp. 89–99.

19. D. Sánchez, M. Tentori, and J. Favela, "Activity recognition for the smart hospital," *IEEE Intelligent Systems*, vol. 23, no. 2, pp. 50–57, Mar. 2008.

20. G. Biagetti, P. Crippa, L. Falaschetti, S. Orcioni, and C. Turchetti, "Human activity recog- nition using accelerometer and photoplethysmographic signals," in *Intelligent Decision Technologies 2017*, vol. 73. Cham, Switzerland: Springer, 2018, pp. 53–62.

21. G. Biagetti, P. Crippa, S. Orcioni, and C. Turchetti, "Surface EMG fatigue analysis by means of homomorphic deconvolution," in *Mobile Networks for Biometric Data Analysis*. Cham, Switzerland: Springer, 2016, pp. 173–188.

22. S. C. Mukhopadhyay, "Wearable sensors for human activity monitoring: A review," *IEEE Sensors Journal*, vol. 15, no. 3, pp. 1321–1330, Mar. 2015.

23. P. Crippa, A. Curzi, L. Falaschetti, and C. Turchetti, "Multi-class ECG beat classification based on a Gaussian mixture model of Karhunen-Loève transform." *International Journal of Simulation–Systems, Science & Technology*, vol. 16, no. 1, 2015.

24. D. Brunelli, A. M. Tadesse, B. Vodermayer, M. Nowak, and C. Castellini, "Low-cost wearable multichannel surface EMG acquisition for prosthetic hand control," in *6th Int. Workshop on Advances in Sensors and Interfaces*, June 2015, pp. 94–99.

25. G. Biagetti, P. Crippa, A. Curzi, S. Orcioni, and C. Turchetti, "A multi-class ECG beat classifier based on the truncated KLT representation," in *2014 European Modelling Symposium*, Oct. 2014, pp. 93–98.

26. G. Biagetti, P. Crippa, L. Falaschetti, and C. Turchetti, "Classifier level fusion of accelerometer and sEMG signals for automatic fitness activity diarization," *Sensors*, vol. 18, no. 9, p. 2850, Aug 2018.

27. G. Biagetti, P. Crippa, L. Falaschetti, S. Orcioni, and C. Turchetti, "A portable wireless sEMG and inertial acquisition system for human activity monitoring," *Lecture Notes in Computer Science*, vol. 10209 LNCS, pp. 608–620, 2017.

28. Y.-Z. Wang, Y.-P. Wang, Y.-C. Wu, and C.-H. Yang, "A 12.6 mW, 573–2901 kS/s recon-figurable processor for reconstruction of compressively sensed physiological signals," *IEEE Journal of Solid-State Circuits*, vol. 54, no. 10, pp. 2907–2916, 2019.

29. Y.-H. Tu, K.-W. Yao, M.-H. Huang, Y.-Y. Lin, H.-Y. Chi, P.-M. Cheng, P.-Y. Tsai, M.-T. Shiue, C.-N. Liu, K.-H. Cheng, and J.-S. Fu, "A body sensor node SoC for ECG/EMG applications with compressed sensing and wireless powering," in *Int. Symp. on VLSI Design, Automation and Test (VLSI-DAT 2017)*, 2017, pp. 1–4.

30. H.-K. Dow, I.-J. Huang, R. Rieger, K.-C. Kuo, L.-Y. Guo, and S.-J. Pao, "A bio-sensing system-on-chip and software for smart clothes," in *IEEE Int. Conf. on Consumer Electronics, (ICCE 2019)*, 2019.

31. T. Isezaki, H. Kadone, A. Niijima, R. Aoki, T. Watanabe, T. Kimura, and K. Suzuki, "Sock-type wearable sensor for estimating lower leg muscle activity using distal emg signals," *Sensors*, vol. 19, no. 8, 2019.

32. E. Doheny, C. Goulding, M. Flood, L. McManus, and M. Lowery, "Feature-based eval-uation of a wearable surface EMG sensor against laboratory standard EMG during force-varying and fatiguing contractions," *IEEE Sensors Journal*, vol. 20, no. 5, pp. 2757–2765, 2020.

33. R. Di Giminiani, M. Cardinale, M. Ferrari, and V. Quaresima, "Validation of fabric-based thigh-wearable EMG sensors and oximetry for monitoring quadricep activity during strength and endurance exercises," *Sensors*, vol. 20, no. 17, pp. 1–13, 2020.

34. G. Rescio, A. Leone, L. Giampetruzzi, and P. Siciliano, *Fall Risk Assessment Using New sEMG-Based Smart Socks*. Cham: Springer, 2021, pp. 147–166.

35. S. Liu, X. Liu, Y. Jiang, X. Wang, P. Huang, H. Wang, M. Zhu, J. Tan, P. Li, C. Lin, G. Zhang, S. Chen, and G. Li, "Flexible non-contact electrodes for bioelectrical signal monitoring," in *Proc. of the Annual International Conference of the IEEE Engineering in Medicine and Biology Society, (EMBS)*, vol. 2018-July, 2018, pp. 4305–4308.

36. M. Magno, L. Benini, C. Spagnol, and E. Popovici, "Wearable low power dry surface wireless sensor node for healthcare monitoring application," in *2013 IEEE 9th International Conference on Wireless and Mobile Computing, Networking and Communications (WiMob)*, Oct 2013, pp. 189–195.

37. S. Y. Lee, K. H. Koo, Y. Lee, J. H. Lee, and J. H. Kim, "Spatiotemporal analysis of EMG signals for muscle rehabilitation monitoring system," in *IEEE 2nd Global Conference on Consumer Electronics (GCCE)*, Oct. 2013, pp. 1–2.

38. K.-M. Chang, S.-H. Liu, and X.-H. Wu, "A wireless sEMG recording system and its application to muscle fatigue detection," *Sensors*, vol. 12, no. 1, pp. 489–499, 2012.

39. T. Y. Fukuda, J. O. Echeimberg, J. E. Pompeu, P. R. G. Lucareli, S. Garbelotti, R. Gimenes, and A. Apolinário, "Root mean square value of the electromyographic signal in the isometric torque of the quadriceps, hamstrings and brachial biceps muscles in female subjects," *Journal of Applied Research*, vol. 10, no. 1, pp. 32–39, 2010.

40. A. Pantelopoulos and N. Bourbakis, "A survey on wearable biosensor systems for health monitoring," in *30th Annual Int. Conf. IEEE Engineering in Medicine and Biology Society*, Aug 2008, pp. 4887–4890.

41. T. Castroflorio, L. Mesin, G. Tartaglia, C. Sforza, and D. Farina, "Use of electromyographic and electrocardiographic signals to detect sleep bruxism episodes in a natural environment," *IEEE Journal of Biomedical and Health Informatics*, vol. 17, no. 6, pp. 994–1001, Nov 2013.

42. J. Oh, M. Kwon, Y. Kim, J. Kim, S. Lee, and J. Kim, "Development and evaluation of myoelectric driving interface," in *2013 IEEE International Conference on Consumer Electronics*, Jan 2013, pp. 248–249.

43. G. Biagetti, P. Crippa, S. Orcioni, and C. Turchetti, "Homomorphic deconvolution for MUAP estimation from surface EMG signals," *IEEE Journal of Biomedical and Health Informatics*, vol. 21, no. 2, pp. 328–338, Mar. 2017.

44. G. Biagetti, P. Crippa, L. Falaschetti, S. Orcioni, and C. Turchetti, "Wireless surface electromyograph and electrocardiograph system on 802.15.4," *IEEE Trans. Consum. Electron.*, vol. 62, no. 3, pp. 258–266, Aug. 2016.

45. S. Orguc, H. Khurana, H.-S. Lee, and A. Chandrakasan, "0.3V ultra-low power sensor interface for EMG," in *ESSCIRC 2017 - 43rd IEEE European Solid State Circuits Conference*, 2017, pp. 219–222.

46. S.-W. Yuk, I.-H. Hwang, H.-R. Cho, and S.-G. Park, "A study on an EMG sensor with high gain and low noise for measuring human muscular movement patterns for smart healthcare," *Micromachines*, vol. 9, no. 11, 2018.

47. H. Kobayashi, "EMG/ECG acquisition system with online adjustable parameters using ZigBee wireless technology," *Electronics and Communications in Japan*, vol. 96, no. 5, pp. 1–10, 2013.

48. A. Burns, E. P. Doheny, B. R. Greene, T. Foran, D. Leahy, K. O'Donovan, and M. J. McGrath, "SHIMMER: An extensible platform for physiological signal capture," in *2010 Annual Int. Conf. IEEE Engineering in Medicine and Biology Society*, Aug 2010, pp. 3759–3762.

49. C. J. De Luca, L. Donald Gilmore, M. Kuznetsov, and S. H. Roy, "Filtering the surface EMG signal: Movement artifact and baseline noise contamination," *Journal of Biomechanics*, vol. 43, no. 8, pp. 1573–1579, 2010.

50. "IEEE Standard for Local and metropolitan area networks—part 15.4: Low-Rate Wireless Personal Area Networks (LR-WPANs)," *IEEE Std 802.15.4–2011*, Sept 2011.

51. B. Latré, P. De Mil, I. Moerman, N. Van Dierdonck, B. Dhoedt, and P. Demeester, "Maximum throughput and minimum delay in IEEE 802.15.4," in *Mobile Ad-hoc and Sensor Networks*, ser. Lecture Notes in Computer Science, X. Jia, J. Wu, and Y. He, Eds. Springer Berlin Heidelberg, 2005, vol. 3794, pp. 866–876.

52. J.-S. Lee, "Performance evaluation of IEEE 802.15.4 for low-rate wireless personal area networks," *IEEE Trans. Consum. Electron.*, vol. 52, no. 3, pp. 742–749, Aug 2006.

53. H. Kobayashi, "A ZigBee based wireless EMG/ECG streaming system for the universal interface," in *IECON 2011 - 37th Annual Conf. IEEE Industrial Electronics Society*, Nov 2011, pp. 2094–2099.

54. H. Kobayashi, "Intelligent wireless EMG/ECG electrode employing ZigBee technology," in *2011 Proc. SICE Annual Conference*, Sept 2011, pp. 2856–2861.

55. A. W. Rohankar, S. Pathak, M. K. Naskar, and A. Mukherjee, "Audio streaming with silence detection using 802.15.4 radios," *ISRN Sensor Networks*, vol. 2012, no. Article ID 590651, p. 5, 2012.

56. H. Y. Tung, K. F. Tsang, H. C. Tung, K. T. Chui, and H. R. Chi, "The design of dual radio ZigBee homecare gateway for remote patient monitoring," *IEEE Trans. Consum. Electron.*, vol. 59, no. 4, pp. 756–764, Nov. 2013.

57. W. Youn and J. Kim, "Development of a compact-size and wireless surface EMG measurement system," in *2009 ICCAS-SICE*. IEEE, Aug 2009, pp. 1625–1628.

58. G. Biagetti, P. Crippa, S. Orcioni, and C. Turchetti, "An analog front-end for combined EMG/ECG wireless sensors," in *Mobile Networks for Biometric Data Analysis*. Cham, Switzerland: Springer, 2016, pp. 215–224.

59. A. G. Pinto, G. Dias, V. Felizardo, N. Pombo, H. Silva, P. Fazendeiro, R. Crisóstomo, and N. Garcia, "Electrocardiography, electromyography, and accelerometry signals collected with BITalino while swimming: Device assembly and preliminary results," in *IEEE 12th Int. Conf. on Intelligent Computer Communication and Processing (ICCP)*, 2016, pp. 37–41.

60. S. Said, S. AlKork, T. Beyrouthy, and M. F. Abdrabbo, "Wearable bio-sensors bracelet for driveras health emergency detection," in *2017 2nd International Conference on Bioengineering for Smart Technologies (BioSMART)*, 2017, pp. 1–4.

61. A. Jani, R. Bagree, and A. Roy, "Design of a low-power, low-cost ECG & EMG sensor for wearable biometric and medical application," in *Proceedings of IEEE Sensors*, vol. 2017-December, 2017, pp. 1–3.
62. M. Tanweer and K. A. I. Halonen, "Development of wearable hardware platform to measure the ECG and EMG with IMU to detect motion artifacts," in *IEEE 22nd Int. Symp. on Design and Diagnostics of Electronic Circuits Systems*, 2019, pp. 1–4.
63. D. M. Das, A. Vidwans, A. Srivastava, M. Ahmad, S. Vaishnav, S. Dewan, and M. S. Baghini, "Design and development of an Internet-of-Things enabled wearable ExG mea- suring system with a novel signal processing algorithm for electrocardiogram," *IET Circuits, Devices Systems*, vol. 13, no. 6, pp. 903–907, 2019.
64. X. Zhang, X. Chen, Y. Li, V. Lantz, K. Wang, and J. Yang, "A framework for hand gesture recognition based on accelerometer and EMG sensors," *IEEE Trans. Syst., Man, Cybem. A, Syst., Humans*, vol. 41, no. 6, pp. 1064–1076, 2011.
65. J. Wu, L. Sun, and R. Jafari, "A wearable system for recognizing american sign language in real-time using IMU and surface EMG sensors," *IEEE Journal of Biomedical and Health Informatics*, vol. 20, no. 5, pp. 1281–1290, 2016.
66. A. Mumtaz, S. Gobee, and C. Venkatratnam, "Development of low-cost wireless EMG sensor network," *ARPN Journal of Engineering and Applied Sciences*, vol. 12, no. 10, pp. 3179–3182, 2017.
67. C. Wu, Y. Yan, Q. Cao, F. Fei, D. Yang, and A. Song, "A low cost surface EMG sensor network for hand motion recognition," in *IEEE 1st Int. Conf. on Micro/Nano Sensors for AI, Healthcare, and Robotics, NSENS 2018*, 2018, pp. 35–39.
68. X. Ding, M. Wang, W. Guo, X. Sheng, and X. Zhu, "Hybrid sEMG, NIRS and MMG sensor system," in *2018 25th International Conference on Mechatronics and Machine Vision in Practice (M2VIP)*, 2018, pp. 1–6.
69. A. Ke, J. Huang, L. Chen, Z. Gao, and J. He, "An ultra-sensitive modular hybrid EMG–FMG sensor with floating electrodes," *Sensors*, vol. 20, no. 17, pp. 1–15, 2020.
70. G. Biagetti, P. Crippa, L. Falaschetti, and C. Turchetti, "A multi-channel electromyography, electrocardiography and inertial wireless sensor module using Bluetooth low-energy," *Electronics*, vol. 9, no. 6, pp. 1–27, 2020.
71. G. Biagetti, P. Crippa, L. Falaschetti, A. Mansour, and C. Turchetti, "Energy and performance analysis of lossless compression algorithms for wireless EMG sensors," *Sensors*, vol. 21, no. 15, 2021.
72. "BTS bioengineering FREEEMG," http://www.btsbioengineering.com/products/freeemg/ (Accessed July 26, 2017), 2017.
73. G. Biagetti, P. Crippa, L. Falaschetti, S. Orcioni, and C. Turchetti, "Human activity monitoring system based on wearable sEMG and accelerometer wireless sensor nodes," *BioMedical Engineering Online*, vol. 17, no. 1, p. 132, Nov. 2018.
74. A. Yousefian, S. Roy, and B. Gosselin, "A low-power wireless multi-channel surface EMG sensor with simplified ADPCM data compression," in *IEEE Int. Symp. on Circuits and Systems (ISCAS)*, May 2013, pp. 2287–2290.
75. S. Bangaru, C. Wang, and F. Aghazadeh, "Data quality and reliability assessment of wearable EMG and IMU sensor for construction activity recognition," *Sensors*, vol. 20, no. 18, pp. 1–24, 2020.

4 ECG Signal Monitoring and Processing in the Operating Room

Kahina Bensafia
University of Mouloud Mammeri
ENSTA-Bretagne

Ali Mansour
ENSTA-Bretagne

Abdel Ouahab Boudraa
Lanvéoc-Poulmic

Sallah Haddab
University of Mouloud Mammeri

Wit Heartlè
Centre Hospitalier Hôtel Dieu

Philippe Aries
Military Teaching Hospital "Clermont Tonnerre"

Benoit Clement
ENSTA-Bretagne

4.1 INTRODUCTION

Health care systems are currently undergoing major changes around the world due to advances in biomedical technologies. At the same time, in the current economic context, many countries are struggling to reduce public spending on all kinds of services, including health care. For this reason, many researchers are looking for ways to use monitoring technologies to improve the efficiency of health systems by reducing their costs. As a result, wireless technologies are increasingly used in hospitals [19,27,37]. Among these technologies are wireless electrocardiogram (ECG) monitoring devices that have been evaluated for ambulatory and in-hospital use [15,17,20,33,25].

DOI: 10.1201/9781003346678-4

Wireless technologies are still not regularly used in operating rooms, mainly due to concerns about possible interference of the wireless transmission by medical equipment or vice versa. To avoid these effects, a separation distance of 1 metre is generally recommended between different devices [9] in operating rooms. Among the available wireless technologies, Bluetooth with its low transmission power is considered to have a low risk of interference with medical equipment [9]. However, electrosurgical units (ESU) used in operating rooms can generate electromagnetic interference (EMI) [22], which could disrupt wireless communications and have an impact on information security.

To ensure patient safety, physiological parameters are monitored during surgery. This chapter presents a wireless device based on the Bluetooth protocol that is designed for continuous ECG signal monitoring in operating rooms. To validate this device, a clinical study is carried out in the Hôpital d'Instruction des Armées Clermont Tonnerre in Brest. The concordance between the ECG signals captured by our wireless device (bt-ECG) and those obtained by the classical device (c-ECG) is established by statistical studies on several volunteer patients. ECG morphology, heart rate, and the durations of P, T waves and QRS complexes of the ECG signal are important features for these statistical studies. The impact of EMI on the ECG signals recorded and transmitted by our system is also studied. As ESU generates electrical artifacts (EAs) that disturb the ECG signals. To remove EAs, we proposed different methods and an extensive experimental study was performed.

4.2 ECG SIGNAL: FROM GENERATION TO ANALYSIS

To enable the heart to fulfil its crucial pumping role, cells of the heart tissue generate electrical impulse more or less regular in the sinus node (SN) [10]. This impulse propagates along the muscle fibres, causing the heart contraction. In fact, each cell is a membrane exchange center in which different ions are involved, such as sodium (Na^+), potassium (K^+), calcium (Ca^{++}) and chlorine (Cl^-). In the resting state, a membrane is only slightly permeable to Na^+ and quite permeable to K^+ or Cl^-. The permeability of the membrane to the K^+ ion is about 50–100 times greater than its permeability to the Na^+ ion. In general, the K^+ concentration in the inner cell medium is about 140 mmole/liter (mmol/L), while that of the outer medium is 2.5 mmol/L [36]. This difference in concentration creates a diffusion gradient that is directed outwards through the membrane. The movement of the K^+ ion along this diffusion gradient respect that direction, so the inside of the cells becomes more negative compared to the external medium. A transmembrane electrical potential difference is thus established −90 mV [13]. To measure and record this potential and thus the currents in the body, biopotential electrodes are provided as an interface between the body and the measuring instrument. Therefore, the electrodes play the role of transducers between the current carried by the ions in the body and the current carried by the electrons in the electrode and its conducting cable. In general, the Silver/Silver Chloride (Ag/AgCl) electrodes are commonly used, see Figure 4.1. The potentials measured on the outer surface of the chest through the electrodes are recorded on moving chart paper in the form of electrocardiograms (ECG).

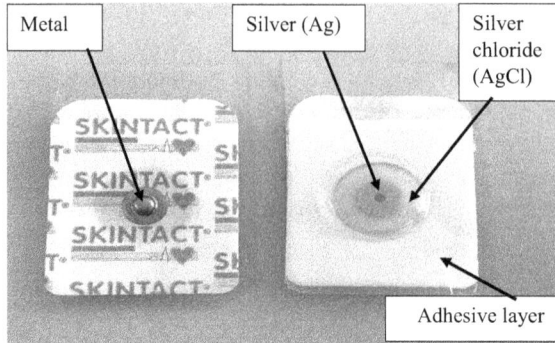

FIGURE 4.1   Cutaneous Ag/AgCl electrodes.

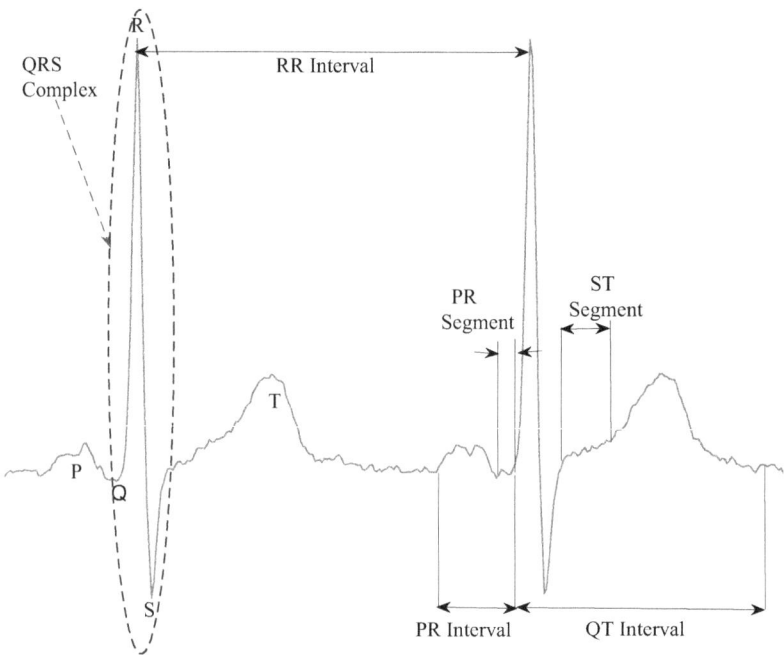

FIGURE 4.2   Waves, segments and intervals of an ECG signal.

The continuous contraction and relaxation of the heart generates a rhythm. The cells in the SN (our natural pacemaker) contribute to this rhythm. The SN generates an electrical impulse which propagates to the atria and depolarizes them, the first deviation of the ECG trace appears, it is named the P-wave (Figure 4.2). The impulse then propagates to the ventricles through the bundle of His[1] and Purkinje fibres[2] so that they contract simultaneously [14]; this is the ventricular depolarization that corresponds to the QRS complexes in the tracing. Followed by the T-wave which represents the ventricular repolarization, i.e. the recuperation of the potential of the

TABLE 4.1
Normal Values of Different Parts of ECG Signals

Duration and Interval	Normal Values
P wave	≤120 ms, amplitude ≤ 0.12 mV
QRS complex	≤120 ms
T wave	≤250 ms,
PR interval	In [120, 200] ms
QT interval	Inversely proportional to heart rate
ST segment	It is isoelectric

membrane at rest. In about a quarter of the population, a U-wave can be observed after the T-wave. The U-wave generally has the same polarity as the T-wave before it. Inverted U waves may occur in the presence of left ventricular hypertrophy or ischemia. The succession of different reactions (P, QRS, T) that occur forms a cardiac cycle. Several cycles make up the heartbeat.

Typical ECG durations and amplitudes of the predefined parts of the cardiac cycle are summarized in Table 4.1 [18]. Abnormal P waves can result from an enlarged and hypertrophied atria or atrial depolarization generated outside the sinus node. QRS complexes with a duration >120 seconds are most likely due to asynchronous depolarization of both ventricles. There are other cardiac pathologies, such as myocardial ischemia, myocardial infarctus, bradycardia, tachycardia, atrial fibrillation (AF) and ventricular fibrillation (VF). For further details on different cardiac pathologies, readers can refer to Refs. [2,14].

4.3 WIRELESS TECHNOLOGY IN THE OPERATING ROOMS

For the patient's safety, various physiological parameters should be monitored during surgery. Ventilator tubing, capnometric tubes and intravenous catheters should be added to the complex network of tubes enclosing the patient and the central anesthesia station. The anesthesiologist often needs to disconnect some of the tubes and cables before and after the surgery and during the transport of the patient to another room. Reducing this network of cables could save time, improve hygiene and facilitate continuous monitoring, even during patient transfer periods. Nowadays, wireless technologies are increasingly introduced in hospitals [19,27,37]. A major challenge is to facilitate, improve monitoring or evaluate in a non-invasive way the physiological parameters that occur in different internal organs of the human body.

Although much work has been done on wireless technology in the hospital environment, it has not yet been regularly used in operating rooms, mainly due to concerns about possible interference from medical devices or vice versa. The Bluetooth wireless technology is considered to have a low risk of interfering with the medical equipment [9]. However, electrosurgical units (ESU) used in the operating room can generate the electromagnetic interference (EMI) [22], which could potentially disrupt wireless signals and impact patient safety. Assessing the reliability of wireless monitoring in the operating room environment becomes essential for the use of

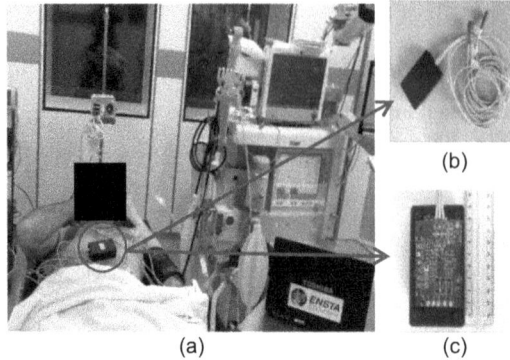

FIGURE 4.3 bt-ECG monitoring device. (a) bt-ECG device in an operating room, (b) bt-ECG with its electrodes, and (c) the circuit.

such technologies. Therefore, we first developed an ECG wireless monitoring device based on Bluetooth that can be safely used during surgery [5], see Figure 4.3. Then, we conducted a study to evaluate the reliability of our Bluetooth ECG device (bt-ECG) in the sensitive and delicate environment of operating rooms and to examine the impact of EMI on wireless signals [1]. To enhance the outcomes of our project, we simultaneously recorded on several patients, during surgeries, ECG signals using bt-ECG and the conventional anaesthesia machine (c-ECG). Hereinafter, we briefly present the steps involved in the acquisition of ECG signals by the two devices, as well as the statistical approaches implemented to compare both recorded signals.

4.3.1 ECG Signal Acquisition by bt-ECG Device and c-ECG Apparatus and Their Synchronization

To record ECG signals using the c-ECG unit, we used a Pico Technology oscilloscope (PicoScope) [34] connected to a PC via a USB port. At the same time, ECG signals have been also acquired and transmitted using our bt-ECG. It should be noted that the sampling frequency used to record the signal of the c-ECG is in 3–60 kHz. To avoid saturation in the transmission of the Bluetooth unit, we used a sampling frequency of 200 Hz. To compare the signals of the two modules bt-ECG and c-ECG, we down-sampled the c-ECG signals. ECG signals are recorded simultaneously by both devices (bt-ECG and c-ECG). However, the synchronization of both devices is almost impossible, as we are not allowed to interfere with the conventional monitoring machine. To create a sort of an artificial synchronization, we have provoked an artifact by tapping lightly near or at the electrodes. This mechanical pressure creates artifact in both sets of recorded signals which should be considered as our starting time reference, see Figure 4.4.

4.3.2 Statistical Methods of Data Comparison

The comparison among obtained signals by the two devices is firstly based on the morphology of the ECG signals. Then, the durations of different P and T waves, as

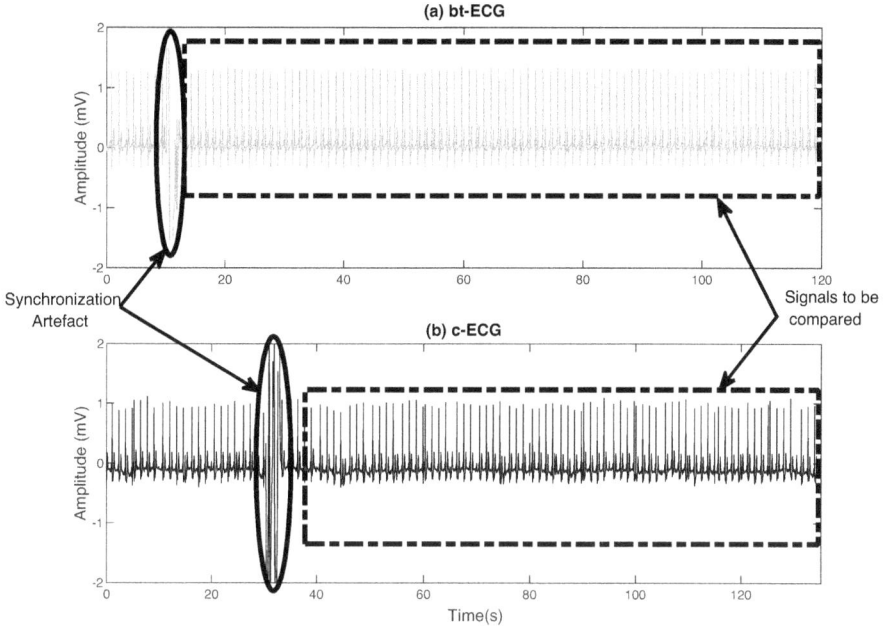

FIGURE 4.4   Synchronization of the two ECG signals (bt-ECG and c-ECG).

well as, the QRS complexes were measured in pairs. Several measurements were made per patient for each wave. A total of 117 measurements were taken per wave. We consolidated our study by further comparing the position of the ST segment with respect to the baseline (20 ST segments per patient). The agreement among the measurements performed by the two devices was confirmed by medical experts and by statistical approaches, such as Bland–Altman's method [6].

4.3.2.1   Bland–Altman's Method

Bland and Altman [6] introduced the Bland–Altman's graph to describe the concordance between two quantitative measurements. Briefly, the resulting graph is a scatter plot, in which the coordinate axis shows the differences among pairs of measurements and the abscissa axis represents the averages of these pairs. We give below the required steps to plot the graph and to properly analyze the data (Figure 4.5).

Let x_{ij} (resp. y_{ij}) be the jth measured value on the ith patient by the c-ECG (resp. bt-ECG) device. Let n be the patient number and m_i the number of measurements made on the ith patient. The Bland–Altman's graph is obtained by plotting the differences d_{ij} on the y-axis and the mean a_{ij} on the x-axis (Figure 4.6):

$$d_{ij} = x_{ij} - y_{ij} \tag{4.1}$$

$$a_{ij} = \frac{x_{ij} + y_{ij}}{2} \tag{4.2}$$

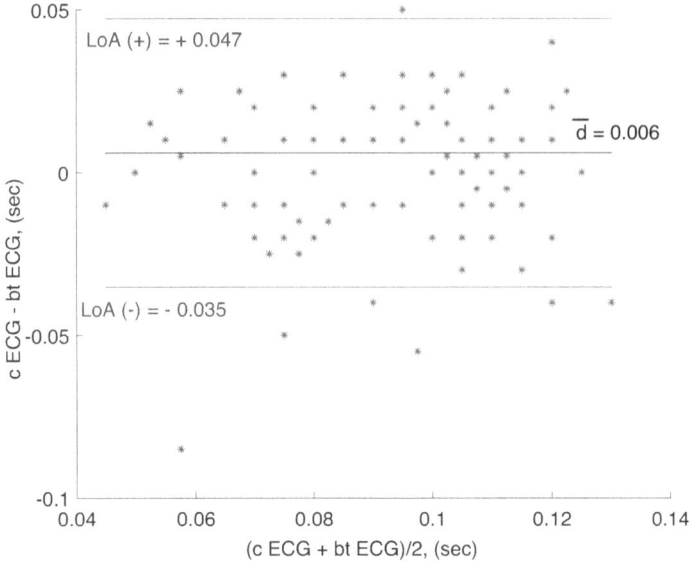

FIGURE 4.5   Diagram illustrating the average difference in P-wave duration with the two ECG devices using the Bland and Altman method.

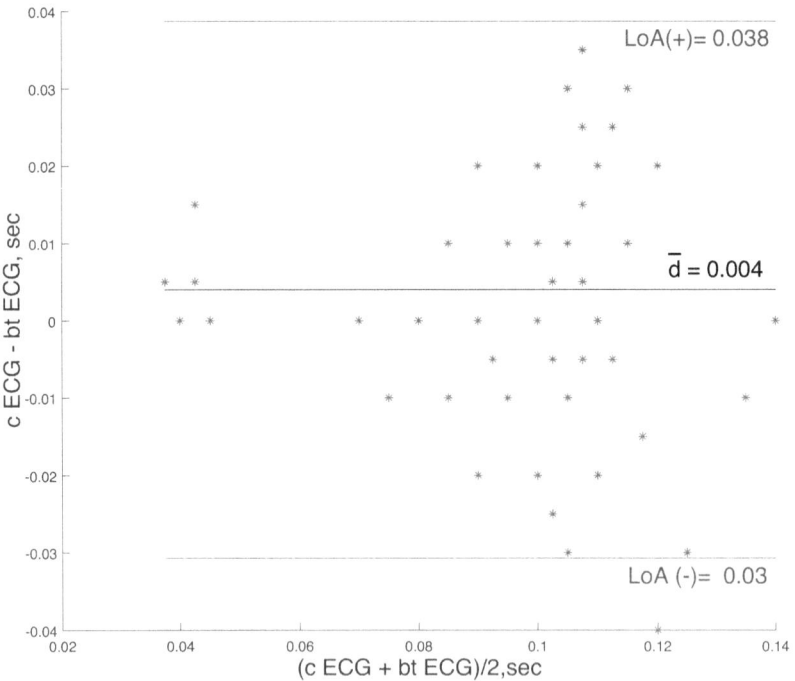

FIGURE 4.6   Diagram showing the average difference in QRS complex duration with the two ECG devices using the Bland–Altman method.

The delay between the measurements made by the two devices $\bar{d}$ is represented by a horizontal line in the graph and is calculated by averaging the differences using the following equation:

$$\bar{d} = \frac{1}{n} \sum_{i=1}^{n} d_{ij} \tag{4.3}$$

Let m_i be to the number of pairs of measurements acquired from the ith patient, then the average of differences for that patient is given by:

$$\bar{d_i} = \frac{1}{m_i} \sum_{j=1}^{m_i} d_{ij} \tag{4.4}$$

Two additional horizontal lines called "Limits of Agreement" (**LoA**) are added to the graph at $\bar{d} \pm 1.96 * s_d$, where s_d stands for the standard deviation of the differences determined by:

$$s_{\bar{d}} = \sqrt{\frac{1}{n-1} \sum_{i=1}^{n} \left(d_{ij} - \bar{d}\right)^2} \tag{4.5}$$

Equations (4.3)–(4.10) give the steps to calculate the **LoA**. The variance of the differences (intrapatient variance) for the ith patient is calculated by:

$$s_i^2 = \frac{1}{m_i - 1} \sum_{j=1}^{m_i} \left(d_{ij} - \bar{d}\right)^2 \tag{4.6}$$

Let s_{dw}^2 be the joint estimate of the intra-subject random error:

$$s_{dw}^2 = \frac{1}{N-n} \sum_{i=1}^{n} (m_i - 1) s_i^2 \tag{4.7}$$

where $N = \sum_{i=1}^{n} m_i$ is the total number of pairs measured on all patients.

The standard deviation of the differences between subjects is therefore given by:

$$s_d^2 = s_{\bar{d}}^2 + \left(1 - \frac{1}{m_h}\right) s_{dw}^2 \tag{4.8}$$

where

$$m_h = \frac{n}{\sum_{i=1}^{n} \frac{1}{m_i}}$$

Finally, the **LoA** are then defined as follows:

$$\text{LoA}_{(-)} = \bar{d} - 1.96 s_d \qquad (4.9)$$

$$\text{LoA}_{(+)} = \bar{d} + 1.96 s_d \qquad (4.10)$$

4.3.3 STATISTICAL RESULTS

Our results revealed a high degree of agreement between the bt-ECG and c-ECG traces, a good correlation between the morphologies is confirmed by physicians. As shown in Figures 4.5, 4.7 and 4.9, the majority of the points are located between the two straight lines of the **LoA**. This statement is satisfied for the durations of P and T waves and QRS complexes. The average of the differences in the durations of P and T waves and QRS complex are 6, −17 and 4 ms respectively. The **LoA** limits are (−0.035, 0.047) ms, (−0.030, 0.038) ms and (−0.011, 0.078) ms for the durations of P waves, QRS complex and T waves, respectively. Regarding the analysis of the ST

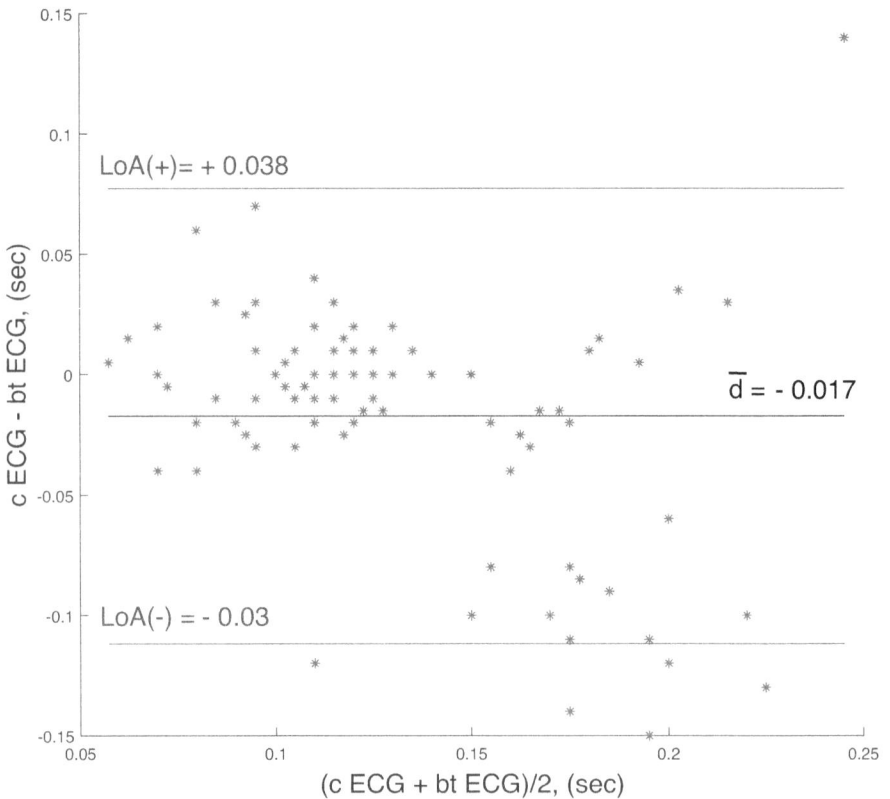

FIGURE 4.7 Diagram showing the average difference in T wave complex duration with the two ECG devices using the Bland–Altman method.

segment, the average of the differences is 0.02 mV. Indeed, these **LoA** were so narrow that differences in P-wave, QRS complex and T-wave duration would be represented by approximately less than one or two small squares (1 mm) on standard ECG traces, which is three times less than the maximum normal duration of each wave (0.12, 0.90 and 0.25 seconds, respectively). For the position of the ST segment, the LoA was <2 mV, i.e. <2 mm which represents a significant value for an electrical elevation for a front lead (D_2).

To confirm that the bt-ECG device is not affected by electromagnetic interference, we recorded an ECG signal on a healthy volunteer subject ~2 m away from the ESU, while membrane surgery is being performed on a patient using the ESU.

4.4 ECG SIGNAL PROCESSING AND REMOVAL OF ELECTRICAL ARTIFACTS

ESU [26] is indispensable in operating rooms because of its ability to cut the skin and coagulate the blood. However, we showed in our clinical study conducted in operating rooms that the ESU generates electrical artifacts that penetrate the patient's skin (surface propagation) and then disrupt the ECG signal. The ESU-generated noise covers a wide range of frequencies [29]. These frequencies can be high-powered and may mask the electrical trace of cardiac activity, as illustrated in Figure 4.8. In this section, we describe our proposed approaches to reduce the EA in the ECG signals.

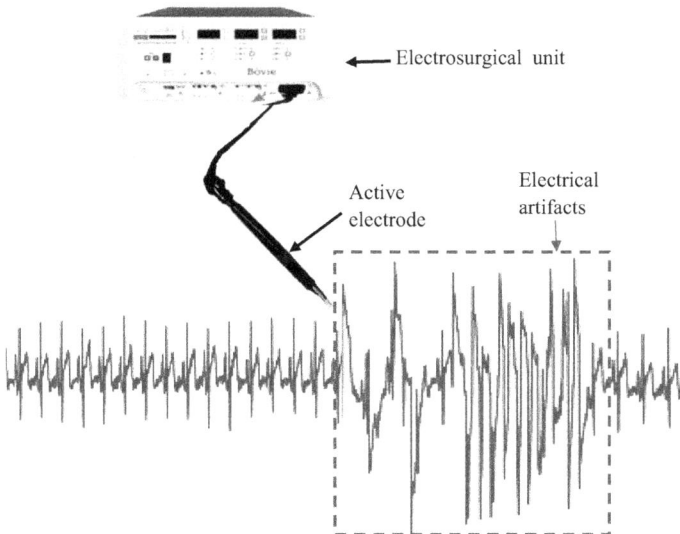

FIGURE 4.8   Impact of the electrosurgical unit on the ECG signal.

4.4.1 PROBLEM STATEMENT

The ECG signal is recorded using surface electrodes to predict abnormalities of the heart. The transfer from the bioelectric current source to the skin electrode can be considered linear [28]. On the other hand, the frequency of bioelectric sources is relatively narrow compared to the high propagation speed of electrical signals [12]. Consequently, the signals picked up by the electrodes can be approximated by instantaneous linear mixing of the sources generated by bioelectric phenomena. Thus, the observed signal can be formulated as follows:

$$y(t) = s(t) + b(t) \tag{4.11}$$

where $s(t)$ is the original source signal (ECG), $b(t)$ is the noise of the electrical artifacts (EAs) and $y(t)$ is the observed signal.

4.4.2 TIME-FREQUENCY REPRESENTATION

Based on the spectrum (obtained by a Fast Fourier Transform, FT) of the observed signal $y(t)$, one can identify the frequency impact of the noise $b(t)$. Figure 4.9c shows the FT of noisy ECG signals and clean ones. We find that the frequency bands of the two signals overlap, therefore making any kind of conventional filtering technique useless. A time-frequency representation can be a good way to better represent the EAs and the ECG. We trace the Smoothed Pseudo Weigner Ville (SPWV) distribution of both ECG signals. Figure 4.9d shows the SPWV applied for noisy ECG signal with EA and Figure 4.9e shows a part of the signal containing pure ECG without EA. We note that the energy of the AE is much higher than the ECG signal, see Figure 4.9d.

4.4.3 THIN SINGULAR VALUES DECOMPOSITION

Applying the singular value decomposition (SVD) to a matrix results in a product of three matrices: $U(m \times m)$, Σ $(m \times n)$ and V $(n \times n)$ [16].

$$A = U \Sigma V^T = \sum_{i=1}^{r} \sigma_i u_i v_i^T \tag{4.12}$$

where $\Sigma = \mathrm{diag}(\sigma_1, \sigma_2, \ldots, \sigma_r, 0, \ldots, 0)$, $\sigma_1 > \sigma_2 > \ldots \sigma_r \geq 0$, $r = \min\{m, n\}$. U is an orthogonal matrix $UU^T = I_m$ and contains the left singular vectors. V is an orthogonal matrix $VV^T = I_n$ and contains the right singular vectors.

Thin SVD (T-SVD) is proposed in Ref. [4] to reduce impulsive noise, similar to electrical artifacts appearing in the ECG signal. The proposed T-SVD is defined as follows:

$$\Sigma_b = \begin{pmatrix} \sigma_1 & 0 & \cdots & 0 \\ 0 & 0 & \cdots & 0 \\ \vdots & \vdots & \ddots & \vdots \\ 0 & 0 & \cdots & 0 \end{pmatrix} \quad \Sigma_s = \begin{pmatrix} 0 & 0 & \cdots & 0 \\ 0 & \sigma_2 & \cdots & 0 \\ \vdots & \vdots & \ddots & \vdots \\ 0 & 0 & \cdots & \sigma_r \end{pmatrix} \tag{4.13}$$

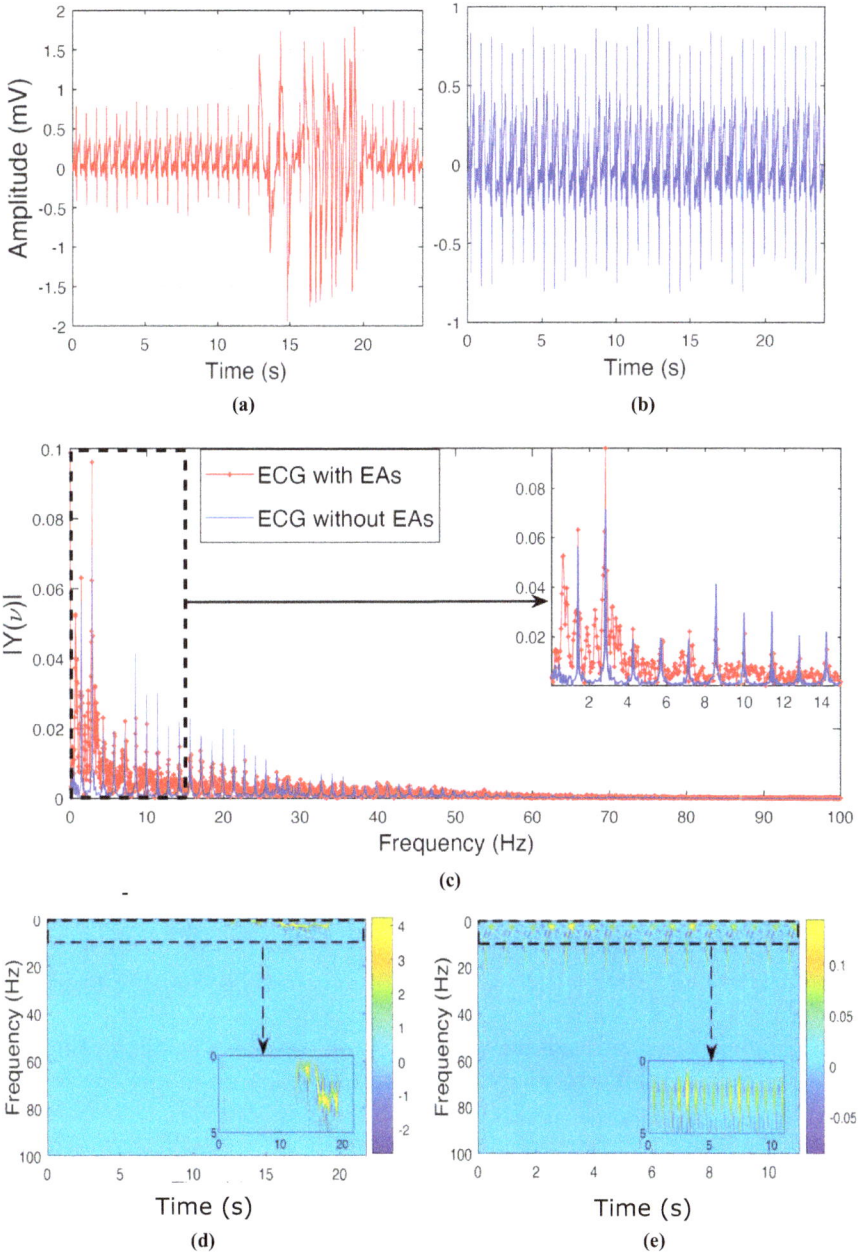

FIGURE 4.9 Time and Frequency representations of ECG signal affected by electrical artifacts (EAs): a) ECG signal source with EAs; b) ECG signal without EAs; c) frequency representation of both signals; d) Time-frequency representation of panel (a); and e) time-frequency representation of panel (b).

The matrix $A_b = U \, \Sigma_b V^T$ and the corresponding vectors (u_1 and v_1^T) give the estimate of the EAs, while the matrix $A_s = U \, \Sigma_s V^T$ and the corresponding vectors form the estimate of the ECG signal. Note that $A = A_b + A_s$.

4.4.4 DISCRETE WAVELET TRANSFORM

Wavelet analysis offers the possibility to analyse stationary or non-stationary signals at different frequencies with different resolutions. Decomposition and reconstruction by Wavelets require the selection of a mother wavelet, a thresholding technique and a level of decomposition. In our case, we opted for the Daubechies wavelets (db4) because its shape is similar to an ECG signal [24] and a decomposition level of 4 was empirically chosen. The db4 is therefore used to decompose the observed ECG signal ($y(t)$) into approximation (based on low frequency) and in detail part (with high frequency).

Since the ECG signal contains abrupt changes (QRS complex), the wavelet decomposition details contain important information. The reconstruction of the denoised ECG signal is achieved by summing all details obtained by the wavelet decomposition, while the last approximation is more related to the EAs.

4.4.5 CEEMDAN BASED ALGORITHM FOR ELECTRICAL ARTIFACTS REDUCTION

The Empirical Mode Decomposition (EMD) method was first introduced by Huang et al. in 1996 [21] to study nonstationary and/or nonlinear signals. Later on, it was further developed by other authors [30,31] and is applied as kind of filters [7,8]. It has also been used in biomedical signal processing, to remove muscle artifacts from EEG data [11,32], and to reduce the artifacts in electrogastrograms [23].

EMD decomposes a signal $y(t)$ into zero-mean oscillating components called the Intrinsic Mode Functions (IMFs). Each decomposition is defined as an IMF, if and only if the resulting signal complies with two conditions:

1. The number of extrema and the number of zerocrossings should be equal or differ by one at most.
2. The local mean values of the upper and lower envelopes determined their maxima and minima respectively must be zero.

Through the sifting process, EMD calculates an IMF which can be taken as details of the original signal and the associated residual. EMD follows the subsequent steps:

- Compute by interpolation the envelope of minima (resp. maxima), $E_{\min}$ (resp. $E_{\max}$) of $y(t)$.
- Calculate the local mean

$$m(t) = \frac{E_{\min}(t) + E_{\max}(t)}{2}$$

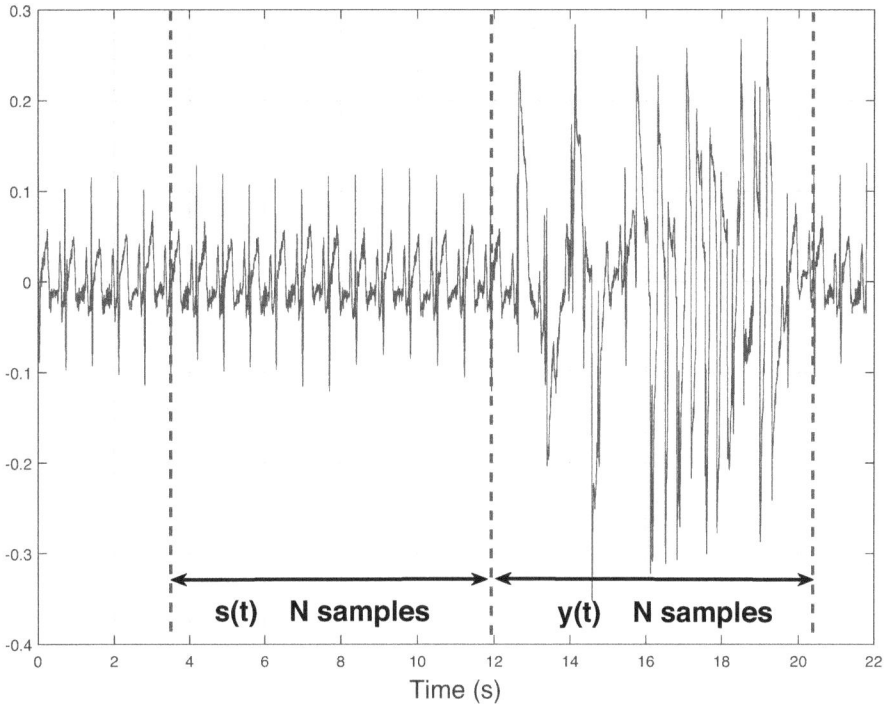

FIGURE 4.10 Observed ECG signal: $s(t)$ clean ECG, $y(t)$ ECG with electrosurgical artifacts (EAs).

If $m(t) = 0$, $y(t)$ is an IMF. Otherwise, one should evaluate the residual $r(t) = y(t) - m(t)$ and repeat the process till one finds all required IMF and reaches insignificant final $r(t)$.

The Empirical Mode Decomposition with Adaptive Noise (CEEMDAN) is an improvement of the EMD algorithm and was proposed by Torres et al. [35]. To apply the proposed algorithms on recorded ECG signals, first, the affected parts by EAs were selected. Two signals are used in our simulations, one is a clean ECG signal $s(t)$ recorded during surgery before the use of EAs, and the second signal consists of an observed noisy signal $y(t)$ as shown in Figure 4.10.

To reduce EAs, we have proposed a method based on the CEEMDAN algorithm [3]. The CEEMDAN decomposition of a given signal generates a set of IMFs. After the decomposition of the observed signal, several reconstructions are performed after selecting some IMFs. The selection has been done based on a cumulative energy criterion.

Algorithm 4.1: Our Proposed Algorithm

1. Calculate the cumulative energy of the ECG signal without EAs ($s(t)$):

$$C_E(n) = \sum_{t=1}^{n} s^2(t) \quad \text{for} \quad n = 1, \ldots, N.$$

$$C_E = \left[C_E(1), C_E(2), \ldots, C_E(N) \right].$$

where N is the length of $s(t)$.

2. Apply CEEMDAN on the observation $y(t)$ to obtain the IMFs:
3. Perform multiple reconstructions $\tilde{s}(t)$:

$$\tilde{s}_j(t) = \sum_{k=1}^{j} \text{IMF}_k(t), \quad \text{for} \quad j = 2, \ldots, K-1$$

where K is the total number of IMFs obtained by CEEMDAN decomposition.

4. Compute the cumulative energies of different reconstructions:

$$\tilde{C}_E^{(j)}(t) = \sum_{t=1}^{n} \tilde{s}_j(t) = \sum_{t=1}^{n} \left(\sum_{k=1}^{j} \text{IMF}_k(t) \right)^2, \quad \text{for} \quad n = 1, \ldots, N$$

5. Determine the differences between the cumulative energies found in steps 1 and 4:

$$DC_E^{(j)}(n) = \left| \tilde{C}_E^j(n) - C_E(n) \right|, \quad j = 2, \ldots, K-1$$

6. Calculate the averages of the differences:

$$M_D^{(j)} = \frac{1}{N} \sum_{n=1}^{N} DC_E^{(j)}(n)$$

7. Find the threshold index j_s which minimizes $M_D^{(j)}$:

$$j_s = \arg\min_j \left(M_D^{(j)} \right)$$

8. The estimated signal becomes:

$$\tilde{s}(t) = \sum_{k=1}^{j_s} \text{IMF}_k(t)$$

9. Figure 4.11 illustrates the differences in the cumulative energies in step 5 in the previous algorithm, e.g. $DC_E^{(2)}$ is the difference between the cumulative energy of the original signal and the cumulative energy of the partial reconstruction $\left(\tilde{s}_2(t) = \sum_{k=1}^{2} \text{IMF}_k(t) \right)$. The closer the difference is to zero, the better the reconstruction. Therefore, in Figure 4.11, as the difference $DC_E^{(4)}$

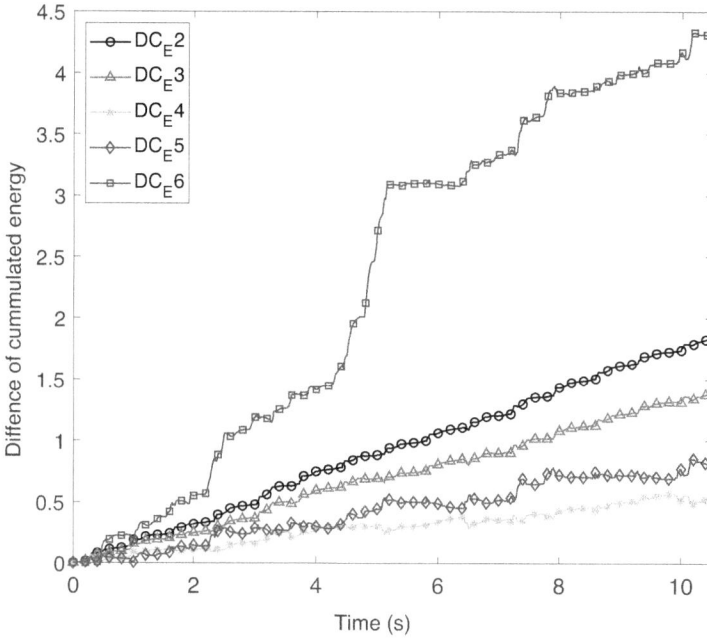

FIGURE 4.11 Differences in the cumulative energies.

is closest to zero, then $\tilde{s}_4(t) = \sum_{k=1}^{4} \mathrm{IMF}_k(t)$ becomes the best approximation of the ECG signal, in this case $j_s = 4$.

To compare the IMFs obtained by CEEMDAN on the source and the observation, then the SPWV distribution is applied the IMFs. Figure 4.12 shows the SPWV of the IMFs, the energy is distributed regularly for the first four IMFs. Whereas with the observation IMFs, this regularity of energy is non-existent as illustrated in Figure 4.13. Therefore, we applied SVD on these IMFs before obtaining the reconstructions.

Algorithm 4.2

1. Apply CEEMDAN on $y(t)$ and obtain IMFs (K is the total number of IMF):

$$y(t) \xrightarrow{\text{CEEMDAN}} \left(\mathrm{IMF}_1(t) \cdots \mathrm{IMF}_K(t) \right)$$

2. Reshape the obtained IMFs into matrices: $\mathrm{IMF}_i(t)(1 \times N) \rightarrow A_i(p \times q)$, where $p \times q = N$ and $i = 1:K$.
3. Apply SVD to the resulting matrices:

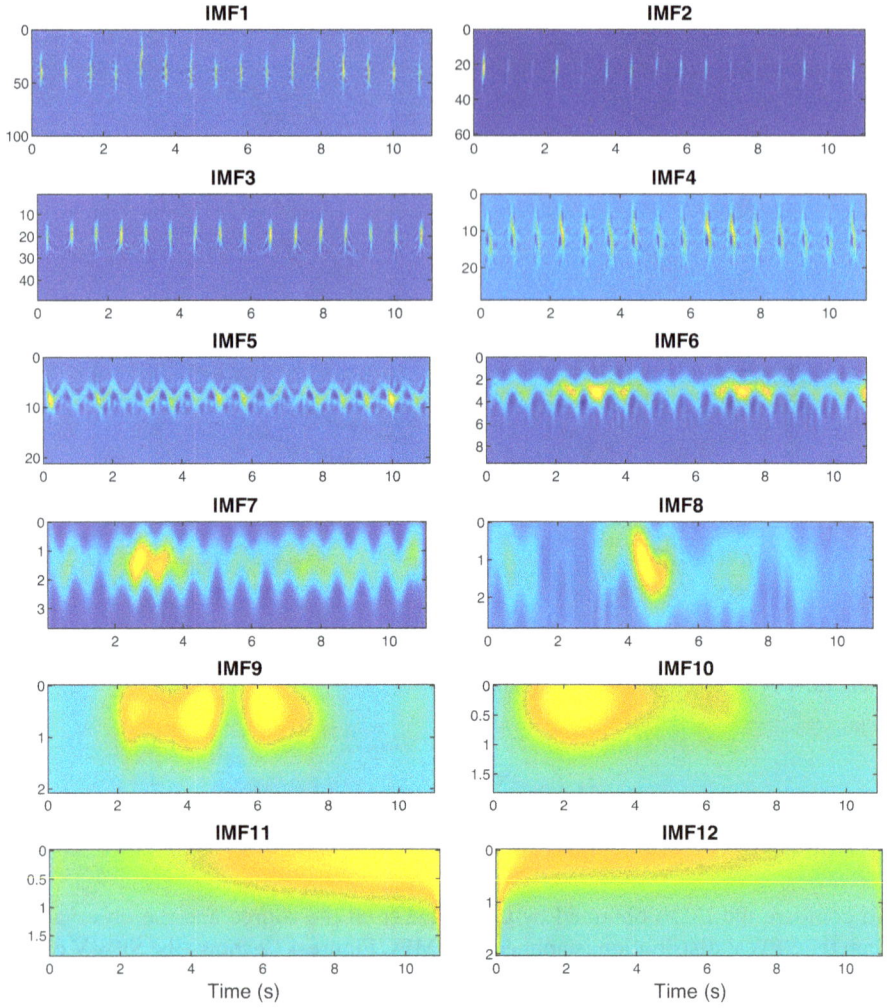

FIGURE 4.12 Smoothed Pseudo Weigner Ville of all IMFs obtained from the decomposition of the source $(s(t))$ by CEEMDAN.

$$
A_i\left(p \times q\right) = U_i \begin{pmatrix} \sigma_{i1} & 0 & 0 & \cdots & 0 \\ 0 & \sigma_{i2} & 0 & \cdots & 0 \\ 0 & 0 & \sigma_{i3} & \cdots & 0 \\ \vdots & \vdots & \vdots & \ddots & \vdots \\ 0 & 0 & 0 & \cdots & \sigma_{ir} \end{pmatrix} V_i^T
$$

where $i = 1:K$ and $r = \min\{p, q\}$.

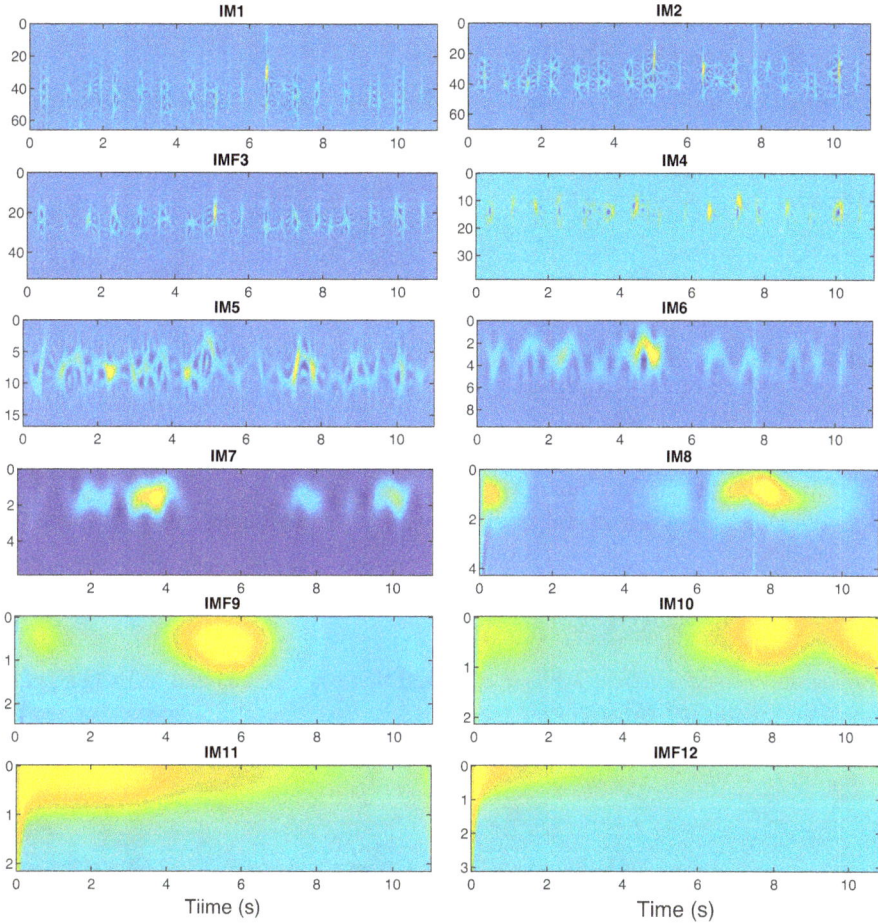

FIGURE 4.13 Smoothed Pseudo Weigner Ville of all IMFs obtained from the decomposition of the observation $(y(t))$ by CEEMDAN.

4. Smooth the first singular values for all matrices $A_i\,(p \times q)$:

$$A_i^{\text{sth}}\,(p \times q) = U_i \begin{pmatrix} \sigma_{i1}/2 & 0 & 0 & \cdots & 0 \\ 0 & \sigma_{i2}/2 & 0 & \cdots & 0 \\ 0 & 0 & \sigma_{i3} & \cdots & 0 \\ \vdots & \vdots & \vdots & \ddots & \vdots \\ 0 & 0 & 0 & \cdots & \sigma_{ir} \end{pmatrix} V_i^T$$

5. Reshape the obtained matrices $A_i^{\text{sth}}\,(p \times q)$ into vectors $\text{IMF}_i^{\text{sth}} \times (1 \times N)$.
6. Then Algorithm 4.1 in Section 4.4.5 is applied on $\text{IMF}_i^{\text{sth}}$ to estimate the denoised ECG signal.

4.4.6   CEEMDAN AND SVD BASED ALGORITHM FOR ELECTRICAL ARTIFACTS REDUCTION

To improve the results, SVD is applied on each IMF, then we adjust the first singular values by dividing them by a factor. We thus perform a soft thresholding on MFIs. The Algorithm 4.2 details this idea. Figure 4.14 shows the SPWV distribution of the IMFs after the application of SVD.

4.4.7   RESULTS AND DISCUSSION

The experimental results were obtained using real ECG signals recorded during several surgeries in operating rooms of the Military Hospital "Clermont Tonnerre" in Brest, France. The objectives of our project are:

1. To overcome the current constraints related to wired monitoring in operating rooms, a wireless monitoring device was designed.
2. Validate this device in real applications, and complex and sensitive environments such as in operation rooms.
3. Comparing the signal of the classical and wireless devices.
4. Clean ECG signal from electrical artifacts.

During our clinical studies, we have observed that both c-ECG and bt-ECG signals are highly disturbed by artefacts, when the electrosurgical unit (ESU) is used at

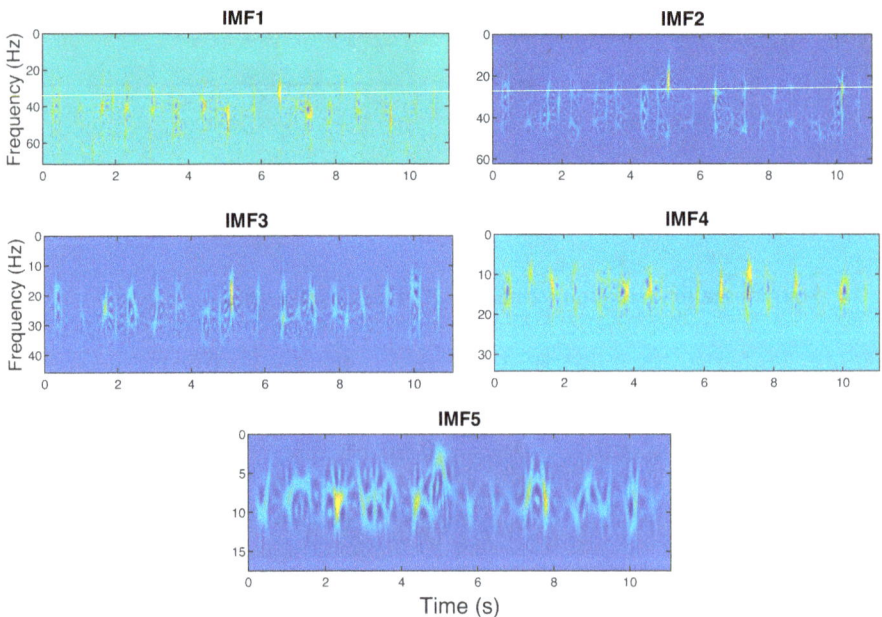

FIGURE 4.14   SPWV distribution of the first IMFs thresholded by SVD.

the trunk (area near the electrodes). However, both signals are minimally disturbed when the surgery is on the limbs. To verify that these artifacts come from an electrical origin and not an electromagnetic effect, we recorded an ECG signal using the bt-ECG device on a healthy subject volunteer, while a membrane surgery was being performed on the patient using ESU. As a result, we didn't notice any kind of artifact in bt-ECG signals recorded on the healthy volunteer while using the ESU on the operated patient. Therefore, we can confirm that the bt-ECG device was not affected by any electromagnetic interference related to the ESU (Figure 4.15).

To eliminate electrosurgical artefacts (EAs) generated by an electrosurgical unit (ESU), several signals were recorded during the clinical phase. The length of the observed signal ($y(t)$) was about 2,000 samples. To achieve our aims, various methods have been proposed, implemented and tested.

- The first one is based on singular values decomposition (T-SVD). Since the energy of the EAs is greater than that of the ECG signal, the SVD decomposition is an appropriate method, in which singular values are represented in decreasing order. The first singular value and the corresponding singular vector are therefore considered to be only related to electrical artifacts. The remaining singular values and the corresponding singular vectors form the estimated ECG signal without noise.
- The second method is based on discrete wavelets decomposition of the real ECG signal into approximation components (low frequency) and details (high frequency). Since QRS complexes are very important and characterized by high frequencies, the preservation of details is therefore essential.

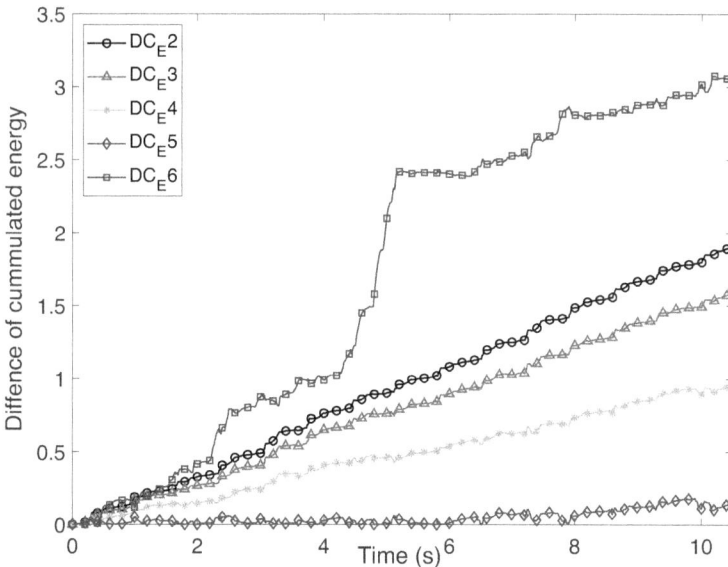

FIGURE 4.15   Differences in the cumulative energies after threshold of IMFs.

- The third method is based on the CEEMDAN algorithm. This method selects IMFs obtained by the CEEMDAN decomposition according to an energy criterion.
- The last method can be considered as an improvement of the previous method, by acting on IMFs. A soft threshold is then applied on IMFs by using the SVD. Afterward, the reconstruction which has a cumulative energy close to the original signal is considered as an estimated ECG signal.

Figure 4.16 shows the estimated ECG signals. Figure 4.16a represents the observation, Figure 4.16b illustrates the result obtained by applying T-SVD, Figure 4.16c shows the outcome of the wavelets-based method, Figure 4.16d depicts the estimated ECG signal by using the CEEMDAN-based method and Figure 4.16d refers to the estimated ECG obtained by applying both SVD and CEEMDAN based method. The last estimation has significantly fewer electrical artefacts compared to the others.

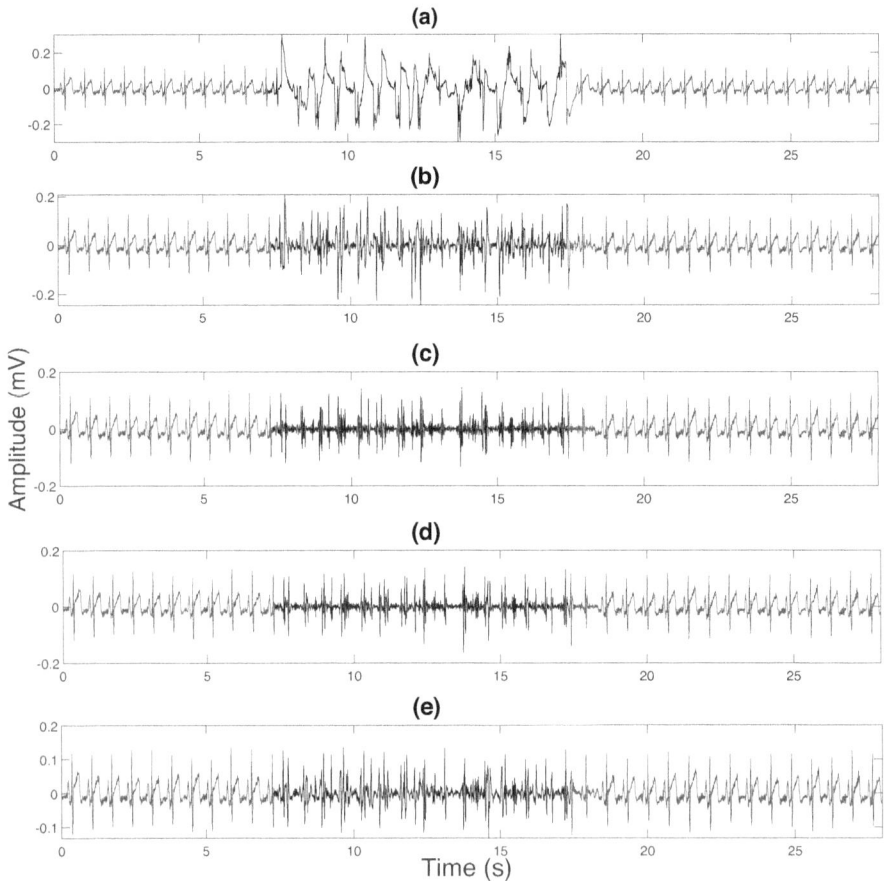

FIGURE 4.16 Estimated ECG signals using: (a) observed signal, (b) thin SVD, (c) wavelets, (d) CEEMDAN-based method, (e) combined method of EMD and SVD.

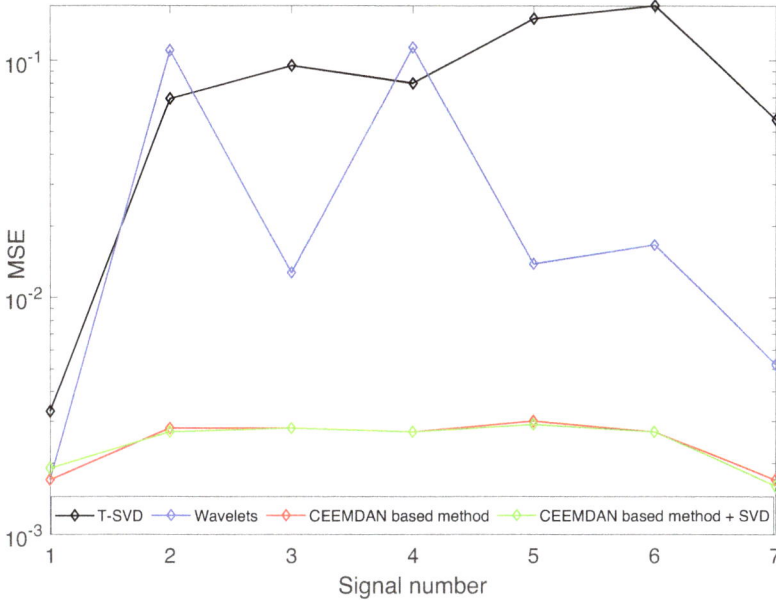

FIGURE 4.17   Mean square error of the results of the applied methods.

Our study is completed by the calculation of performance indices on several signals. The mean square errors (MSE) between the original ECG signals and the signals estimated by different methods are calculated. The method based on both SVD and CEEMDAN offered the best results for major signals. Indeed, the MSE values obtained by applying the proposed method are lower than those obtained by other methods and that was observed over several signals, as shown in Figure 4.17.

4.5   CONCLUSION

To help physicians and surgeons in their duties, to better monitor a patient in real-time during and after surgery. We have been asked by doctors to design a wireless ECG monitor (bt-ECG). This device ensures a continuous monitoring of the patient, during his surgery and after his transfer to the intensive care unit or another room. We have demonstrated that our bt-ECG prototype can provide reliable ECG monitoring in the operating rooms. Various real experiments were conducted during a surgical operation with the help of the medical team, these experiments show the reliability of our device, and prove the safety of using Bluetooth devices in the operating room. Indeed, the presence of multiple electrical medical devices did not disturb the Bluetooth. Electromagnetic interference in the operating room did not disrupt the bt-ECG signal and it is not an obstacle to the development of wireless technology in the operating room.

Regarding the removal of electrical artifacts, several methods are proposed and implemented; discrete wavelet decomposition, thin SVD, CEEMDAN algorithm and a hybrid method based on both SVD and CEEMDAN algorithm. The MSE reveals that both methods based on the CEEMDAN algorithm give better results. Due to the sensitivity of the ECG signal, accurate monitoring is important to diagnose the patient's condition. Thus, quantitative results are not sufficient to determine the best method, a visual analysis of the estimated signal is very important. Cardiologists consider the signals obtained by the hybrid method combining CEEMDAN-based method and SVD to be the best one.

NOTES

1 The bundle of His is a heart muscle that takes part in electrical conduction in the heart.
2 Purkinje fibres are branched fibres that carry the electrical impulse to the ventricles.

REFERENCES

1. Philippe Ariès, Kahina Bensafia, Ali Mansour, Benoît Clèment, Jean-louis Vincent, and Ba Vinh Nguyen. Design and evaluation of a wireless electrocardiogram monitor in an operating room: A pilot study. *Anesthesia & Analgesia*, 129(4):991–996, 2018.
2. Jean-Yves Artigou, *Jean-Jacques Monsuez,* et al. *Cardiologie et maladies vasculaires.* Sociètè fran caise de cardiologie, Elsevier Masson, Paris, 2020.
3. Kahina Bensafia, Ali Mansour, Abdel-Ouahab Boudraa, Salah Haddab, Philippe Ariès, and Benoit Clement. Blind separation of ECG signals from noisy signals affected by electrosurgical artifacts. *Analog Integrated Circuits and Signal Processing*, 104(2):191–204, 2020.
4. Kahina Bensafia, Ali Mansour, and Salah Haddab. Blind elimination of electrical artifacts caused by the electrosurgical units (ESU) for ECG signals. In *European Conference on Electrical Engineering and Computer Science*, pp. 290–295. IEEE, Bern, Switzerland, December 2018.
5. Kahina Bensafia, Ali Mansour, Gilles Le Maillot, Benoit Clement, Olivier Reynet, Philippe Ariès, and Salah Haddab. Wireless based system for continuous electrocardiography monitoring during surgery. *World Academy of Science, Engineering and Technology, International Journal of Medical, Health, Biomedical, Bioengineering and Pharmaceutical Engineering*, 11(10):560–566, 2017.
6. John Martin Bland and Douglas G. Altman. Measuring agreement in method comparison studies. *Statistical Methods in Medical Research*, 8(2):135–160, 1999.
7. Abdel-Ouahab Boudraa and Jean-Christophe Cexus. EMD-based signal filtering. *IEEE Transactions on Instrumentation and Measurement*, 56(6):2196–2202, 2007.
8. Abdel-Ouahab Boudraa, Jean-Christophe Cexus, and Zazia Saidi. EMD-based signal noise reduction. *International Journal of Signal Processing*, 1(1):33–37, 2004.
9. Justin Boyle. Wireless technologies and patient safety in hospitals. *Telemedicine Journal and e-Health*, 12(3):373–382, 2006.
10. *Virual Medical Centre. Cardiovascular system (heart) anatomy.* https://www.myvmc.com/anatomy/cardiovascular-system-heart/, May 2018.
11. Marcelo A. Colominas, Gaston Schlotthauer, and María E. Torres. Improved complete ensemble EMD: A suitable tool for biomedical signal processing. *Biomedical Signal Processing and Control*, 14(1):19–29, 2014.
12. Lieven De Lathauwer, Bart De Moor, and Joos Vandewalle. Fetal electrocardiogram extraction by blind source subspace separation. *IEEE Transactions on Biomedical Engineering*, 47(5):567–572, 2000.
13. Adam Gacek and Witold Pedrycz. *ECG Signal Processing, Classification and Interpretation: A Comprehensive Framework of Computational Intelligence.* Springer Science & Business Media, Berlin, Germany, 2011.

14. Henry B. Geiter Jr. *EZ ECG Rhythm Interpretation*. F. A Davis Company, Philadelphia, PA, 2006.
15. Muhammad Wildan Gifari, Hasballah Zakaria, and Richard Mengko. Design of ECG homecare: 12-lead ECG acquisition using single channel ECG device developed on AD8232 analog front end. In *2015 International Conference on Electrical Engineering and Informatics (ICEEI)*, pp. 371–376. IEEE, Denpasar, Bali,10–11 August, 2015.
16. Gene H. Golub and Charles F. Van Loan. *Matrix Computations*, volume 3. JHU Press, Baltimore, MD, 2012.
17. Salah Haddab and Mourad Laghrouche. Microcontroller-based system for electrogastrography monitoring through wireless transmission. *Measurement Science Review*, 9(5):122–126, 2009.
18. John Hampton. *The ECG in Practice*, 4th ed. Elsevier Health Sciences, Oxford, 2013.
19. Ira Hofer and Maxime Cannesson. Is wireless the future of monitoring? *Anesthisia & Analgesia*, 122(2):305–306, 2016.
20. Chin-Tang Hsieh, Guang-Lin Hsieh, Eugene Lai, Zong-Ting Hsieh, and Guo-Ming Hong. A holter of low complexity design using mixed signal processor. In *Fifth IEEE Symposium on Bioinformatics and Bioengineering (BIBE'05)*, pp. 316–319. IEEE, Minneapolis, MN, 19–21 October, 2005.
21. Norden E. Huang, Zheng Shen, Steven R. Long, Manli C. Wu, Hsing H. Shih, Quanan Zheng, Nai Chyuan Yen, Chi Chao Tung, and Henry H. Liu. The empirical mode decomposition and the hilbert spectrum for nonlinear and non-stationary time series analysis. *Proceedings of the Royal Society of London A: Mathematical, Physical and Engineering Sciences*, 454:903–995, The Royal Society, 1998.
22. Stephen E. Lapinsky and Anthony C. Easty. Electromagnetic interference in critical care. *Journal of Critical Care*, 21(3):267–270, 2006.
23. Hualou Liang, Zhiyue Lin, and Richard W. McCallum. Artifact reduction in electrogastrogram based on empirical mode decomposition method. *Medical and Biological Engineering and Computing*, 38(1):35–41, 2000.
24. Stèphane Mallat. *A Wavelet Tour of Signal Processing*. Elsevier, Amsterdam, Netherlands, 1999.
25. Diego P. Morales, Antonio Garcia, Encarnacion Castillo, Miguel A. Carvajal Rodriguez, Jesus Banqueri, and Alberto J. Palma. Flexible ECG acquisition system based on analog and digital reconfigurable devices. *Sensors and Actuators A: Physical*, 165(2):261–270, 2011.
26. Malcolm G. Munro. Fundamentals of electrosurgery part I: Principles of radiofrequency energy for surgery. In Liane Feldman, Pascal Fuchshuber, and Daniel B. Jones (eds.), *The SAGES Manual on the Fundamental Use of Surgical Energy (FUSE)*, pp. 15–59. Springer, Berlin, Heidelberg, 2012.
27. Rita Paradiso, Gianni Loriga, Nicola Taccini, Maria Pacelli, and Roberto Orselli. Wearable system for vital signs monitoring. In *International IEEE EMBS Special Topic Conference on Information Technology Applications in Biomedicine*, pp. 253–259, Birmingham, UK, April 2004.
28. *Robert Plonsey. Bioelectric Phenomena*. Wiley & Sons, Hoboken, NJ, 1999.
29. Kenneth M. Riff. *Electrosurgery detection*. US Patent 8,961,505, February 24 *2015*.
30. Gabriel Rilling, Patrick Flandrin, Paulo Goncalves, et al., On empirical mode decomposition and its algorithms. In *IEEE-EURASIP Workshop on Nonlinear Signal and Image Processing*, NSIP-03, vol. 3, pp. 8–11, Trieste, 8–11 June, 2003.
31. Gabriel Rilling, Patrick Flandrin, Paulo Goncalves, and Jonathan M. Lilly. Bivariate empirical mode decomposition. *IEEE Signal Processing Letters*, 14(12):936–939, 2007.
32. Doha Safieddine, Amar Kachenoura, Laurent Albera, Gwènaël Birot, Ahmad Karfoul, Anca Pasnicu, Arnaud Biraben, Fabrice Wendling, Lotfi Senhadji, and Isabelle Merlet. Removal of muscle artifact from EEG data: Comparison between stochastic (ICA and CCA) and deterministic (EMD and wavelet-based) approaches. *EURASIP Journal on Advances in Signal Processing*, 2012(1):127, 2012.

33. Maryam Shojaei-Baghini, Rakesh K. Lal, and Dinesh K. Sharma. A low-power and compact analog CMOS processing chip for portable ECG recorders. In *2005 IEEE Asian Solid-State Circuits Conference*, pp. 473–476. IEEE, Hsinchu, 1–3 November, 2005.

34. Pico Technology. *Pico oscilloscope range*.

35. Maria E. Torres, Marcelo A. Colominas, Gaston Schlotthauer, and Patrick Flandrin. A complete ensemble empirical mode decomposition with adaptive noise. In *2011 IEEE International Conference on Acoustics, Speech and Signal Processing (ICASSP)*, pp. 4144–4147. IEEE, Prague, 22–27 May, 2011.

36. John G. Webster. *Medical Instrumentation: Application and Design*. John Wiley & Sons, Hoboken, NJ, 2009.

37. Jiewen Zheng, Congying Ha, and Zhengbo Zhang. Design and evaluation of a ubiquitous chest-worn cardiopulmonary monitoring system for healthcare application: A pilot study. *Medical & Biological Engineering & Computing*, 55(2):283–294, 2017.

5 Photoplethysmography and Inertial Sensors in Wearable Devices for Healthcare

Multimodal Signal Processing for Increasing Accuracy

Giorgio Biagetti, Paolo Crippa,
Laura Falaschetti, and Claudio Turchetti
Università Politecnica delle Marche

5.1 INTRODUCTION

As the miniaturization of sensors and processing electronics progresses, the possibilities of integrating multiple data acquisition chains, thus exploiting a variety of signals originating in different physical domains, into the same small, wearable device, are also dramatically increasing.

This allowed the realization and mass-, low-cost production of devices such as smartwatches, fitness trackers, or similar apparatuses that can easily be worn e.g. on the wrist or arm, providing a comfortable, unobtrusive way of collecting biometric data from the person wearing it.

Although originally, and still mostly, developed for recreational usage in the context of fitness or sports activities, these devices collect data that can be fruitfully processed and applied in many other scenarios. Remote monitoring of a patient's conditions and activities in healthcare applications, where there might be the need to ensure personal safety of persons that, maybe due to cognitive impairments, might engage in risky activities, is certainly one of these. Keeping track of the person's vital parameters, such as heart rate (HR) and blood oxygenation (SpO2), or even breathing rate (BR), while still allowing the person as much freedom of movement as possible, is another one.

In all these contexts, the combination of data coming from heterogeneous physical domains is collected, and by processing them all together it is possible to extract much more meaningful and accurate information than by using only one type at once.

DOI: 10.1201/9781003346678-5

In this chapter, we will thus briefly illustrate two of the most common types of data collected by wearable devices, i.e., the photoplethysmographic (PPG) signal, and motion signals obtained through an inertial platform (usually comprising at least a linear accelerometer, often a gyroscope, and sometimes an electronic compass for absolute orientation).

A survey of the literature on the topic will next be presented, and we will finally dig into a few details of some recently proposed algorithms that take advantage of this multimodal signal processing paradigm to improve the accuracy of heart rate estimation and activity detection.

5.1.1   INERTIAL SENSORS

The most common and ubiquitous type of inertial sensor deployed in wearable devices is the linear accelerometer. It measures the force to which the sensor is subject, including the gravitational pull, and can thus be used during rest conditions to determine the orientation of the device along two axes by computing the angles between the device reference frame and the vertical direction along which the gravitational pull exerts its force. Of course, the angle along the vertical direction cannot be determined by the accelerometer alone. For that measure, an absolute sensor, such as an electronic compass, must be used.

Nevertheless, for wearable devices, the orientation with which the device is worn on the body is not usually known, as it is often a matter of user preference. Hence, knowing the complete, absolute orientation of the device is not of much use, and in most applications, it is usually possible to dispense with the electronic compass altogether.

Of course, when a person moves, the accelerometer registers (the second derivative of) these movements, allowing a suitable algorithm to infer information about the type of activity that is being performed.

Sometimes, linear accelerometers are augmented with gyroscopes. In the form commonly employed in embedded systems, gyroscopes measure the angular velocity of the device along the three axes. This information can be used to increase the accuracy of an electronic compass, which is typically a slow-responding and noisy sensor, during sudden movements, and can also improve the determination of the orientation done with the accelerometer alone. Nevertheless, their use is not as widespread as that of the accelerometer, and many readily available datasets [1,2] do not include gyroscope data. Hence, we will not make use of this type of data in the remainder of this chapter.

5.1.2   PHOTOPLETHYSMOGRAPHY BASICS

A photoplethysmograph works by shining light, possibly at different wavelengths, through the skin of the person wearing the device, and by detecting the amount of light reflected back, or passing through, depending on the mechanical arrangement of the light source and detector. The least obtrusive types, i.e., the wrist-worn or arm-wearable ones, collect the reflected light. Other types, usually worn on a finger or on an earlobe, collect the transmitted light. They usually allow greater accuracy, but are a little more clumsy to wear, so are mostly used in controlled clinical settings.

The blood, with which the tissues are perfused, absorbs or reflects light of different wavelengths in different manners according to how much oxygen it carries. So, by an appropriate choice of the wavelength (color) of the light, it is possible to detect both the amount of blood present and its level of oxygenation. For the first task, the total amount of blood, a green light source is commonly employed as it usually gives the highest signal-to-noise ratio. Since the volume of blood varies during each heart pulse (increasing during the systolic phase), this technique allows the measurement, or at least estimation, of the heart rate. If oxygenation is also to be measured, then two different light sources are needed, one in the red region, and one in the infrared region, as the oxygenated hemoglobin absorbs more in the infrared region and less in the red region, the opposite of the deoxygenated hemoglobin. Sensors that integrate all three types of light sources are thus not uncommon [3].

5.2   RELATED WORKS

In recent years, there has been an increasing interest in the photoplethysmography, driven by the demand for low-cost, simple and portable technology for healthcare and fitness, the availability of low-cost and small semiconductor components, and the advancement of computer-based pulse wave analysis techniques that all allow a variety of multimodal biosignal recording and processing [4]. As an example in Ref. [5] a light-to-digital converter (LDC) for the long-time continuous monitoring of the PPG was designed. The overall readout circuit implements a transimpedance amplifier, a delta-sigma modulator, and a decimation filter on a silicon chip of $2.76 \, mm^2$ and consumes less than $100 \, \mu m$.

PPG-based technology [6–8] and its derivatives [9,10] can be found in a wide range of commercially available medical devices to provide useful information about HR and HR variations, as well as to measure oxygen saturation and blood pressure.

In clinical applications, the PPG signal is rarely affected by motion artifacts (MAs) and in general the patient movements are of reduced entity. In these cases, custom techniques based on wavelets or on different time-frequency algorithms have been proposed to reduce the effects of MAs on the PPG signal, e.g. for estimating the oxygen saturation.

As an example, in Ref. [11] a study has demonstrated that the patient's horizontal, vertical and bending finger movements, can be filtered out by using techniques based on different types of wavelets such as Daubechies, biorthogonal, reverse biorthogonal, symlet and Coiflet. Similarly, the empirical mode decomposition (EMD) time-frequency technique based on the Hilbert-Huang transform [12], and a modified version called the Ensemble EMD (E^2MD) method [13] have been proposed for PPG data affected by different types of MAs due to finger movement in pulse oximetry applications. Combinations of several algorithms have also been studied in Ref. [14], where a simple and efficient approach based on adaptive step-size least mean squares (AS-LMS) adaptive filter has been proposed for the same purpose.

Outside clinical settings, e.g. in fitness and sporting applications, the HR estimation from PPG signals is a challenging task due to the fact that capturing these signals is highly sensitive to the subject body movements. Indeed, if the subject is not perfectly still, as in a clinical setting, but instead is performing a physical activity, it can be expected that MAs get superimposed to the heart-beat-related signal, making its extraction very difficult.

To cope with this issue, a plethora of different approaches to reduce the effects of MAs have been proposed in recent years [15]. In the following, some of the most relevant techniques proposed in the literature are reported and discussed.

Firstly, some of the traditional signal processing algorithms have been used to this end. In Ref. [16] a method is proposed, based on the combination of independent component analysis and block interleaving with low-pass filtering, as a signal enhancement step to extract the true PPG signal from the MA-contaminated measured signal.

Some techniques are based on adaptive filtering [17,18]. Among them, in Ref. [19] the MA removal from PPG signal during jogging was performed by an efficient method based on a reference generation using singular value decomposition and then the multistage application of filtered X-LMS algorithm. Simultaneously, three-axis accelerometer data is used to supply a reference signal to estimate time and extent of MA in the PPG signal. This is followed by an application of the slope sum method to track peaks, and thus determine the HR.

Similarly, [20] presented a methodology called MURAD for HR estimation from wrist-type PPG signals, using a multiple reference adaptive noise cancellation algorithm. The MURAD uses four reference noise signals (RNSs) (i.e. the three-axis accelerometer data and the difference signal between the two PPG signals) to the adaptive filter for obtaining four different versions of the cleaned PPG signal. Then, an intelligent peak tracking process selects the most suitable HR for the current window from the periodograms of these cleaned PPG signals.

Kalman filtering was used in Refs. [21,22]. In particular, in Ref. [21] the tracking of HR by PPG signal was performed in three sequential signal processing stages: (i) applying the principal component analysis to both the PPG and accelerometer signals for denoising, (ii) applying the spectral analysis to the denoised PPG signal, obtained by a Discrete Fourier Transform (DFT) that provides the initial HR measurement; (iii) a Kalman filtering of the raw DFT-based measurement, that allows to track HR temporal evolution and produce a refined and smoothed estimation of HR.

In Ref. [1] a general framework, called TROIKA, for HR estimation using wrist-type PPG signals when MAs are extremely strong due to intensive physical activity, e.g. in wearable devices during sports or fitness activities, is presented. It is based on (i) signal decomposition to partially remove MA components, overlapping with the heartbeat-related PPG components in the same frequency band, (ii) a sparsity-based high-resolution spectrum estimation, (iii) and a spectral peak tracking with verification to correctly select the spectral peaks corresponding to the HR. Different variants of this technique can be derived from it, by choosing specific algorithms for performing its key steps, according to specific requirements in hardware design. Thus, it is of great value to any wearable device.

A modified version of TROIKA has been proposed in Ref. [23]. The method is based on JOint Sparse Spectrum reconstruction (JOSS). It exploits the fact that the spectra of acceleration and PPG signals have some common spectral structures and, thus, formulates the spectrum estimation of these signals into a joint sparse signal recovery model called the multiple measurement vector (MMV) model [24]. Due to the common sparsity constraint on the spectral coefficients, identifying and removing spectral peaks of MA in the PPG spectra is easier. As a result, no extra signal processing is needed to remove MA. The algorithm's simplicity and its low sampling rates requirement allow significant energy saving in data acquisition and wireless transmission, which is welcomed in wearable devices.

Another approach [25], makes use of a cascade of adaptive, recursive-least-squares filters, aided by singular spectrum analysis, to perform HR tracking. Spectrum subtraction techniques can also be used with E^2MD to track HR changes [26]. These two methods have also been combined together and with a higher-accuracy wavelet-Fourier frequency estimator [27]. In a few cases, the possibility of tracking HR without using data from accelerometers to clean the MAs has also been investigated, such as in Ref. [28], where a short-time Fourier transform feeds tentative estimations to a sophisticated statistical-based tracking algorithm, or in Ref. [29], where a technique based on convolutional neural network for human activity recognition (HAR) and HR estimation using wrist-worn sensors without the aid of inertial data, has been proposed.

Specialized approaches [30] have also been developed for specific use cases, e.g. for HR monitors for running, that trade accuracy for versatility by exploiting the quasi-periodic nature of the artifacts. Time-domain techniques may pose less computational burden on resource-constrained devices [31], while offline processing can provide increased accuracy, as is shown in Ref. [32] which exploits a Viterbi-based tracker. And, recently, in Ref. [33] a model for noise components of motion artifacts using an open database of PPG corrupted data was created with the purpose of heart-monitoring using wearable devices.

Approaches based on the geometric separation of signal subspaces have also been proposed [34–36]. These have been designed to have a lower computational complexity than [1], while still offering comparable accuracy. Indeed, the dominant computational burden of these approaches lies on the SVD, which can be implemented leveraging any of many specialized algorithms [37,38], some of which are optimized for frequency tracking and so especially efficient in this context. These techniques will be discussed in detail later on in this chapter.

Another field where the PPG signal can be exploited is in the detection, or recognition, of human activity. In this area, recently, a robust technique for HAR, using only the PPG signal has been proposed [39]. There, the features are extracted from raw PPG segments by Hilbert transform and then classified by the k-nearest neighbor algorithm, naive Bayes, and decision tree algorithms, to achieve an average classification accuracy of 89.39% on the test data derived by resting, squat, and stepper exercises.

Of course, HAR was conventionally done using accelerometer data [40], but it has been demonstrated that using other information, e.g. the PPG signal, can improve the accuracy [41,42]. Moreover, when combined with modern classifiers based on deep-learning techniques such as recurrent neural networks, it also lends itself to an efficient implementation on embedded platforms [43]. But more about this combined data processing will be discussed in detail in the remainder of this chapter.

5.3 PROCESSING OF INERTIAL DATA

The data acquired from the inertial sensors usually need some preprocessing before it can be fruitfully applied to tasks such as activity recognition or motion artifact removal. This preprocessing has two objectives: reduce the dimensionality of the data, and put it into a format that is less sensitive to the particular circumstances of the acquisition environment.

One technique that simultaneously achieves both objectives is described in the following.

Let $x[n]$, $y[n]$, $z[n]$ be the signals coming from the accelerometer, representing the measured acceleration along the three axes of the sensor (and hence relative to the device reference frame). Depending on the bandwidth of the accelerometer and the type of activity that the person is expected to do, a low-pass filter might be employed to remove high-frequency noise. Usually, a cut-off frequency of around 4 Hz is enough to filter out most of the noise and keep the relevant information even for high-intensity activities such as running.

As most classification algorithms work on a frame-by-frame basis, it is necessary to collect a number of samples over a sliding window, let us call W such number. It must encompass enough time to allow a particular movement to be contained within the window, so a duration of about 8 seconds is considered. With a sampling frequency of, e.g., 125 Hz, we thus process windows of $W = 1{,}000$ samples. Of course, these windows do overlap in time and can be shifted by 1 or 2 seconds to provide higher time resolution in the output of the classifiers or other extracted parameters.

The acquired signals can be represented as $N \times L$ Hankel data matrices X, Y, Z, which are matrices whose rows i are sub-windows of the signal starting from sample i and of length L. These can mathematically be defined as

$$[X]_{ij} = x[i + j - 2], \quad i = 1, \ldots, N, j = 1, \ldots, L \tag{5.1}$$

$$[Y]_{ij} = y[i + j - 2], \quad i = 1, \ldots, N, j = 1, \ldots, L \tag{5.2}$$

$$[Z]_{ij} = z[i + j - 2], \quad i = 1, \ldots, N, \ j = 1, \ldots, L \tag{5.3}$$

and basically comprise N replica of time-shifted versions of the signal. Of course, if W is the total number of available samples, $N + L$ must equal $W + 1$, and a suitable choice can be $N = 400$, $L = 601$.

Using Hankel matrices allow the further processing of data to be carried out independently of the phase of the window with respect to the movement being performed, hence avoiding a synchronization step.

Moreover, since, as previously mentioned, these devices are usually worn without a specific orientation, the three matrices just defined can be adjoined together to form a single data matrix $H \in \mathbb{R}^{N \times 3L}$, as

$$H = \begin{bmatrix} X & Y & Z \end{bmatrix} \tag{5.4}$$

so that all the three axes are represented in each row.

This matrix is then further processed by singular value decomposition (SVD), which is used to obtain a more compact representation. Indeed, every matrix possesses a representation of the form

$$H = U\Lambda V^T \tag{5.5}$$

where Λ is a diagonal matrix that contains the so-called singular values in its diagonal entries, usually sorted in decreasing order of magnitude, and U and V are matrices of singular vectors normalized to unit norm.

In many real-world-derived signals, it happens that most of the singular values are usually negligible in magnitude and are often associated with noise instead of to a significant feature of the signal. This is depicted in Figure 5.1, where the first 100 singular values λ_n (the diagonal elements of Λ) are shown in normalized units $\left(\lambda_n^* = \lambda_n \middle/ \sqrt{\sum_i \lambda_i^2} \right)$ for the acceleration data measured during running.

As can be seen, after the first three pairs, the amplitude of the singular values has already dropped by an order of magnitude, and another order of magnitude at about index 20.

By actually zeroing these small singular values, little or no information is lost, but (5.5) can be simplified because the rightmost columns of U and V can be removed, keeping only the first P columns that correspond to the highest P singular values. This procedure is akin to principal component analysis, and it can be proven to provide optimal compression of data [44].

In this context, we can interpret the first P columns of U as a basis for a subspace that represents the motion being performed. An example of a few components of such basis is shown in Figure 5.2.

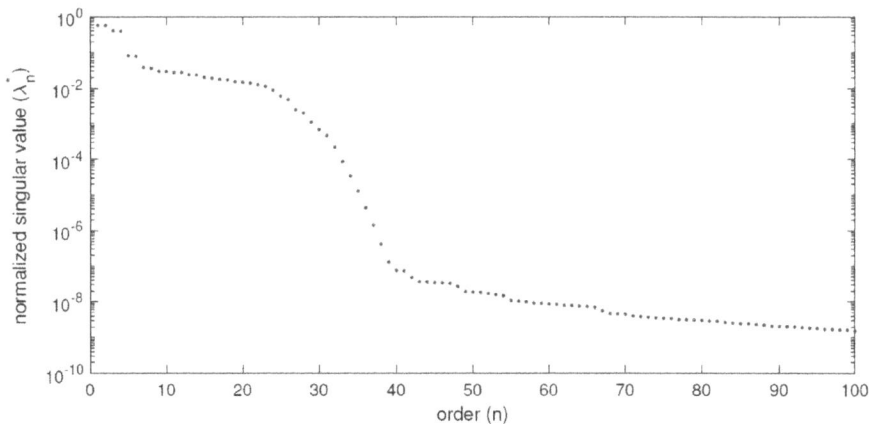

FIGURE 5.1   Spectrum of the normalized singular values, extracted from the Hankel matrix composed of acceleration data measured on the wrist of a subject during running.

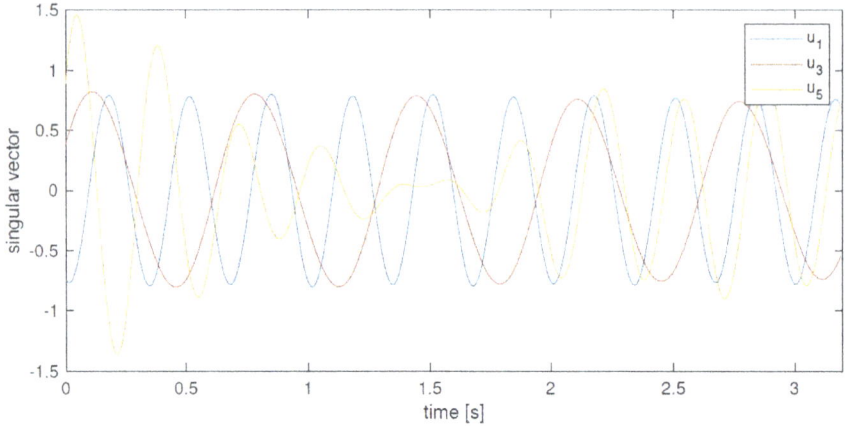

FIGURE 5.2   Three singular vectors. As can be seen, each one is characterized by a dominant frequency. Indeed, the corresponding even-numbered vectors are almost 90° phase-shifted versions of these. That is because the Hankel matrix contains all possible phase shifts of the data segment, hence to form a complete basis two differently phased (nearly) sinusoidal components are needed for each frequency.

5.4   PROCESSING OF THE PPG SIGNAL

The PPG signal can be acquired from different positions on the body, but for health-care applications, it is customarily sensed either on the finger or on the wrist. Of course, to limit the discomfort to the user and allow the subject more freedom of movement, it is better to leave the fingers free and to acquire the PPG signal from the wrist. Unfortunately, the signal acquired has a lower signal-to-noise ratio. And in either case, the PPG signal is always contaminated by so-called motion artifacts, when acquired from a subject that is not perfectly still. These are caused by variations in the amount of subcutaneous blood resulting from inertial forces that act on the blood itself, which might even be of greater amplitude than the fluctuations due to heart pulses. An example is shown in Figure 5.3.

There, the signal was acquired while the subject climbed a few steps. In such a case, it is quite easy to discriminate the artifacts from the useful signal, as the frequencies are quite different. In other activities, such as running or other fast-moving exercises, this frequency discrimination is not possible as the cadence of the movement and the heart rate are often very similar and can cross each other frequently.

5.4.1   MOTION ARTIFACT REDUCTION

For HR detection applications, it is thus necessary to remove these artifacts from the PPG signal, as simple frequency-based filtering cannot be applied in general due to overlapping bands of the useful and interfering signals.

On the other hand, having defined a subspace that contains the motion signal in the previous sections, this concept can be applied to artifact removal quite easily. Indeed, if we subject the PPG signal to the same preprocessing steps as the acceleration

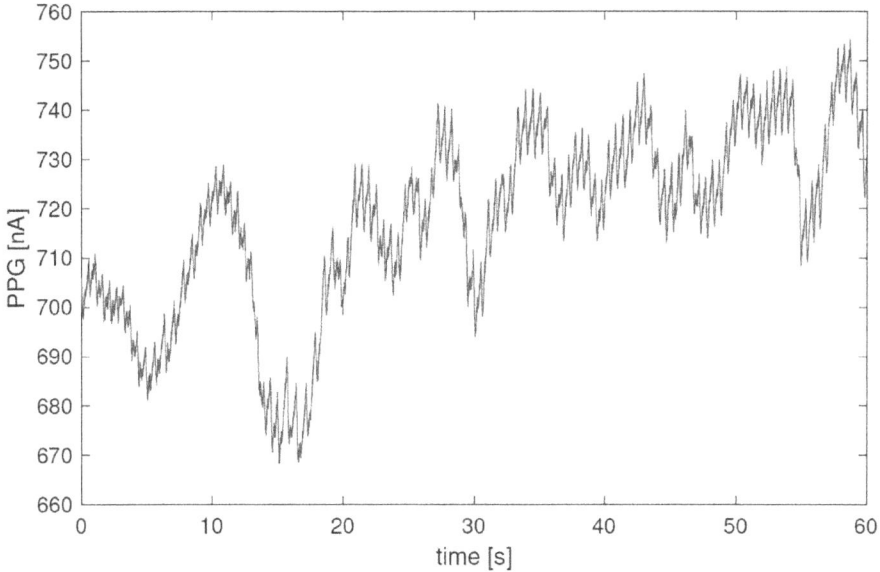

FIGURE 5.3   Example of a PPG signal acquired for 1 minute. Vertical units denote the current as measured by the photodetector, a MAXREFDES100 device by Maxim Integrated [2]. The "high-frequency" component is due to the heart pulses, the "slowly-varying", but wider, oscillations, are due to motion artifacts.

signal, i.e. filtering (it can be band-pass for the PPG to also remove the DC bias) and principal component analysis of its Hankel matrix, we can try to remove the components of the PPG signal that lies in the same subspace that defines the motion signal. If, as it is reasonable to assume, heart rate and motion are independent of each other, the remaining components should only contain frequencies that are due to the (quasi-) periodic nature of the heart pulses.

So, a suitable algorithm for HR detection can include a motion artifact removal stage that is based on the comparison of the basis vectors for the PPG signal to those of the acceleration signal. Similar vectors are discarded. The remaining ones are characterized in the frequency domain to extract possible HR candidates (HR pulses are of course not sinusoidal in nature, so many harmonics will surely be found).

A block diagram summarizing the data flow of the whole procedure is reported in Figure 5.4, where the main steps previously mentioned are reported. Specifically, the first four stages in the dashed box (Windowing, BP Filtering, Hankel, SVD) perform the computations expressed in Equations (5.1)–(5.5), while HR candidates and MA frequencies are computed by the FFT Peak Finder blocks.

Indeed, as previously mentioned, each singular vector is typically characterized by a dominant frequency. Its value is thus extracted by means of an accurate peak interpolation of its Fourier transform (the interpolation is necessary to improve frequency resolution, as with the employed window length the natural frequency resolution would be 18 BPM, way too coarse, and windows cannot be extended without unduly increasing the processing latency). Figure 5.5 shows such dominant frequencies for

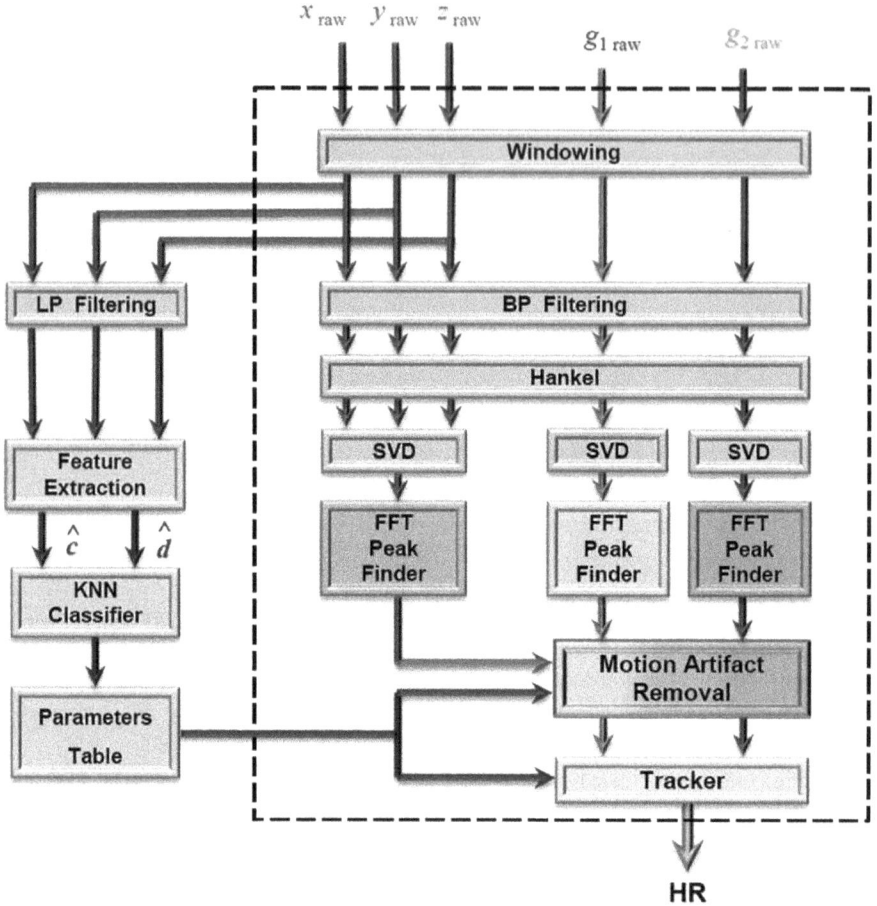

FIGURE 5.4 Block diagram of the signal processing and algorithms used to extract HR information from a PPG signal contaminated by motion artifacts, as described in Ref. [36].

some of the most important singular vectors, obtained from a subject that performed a set comprised of walking, running, walking, running, and walking again, on a treadmill, using the dataset available from Ref. [1]. As can be seen, there are vectors with dominant frequencies very close to the true HR, but those are not always the strongest, especially during the high-intensity phases such as running (these phases can easily be spotted by the higher cadence, i.e., the higher frequency of the motion component, of which the first two harmonic components are clearly identifiable). While walking, the strongest PPG component usually match HR.

To detect the proper component, the ones that are most likely due to motion artifact are removed by the Motion Artifact Removal block. This is shown in Figure 5.6, where the results of this elimination can be seen. After discarding the components too close to MA, the strongest remaining one is chosen, and the others are simply left unused.

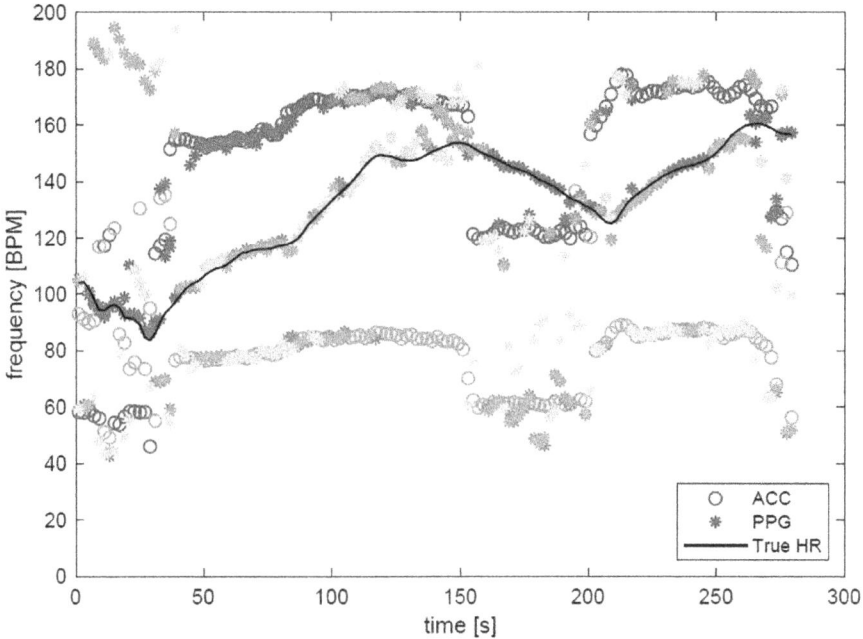

FIGURE 5.5 Example of the pre-processing to remove MA from a PPG signal. Circles denote the dominant frequency of the singular vectors extracted from the inertial sensor, darker color indicating stronger (higher λ) components. Only the first and third components are shown to avoid cluttering the figure. Stars denote the dominant frequency of the singular vectors extracted from one PPG channel. Again, darker colors indicate stronger (higher λ) components. Here, the first 5 are shown. For reference, the true HR as computed from simultaneous ECG is also reported in a solid black line.

This procedure leaves with one estimate per PPG channel, but it can be that it removes the correct candidate. To deal with such cases, the harmonics and subharmonics of the candidate are also considered. These candidates will then be filtered and smoothed by a tracking algorithm (Tracker block) that tries to pick up the ones with the highest likelihood (also considering the previous estimates for continuity), which yields the final estimate shown in green on the same figure.

These last two stages, MA removal and HR tracking, need some parameters to operate, like the radius of MA cancellation and tracking time constants, that can be hand-tuned and optimized for different physical activities in order to improve accuracy. But once this tuning is done, a means of selecting the optimal parameter set based on the automatic detection of the activities currently being performed can be useful. To that end, the left part of Figure 5.4 shows a possible flow, where the results of an activity classifier (a KNN-based one, in this case) are used to select from different sets of pre-tuned parameters, but more on this topic will be discussed in the next section.

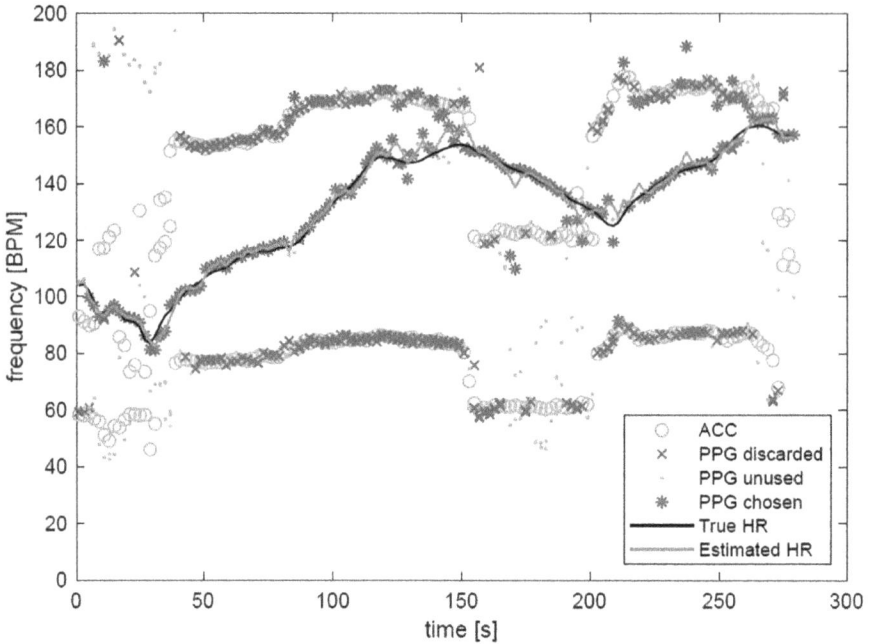

FIGURE 5.6 Example of the result of MA removal from a PPG signal. The PPG data points from Figure 5.5 that lie in close proximity to motion-artifact frequencies (gray circles) are discarded (gray crosses). The strongest remaining components (bright orange stars) are chosen and smoothed by a suitable tracking filter to provide the estimated HR (solid green line).

5.5 ACTIVITY CLASSIFICATION

Another application area where multimodal signal processing helps in increasing the accuracy of the result can be found in the human activity classification, i.e., the task of recognising what kind of activity, usually from a predefined set, a person wearing a suitable tracking device is performing.

Besides obvious uses in sports applications, this can also be quite useful in healthcare applications, where the need to monitor a person with cognitive impairments, without needing constant human supervision, might ease the task of caregivers, by having an automated system that raises alarms when the subject attempts to perform activities that are classified as dangerous.

5.5.1 FEATURE EXTRACTION

There are many well-known classification algorithms that can be applied to the scenario at hand. Although some can directly deal with the raw data sampled by the sensor [42], it is customary to apply some form of feature extraction before feeding the data to the classifier.

Indeed, the preprocessing examined in the previous sections proved to be quite useful as a feature extraction step. The normalized spectrum of the singular values, like that shown in Figure 5.1, contains information on the "complexity" of the signal: the more complex the signal (and hence the movement), more different singular vectors are needed to represent it, and so more high-valued singular values will show up in the spectrum.

This property has been successfully exploited [41] to accurately classify several different activities, drawn from the sports scenario in this case, such as high- and low-intensity biking, running, and walking.

An example of the extracted features is shown in Figure 5.7. Since, in general, the number of singular values that compose the spectra is large, depending on the window length used to segment the signal, and we concatenate two vectors together (one for the PPG signal and one for the acceleration signal), it is necessary to compress the data before further processing.

This can be done by principal component analysis (PCA), which performs another SVD of the data to estimate the eigenvectors of the data's covariance matrix, in order to perform optimal compression. The projections of the train data spectra along the first three of these eigenvectors are thus shown in the figure.

As can be seen very easily in the bottom panel, the three different activities (biking, running, and walking) cluster themselves in well-separated zones of the plane, while, as could probably be expected, high-resistance biking and low-resistance biking are harder to separate at first inspection. But with the aid of an automatic classifier, they can also be distinguished with reasonable accuracy.

5.5.2 CLASSIFICATION

A simple Bayesian classifier was used for the task. A Bayesian classifier assumes that the input to be classified belongs to a random distribution which differs according to which class the input belongs to. It then uses a model of the probability density function (pdf) of the input data to determine the most probable class the unknown sample was drawn from.

It thus needs a training phase, whose output is a set of pdf models, which usually take the form of a Gaussian mixture. If k is a feature vector, the Gaussian mixture can be written as

$$p_\gamma(k) = \sum_{i=1}^{F_\gamma} \alpha_i^{(\gamma)} \mathcal{N}\left(k | \mu_i^{(\gamma)}, C_i^{(\gamma)}\right) \tag{5.6}$$

where γ is an index denoting the class, F_γ is the number of Gaussians composing the mixture that describes class γ, and $\left\{\alpha_1^{(\gamma)}, \mu_1^{(\gamma)}, C_1^{(\gamma)}, ..., \alpha_{F_\gamma}^{(\gamma)}, \mu_{F_\gamma}^{(\gamma)}, C_{F_\gamma}^{(\gamma)}\right\}$ is the set of parameters to be estimated for each class during training. There, α are the weighting factors, μ are the vectors of mean values, and C the covariance matrices, one for each Gaussian that composes the mixture. All these parameters can be found using an algorithm known as Expectation-Maximization [45], which can be quite computationally demanding, but needs only be performed during training.

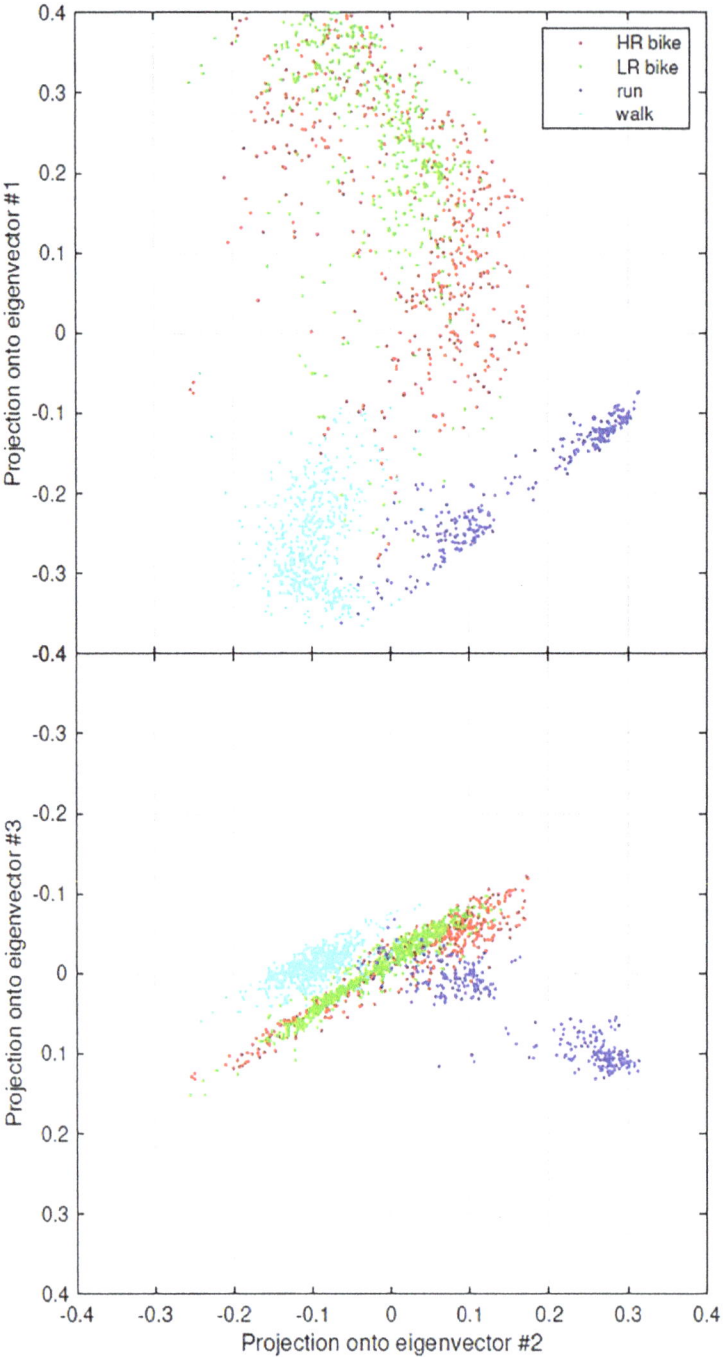

FIGURE 5.7 Projections of the training set over the first three eigenvectors, shown at two different angles (#1 vs. #2 and #2 vs. #3), to show how data cluster together in different ways along the various axes.

The recognition phase is more straightforward. Given an unknown feature vector k, Equation (5.6) is evaluated for all γ, and the one that gives the highest result is chosen

$$\hat{\gamma} = \arg\max_\gamma p_\gamma (k) \tag{5.7}$$

Using this framework it is not always true that the more features are extracted, the better the results are. Indeed, estimating a GMM in a very high-dimensional space is not a trivial task and might require huge amounts of training material. Hence compression of the data, by means e.g. of PCA, as employed in the cited example, is of paramount importance. With these precautions, as was shown, while using acceleration data alone or PPG data alone yielded accuracies of 65.7% and 44.7% respectively, the concatenation of the two spectra allowed the overall accuracy to reach 78.0%, which is a huge improvement made possible by the multimodal signal processing employed.

5.6   CONCLUSIONS

In this chapter, a brief overview of the usefulness of combining PPG and inertial sensors to monitor human physical conditions and activities was presented. We discussed some conventional applications, such as using the PPG signal to estimate the HR, estimation that almost always also needs information from an accelerometer to clean the optical signal from artifacts due to motion. A simple approach based on a subspace decomposition of the artifact from the useful signal is discussed as an example of multimodal signal processing applied to this context. Some novel trends, as the automatic classification of activities, had also been discussed, together with examples of how a multimodal signal processing approach can again be fruitfully applied to increase the accuracy of these techniques.

REFERENCES

1. Z. Zhang, Z. Pi, and B. Liu, "TROIKA: A general framework for heart rate monitoring using wrist-type photoplethysmographic signals during intensive physical exercise," *IEEE Transactions on Biomedical Engineering*, vol. 62, no. 2, pp. 522–531, 2015.
2. G. Biagetti, P. Crippa, L. Falaschetti, L. Saraceni, A. Tiranti, and C. Turchetti, "Dataset from PPG wireless sensor for activity monitoring," *Data in Brief*, vol. 29, p. 105044, 2020.
3. S. K. Longmore, G. Y. Lui, G. Naik, P. P. Breen, B. Jalaludin, and G. D. Gargiulo, "A comparison of reflective photoplethysmography for detection of heart rate, blood oxygen saturation, and respiration rate at various anatomical locations," *Sensors*, vol. 19, no. 8, p. 1874, 2019.
4. Q. Lin, S. Song, I. Castro, H. Jiang, M. Konijnenburg, R. Van Wegberg, D. Biswas, S. Stanzione, W. Sijbers, C. Van Hoof, F. Tavernier, and N. Van Helleputte, "Wearable multiple modality bio-signal recording and processing on chip: A review," *IEEE Sensors Journal*, vol. 21, no. 2, pp. 1108–1123, 2021.
5. E. Pribadi, R. Pandey, and P.-P. Chao, *"Design and implementation of a new light to digital converter for the PPG sensor," Microsystem Technologies*, vol. 27, no. 3, pp. 1–23, 2021.

6. J. Allen, "Photoplethysmography and its application in clinical physiological measurement," *Physiological Measurement*, vol. 28, no. 3, pp. R1–R39, 2007.
7. E. Gil, M. Orini, R. Bailón, J. Vergara, L. Mainardi, and P. Laguna, "Photoplethysmography pulse rate variability as a surrogate measurement of heart rate variability during non-stationary conditions," *Physiological Measurement*, vol. 31, no. 9, pp. 1271–1290, 2010.
8. E. Gil, P. Laguna, J. P. Martínez, O. Barquero-Pérez, A. García-Alberola, and L. Soĺrnmo, "Heart rate turbulence analysis based on photoplethysmography," *IEEE Transactions on Biomedical Engineering*, vol. 60, no. 11, pp. 3149–3155, 2013.
9. M. Elgendi, M. Jonkman, and F. DeBoer, "Heart rate variability measurement using the second derivative photoplethysmogram," In *Proceedings of 3rd International Conference on Bio-Inspired Systems and Signal Processing (BIOSIGNALS 2010)*, Valencia, Spain, 2010, pp. 82–87.
10. M. Elgendi, M. Jonkman, and F. De Boer, "Applying the APG to measure heart rate variability," In *Proceedings of 2nd International Conference on Computer and Automation Engineering, (*ICCAE 2010), vol. 3, Singapore, 2010, pp. 514–517.
11. M. Raghuram, K. V. Madhav, E. H. Krishna, and K. A. Reddy, "Evaluation of wavelets for reduction of motion artifacts in photoplethysmographic signals," In *10th International Conference on Information Sciences Signal Processing and their Applications (ISSPA)*, Kuala Lumpur, Malaysia, May 2010, pp. 460–463.
12. M. Raghuram, K. V. Madhav, E. H. Krishna, N. R. Komalla, K. Sivani, and K. A. Reddy, "HHT based signal decomposition for reduction of motion artifacts in photoplethys mographic signals," In *IEEE International Instrumentation and Measurement Technology Conference (I2MTC)*, Graz, Austria, May 2012, pp. 1730–1734.
13. M. Raghuram, K. Sivani, and K. A. Reddy, "E2MD for reduction of motion artifacts from photoplethysmographic signals," In *International Conference on Electronics and Communication Systems (ICECS)*, Coimbatore, India, February 2014, pp. 1–6.
14. M. R. Ram, K. V. Madhav, E. H. Krishna, N. R. Komalla, and K. A. Reddy, "A novel approach for motion artifact reduction in PPG signals based on AS-LMS adaptive filter," *IEEE Transactions on Instrumentation and Measurement*, vol. 61, no. 5, pp. 1445–1457, 2012.
15. S. Ismail, U. Akram, and I. Siddiqi, "Heart rate tracking in photoplethysmography signals affected by motion artifacts: A review," *Eurasip Journal on Advances in Signal Processing*, vol. 2021, no. 1, pp. 1–27, 2021.
16. B. S. Kim and S. K. Yoo, "Motion artifact reduction in photoplethysmography using independent component analysis," *IEEE Transactions on Biomedical Engineering*, vol. 53, no. 3, pp. 566–568, 2006.
17. J. Y. A. Foo, "Comparison of wavelet transformation and adaptive filtering in restoring artefact-induced time-related measurement," *Biomedical Signal Processing and Control*, vol. 1, no. 1, pp. 93–98, 2006.
18. P. T. Gibbs, L. B. Wood, and H. H. Asada, "Active motion artifact cancellation for wearable health monitoring sensors using collocated MEMS accelerometers," In M. Tomizuka (ed.), *Smart Structures and Materials*, vol. 5765. Bellingham, WA: International Society for Optics and Photonics, 2005, pp. 811–819.
19. K. T. Tanweer, S. R. Hasan, and A. M. Kamboh, "Motion artifact reduction from PPG signals during intense exercise using filtered X-LMS," In *2017 IEEE International Symposium on Circuits and Systems (ISCAS)*, Baltimore, MD, May 2017, pp. 1–4.
20. S. S. Chowdhury, R. Hyder, M. S. B. Hafiz, and M. Haque, "Real-time robust heart rate estimation from wrist-type PPG signals using multiple reference adaptive noise cancellation," *IEEE Journal of Biomedical and Health Informatics*, vol. 22, no. 2, pp. 450–459, 2018.

21. A. Galli, G. Frigo, C. Narduzzi, and G. Giorgi, "Robust estimation and tracking of heart rate by PPG signal analysis," In *IEEE International Instrumentation and Measurement Technology Conference (I2MTC)*, Torino, Italy, May 2017, pp. 1–6.

22. B. Lee, J. Han, H. J. Baek, J. H. Shin, K. S. Park, and W. J. Yi, "Improved elimination of motion artifacts from a photoplethysmographic signal using a Kalman smoother with simultaneous accelerometry," *Physiological Measurement*, vol. 31, no. 12, p. 1585, 2010.

23. Z. Zhang, "Photoplethysmography-based heart rate monitoring in physical activities via joint sparse spectrum reconstruction," *IEEE Transactions on Biomedical Engineering*, vol. 62, no. 8, pp. 1902–1910, 2015.

24. S. F. Cotter, B. D. Rao, Kjersti Engan, and K. Kreutz-Delgado, "Sparse solutions to linear inverse problems with multiple measurement vectors," *IEEE Transactions on Signal Processing*, vol. 53, no. 7, pp. 2477–2488, 2005.

25. M. T. Islam, I. Zabir, S. T. Ahamed, M. T. Yasar, C. Shahnaz, and S. A. Fattah, "A time-frequency domain approach of heart rate estimation from photoplethysmographic (PPG) signal," *Biomedical Signal Processing and Control*, vol. 36, pp. 146–154, 2017.

26. Y. Zhang, B. Liu, and Z. Zhang, "Combining ensemble empirical mode decomposition with spectrum subtraction technique for heart rate monitoring using wrist-type photoplethysmography," *Biomedical Signal Processing and Control*, vol. 21, pp. 119–125, 2015.

27. M. S. Islam, M. Shifat-E-Rabbi, A. M. A. Dobaie, and M. K. Hasan, "PREHEAT: Precision heart rate monitoring from intense motion artifact corrupted PPG signals using constrained RLS and wavelets," *Biomedical Signal Processing and Control*, vol. 38, pp. 212–223, 2017.

28. D. Zhao, Y. Sun, S. Wan, and F. Wang, "SFST: A robust framework for heart rate monitoring from photoplethysmography signals during physical activities," *Biomedical Signal Processing and Control*, vol. 33, pp. 316–324, 2016.

29. E. Brophy, W. Muehlhausen, A. Smeaton, and T. Ward, "CNNs for heart rate estimation and human activity recognition in wrist worn sensing applications," In *2020 IEEE International Conference on Pervasive Computing and Communications Workshops, PerCom Workshops*, Austin, TX, 2020, 2020.

30. H. Dubey, R. Kumaresan, and K. Mankodiya, "Harmonic sum-based method for heart rate estimation using PPG signals affected with motion artifacts," *Journal of Ambient Intelligence and Humanized Computing*, vol. 9, no. 1, pp. 137–150, 2018.

31. M. Wójcikowski and B. Pankiewicz, "Photoplethysmographic time-domain heart rate measurement algorithm for resource-constrained wearable devices and its implementation," *Sensors (Switzerland)*, vol. 20, no. 6, pp. 1–16, 2020.

32. A. Temko, "Accurate heart rate monitoring during physical exercises using PPG," *IEEE Transactions on Biomedical Engineering*, vol. 64, no. 9, pp. 2016–2024, 2017.

33. S. Cajas, M. Landínez, and D. López, "Modeling of motion artifacts on PPG signals for heart-monitoring using wearable devices," In *Proceedings of SPIE: 15th International Symposium on Medical Information Processing and Analysis,* 2019, Medelin, vol. 11330, 2020.

34. A. Bacà, G. Biagetti, M. Camilletti, P. Crippa, L. Falaschetti, S. Orcioni, L. Rossini, D. Tonelli, and C. Turchetti, "CARMA: A robust motion artifact reduction algorithm for heart rate monitoring from PPG signals," In *23rd European Signal Processing Conference (EUSIPCO 2015)*, Nice, France, September 2015, pp. 2696–2700.

35. G. Biagetti, P. Crippa, L. Falaschetti, S. Orcioni, and C. Turchetti, "Motion artifact reduction in photoplethysmography using Bayesian classification for physical exercise identification," In *Proceedings of 5th Internatinal Conference on Pattern Recognition Applications and Methods (ICPRAM 2016)*, Rome, Italy, February 2016, pp. 467–474.

36. G. Biagetti, P. Crippa, L. Falaschetti, S. Orcioni, and C. Turchetti, "Reduced complexity algorithm for heart rate monitoring from PPG signals using automatic activity intensity classifier," *Biomedical Signal Processing and Control*, vol. 52, pp. 293–301, 2019.

37. R. Badeau, G. Richard, and B. David, "Sliding window adaptive SVD algorithms," *IEEE Transactions on Signal Processing*, vol. 52, no. 1, pp. 1–10, 2004.

38. P. Strobach, "Sliding window adaptive SVD using the unsymmetric householder partial compressor," *Signal Processing*, vol. 90, no. 1, pp. 352–362, 2010.

39. T. Aydemir, M. Şahin, and O. Aydemir, "A new method for activity monitoring using photoplethysmography signals recorded by wireless sensor," *Journal of Medical and Biological Engineering*, vol. 40, no. 6, pp. 934–942, 2020.

40. G. Biagetti, P. Crippa, L. Falaschetti, S. Orcioni, and C. Turchetti, "An efficient technique for real-time human activity classification using accelerometer data," In I. Czarnowski, A. M. Caballero, R. J. Howlett, and L. C. Jain (eds.), *Intelligent Decision Technologies 2016*. Cham: Springer International Publishing, 2016, pp. 425–434.

41. G. Biagetti, P. Crippa, L. Falaschetti, S. Orcioni, and C. Turchetti, "Human activity recognition using accelerometer and photoplethysmographic signals," In , I. Czarnowski, R. J. Howlett, and L. C. Jain (eds.), *Intelligent Decision Technologies 2017*. Cham: Springer International Publishing, 2018, pp. 53–62.

42. G. Biagetti, P. Crippa, L. Falaschetti, E. Focante, N. M. Madrid, R. Seepold, and C. Turchetti, "Machine learning and data fusion techniques applied to physical activity classification using photoplethysmographic and accelerometric signals," *Procedia Computer Science*, vol. 176, pp. 3103–3111, 2020.

43. M. Alessandrini, G. Biagetti, P. Crippa, L. Falaschetti, and C. Turchetti, "Recurrent neural network for human activity recognition in embedded systems using PPG and accelerometer data," *Electronics*, vol. 10, no. 14, pp. 1–18, 2021.

44. G. Biagetti, P. Crippa, L. Falaschetti, S. Orcioni, and C. Turchetti, "An investigation on the accuracy of truncated DKLT representation for speaker identification with short sequences of speech frames," *IEEE Transactions on Cybernetics*, vol. 47, no. 12, pp. 4235–4249, 2017.

45. M. A. T. Figueiredo and A. K. Jain, "Unsupervised learning of finite mixture models," *IEEE Transactions on Pattern Analysis and Machine Intelligence*, vol. 24, no. 3, pp. 381–396, 2002.

6 Non-Invasive System for Measuring Parameters Relevant to Sleep Quality and Detecting Sleep Diseases
The Data Model

Daniel Vélez Gutiérrez
HTWG Konstanz

Natividad Martínez Madrid
Reutlingen University

Ralf Seepold
HTWG Konstanz

6.1 INTRODUCTION

Medicine is nowadays in the middle of a digital revolution. Digitalized patient information supports the development of clinical decisions, and new digital interventions are being created and improved to promote healthy behaviors [1]. Advances in electronics have developed more accurate and sensible components that can detect and extract useful information. This has contributed to creating different mechanisms for remote patient assistance [2], making telemedicine, remote healthcare support, and home healthcare a new trend in modern society.

In the field of sleep medicine, the measurement of accurate vital signs has always been a challenge; the current standard method for recording physiological parameters during sleep is polysomnography (PSG), which uses, among others, electroencephalogram, electrooculogram, electromyogram, electrocardiogram, pulse oximetry, airflow, and respiratory effort to diagnose and determine the causes of sleep disturbances and pathologies [3]. Despite its utility in clinical sleep medicine, applying such a set of sensors in the patient's body can cause discomfort, stress, and anxiety and, under certain circumstances, compromise the results obtained, making this device not the

DOI: 10.1201/9781003346678-6

most suitable candidate for long-term sleep monitoring. This has motivated the development of portable devices capable of measuring sleep at home [4,5].

6.1.1   SLEEP MONITORING DEVICES

In recent years, there has been increasing interest in developing commercial and research-oriented non-invasive sleep measurement devices that provide promising results in detecting and measuring vital signs and sleep events [6]. Some of these devices, nonetheless, tend to focus on individual aspects of the sleep monitoring process, such as collecting the data on a cloud-based platform using commercially available low-cost sensors [7] or classifying sleep stages based on non-invasive and wireless sensors [8,9]. More comprehensive approaches can be found in available commercial implementation [10,11], but these applications restrain the information about the data and methods employed, creating a liability that prevents their use in large-scale sleep studies [12]. Other approaches go further; Nam et al. [13] propose a method to assess sleep quality using a device with two non-invasive sensors: a three-axis accelerometer and a pressure sensor. The device is used to determine sleeping pose, movement, HR, and respiration (RR), and from these parameters, determine a sleep quality measurement based on the sleep time and the prediction of the sleep stages and sleep-apnea events. Yacchirema et al. [14] describe a more elaborated system that provides real-time alerts on apnea events to medical personnel using different IoT devices to measure environmental factors and patient biosignals and behavior.

6.1.2   THE MORPHEUS PERSPECTIVE

The implementation proposed as part of the MORPHEUS project differs from previous systems and devices by covering multiple aspects with a systematic and open-access approach in both the hardware and the software implementation, providing a baseline for further developments and analysis in the field. The MORPHEUS Box (MoBo) device aims to monitor body signals over long periods from patients during sleep, using exclusively non-invasive technologies. This includes the absence of privacy-invasive technologies such as cameras and microphones. The software interfaces focus on providing sharing capabilities and privacy compliance, and the user application aims to deliver support in treating sleep-related pathologies to patients and physicians. It is now a known fact that the frequency of medically diagnosed sleep disorders increases from the age of 40 [15], and different sleep parameters can be used to monitor actual health status changes in older adults [16]. Thus in this sense, the proposed system offers an opportunity to enhance the quality of life of an aging society by putting the collection of sleep data at the center of the analysis and making it available for medical evaluations.

6.1.3   SECTION STRUCTURE

The following sections describe the project's objective, the architecture of the proposed system, and its components using SysML and UML diagrams. Therefore, each topic is detailed in a section: Section 6.2 describes the conceptualization behind the project. Section 6.3 depicts the project's general architecture, including hardware and software. Section 6.4 outlines the components designed in the backend. Section 6.5 explains the dataflow designed for the system from the hardware to the user application. Section 6.6 ends with a conclusion.

6.2 CONCEPT OF MORPHEUS

The MORPHEUS project aims to monitor patients' vital signs over long periods of sleep without causing additional discomfort by developing a system (including both the software and the hardware).

The core of the project is the development of a system for automatically collecting sleep-related data, providing both the software and the hardware. The system will make recommendations (e.g., CBT-I based) and verify compliance. Continuously measured values (e.g., sleep/wake state, respiration, or heart rate during sleep) will be compared with the recommendations and readjusted if necessary. The system uses exclusive, non-invasive technologies deployed in a patient's home environment. The MORPHEUS Box is at the center of the development (MoBo, see Figure 6.1). Conceptually, the development can be divided into five components: The MoBo core, as the core component, provides the basic functionality of an embedded system and implements the interfaces to the other components. The MoBo algorithm receives the signal data stream from the MoBo HW and processes it using specific evaluation

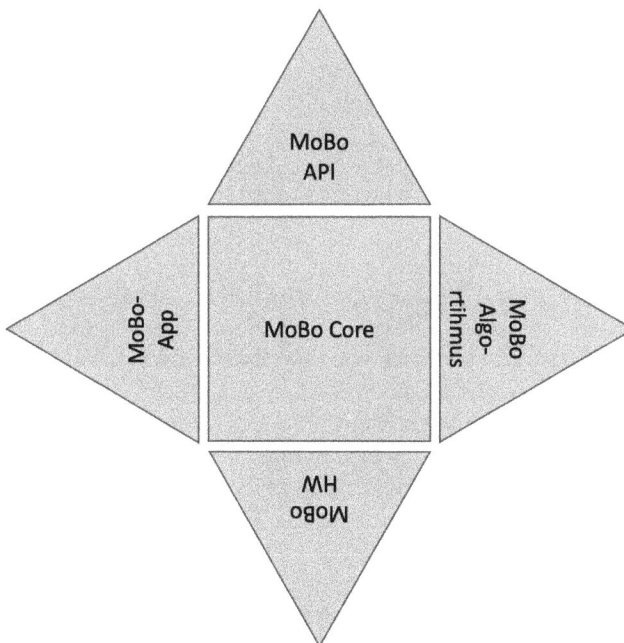

FIGURE 6.1 MORPHEUS box.

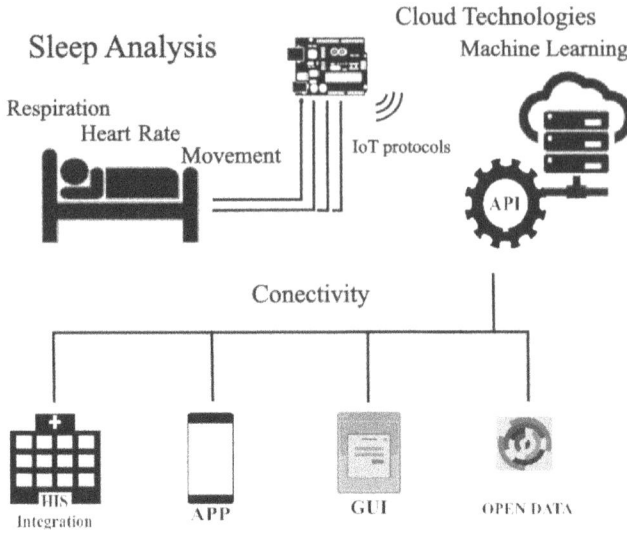

FIGURE 6.2   Layout diagram of final prototype.

algorithms. Different intelligent algorithms are used to personalize the process-ing of sleep quality and relevant vital data (e.g., respiration or apnea events). The MoBo-HW is a network of sensors that captures, in a non-invasive way, a patient's vital data and pre-processes it. Later it passes it to a downstream rule-based or learn-ing algorithm. The MoBo API provides an open interface of the system to external platforms to connect it to hospital or practice information systems. The MoBo App will provide and visualize the data suitably for the target groups (like patients or physicians), e.g., via a smartphone.

MORPHEUS is designed to explore various aspects of sleep medicine and remote medical care. Figure 6.2 depicts the basic layout of the product and the fields related to its components.

6.3   GENERAL SYSTEM ARCHITECTURE

The engineering of the system has been developed using SysML and UML dia-grams. The standard process has been conducted, creating corresponding use cases and requirements documentation. Figure 6.3 describes the basic structure of the designed system using a SysML block diagram. It comprises a set of physical com-ponents (bed, sensors, and microcontroller) and a software infrastructure deployed using cloud technologies that will provide the connectivity and processing features required to endow the system with intelligent capabilities to generate the recom-mendations and deploy helpful information to the final users. It also includes an artifact providing passive connection capabilities with standardized health infor-mation systems.

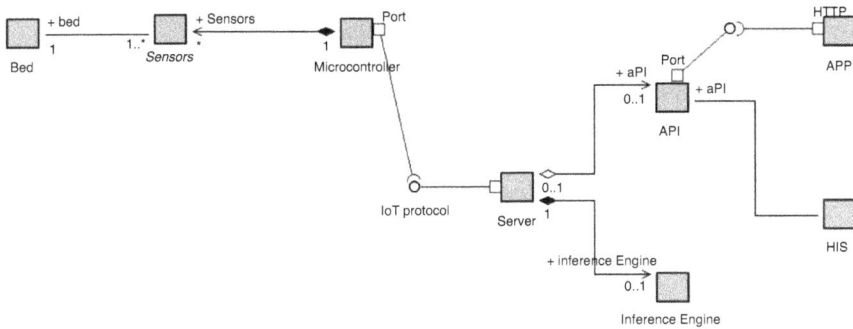

FIGURE 6.3 Block diagram of MORPHEUS.

6.3.1 BED AND SENSOR NETWORK

The system focuses on implementing current non-invasive sleep measurement devices, their improvement, and the proposal of new mechanisms. The current design supports the inclusion of multiple sensors and wearable devices to improve the accuracy of actigraphy-based sensors.

Initial implementations are currently using force-sensitive resistors (FSR) and accelerometers to register heart, respiratory, and movement signals, taking as a starting point prior research conducted [17,18]. Initial research on these devices has confirmed the applicability of the non-intrusive sensors and ballistocardiography (BCG) techniques in diagnosing sleep-related pathologies [19]. Additional non-invasive technologies and sensors are being tested to be included in the final implementation.

Environmental sensors are also considered to automate the collection of metrics from patient surroundings, such as the intensity of illumination, the status of home appliances, or even patient and device location within the home.

6.3.2 EDGE COMPUTING

The signal processing is performed on a customized electronic board (Figure 6.4), also supported by a Raspberry microcomputer to help connect and transmit data. The Raspberry works as an edge computer that would eventually run tailored ML algorithms based on the usage of the MORPHEUS box. Current local implementation is designed to treat data in a structured way [20]. Figure 6.5 depicts the entities modeled to handle the information during communications and local storage.

The metrics collected by wearable devices are treated as separate sensors, thus supporting the interaction and recording with multiple ready-to-use devices and individual sensors. Figure 6.6 describes the data transmission activities. The device is designed to support local data storage in case the Cloud services are unavailable or the Internet connection fails during a recording. Other synchronization processes with the server are also considered (although not included in the activity diagram).

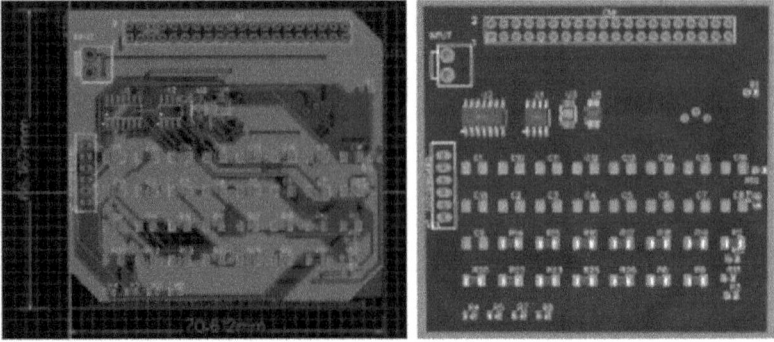

FIGURE 6.4   Customised electronic board.

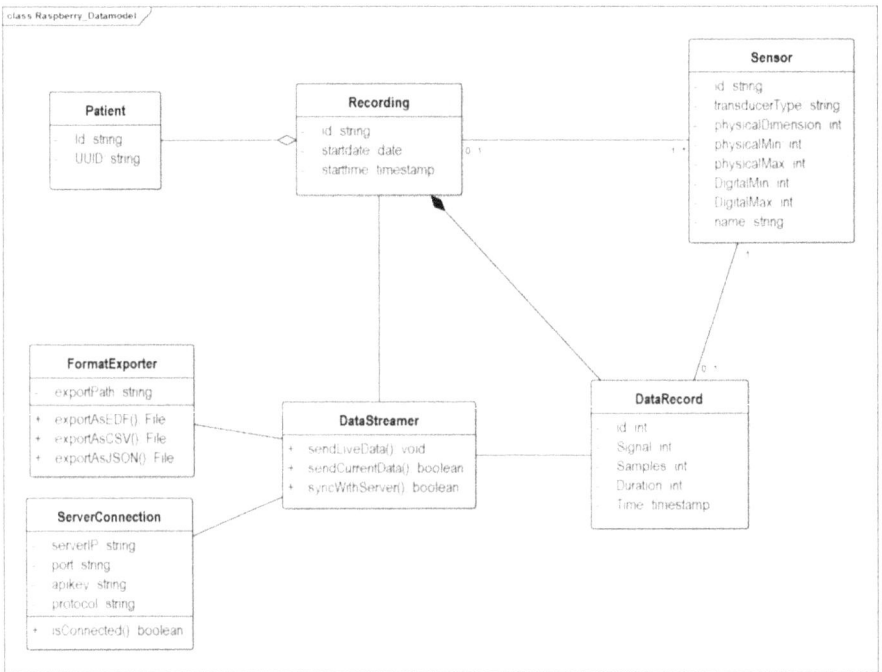

FIGURE 6.5   Edge computer data entity model.

6.3.3   SERVER

The cloud server is where further data processing and permanent storage take place. The technologies implemented in this block will allow the system to create intelligent recommendations, predictions, and sleep-disorder detections. All this depends on the algorithms inserted into the backend system. Since the data does not need a real-time analysis, the cloud server can be designed in a way taking this into account.

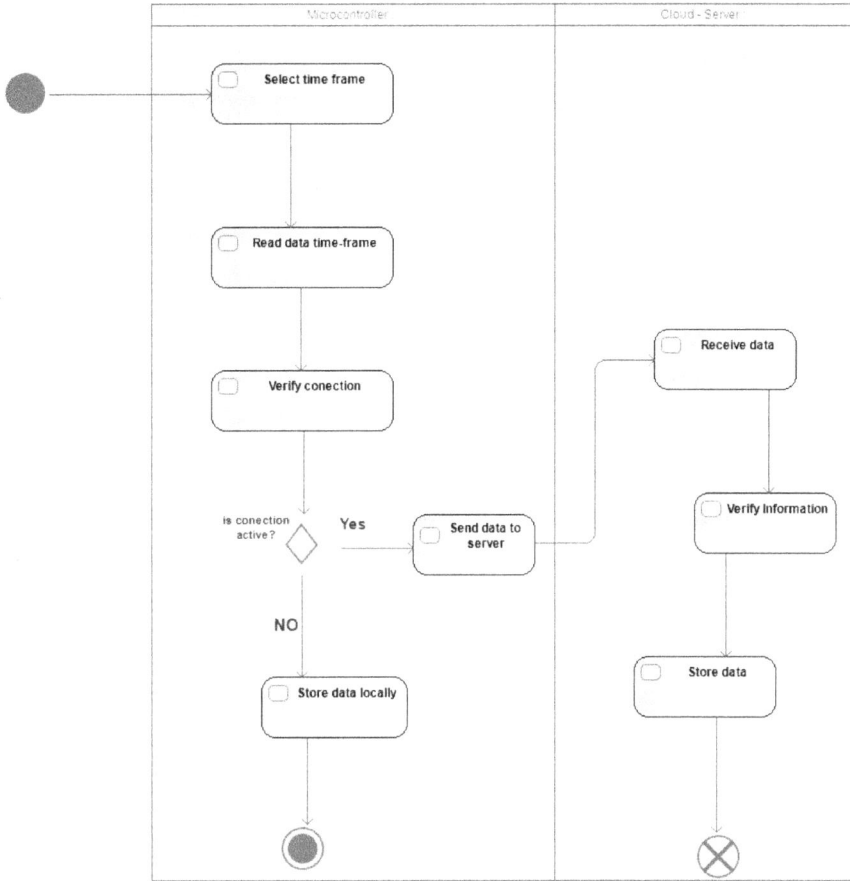

FIGURE 6.6   Data transmission activity diagram.

The specification of the current design for backend software is described in the following section.

6.4   BACKEND SOFTWARE ARCHITECTURE

The internal architecture of the software is designed as a service-oriented (SOA) system. The design includes several predefined components that belong to the Fiware initiative [21]. Figure 6.7 describes the components and layers included.

Depicted components from the physical layer were described in the previous section. Regarding the service layer, the main components are predefined Fiware components that create a technological base for the software infrastructure.

The Fiware Community [21] is an independent Open Community whose members and partners contribute to a trusted brand and technology. It aims to build an open, sustainable ecosystem around public, royalty-free, and implementation-driven

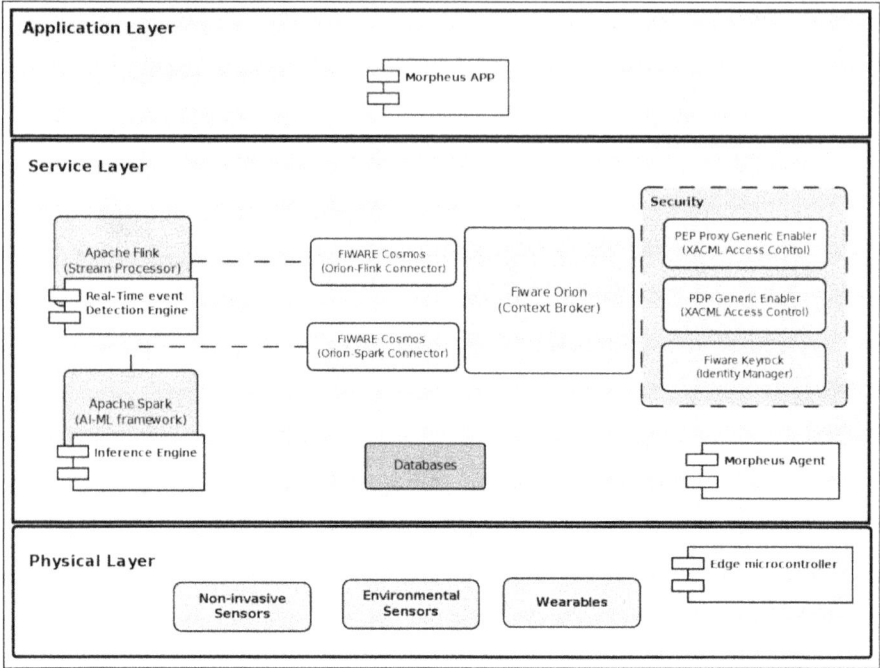

FIGURE 6.7   Components of intended infrastructure.

software platform standards that ease the development of new smart applications in multiple sectors.

6.4.1 FIWARE ORION

The Fiware Orion (context broker) is a C++ implementation of the NGSI information model and API. NGSI stands for "Next Generation Service Interfaces" and is a model for managing digital context information. It was formulated by the European Telecommunications Standardization Institute (ETSI) as part of the "Rolling plan for ICT standardisation" [22] to provide a helpful bridge between EU policies and standardization activities in information and communication technologies. Fiware Context Broker facilitates the management of context information with the API and allows easy creation, updating, and querying of context elements, and subscribing to custom alerts. All usage scenarios are described in the documentation [23].

By including the Fiware context broker, the project infrastructure will be able to support current and future implementations guided by the European Commission regarding the treatment and transmission of health information.

The European Commission announced the creation of a European Health Data Space in 2019 that will enable secure and trustworthy accessibility to health data across EU borders. To facilitate health data interoperability in the EU, the Commission adopted a Recommendation for a European Electronic Health Record exchange format (EEHRxF), which sets out the consensus concerning the technical specifications for patient summary reports, and e-prescriptions [22].

6.4.2 SECURITY COMPONENTS

In the proposed design, a security-related set of components take care of the control access and authorization protocols. These components are also provided in the Fiware ecosystem, including its security, authorization, and API access control catalog, and comply with the OASIS XACML (eXtensible Access Control Markup Language) standard [24].

In a standardized XACML architecture, a *Policy Decision Point* (PDP) evaluates policies against access requests provided by a *Policy Enforcement Point* (PEP). It is designed to support large-scale environments where resources are distributed, and policy administration is federated. For this, Keyrock Identity Manager provides a bridge between different IDM systems (at the connectivity and application levels) for authorizing foreign services into the secure stored data. This component complies with existing standards for user authentication and provides access to services acting as a Single Sign-On platform. By including this component, the system can provide secure and private authentication for users to devices, networks, and services, user profile management, privacy-preserving disposition of personal data, sign-in to service domains, and identity federation towards different services.

The PEP Proxy Generic Enabler works with OAuth2 and XACML protocols and provides authentication and authorization security features to the backend applications. At the same time, the PDP Generic Enabler provides an API to confirm access decisions based on authorization policies and evaluate authorization requests from PEPs. As with the other Fiware security components, the provided API follows the REST architecture style and complies with XACML v3.0.

6.4.3 MORPHEUS AGENT

The communication between the physical and the service layer is controlled by the MORPHEUS-Agent component, which contains the processing logic to store and share information with the context broker (Fiware-Orion). The services provided by this agent control the data flow emitted by the microcontroller, ensuring proper communication between the two layers. When needed, this component will also perform practical transformations and data extraction from the communication channel to fulfill the requirements for the intelligent behavior in the platform. The data transmission is managed by creating a structured replica of the data entities provided by the microcontroller, with enhanced capabilities for storing historical data.

Figure 6.8 describes the data model used by the agent. It comprises a set of entities that register the sensor measurements and the information from their physical origins. Patient identification is also registered to support multiple device connections through the same agent.

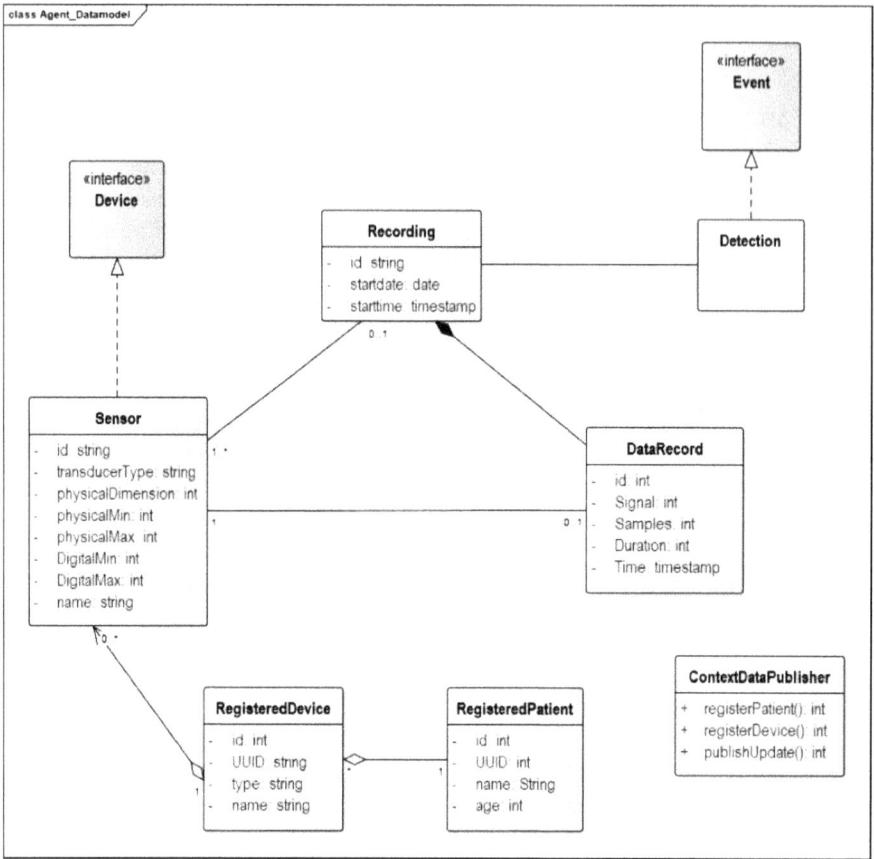

FIGURE 6.8   MORPHEUS agent data model.

6.4.4   INTELLIGENT BEHAVIOR

The intelligent behavior of the system is designed to be integrated with Open source components. There are two main implementations: one is supported on the Apache Flink Framework to drive real-time processing of data streams (Detection Engine), and the second is built on Apache Spark Framework to use machine learning libraries that provide intelligent recommendations based on pre-trained models (Inference Engine).

6.4.4.1   Apache Flink: RT Detection Engine

Apache Flink is a unified runtime environment that runs internal programs as structured directed graphs (JobGraphs) with parallelized operations [25]. A distributed engine can adapt the execution depending on the cluster environment, adapting each run depending on the data distribution. Queries can be specified in various programming languages. It provides an API with generic data operations like cross, join, map, reduce and filter, and user-defined operations.

A custom implementation within this framework seeks to provide the platform with capabilities to process real-time information and detect sleep events even during data transmission.

6.4.4.2 Apache Spark: Inference Engine

The Inference Engine is based on Apache Spark Framework, an analytics engine for distributed data processing and machine learning workloads. It is a powerful big data processing platform supporting batch and stream processing capabilities. Batch processing is performed with fast in-memory computations using Resilient Distributed Datasets (RDD), a specialized type of read-only data structure maintained in memory, ensuring Spark fault-tolerant capabilities and avoiding disk writing after every operation. Similarly, stream processing in Spark uses a micro-batch strategy which treats data streams as a group of tiny batches that are then handled in turns by the batch engine [26]. This strategy, nevertheless, leads to performance disadvantages compared with Flink and is the reason for implementing both as separate service components.

Spark services are included to endow the system with enterprise-level machine learning and Artificial Intelligence processing capabilities. Using Apache-Spark libraries, the system will generate intelligent recommendations based on the data in compliance with current medical procedures and recommendations.

6.4.5 MORPHEUS App

MORPHEUS APP is still being designed. Figure 6.9 describes the use cases identified. This component comprises the final user interface that allows patients and physicians to interact with the device and the platform, read and visualize sleep data and medical recommendations, and manage profile information. It is also intended as a tool for facilitating patient-doctor interaction.

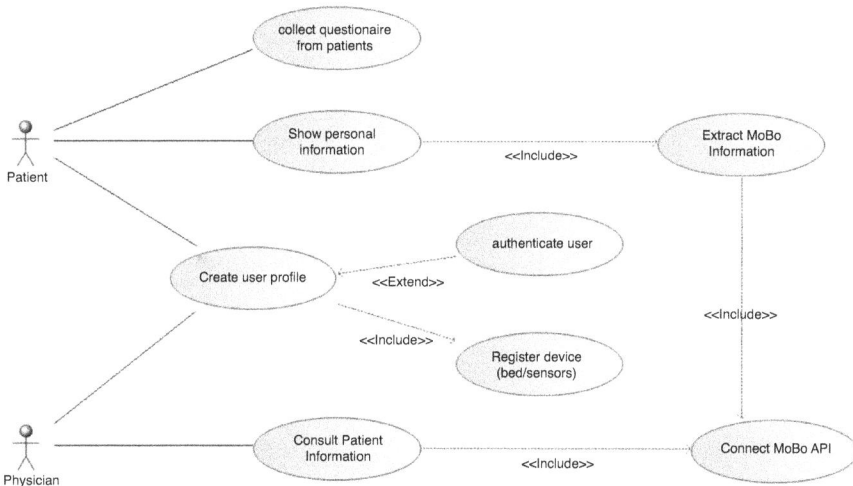

FIGURE 6.9   MORPHEUS App use cases.

6.5   DATA FLOW AND COMPONENT INTERACTIONS

The system was developed to provide interoperability of services using TCP/IP and HTTPS protocols. The following sections offer an extended description of the information flows considered in the proposed solution. Figure 6.10 depicts the designed interaction between the components.

6.5.1   From Sensors to the Context Broker

The actigraphic signals are detected by a series of sensors in the bed and sent to the microcomputer via electronic bus wired connections via the I2C protocol. A program deployed in a Raspberry establishes the connection with the MORPHEUS Agent using the HTTPS protocol over an internet connection. The Agent has TCP/IP port connections to a MySql database to store relevant information. It sends the context information to the broker using the HTTP protocol. The broker will authenticate any connection using the associated Identity Manager and store the context information in a MongoDB accessed by a TCP/IP secure port connection.

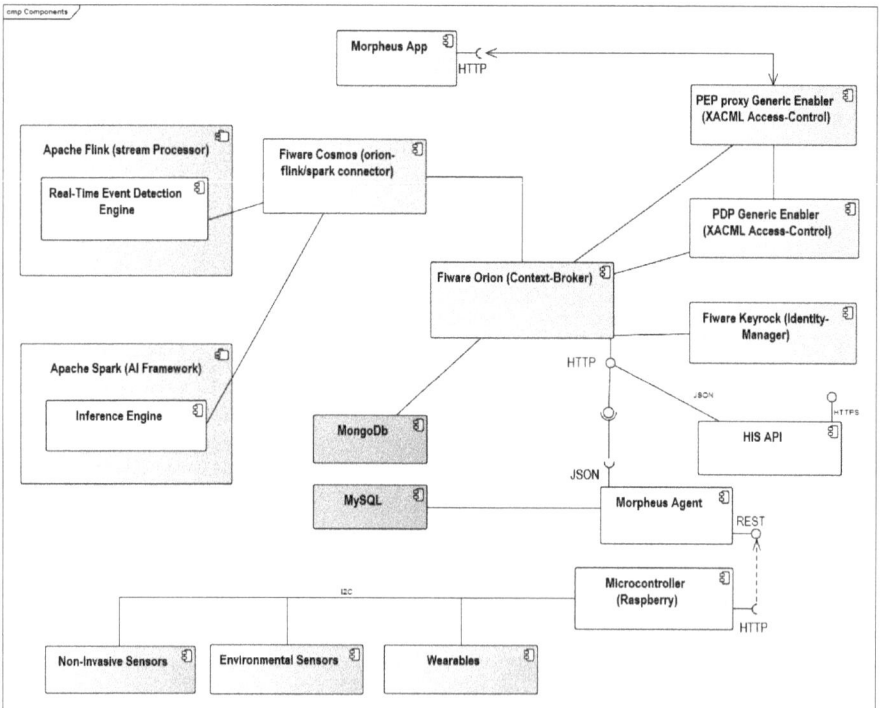

FIGURE 6.10   Connections between system components.

6.5.2 INTELLIGENT ENGINES INFORMATION FLOW

The information sources to run the intelligent algorithms are mainly three: The context broker, the system databases, and the data stream. Training the algorithms is a separate process that is not included in the system's workload. The data used for the training process can be extracted from the system, but it will also consider additional external sources. The Real-Time event Detection Engine will extract the information by registering for data events in the context-broker. The Inference Engine will extract the information from both the broker and directly from the data sources, as this can improve the performance of the system. Results from both Intelligent Engines will be sent to the context broker using RestFul request over HTTPS.

6.5.3 FROM CONTEXT BROKER TO USER INTERFACE

The user interface and mobile applications retrieve information using the RestFul API from Fiware-Orion. The Keyrock Identity Manager handles the user APP authentication, and the data querying and security from the server side will be granted by interacting with the PDP and PEP generic enablers. From the user application side, the security will be handled using HTTPS connections and token-based authentication.

6.6 CONCLUSION

Non-invasive measurements are key to future developments in sleep quality and disease detection. The infrastructure behind this takes advantage of new trends in software development, as improved connectivity is required to work with cloud computing and Artificial Intelligent approaches. In this chapter, we have seen an example of these technologies and the features that bind to this type of implementation. Service-oriented approaches and methodologies in software development improve the ability of back-end systems to respond to these requirements, and open-source implementations and initiatives make it possible to accelerate the construction of complete complex solutions that are reliable enough to be implemented in end-product solutions. The components provided by FIWARE provide several standard features needed in service-oriented solutions, granting a secure environment for processing sensor data. The Apache initiatives and products provide powerful platforms on which to base the custom implementations needed to generate smart predictions and include the system's intelligent capabilities.

With the specification provided in this chapter, the MORPHEUS project set a good platform, data model, and transmission strategies to bring an innovative proposal to measure sleep quality and detect sleep diseases from non-invasive sensors.

ACKNOWLEDGMENTS

Carl-Zeiss-Foundation funded this research with the MORPHEUS-Project "Non-invasive system for measuring parameters relevant to sleep quality" (project number: P2019-03-003).

REFERENCES

1. X. Hu, "Physiological measurement: A vital piece of the puzzle of precision medicine and health," *Physiological Measurement*, vol. 41, no. 2, p. 020401, 2020. [Online]. Available: https://doi.org/10.1088/1361-6579/ab6900.

2. R. A. Khan and A.-S. K. Pathan, "The state-of-the-art wireless body area sensor networks: A survey," *International Journal of Distributed Sensor Networks*, vol. 14, no. 4, p. 1550147718768994, 2018. [Online]. Available: https://doi.org/10.1177/1550147718768994.

3. J. V. Rundo and R. Downey, "Polysomnography," In L. S. de Vries and H. C. Glass (Eds.), *Handbook of Clinical Neurology*, vol. 160, pp. 381–392. Elsevier: Amsterdam, The Netherlands, 2019. [Online]. Available: https://doi.org/10.1016/B978-0-444-64032-1.00025-4.

4. J. M. Kelly, R. E. Strecker, and M. T. Bianchi, "Recent developments in home sleep-monitoring devices," *International Scholarly Research Notices*, vol. 2012, p. e768794, 2012. [Online]. Available: https://www.hindawi.com/journals/isrn/2012/768794/.

5. M. Conti, S. Orcioni, N. Martínez Madrid, M. Gaiduk, and R. Seepold, "A review of health monitoring systems using sensors on bed or cushion," In *International Conference on Bioinformatics and Biomedical Engineering,* pp. 347–358. Springer International Publishing: Cham, 2018. [Online]. Available: https://link.springer.com/chapter/10.1007%2F978-3-319-78759-6_32.

6. S. F. Green, T. Frame, L. V. Banerjee, A. Gimson, J. Blackman, H. Morrison, K. Lloyd, S. Rudd, W. G. Frederick Fotherby, U. Bartsch, S. Purcell, M. Jones, and L. Coulthard, "A systematic review of the validity of non-invasive sleep-measuring devices in mid-to-late life adults: Future utility for Alzheimer's disease research," *Sleep Medicine Reviews*, vol. 65, p. 101665, 2022. [Online]. Available: https://www.sciencedirect.com/science/article/pii/S1087079222000788.

7. K. Rajguru, P. Tarpe, V. Aswar, K. Bawane, S. Sorte, and R. Agrawal, "Design and implementation of IoT based sleep monitoring system for Insomniac people," In *2022 Second International Conference on Artificial Intelligence and Smart Energy (ICAIS),* February 2022, pp. 1215–1221. [Online]. Available: https://doi.org/10.1109/ICAIS53314.2022.9742803.

8. B. Yu, Y. Wang, K. Niu, Y. Zeng, T. Gu, L. Wang, C. Guan, and D. Zhang, "WiFi-sleep: Sleep stage monitoring using commodity Wi-Fi devices," *IEEE Internet of Things Journal*, vol. 8, no. 18, pp. 13 900-13 913, 2021. [Online]. Available: https://doi.org/10.1109/JIOT.2021.3068798.

9. X. Liu, J. Cao, S. Tang, J. Wen, and P. Guo, "Contactless respiration monitoring via off-the-shelf WiFi devices," *IEEE Transactions on Mobile Computing*, vol. 15, no. 10, pp. 2466–2479, 2016. [Online]. Available: https://doi.org/10.1109/TMC.2015.2504935.

10. A. Tal, Z. Shinar, D. Shaki, S. Codish, and A. Goldbart, "Validation of contact-free sleep monitoring device with comparison to polysomnography," *Journal of Clinical Sleep Medicine*, vol. 13, no. 03, pp. 517–522, 2017. [Online]. Available: https://jcsm.aasm.org/doi/full/10.5664/jcsm.6514.

11. J. Mantua, N. Gravel, and R. M. C. Spencer, "Reliability of sleep measures from four personal health monitoring devices compared to research-based actigraphy and poly-somnography," *Sensors*, vol. 16, no. 5, p. 646, 2016. [Online]. Available: https://www.mdpi.com/1424-8220/16/5/646.

12. S. Roomkham, D. Lovell, J. Cheung, and D. Perrin, "Promises and challenges in the use of consumer-grade devices for sleep monitoring," *IEEE Reviews in Biomedical Engineering*, vol. 11, pp. 53–67, 2018. [Online]. Available: https://ieeexplore.ieee.org/document/8309286.

13. Y. Nam, Y. Kim, and J. Lee, "Sleep monitoring based on a tri-axial accelerometer and a pressure sensor," *Sensors*, vol. 16, no. 5, p. 750, 2016. [Online]. Available: https://www.mdpi.com/1424-8220/16/5/750.

14. D. C. Yacchirema, D. Sarabia-JáCome, C. E. Palau, and M. Esteve, "A smart system for sleep monitoring by integrating IoT with big data analytics," *IEEE Access*, vol. 6, pp. 35 988–36 001, 2018. [Online]. Available: https://doi.org/10.1109/ACCESS.2018.2849822.

15. C. J. Lavoie, M. R. Zeidler, and J. L. Martin, "Sleep and aging," *Sleep Science and Practice*, vol. 2, no. 1, p. 3, 2018. [Online]. Available: https://doi.org/10.1186/s41606-018-0021-3.

16. N. Schütz, H. Saner, A. Botros, B. Pais, V. Santschi, P. Buluschek, D. Gatica Perez, P. Urwyler, R. M. Müri, and T. Nef, "Contactless sleep monitoring for early detection of health deteriorations in community-dwelling older adults: Exploratory study," *JMIR mHealth and uHealth*, vol. 9, no. 6, p. e24666, 2021. [Online]. Available: https://mhealth.jmir.org/2021/6/e24666.

17. M. Gaiduk, J. J. Perea Rodríguez, R. Seepold, N. Martínez Madrid, T. Penzel, M. Glos, and J. A. Ortega, "Estimation of sleep stages analyzing respiratory and movement signals," *IEEE Journal of Biomedical and Health Informatics*, vol. 26, no. 2, pp. 505–514, 2022. [Online]. Available: https://doi.org/10.1109/JBHI.2021.3099295.

18. A. Serrano Alarcón, N. Martínez Madrid, and R. Seepold, "A minimum set of physiological parameters to diagnose obstructive sleep apnea syndrome using non-invasive portable monitors: A systematic review," *Life*, vol. 11, no. 11, p. 1249, 2021. [Online]. Available: https://doi.org/10.3390/life11111249.

19. M. Conti, C. Aironi, S. Orcioni, R. Seepold, M. Gaiduk, and N. Martínez Madrid, "Heart rate detection with accelerometric sensors under the mattress," In *2020 42nd Annual International Conference of the IEEE Engineering in Medicine & Biology Society (EMBC)*, 2022, pp. 4063–4066. [Online]. Available: https://doi.org/10.1109/EMBC44109.2020.9175735.

20. A. Asadov, A. Boiko, M. Gaiduk, W. D. Scherz, R. Seepold, and N. Martínez Madrid, "Evaluation of a prototype for early active patient mobilization," *Procedia Computer Science*, vol. 207, pp. 2223–2231, 2022. [Online]. Available: https://doi.org/10.1016/j.procs.2022.09.282.

21. "About FIWARE | FIWARE." [Online]. Available: https://www.fiware.org/about-us/.

22. "Rolling Plan 2021 | Joinup." [Online]. Available: https://joinup.ec.europa.eu/collection/rolling-plan-ict-standardisation/rolling-plan-2021.

23. "Orion Context Broker," November 2022, original-date: 2013-09-13T11:55:25Z. [Online]. Available: https://github.com/telefonicaid/fiware-orion/blob/9e0b1ea18dc4a7 62c94c1d2315a0de84f0dd99bd/README.md.

24. "eXtensible Access Control Markup Language (XACML) version 3.0" [Online]. Available: https://www.okbsapr.ru/upload/iblock/0ac/xacml-3.0-core-spec-os-en.pdf.

25. T. Rabl, J. Traub, A. Katsifodimos, and V. Markl, "Apache Flink in current research," it - Information Technology, vol. 58, 2016. [Online]. Available: https://doi.org/10.1515/itit-2016-0005.

26. E. Shaikh, I. Mohiuddin, Y. Alufaisan, and I. Nahvi, "Apache spark: A big data processing engine," In *2019 2nd IEEE Middle East and North Africa COMMunications Conference (MENACOMM)*, 2019, pp. 1–6. [Online]. Available: https://doi.org/10.1109/MENACOMM46666.2019.8988541.

7 Predicting Incidence of Stroke via Supervised Machine Learning Methods on Class Imbalanced Data

Sulaf Assi, Manoj Jayabalan, and Vaibhav Parakh
Liverpool John Moores University

Jolnar Assi
Traders Island Ltd

Abdullah Al Hamid
King Faisal University

Dhiya Al-Jumeily OBE
Liverpool John Moores University

7.1 INTRODUCTION

Stroke is a cardiovascular disease that contributes to morbidities and mortalities globally. Stroke is the second largest cause of death and the third most common cause of disability worldwide (WHO, 2021a, 2021b). According to the World Health Organization (WHO), there are 12.2 million new strokes per year with the majority occurring in low- and middle-income (LMIC) countries (WHO, 2021b; WSO, 2022). Nonetheless, the increase in desk-based jobs and working from home post-COVID have led to an increase in chronic conditions including stroke (Lippi et al., 2020).

A stroke is defined as the cessation of blood flow to a part of the brain, leading to cell death and loss of function. Strokes are classified into two types: ischaemic stroke and haemorrhagic stroke. Ischaemic stroke represents the major type of stroke and contributes to 87% of total stroke cases (Johnson et al., 2016). In most cases, blood clots of the carotid arteries are responsible for ischaemic strokes. However, haemorrhagic stroke occurs when a brain artery bursts and is classified into intracerebral haemorrhage and subarachnoid haemorrhage (Chatterjee et al., 2017). Intracerebral haemorrhage is the most common type of haemorrhage and occurs when blood

DOI: 10.1201/9781003346678-7

seeps out of a brain aneurysm and into the surrounding tissue causing this disease. Subarachnoid haemorrhage is rare and refers to a haemorrhage in the space between the brain and the surrounding tissue. When mentioning stroke, it is worth noting that 'mini-strokes', also known as transient ischemic attacks (TIA), occur when blood flow to the brain is disrupted (Sposato et al., 2015). TIAs usually serve as an early warning sign of stroke. Hence, one in three patients who suffer from TIA and do not seek medical treatment experience a stroke within a year after the TIA incidence (CDC, 2022). Multiple risk factors contribute to stroke and include medical and patient and lifestyle risk factors. Medical risk factors include history of stroke and/or TIA, high blood pressure, high cholesterol level, heart disease, diabetes and/ or sickle cell disease. In addition, patient and lifestyle risk factors include gender, genetic predisposition, unhealthy diet, obesity, physical inactivity, smoking and alcohol intake (Prior et al., 2011).

Stroke patients are usually admitted to a stroke unit or a clearly defined area or ward within the hospital where they are looked after by a team of specialist doctors and nurses (Khurana et al., 2018). The earlier patients arrive to stroke unit, the better prognosis they have (Pandian and Sudhan, 2013). However, not all countries have stroke units or team of specialists dedicated to stroke patients. This latter problem is more prevalent in LMIC which also suffer from lack of service integration between authorities, healthcare professionals and stakeholders (Bernhardt et al., 2020). In this respect, early prediction of stroke can help mitigate it, especially in LMIC.

Interpretable machine learning (IML) has been an area of interest for researchers as it addresses a persisting issue in relation to models' transparency and interpretability (Watson, 2022). Hence, IML enables humans to understand how ML models have achieved a certain decision. Conventional ML models such as ensemble models and deep neural networks (DNNs) have been driving ML forward at an incredible pace (Al-Tameemi et al., 2021). A wide range of real-world applications use these ML models including Amazon, Apple, Flipkart and Google Cloud. Despite the advantages of these ML models in decision-making, they lack transparency in how a decision has been achieved. The complex 'black box character' of such a model has flagged concerns, especially in important decision-making such as the implementation of AI in the medical field. To solve the aforementioned issues, interpretable ML emerged which encompasses the ability of ML models to explain or portray their behaviours to humans in a comprehendible manner (Doshi-Velez and Kim, 2017).

7.2 METHODS

It is common for enormous amounts of medical data to be generated quickly. Preprocessing techniques such as missing value imputation and approaches such as Synthetic Minority Oversampling Technique (SMOTE) or Adaptive Synthetic Sampling Technique (ADASYN), and their influence on predictive modelling should be examined before continuing (Kumar et al., 2019). Subsequently, both SMOTE and ADASYN have been used for pre-processing the dataset used in this research.

The dataset was retrieved from the Kaggle platform and comprised 5,110 observations with 12 attributes (Kaggle, 2022). Ten of these attributes were clinical features commonly used for predicting incidences of stroke and included:

- **Patient demographic markers**: Age, gender, marital status, type of work, type of residence
- **Patient health markers**: Average glucose level, body mass index (BMI), heart disease, hypertension and smoking
- **Target label (stroke: yes, no)**: Indicated what can be observed from the dataset. Out of 5,110 observations, there were 249 events of stroke and 4,861 events of a non-stroke that yielded an imbalanced dataset.

7.2.1 DATASET PRE-PROCESSING AND EXPLORATORY DATA ANALYSIS

All clinical features had data points apart from BMI. There were missing data (201 rows) but the choice was made to impute them rather than drop them completely. Categorial feature variables (e.g. type of work and type of residence) were one-hot encoded to prepare the dataset for ML processing. Moreover, exploratory data analysis was applied, which included univariate and/or bivariate analysis to uncover relationships/correlations between feature variables and target variables.

7.2.2 CLASS BALANCING

Due to the dataset class imbalance, appropriate sampling techniques such as SMOTE, ADASYN and/or resampling were applied to rebalance the classes. SMOTE works by taking instances that are in proximity in the feature space, drawing a line between such points and then plotting new samples along that line. For the minority class, this method could be used to build as many synthetic instances as needed. The approach has the disadvantage of constructing synthetic instances without taking into consideration the majority class, which might result in ambiguous examples if the classes overlap a lot (Fernández et al., 2018). ADASYN is based on the idea of adaptively generating minority data samples according to their distributions using K-nearest neighbour (KNN). There are no assumptions about the underlying distribution of the data because the algorithm adjusts the distribution dynamically. For the KNN algorithm, Euclidean distance was used. A density distribution was used to determine the number of synthetic samples that must be generated for each minority sample by adaptively changing the weights of different minority samples to compensate for skewed distributions in ADASYN. Nevertheless, in SMOTE, the weights of each minority sample were fixed. For each original minority sample, the latter generated the equivalent number of synthetic samples (Fernández et al., 2018).

7.2.3 EVALUATION

Following the training of the various models on the train dataset, the model was evaluated using a test dataset. The Confusion Matrix played a key role in determining the model's ability to accurately predict the correct classes. Using the confusion matrix, the performance of various models could be compared, and based on the results, the best performing model could be determined. Predicted classes were then classified into True positive (TP), False positive (FP), True Negative (TN), and False Negative (FN). The four aforementioned classes are explained below:

- **TP**: The predicted output is positive (incidence of stroke) and the ground truth is also positive (incidence of stroke).
- **FP**: The predicted output is positive (incidence of stroke) but the ground truth is negative (non-incidence of stroke).
- **TN**: If the predicted output is negative (non-incidence of stroke) and the ground truth is also negative (non-incidence of stroke).
- **FN**: If the predicted output is negative (non-incidence of stroke) but the ground truth is positive (incidence of stroke).

Moreover, six statistics criteria were used to evaluate the models:

- **Accuracy**: compared the models' overall performance in detecting both default classes and gave an indication of how well the model accurately recognised all data points belonging to a given class when compared to the actual class in the test data.
- **Precision**: compared the models' overall performance in accurately detecting correct instances out of all the positive insurances.
- **Sensitivity/recall**: When compared to the actual occurrences in the dataset, sensitivity/recall indicated how many stroke incidents were accurately predicted by the model. This is an important parameter since higher sensitivity indicated that the algorithm was capable of properly identifying all stroke-related incidents.
- **Specificity**: was calculated as the percentage of genuine negatives that were predicted to be negatives (or true negative). This means that a portion of actual negatives were predicted as positives, resulting in false positives.
- **F1-score**: This metric was particularly helpful in a class imbalanced dataset. This metric was the harmonic mean of precision and recall. A higher F1-score indicated the model had correctly predicted TP and TN. This was especially useful where the emphasis was on having a lower FN and an overall lower FP.
- **ROC-AUC curve**: This metric showed the TP rate and FP rate for every probability threshold of a binary classifier.

7.2.4 DATA PREPARATION

Table 7.1 shows the first three records of data transposed and that corresponded to 12 variables that comprised: ten clinical features (independent features) and stroke (dependent feature).

Using the info method, we could get a basic description of the dataset which informed us of the number of rows as well as the data type. We could see that we are dealing with 3 float 64 data type features, 4 int 64 data type features, and five object (string) data type features. We could also observe that BMI has only 4,909 non-null records as against 5,110 non-null records in the other features which indicated missing data (Figure 7.1).

Using the value_counts method, we have attempted to look at what all values were under the object data type features which would form an important base for our

TABLE 7.1

Dependent and Independent Features in Stroke Patients

Code	9046	51676	31112
Gender	Male	Female	Male
Age (years)	67	61	80
Hypertension	0	0	0
Heart disease	1	0	1
Marital status	Yes	Yes	Yes
Occupation	Private	Self-employed	Private
Residence type	Urban	Rural	Rural
Average glucose level	228.69	202.21	105.95
BMI	36.6	NaN	32.5
Smoking	Formerly smoked	Never smoked	Never smoked
Stroke	1	1	1

```
<class 'pandas.core.frame.DataFrame'>
RangeIndex: 5110 entries, 0 to 5109
Data columns (total 12 columns):
 #   Column             Non-Null Count   Dtype
---  ------             --------------   -----
 0   id                 5110 non-null    int64
 1   gender             5110 non-null    object
 2   age                5110 non-null    float64
 3   hypertension       5110 non-null    int64
 4   heart_disease      5110 non-null    int64
 5   ever_married       5110 non-null    object
 6   work_type          5110 non-null    object
 7   Residence_type     5110 non-null    object
 8   avg_glucose_level  5110 non-null    float64
 9   bmi                4909 non-null    float64
 10  smoking_status     5110 non-null    object
 11  stroke             5110 non-null    int64
dtypes: float64(3), int64(4), object(5)
memory usage: 479.2+ KB
```

FIGURE 7.1 Information about clinical features used in this study.

univariate and bivariate analysis. This also provided a high-level scenario of what all values were considered by the model. For example, was the concerned patient employed in a private occupation or a government occupation; was the concerned patient married or unmarried; lastly, what was their smoking status (Figure 7.2).

7.2.5 IDENTIFYING AND TREATING MISSING VALUES

Out of the 5,110 records, 201 records (4%) were missing the BMI clinical feature. These missing values were imputed by using a decision tree regressor and basis other

```
gender
-----------------------------------
Female      2994
Male        2115
Other          1
Name: gender, dtype: int64

ever_married
-----------------------------------
Yes      3353
No       1757
Name: ever_married, dtype: int64

work_type
-----------------------------------
Private            2925
Self-employed       819
Children            687
Govt_job            657
Never_worked         22
Name: work_type, dtype: int64

Residence_type
-----------------------------------
Urban      2596
Rural      2514
Name: Residence_type, dtype: int64

Smoking_status
-----------------------------------
never smoked        1892
Unknown             1544
formerly smoked      885
smokes               789
Name: Smoking_status, dtype: int64
```

FIGURE 7.2   Value-count method applied to clinical features.

two features (age and gender) in order to generate new values. Age and gender were chosen considering the role these two variables play in determining BMI. The operation was confirmed by using the info method on the data frame where we could observe that now there were no more null values. Thus, all clinical features had 5,110 non-null records, which could also be corroborated from Figures 7.4 and 7.5. Post data imputation data information (Figure 7.3).

7.2.6   IDENTIFICATION OF OUTLIERS

For outlier identification and treatment, all records that were beyond the fifth deviation from the normal (i.e., z score > 5) were removed. As per theory, 99.99% of data could be summarised within z score $= 3$ in the case of normally distributed data. But since the statistical distribution of the present data was not normal, data was kept up to the fifth deviation. This was chosen in order to preserve the maximum number of records the model was trained on. Four records were shown to be outlier, and subsequently, 5,106 records remained post-outlier treatment.

```
<class 'pandas.core.frame.DataFrame'>
RangeIndex: 5110 entries, 0 to 5109
Data columns (total 11 columns):
 #   Column             Non-Null Count  Dtype
---  ------             --------------  -----
 0   gender             5110 non-null   object
 1   age                5110 non-null   float64
 2   hypertension       5110 non-null   int64
 3   heart_disease      5110 non-null   int64
 4   ever_married       5110 non-null   object
 5   work_type          5110 non-null   object
 6   Residence_type     5110 non-null   object
 7   avg_glucose_level  5110 non-null   float64
 8   bmi                5110 non-null   float64
 9   smoking_status     5110 non-null   object
 10  stroke             5110 non-null   int64
dtypes: float64(3), int64(3), object(5)
memory usage: 439.3+ KB
```

FIGURE 7.3   Post-data amputation and data information.

7.2.7   SCALING OF DATA

Three copies of the dataset were used for scaling of the data. The first copy of dataset was used 'as-it-is', i.e., preserving the imbalanced nature of the dataset and experimenting with classification algorithms. For the remaining two copies of dataset, one was augmented using SMOTE and other with ADASYN to address the imbalanced nature of the dataset. Scaling was opted during the first run of model building process, but during results gathering, it was observed that if features such as average glucose levels, BMI, and age were scaled then their interpretation would be rendered useless and hence it was a conscious decision to not scale the data in order to let the features be intelligible post modelling for inferences/interpretation.

7.2.8   EXPLORATORY DATA ANALYSIS

Exploratory data analysis had been applied before formally commencing the model in order to obtain insight into the data and identify any key interferences. This was also in line with the model interpretation expectations, where pre-modelling interpretation/inferences enabled understanding post-modelling interpretation/inferences. In all cases, it has been maintained that a class imbalanced dataset was analysed. Thus 4.87% of records showed incidence of stroke; while 95.1% showed non-incidence of stroke. This imbalance showed similarity to real world data, especially in the medical domain.

7.2.9   UNIVARIATE ANALYSIS

Univariate analysis was applied to different combination of clinical features depending on the context they had encountered.

7.2.9.1   Age, Average Glucose Levels, BMI

Univariate analysis commenced with assessing the distribution of float64 data type features including 'age', 'avg_glucose levels' and 'BMI'. The three variables showed a skewed distribution. Age showed tapering off at extremes, whereas, average glucose level and BMI showed positive skew (Figure 7.4). This was also observed for 'incidence of stroke' and 'non-incidence of stroke' (Figure 7.4). Nonetheless, the latter records showed that the incidence of strokes had age concentrated beyond 40 years which indicates that older people tend to have higher probability of experiencing stroke. However, this pattern was not seen for average glucose level and BMI that showed similar distribution for 'incidence of stroke' and 'non-incidence of stroke'.

7.2.9.2   Gender and Age of Population

Univariate analysis of gender and age population showed higher incidence of stroke in women versus men. Distribution of stroke per gender showed 58.6% of records belonged to women; whereas, 41.4% belonged to men (Figure 7.5). This could be linked to multiple factors relating to women that exacerbate with age e.g. menopause and decrease of estrogen levels (Bushnell, 2008), under-diagnosis (Jneid and Thacker, 2001) and under-prescription of medicines (Vogel et al., 2021). These factors have been confirmed by the present datasets that showed the highest incidence of stroke in women was encountered around the age of 80 years; thus, this could be the result of a combination of the factors related to underdiagnosis, underprescription and menopause (Figure 7.5). On the other hand, men showed the highest incidence at adult age (<60 years old).

Numeric Variable Distribution
We see a positive skew in BMI and Glucose Level   (a)

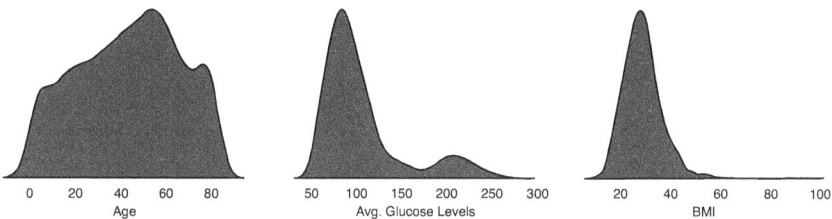

Numeric Variables by Stroke & No Stroke
Age looks to be a prominent factor - this will likely be a salient feautre in our models   (b)

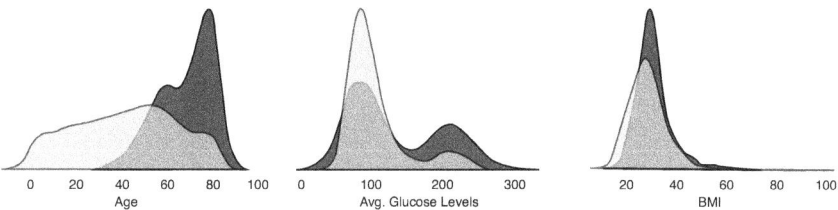

FIGURE 7.4   Distribution of age, average glucose levels and BMI for (a) entire dataset and (b) incidence of stroke (dark blue) and non-incidence of stroke (light blue).

Age-Sex Infrence

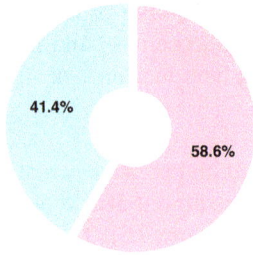

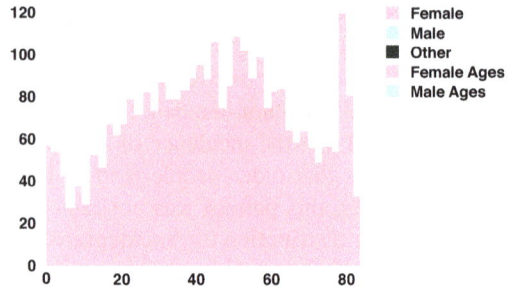

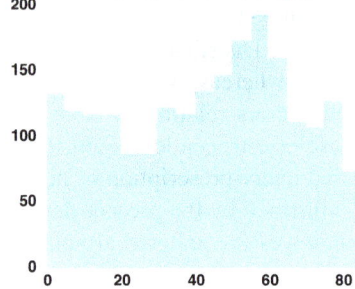

FIGURE 7.5   Distribution of age and gender in relation to incidence of stroke.

7.2.9.3   Socioeconomic Status of Stroke-Positive Patients

The socioeconomic status of stroke-positive patients showed three key variables relating to the type of work, marital status and residence types. For the type of work among stroke-positive patients, 59.4% were privately employed, 26.1% were self-employed and 13.3% were government employees. Moreover, 88.4% of stroke-positive patients were married. As for the residence type, 54.2% of stroke-positive patients lived in urban areas, whereas the remaining lived in rural areas.

7.2.9.4   Prevalence of Smoking among Stroke-Positive Patients

Unexpectedly, the highest smoking category (36.1%) of stroke-positive patients were those who had never smoked prior to the stroke. This was followed by patients who had formerly smoked and had quit smoking prior to the stroke (28.1%) and those with unknown smoking habits (18.9%). Only 16.9% of stroke-positive patients reported themselves as smokers. Whereas, this result contradicts the vast literature relating to smoking as a risk factor for stroke (Shah and Cole, 2010). Nonetheless, the result could be explained by the unreported smoking status of 18.9% of the patients

7.2.10 Bivariate/Multivariate Analysis

7.2.10.1 Bivariate Analysis of Average Glucose Level and BMI with Age

Figure 7.6 shows the relationship between average glucose level and BMI with age. The darker shaded dots represent the incidence of stroke. In this respect, age showed a crucial factor in case of average glucose levels and BMI. Towards the right extreme of age, we see many incidences of stroke which are fairly symmetrically spread in case of average glucose level; and in case of BMI, even at lower BMIs, we can see incidences of stroke if the sample belongs to the elderly population category.

7.2.10.2 Correlation of Features Belonging to Entire Dataset

Two correlation models were applied: the first to the entire dataset and the second to the stroke-positive datasets. The entire dataset correlation model showed low correlation values between individual features and that was a good outcome in terms of model building (Figure 7.7). This is because highly correlated features could be misleading and skew the results of the model. The stroke-positive dataset showed a slight negative correlation (−0.25) between age and BMI and a slight positive correlation (0.32) between average glucose levels and BMI.

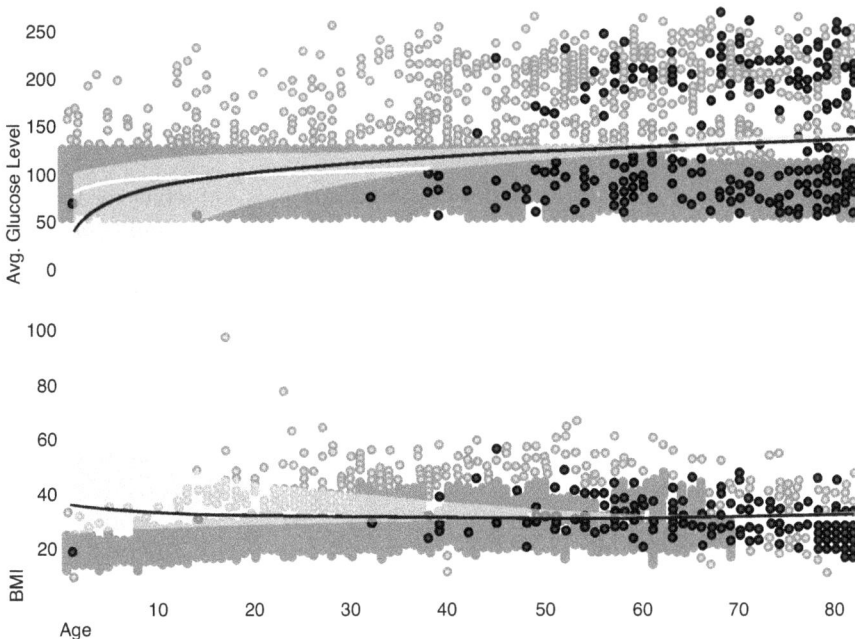

Strokes by Age, Glucose Level, and BMI
Age appears to be a very important factor

FIGURE 7.6 Bivariate analysis of average glucose level and BMI with age.

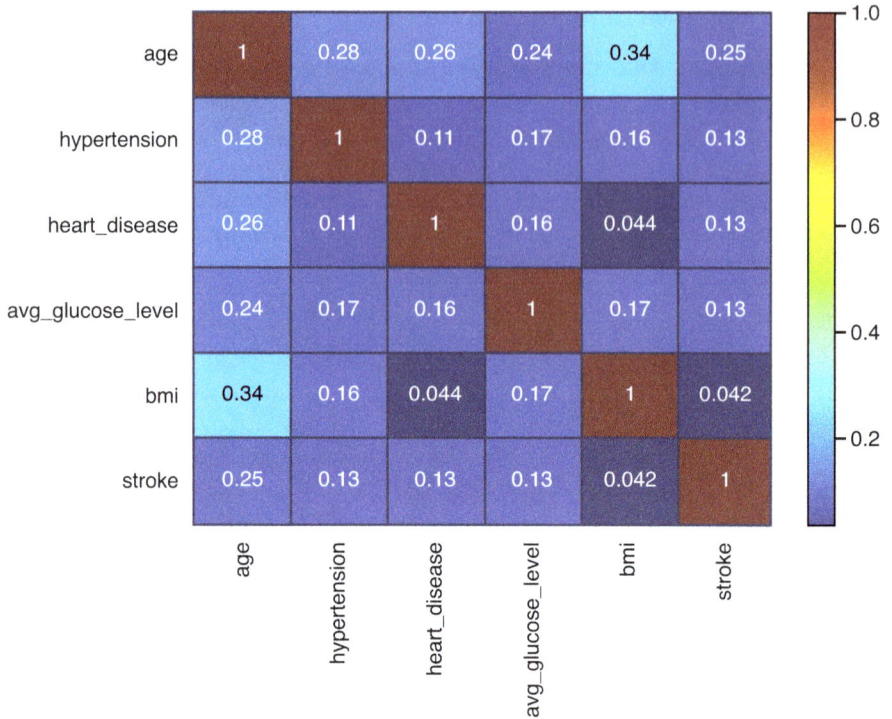

FIGURE 7.7 Correlation of features belonging to the entire dataset.

7.2.11 CLASS BALANCING

For model to be effective at the task of classification, it is important for the dataset to be balanced. A balanced dataset underlies an equal representation of the target variable(s). The present dataset showed 249/5,110 stroke-positive patients and 4,861/5,110 stroke-negative patients and which indicated an imbalanced dataset. Therefore, class imbalance was applied in order to prevent the model from predicting the majority class (i.e. non-stroke). The latter case would render the model ineffective.

Therefore, to improve the model's performance, two techniques that address class imbalance were applied: SMOTE and ADASYN. Both techniques addressed class imbalance by oversampling the minority class to match the number of observations of the majority class.

7.2.11.1 Class Balancing – SMOTE

SMOTE worked by taking instances which were in proximity in the feature space, drawing a line between such points and then plotting new samples along that line. For the minority class, this method could be used to build as many synthetic instances as needed. In this respect, the 5,106 post-outlier treatment records pre-SMOTE application became 9,714 records post-SMOTE application. The 9,714 records comprised 50.00% records of positive (1) classes and 50.00% records of negative (0) classes.

7.2.11.2 Class Balancing – ADASYN

ADAptive SYNthetic (ADASYN) is based on the idea of adaptively generating minority data samples according to their distributions using K-nearest neighbour. There are no assumptions about the underlying distribution of the data because the algorithm adjusts the distribution dynamically. For the KNN Algorithm, Euclidean distance was used. A density distribution was used to determine the number of synthetic samples that needed to be generated for each minority sample by adaptively changing the weights of different minority samples to compensate for skewed distributions in ADASYN, whereas in SMOTE, the weights of each minority sample are fixed. For each original minority sample, the latter generates the equivalent number of synthetic samples.

In this respect, the 5,106 post-outlier treatment records pre-ADASYN application became 9,651 records post-SMOTE application. The 9,714 records comprised 50.32% records of positive (1) classes and 49.67% records of negative (0) classes.

7.2.12 Model Building and Hyper-Parameters Tuning

For the three datasets:

- The first copy of dataset used as-it-is with class imbalance;
- The second copy of dataset post application of SMOTE;
- The third copy of dataset post application of ADASYN.

In all cases, the training size was 75% and test size was 25%. Default random state '42' was used. All the object type variables were converted to dummy variables (numeric format) to facilitate model building process. In order to find the best hyperparameters for the ensemble classification models, GridSearch CV was employed with 10-fold cross-validation. This implied, during the search for the best hyperparameters, out of the 10 folds, the model was trained on nine folds and the remaining 10th fold was employed for testing. Ten-fold cross-validation has been deemed as an optimum number of folds, especially when there was no clarity on how many folds should be employed since 10-fold cross-validation helps in model optimisation by greatly reducing the bias and variance of the mode. Table 7.2 shows the hyperparameters that were obtained during the training process:

7.2.13 Machine Learning Interpretation

ML interpretation was applied in two phases: The first being the pre-modelling phase and the second the post-modelling phase. The pre-modelling phase represented an exploratory phase for interpretation and gaining insights from the data. Post-modelling phase yielded how the machine had determined the results by exploring features' importance and depending on SHAP values. Cooperative game theory underpins SHAP values (SHapley Additive exPlanations), a technique for making machine learning models more understandable and transparent. SHAP explained the individual contribution of each feature on the output of the model.

TABLE 7.2
Hyperparameters Obtained for Various Models during the Training

Algorithm/ Balance	Criterion	Max_ depth	Max_ features	Min_ samples_leaf	Learning Rate	n-Estimators
			Decision Tree			
Imbalanced	Gini	2	-	10	-	-
SMOTE	Gini	14	-	5	-	-
ADASYN	Gini	14	-	5	-	-
			Random Forest			
Imbalanced	-	1	2	5	-	10
SMOTE	-	20	5	5	-	100
ADASYN	-	20	5	5	-	200
			Gradient Boost			
Imbalanced	-	1	-	-	0.3	200
SMOTE	-	9	-	-	0.4	200
ADASYN	-	9	-	-	0.4	200
			XGBoost			
Imbalanced	-	2	-	-	0.1	150
SMOTE	-	9	-	-	0.1	200
ADASYN	-	9	-	-	0.1	150

7.2.14 IMBALANCE XGBOOST

Parallelly, in a separate experiment, Imbalance XGBoost algorithm was trained on class imbalanced dataset (data as-it-is) to determine performance across metrics such as accuracy, precision, recall, and F1-score. All the steps up to Exploratory Data Analysis were followed and then Imbalance XGBoost was applied on imbalanced dataset. For 'Focal' Imbalance XGBoost, 3.0 focal gamma was the best hyperparameter; for 'Weighted' Imbalance XGBoost, 1.5 imbalance alpha was the best hyperparameter.

7.3   RESULTS AND DISCUSSION

The four classifiers were evaluated with and without class imbalance algorithms that including SMOTE and ADASYNE. To evaluate the effect of the class imbalance, different classifiers were used without class imbalance, with SMOTE and with ADASYN.

7.3.1   EVALUATION OF TRAINING AND TEST DATA WITHOUT CLASS IMBALANCE

When used without class imbalance, 4,084 records were used in the training set and 1,022 in the test set. Of the 4,084 records, 3,897 were stroke non-incidence and 187 were stroke incidence. In the cases of the five classifiers (logistic regression, decision tree, random forest, Gradient Boost and XGBoost), all 3,897 stroke non-incidence were correctly classified as true negative; however, 187 were classified as false negative. This was an expected problem with imbalanced data where the machine would predict the outcome of the dominant class (López et al., 2013). Out of the 1,022 test samples, 960 records were stroke non-incidence and 62 were stroke incidence. As in the case of the training set, the test set yielded correct classification for the stroke non-incidence records but false negative for the stroke incidence records. This in turn highlighted the limitations with using imbalanced datasets.

7.3.2 EVALUATION OF TRAINING AND TEST DATA AFTER USING SMOTE

This limitation was further confirmed when SMOTE and ADASYN were applied. Hence, 9,714 records were used for the logistic regression classifier with SMOTE of which 7,285 were used as a training set and 2,429 as a test set. The training set had 3,647 records as non-stroke incidence and 3638 as stroke incidence. Of these records using logistic regression, 3,125 records were correctly classified as true negatives and 3,150 records were correctly classified as true positives. Decision trees showed 3,454 true negatives and 3,476 true positives. Likewise, the number of true negatives and true positives using random forest and XGBoost classifiers were 3,430 and 3,519, respectively. Similar numbers were obtained for Gradient Boost which yielded 3,647 true negatives and 3,638 true positives. The test set contained 1,210 records as non-stroke incidence and 1,219 as stroke incidence. Of the test, 1,036 and 1,050 records were classified as true negatives and true positives, respectively, using logistic regression. Decision trees showed 1,083 true negatives and 1,101 true positives. Moreover, both random forest and XGBoost showed 1,099 and 1,114 records true negatives and true positives, respectively. Gradient Boost gave 1,136 and 1,178 true negatives and true positives, respectively.

7.3.3 EVALUATION OF TRAINING AND TEST DATA AFTER USING ADASYN

The 9,651 records were used for logistic regression classifier with ADASYN of which 7,238 records were used as training set and 2,413 records were used as test set. Of the 7,238 samples in the training set, 3,630 records were used as stroke incidence and 3,608 were used as non-stroke incidence. Of these incidences, 3,089 and 3,101 records were classified correctly as true negatives and true positives, respectively, by using logistic regression. Decision trees showed 3,414 true negatives and 3,448 true positives. Moreover, ADASYN applied to random forest and XGBoost classifiers showed 3,422 true negatives and 3,487 true positives. Similarly, Gradient Boost showed 3,630 and 3,608 true negatives and positives, respectively. Furthermore, the test set sample contained 1,227 records of stroke non-incidence and 1,186 records were stroke incidence. In this respect, 1,063 and 1,034 records were classified as true negative and true positives, respectively, by using logistic regression. Decision trees yielded 1,097 true negatives and 1,079 true positives. Likewise, 1,120 and 1,114 records were correctly classified using random forest and XGBoost. Also, Gradient Boost showed 1,164 and 1,141 true positives and true negatives, respectively.

7.3.4   COMPARISON OF PERFORMANCE OF DIFFERENT MODEL BASIS SAMPLING TECHNIQUES

When the three models for each classifier were compared over the different sampling techniques, the model on class imbalanced dataset failed with the accuracy of 0.5 only (Table 7.3). On the other hand, models with SMOTE or ADASYN showed higher

TABLE 7.3
Performance of Different Classifier Basis Sampling Techniques

Model/Imbalance	Accuracy	Balanced Accuracy	Recall	Precision	F1-Score	ROC-AUC
		Logistic Regression				
TR	0.954	0.5	0	0	0	0.843
TR/SMOTE	0.861	0.861	0.866	0.858	0.862	0.939
TR/ADASYN	0.855	0.855	0.859	0.851	0.855	0.935
TE	0.939	0.5	0	0	0	0.835
TE/SMOTE	0.859	0.859	0.861	0.858	0.86	0.937
TE/ADASYN	0.869	0.869	0.872	0.863	0.867	0.941
		Decision Tree				
TR	0.954	0.5	0	0	0	0.823
TR/SMOTE	0.951	0.951	0.955	0.947	0.951	0.992
TR/ADASYN	0.948	0.948	0.956	0.941	0.948	0.991
TE	0.939	0.5	0	0	0	0.799
TE/SMOTE	0.899	0.899	0.903	0.897	0.9	0.94
TE/ADASYN	0.902	0.902	0.91	0.892	0.901	0.944
		Random Forest				
TR	0.954	0.5	0	0	0	0.731
TR/SMOTE	0.954	0.954	0.967	0.942	0.954	0.993
TR/ADASYN	0.955	0.955	0.966	0.944	0.955	0.993
TE	0.939	0.5	0	0	0	0.736
TE/SMOTE	0.923	0.923	0.938	0.912	0.925	0.977
TE/ADASYN	0.926	0.926	0.939	0.912	0.926	0.979
		Gradient Boost				
TR	0.954	0.5	0	0	0	0.876
TR/SMOTE	1	1	1	1	1	1
TR/ADASYN	1	1	1	1	1	1
TE	0.939	0.5	0	0	0	0.838
TE/SMOTE	0.953	0.953	0.966	0.941	0.953	0.99
TE/ADASYN	0.955	0.955	0.962	0.948	0.955	0.991
		XGBoost				
TR	0.954	0.5	0	0	0	0.851
TR/SMOTE	0.942	0.942	0.973	0.916	0.944	0.986
TR/ADASYN	0.938	0.938	0.977	0.905	0.94	0.983
TE	0.939	0.5	0	0	0	0.827
TE/SMOTE	0.916	0.916	0.939	0.899	0.919	0.974
TE/ADASYN	0.921	0.921	0.947	0.898	0.922	0.972

TR, training set; TE, test set.

accuracy within the range of 0.85-1 with similarities among the training and test sets. The close agreement between training and test sets indicates the robustness of the models. In this respect, the highest accuracy for all models was seen for Gradient Boost which showed exponential performance of all metrics for both training and test sets. This was followed by random forest, XGBoost and decision trees that showed accuracy above 0.9. On the other hand, logistic regression models showed the least accuracy for all. Yet for all classifiers, the performance was massively improved when oversampled with SMOTE or ADASYN. This was further supported by the F1-score that was high for all models indicating a high recall and subsequently high precision.

7.4   CONCLUSION

There is extensive research required in the analysis of the large volumes of patient data that can potentially be obtained, to identify patterns that can aid clinical and social care provision. It is also important to consider the reliability and security of patient data if such data is to support clinical and social care decision-making. In cases of large data, data preparation is crucial, particularly in case of imbalanced data. In this case, oversampling and addressing data imbalance offer an ideal approach in addressing inaccuracies in predictions. This was proven in this work where SMOTE and ADASYN algorithms improved massively imbalanced data for patients with stroke incidence. Nonetheless, there is still a scope for developing deep learning models that may use auto-hyperparameter tuning that can bypass the need for balancing classes in a class imbalanced dataset. There is also a scope to employ other class balancing such as SMOTE-NC (Nominal and Continuous), Borderline-SMOTE, Borderline-SMOTE SVM for the purpose of class balance.

REFERENCES

Abedi, Vida, Nitin Goyal, Georgios Tsivgoulis, Niyousha Hosseinichimeh, Raquel Hontecillas, Josep Bassaganya-Riera, Lucas Elijovich, Jeffrey E. Metter, Anne W. Alexandrov, David S. Liebeskind, Andrei V. Alexandrov, and Ramin Zand, 2017. Novel screening tool for stroke using artificial neural network. *Stroke*, 48(6): 1678–1681.

Al-Tameemi, Ghaith, James Xue, Suraj Ajit, Triantafyllos Kanakis, Israa Hadi, Thar Baker, Mohammed Al-Khafajiy, and Rawaa Al-Jumeily, 2021. A deep neural network-based prediction model for students' academic performance. In *2021 14th International Conference on Developments in eSystems Engineering (DeSE)*, Sharjah, United Arab Emirates, pp. 364–369, IEEE.

Bernhardt, Julie, Gerard Urimubenshi, Dorcas B. C. Gandhi, and Janice J. Eng, 2020. Stroke rehabilitation in low-income and middle-income countries: A call to action. *The Lancet* 396(10260): 1452–1462.

Bushnell, Cheryl D., 2008. Stroke in women: Risk and prevention throughout the lifespan. *Neurologic Clinics* 26(4): 1161–1176.

Centre for Disease Control (CDC), 2022. Available at: https://www.cdc.gov/stroke/about.htm Accessed: 19/09/2022.

Chatterjee, Saurav, Ido Weinberg, Robert W. Yeh, Anasua Chakraborty, Partha Sardar, Mitchell D. Weinberg, Christopher Kabrhel, et al., 2017. Risk factors for intracranial haemorrhage in patients with pulmonary embolism treated with thrombolytic therapy development of the PE-CH score. *Thrombosis and Haemostasis* 117(2): 246–251.

Doshi-Velez, Finale and Been Kim, 2017. *Towards a rigorous science of interpretable machine learning*. arXiv preprint arXiv:1702.08608.

Fernández, Alberto, Salvador Garcia, Francisco Herrera, and Nitesh V. Chawla, 2018. SMOTE for learning from imbalanced data: Progress and challenges, marking the 15-year anniversary. *Journal of Artificial Intelligence Research* 61: 863–905.

Goyal, Nitin, Georgios Tsivgoulis, Shailesh Male, Jeffrey E. Metter, Sulaiman Iftikhar, Ali Kerro, Jason J. Chang, James L. Frey, Sokratis Triantafyllou, Georgios Papadimitropoulos, Vida Abedi, Anne W. Alexandrov, Andrei V. Alexandrov, and Ramin Zand, 2016. FABS: An intuitive tool for screening of stroke mimics in the emergency department. *Stroke*, 47(9): 2216–2220.

Jneid, Hani and Holly L. Thacker, 2001. Coronary artery disease in women: Different, often undertreated. *Cleveland Clinic Journal of Medicine* 68(5): 441–448.

Johnson, Walter, Oyere Onuma, Mayowa Owolabi, and Sonal Sachdev, 2016. Stroke: A global response is needed. *Bulletin of the World Health Organization* 94(9): 634.

Kaggle, 2022. Available at: https://www.kaggle.com/datasets.

Khurana, Dheeraj, Madakasira Vasantha Padma, Rohit Bhatia, Subhash Kaul, Jeyaraj Pandian, and Padmavathyamma Narayanapillai Sylaja, Guideline Core Committee, et al., 2018. Recommendations for the early management of acute ischemic stroke: A consensus statement for healthcare professionals from the Indian Stroke Association. *Journal of Stroke Medicine* 1(2): 79–113.

Kumar Das, Sujit, Arnab Kumar Mishra, and Pinki Roy, 2019. Automatic diabetes prediction using tree based ensemble learners. *International Journal of Computational Intelligence & IoT* 2(2): 6.

Lippi, Giuseppe, Brandon M. Henry, and Fabian Sanchis-Gomar, 2020. Physical inactivity and cardiovascular disease at the time of coronavirus disease 2019 (COVID-19). *European Journal of Preventive Cardiology* 27(9): 906–908.

López, Victoria, Alberto Fernández, Salvador García, Vasile Palade, and Francisco Herrera, 2013. An insight into classification with imbalanced data: Empirical results and current trends on using data intrinsic characteristics. *Information Sciences* 250: 113–141.

Newman-Toker, David E., Ernest Moy, Ernest Valente, Rosanna Coffey, and Anika L. Hines, 2014. Missed diagnosis of stroke in the emergency department: A cross-sectional analysis of a large population-based sample. *Diagnosis* 1(2): 155–166.

Pandian, Jeyaraj Durai and Paulin Sudhan, 2013. Stroke epidemiology and stroke care services in India. *Journal of Stroke* 15(3): 128.

Prior, Peter L., Vladimir Hachinski, Karen Unsworth, Richard Chan, Sharon Mytka, Christina O'Callaghan, and Neville Suskin, 2011. Comprehensive cardiac rehabilitation for secondary prevention after transient ischemic attack or mild stroke: I: Feasibility and risk factors. *Stroke* 42(11): 3207–3213.

Shah, Reena S. and John W. Cole, 2010. Smoking and stroke: The more you smoke the more you stroke. *Expert Review of Cardiovascular Therapy* 8(7): 917–932.

Sposato, Luciano A., Lauren E. Cipriano, Gustavo Saposnik, Estefanía Ruíz Vargas, Patricia M. Riccio, and Vladimir Hachinski, 2015. Diagnosis of atrial fibrillation after stroke and transient ischaemic attack: A systematic review and meta-analysis. *The Lancet Neurology* 14(4): 377–387.

Vogel, Birgit, Monica Acevedo, Yolande Appelman, Cathleen Noel Bairey Merz, Alaide Chieffo, Gemma A. Figtree, Mayra Guerrero et al., 2021. The Lancet women and cardiovascular disease commission: Reducing the global burden by 2030. *The Lancet* 397(10292): 2385–2438.

Watson, David S., 2022. Conceptual challenges for interpretable machine learning. *Synthese* 200(1): 1–33.

World Health Organisation (WHO), 2021a. *World Stroke Day*. Available at: https://www.who.int/southeastasia/news/detail/28-10-2021-world-stroke-day. Accessed 19/09/2022.

World Health Organisation (WHO), 2021b. Available at: https://www.emro.who.int/health-topics/stroke-cerebrovascular-accident/index.html. Accessed 15.11.2022.

World Stroke Organisation (WSO): Global Stroke Fact Sheet, 2022. Available at: https://www.world-stroke.org/assets/downloads/WSO_Global_Stroke_Fact_Sheet.pdf. Accessed: 16.11.2022.

8 Ultrasound Vector Flow Imaging, a Promising Technique towards a New Carotid Atheroma Risk Stratification

Istvàn Defrançais
CHRU de Brest, Blvd T. Prigent

Ali Mansour
ENSTA-Bretagne

Luc Bressollette
CHRU de Brest, Blvd T. Prigent

8.1 INTRODUCTION

Nowadays, ultrasound technology is a very powerful tool for clinicians. In fact, compared to classical imageries (like computed tomography scan or magnetic resonance angiography) ultrasound benefits from many advantages. Compared to other technics, this is relatively inexpensive and efficient to describe in vivo process. It is worth mentioning that ultrasound devices are mobile, so the examination can be carried out under difficult conditions or in the patient's bedside. Unfortunately, the main drawback of ultrasound comes almost entirely from a potential lack of image clarity and the interpretation process. Hereinafter, we describe the current ultrasound status in today's clinical practice before suggesting new approaches. The consequences of carotid atheroma plaques are discussed to support the need for innovative tools.

A major cause of cerebral infarction is atherosclerosis of the carotid arteries, which is also an independent risk marker for myocardial infarction or vascular death. According to the HAS (Haute Autorité de Santé, i.e. the Higher French Health National Authority) carotid bifurcation stenosis is both frequent (5%–10% of subjects over 65 years old have a carotid) and serious, due to a potential infarction risk that they train (stroke). This risk becomes greater than 10% per year in the case of symptomatic stenosis and around 2% per year for asymptomatic stenosis. Their treatment is based on drug and vascular risk factor control. In addition, a revascularization procedure may be indicated. The management of symptomatic stenosis is clearly defined by the diameter reduction, while the actions to be performed when diagnosing asymptomatic stenosis are more difficult to define. According to the Asymptomatic Carotid Atherosclerosis (ACAS) and Asymptomatic Carotid Surgery Trial (ACST) studies, the surgery benefit manifests itself only in a long-term period (1–2 years) in asymptomatic stenosis, whereas for symptomatic forms, it appears early after the intervention. In fact, the benefice of surgery for asymptomatic stenosis appears to be independent of the degree of carotid stenosis beyond 60% of diameter reduction. Given the current data, parameters (fluidic and structural components) have been reported through various studies to help the clinician define his therapeutic strategy.

8.1.1   CURRENT STRATEGY

As proposed in Figure 8.1, a decision algorithm has been created to illustrate the commonly accepted treatment method. An emphasis was made on the clinical situation specifically developed later.

Risk stratification of strokes for patients with asymptomatic carotid artery stenosis is a matter of active research, and despite decades of research, no common

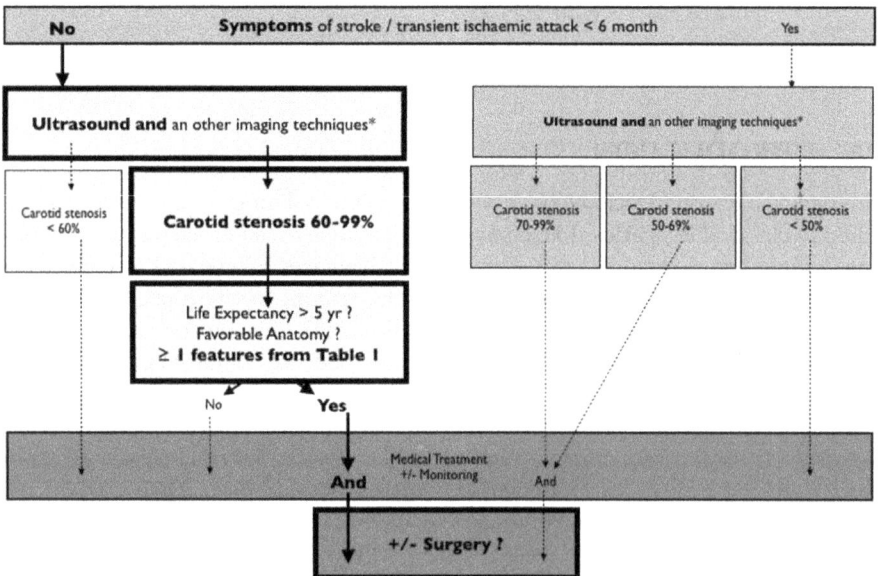

FIGURE 8.1   Algorithm of therapeutic strategy: orange boxes refer to symptomatic patients, yellow ones to asymptomatic patients, green boxes to therapeutics. * Imaging techniques like computer-tomography scan or magnetic resonance imaging. (Inspired from: Aboyans, V., Ricco, J. B., Bartelink, M. L. E. L., Björck, M., Brodmann, M., Cohnert, T., et al. (2017). ESC Guidelines on the Diagnosis and Treatment of Peripheral Arterial Diseases, developed in collaboration with the European Society for Vascular Surgey (ESVS).)

TABLE 8.1
Features Associated with an Increased Risk of Late Stroke

Stenosis Severity (%)	Clinical Criteria	Imaging Criteria
50–99	Contralateral TIA/Stroke	• Stenosis progression (1°) • Intra-plaque haemorrhage on MRI (2°) • Plaque lucency on duplex US (3°) • Juxta-luminal Hypoechogenic Area size (4°) • Spontaneous embolization on TCD (5°)
60–99		1° and 2°, 3°, 4°, 5° & silent infarction on CT (6°)
70–99		2° and 3°, 4°, 5°, 6° &
		• Spontaneous embolization on TCD if associated with echo-lucent plaque on US (7°) • Impaired cerebro-vascular reactivity

Source: Inspired from A.R. Naylor et al. and ESVS recommendations.
TIA, Transient Ischaemic Attack; MRI, Magnetic Resonance Imaging; US, Ultrasound; TCD, Trans Cranial Doppler; CT, Computed Tomography.

agreement on prognostic factors has been reached. Several factors (Table 8.1) are proposed to predict strokes in these patients, including fluidic information (severity of stenosis) and structural information (echo-lucent plaque type, low grayscale median score, presence of discrete white areas without acoustic shadowing, size of juxta luminal black area) as well as clinical risk factors. This information seems insufficient because it is not much used in clinical practice. While no factors were studied from the interaction of fluid (the blood) and this structure (the plaque). This chapter aims to introduce the Ultrasound Vector Flow Imaging (UVFI) technic, as a possible efficient strategy to analyze interactions between fluidic and structural components, as well as a new method to apprehend blood flow and atheroma genesis.

8.1.2 PLAQUE FORMATION

Nowadays, the mechanisms of atherogenesis are still not fully understood; yet, a number of them are now well known, while others remain more hypothetical. It is therefore appropriate, in a first approach, to briefly describe the well-known patterns of atherogenesis before getting into the more recent hypotheses. We can divide the evolution of the atherosclerotic plaque in different stages (Table 8.2).

Throughout its formation, the atherosclerotic plaque experiences a long asymptomatic evolution. After the initial phase of a progressive growth, the plaque experiences an incremental growth phase during acute events. The exposure of the lipid nucleus constituents from the plaque, in particular the release of tissue factor, will be at the origin of the thrombotic process. The occurrence of fractures or cracks in the fibrous screed, results from the conjunction of two types of factors:

TABLE 8.2
Different Stage's Formation of Atheroma Plaque

Stages	Description	Illustration
I	Low Density Lipoprotein (LDL) penetration at the intima level and their oxidation. Activation of endothelial cells; adhesion of monocytes to the endothelium and penetration at the level of the intima.	
II	Formation of foam cells from macrophages and smooth muscle cells (SMCs). Proliferation of smooth muscle cells (SMCs) and migration of these SMCs from the media to the intima.	
III	Secretion of collagen, elastic fibers and proteoglycans by SMCs. Accumulation of connective tissue, lipid, SMCs and foam cells. Formation of the lipid nucleus from the accumulated lipid elements.	
IV	Ulceration/rupture of the vascular wall and exposure of the sub endothelium. Adhesion and platelet activation causing thrombosis.	

Source: Inspired by the works of Dr. J. Léoni, E. Daubrosse at http///smart.servier.com/.

- Intrinsic factors are likely to weaken the plaque. They are linked to the size and the composition of the fibrous screed, as well as the metabolic and catabolic activity.
- Extrinsic factors that can trigger the plaque rupture like mechanical stresses induced by blood flow.

8.1.3 PLAQUE COMPOSITION

Figure 8.2 illustrates intrinsic factors. The main mediators making up an atheroma plaque are shown. The balance governing its stability lies in the equity between the formation and degradation of its components.

The fibrous capsule is composed of smooth muscle cells (SMCs), an extracellular matrix that conditions plaque resistance. The vulnerability of the plaque will depend on screed cellularity and quality of extracellular matrix. Macrophage (Ma) infiltration

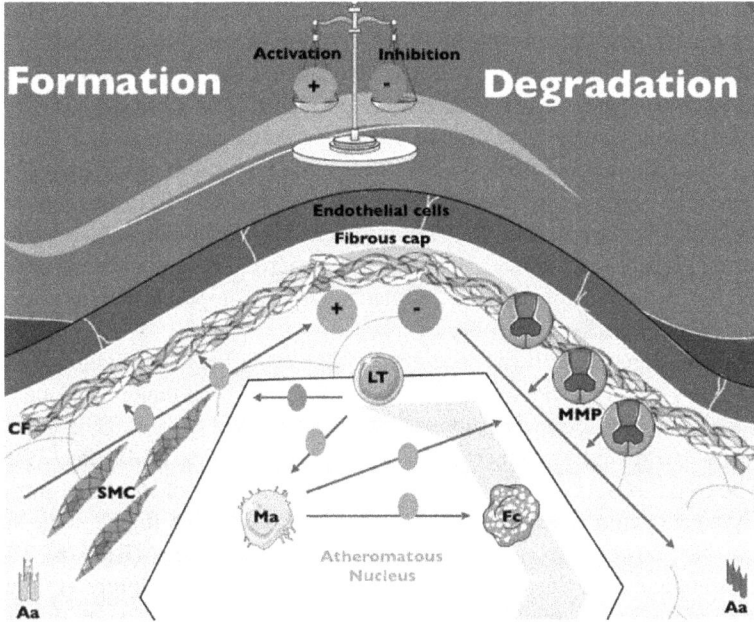

FIGURE 8.2 Atheroma plaque composition and balance between formation and degradation of its components. (Inspired by the works of Dr. J. Léoni, E. Daubrosse at http://smart.servier.com.)

into the fibrous capsule lowers the breaking stress threshold (Lendon et al., 1992). This observation is first linked to collagen (CF) decreased fibrous capsule (Burleigh et al., 1992). This screed is the first barrier to mechanical stress. The plaque composition can be summarized in the following steps:

1. Extracellular matrix of the fibrous cap is essentially composed of collagen and elastin. The decrease in synthesis and degradation of this support tissue will weaken the fibrous capsule. Depletion in smooth muscle cells (SMCs) results in a decreased synthesis of extra-cellular matrix elements by several mechanisms: Apoptosis (programmed death) of SMCs under the action of interleukin one (IL-1), tumor necrosis factor alpha (TNFα) and interferon gamma (INFγ). Degradation of the matrix results from the secretion of a group of enzymatic protein called matrix metalloproteinases (MMP). In fact, it seems that the result of MMPs' activity is due to a tradeoff between the synthesis of MMPs and their inhibitors by macrophages.
2. Smooth Muscle Cells (SMCs) encountered within the fibrous screed have a synthetic phenotype; they secrete components of extracellular matrix. Over time, the proportion of SMCs, which is initially high, tends to decrease, in particular at the level of shoulder zones. This regression of SMCs is essentially linked to the production (by local inflammatory cells) of different pro-apoptotic mediators: IL-1, TNFα and INFγ. The mechano-transduction linked to the stretching of the smooth muscle cells from arterial wall

increases vascular remodeling (Haga et al., 2007). Hemodynamic stimuli induce the adaptive response to preserve a sufficient luminal area.

3. **Lipid core**: plaque stability largely depends on the atheromatous nucleus composition and size. The lipid mass, which has soft consistency, transmits hemodynamic forces to the edges of the plaque (Kakkos, 2013). The nucleus can become necrotic after repeated episodes of macrophage malfunction. Thus, a phenomenon on a microscopic scale can become macroscopically viewable in the form of a necrotic nucleus close to a vascular lumen. Histologic studies showed that the necrotic core is twice as close to the lumen in unstable and symptomatic plaques compared with asymptomatic plaques (Bassiouny, 1997).

4. **Macrophages**: Finally, what determines the evolution of the plaque towards a fibrous lesion or a necrotic lesion is the preserved function of macrophages. Specifically, a larger macrophage presence but with altered function, was related to a higher vulnerability of the plaque (Alegre-Martínez, 2019). The initial observation, which led to the development of the Juxta-luminal Hypoechogenic/Black Area (JBA) criteria, was reported in histologic studies that showed a correlation with symptomatic plaques (Kakkos, 2013). Histological data suggests that macrophages and plaque rupture are higher in the proximal segments of coronary lesions. On the other hand, the distal shoulders of the plaque, where smooth muscle cells are dominant, seem less susceptible to symptoms. Supra-physiological or high wall shear stress (later discussed) has been associated with up-regulation of macrophages and smooth muscle cell apoptosis, which leads to matrix breakdown (Kumar, 2018). In cases where stabilizing factors are predominant, the lesion tends to undergo fibrosis or calcification. A fibrous lesion will develop if most of the apoptotic cells are eliminated by phagocytosis, thus avoiding the release of their content in the plaque. In cases where the sensitizing factors are predominant, the lesion tends to be rich in lipids and to be depleted in collagen. Senescent cells promote elastic fiber degradation and fibrous cap thinning by increasing metalloprotease production. The growth of atherosclerotic plaques is accompanied by the development of a microcirculatory network from the vasa-vasorum. Particularly fragile, these vessels are responsible for bleedings in atherosclerotic plaque heart, which are described as being associated with progression of the plaque. The plaque also undergoes external stresses, which are partly responsible for its rupture.

8.1.4 MECHANICAL CONSTRAINTS

The mechanical stresses are due to the forces exerted by the blood flow. Figure 8.3 illustrates the different forces applied on atheroma plaques.

Blood flowing through a vessel exerts a physical force, which generates two principal stress vectors on the vessel wall. The first one is perpendicular to the vessel wall, defined as the tensile stress, which represents the dilating force exerted by the blood pressure on the vessel wall (extension and pressure). The second one is a parallel stress vector to the vessel wall, defined as shear stress, which represents the frictional force exerted on the endothelial surface. All the forces exerted on the plaque

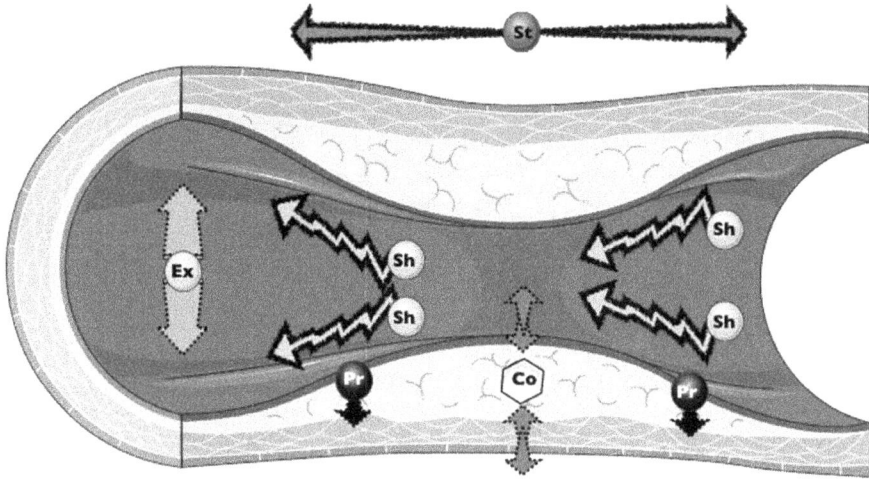

FIGURE 8.3 Hemodynamic simplification of forces acting on atheroma plaque. Two kinds of forces are illustrated: ⟹ parallel to the wall & ⤑: perpendicular to the wall. Pr, Pressure transmitted towards the junction by the lipid core; CO, Compression; Sh, Shear stress; St, Stretch of artery longitudinally; Ex, Extension of artery diameter.

are redistributed by the lipid nucleus towards the junction between the capsule and the healthy wall. The circumferential tension forces exert the most important stresses during an acute event linked with the rupture of the plaque. Longitudinal forces have also been observed. Shear forces appear to act as a long-term influencer on the plaque evolution. In a situation of asymptomatic plaques, the wall shear stress has been specifically highlighted and studied through the rest of this chapter.

8.1.4.1 Defining a Shear Stress

For any real fluid having viscosity, there are shear stresses. Indeed, even if a fluid is in motion, it must have a zero speed in solids contact zone. Any speed differences within a viscous fluid cause shear stress. The fluid particles going faster are braked by those going slower. This is also why it is necessary to exert a certain force to put a fluid in motion. A shear stress is a mechanical constraint applied in a parallel or tangential manner in face of a material, as opposed to classical stresses which are applied perpendicularly. It is the ratio of a force to a surface. It therefore has the dimension of a pressure, expressed in pascals (Pa). By definition, for the same applied force, the pressure increases inversely proportional to the size of the applied area. Thus, the thinning of the fibrous capsule results in an increase in stresses. In practical situation, the measurement of this fibrous capsule could not be validated yet. Plaque rupture refers to the loss of physical integrity of fibrous capsule. This loss occurs following a physical tearing of the tissue, induced by the application of these forces. The erosion, or even the rupture of the fibrous cap covering this plaque will lead to the occurrence of a thrombotic event.

After the rupture, three stages are described: adhesion, activation and platelet aggregation. These three stages lead to the formation of the thrombus. Indeed, during

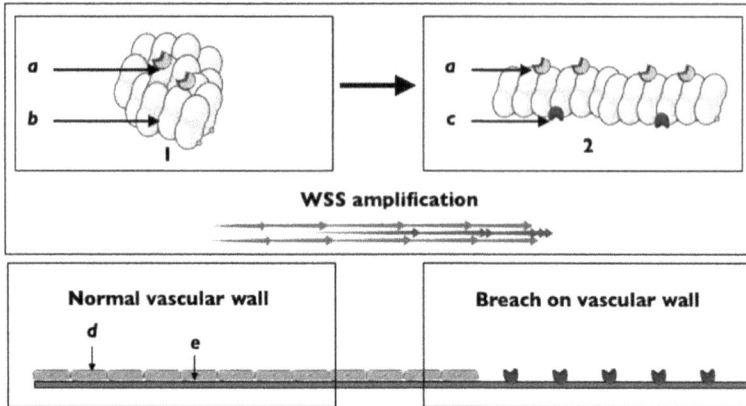

FIGURE 8.4 Willebrand factor conformation modifications under shear stress and its role in hemostasis initiation. **1**: Globular conformation; platelet and sub-endothelial receptor are not available for fixation. **2**: Stretched conformation; platelet and sub-enfothelial receptor are fully available for fixation. **a**: Platelet receptor Gplb-IX-V; **b**: Willebrand multimers; **c**: Subendothelial receptor; **d**: Endothelial cells; **e**: Subendothelial wall. WSS, Wall shear stress.

the rupture of the fibrous capsule, endothelial cells and platelets secrete a multimer of very high molecular weight called Von Willebrand factor. This factor represents the key axis of arterial thrombotic events. The latter having a so-called globular confor-mation in the inactive state, changes of conformation following its attachment to the sub-endothelium or after an increase in shear stresses (Casa et al., 2015). It is now established that a high shear rate favors its change in active conformation (stretched form) as illustrated in Figure 8.4.

The initial fixation of the pads induces a rapid reduction in the diameter of the arteries responsible for an increase in the shear stress as well as a positive feedback to the thrombotic phenomenon. This process is called atherothrombosis. The throm-bus can then embolize to the intracranial arteries or even create a local occlusion of the artery resulting in both cases in brain ischemia (stroke). In fact, instead of the hypoperfusion in severe stenosis, plaque rupture caused by a mechanical destruc-tion acts as the major pathological process of a stroke in mild to moderate stenosis of intracranial atherosclerotic stenosis, and it is therefore, of clinical value to detect the plaque vulnerability (Chen, 2020). Predicting the critical indices of hemody-namic forces such as maximal wall shear stress (WSS) and pressure drop thresholds, according to the progression of stenotic severity, could be beneficial for assessing the functional severity of the atherosclerotic stenosis. In a recent study, a good correla-tion was found between in vitro WSS measurements and the theoretical WSS values. UVFI for vascular explorations (V-Flow) allowed a local and a direct evaluation of the plaque's WSS (Goudot, 2019).

8.1.5 A REVIEW OF CURRENT KNOWLEDGE

New ultrasound technics have emerged in recent years; one of them (Vector Flow Imaging) allows to calculate the wall shear stresses of a given point in near real time with some ease (Asami, 2019; Assi, 2017). In this context, a summary of the current data can support the use of WSS in practice. Anatomically, the average diameter of a proximal coronary artery is around 4 mm (Dodge et al., 1992); comparatively, a common carotid artery diameter seems to be about 7 mm; the wall constraints might be different (Sigala, 2018), but most of the studies show similar correlations.

8.1.6 WALL SHEAR STRESS VARIATIONS AND CONSEQUENCES

Compared to physiological WSS, high or low WSS appears to make different impact on atherosclerotic evolution. Lowered wall shear stress (WSS) appears in most of the studies as being a factor in favor of atheroma plaque progression.

On Coronary Arteries	On Carotid Arteries
According to a study carried out by optical coherence tomography (OCT) (Samady, 2019), a reduction in the parietal shear stresses leads to progression of coronary atheroma plaque by activating inflammation and proatheromatous pathways. These results are consistent with a comparable study carried out using an angiographic technique (Kumar, 2018)	With ultrasound color flow analysis, WSS was becoming lower in an early stage of atherosclerosis (Zhang, 2018). In the same way, significant relationship between lower WSS and internal carotid artery stenosis progression was found with MRI (Magnetic Resonance Imaging) on TOF (Time of Fly) sequences (Chen, 2020). Histological analyses on mice permitted to argue a significant relationship between atherosclerotic lesions development and lowered shear stress or vortices (Cheng, 2006)

In the over hand, higher wall shear stress (WSS) seems to change plaque composition, as well as plaque vulnerability.

On Coronary Arteries	On Carotid Arteries	On Intracerebral Arteries
According to a study carried out by OCT (Samady, 2019), high shear stresses lead to a vulnerable plaque phenotype. These results were consistent with a comparable study carried out using in vivo intravascular ultrasound (IVUS) (Kumar, 2018). The activation of MMP with vulnerability phenotype in the shoulders of plaques was observed (Kumar, 2018). Thin fibrous cap and high cap stress were reported as well to be keys risk factors for plaque rupture (Wang, 2017).	On MRI carotid bifurcations analyses, regions of increased shear stress were at higher risk to develop ulceration prone plaque (Groen, 2008). On the other hand, intraplaque hemorrhage were shown to be associated as well (Beach et al., 1993). Axial location of plaque rupture in a carotid artery would strongly be influenced by changes of the lipid core stiffness and the geometry (Alegre-Martínez, 2019).	With MRI on TOF sequences, high mechanical load generated by hemodynamic forces often revealed a hemodynamic pattern evocating a prone to rupture plaque (Chen, 2020).

Most of the results show plaques modifications for vulnerability profile. Changes in plaque composition such as macrophages or cap thickness are induced by high WSS (Samady, 2019). In fact, we may wonder if changes of WSS along the plaque could have a consequence of the plaque composition. We could therefore deduct the composition of the plaque according to WSS modifications.

8.1.7 AXIAL VARIATIONS OF WALL SHEAR STRESS (WSS)

Can a dynamic mapping could be possible? And what do we already know about this? Here is a cartographic viewpoint (Figure 8.5).

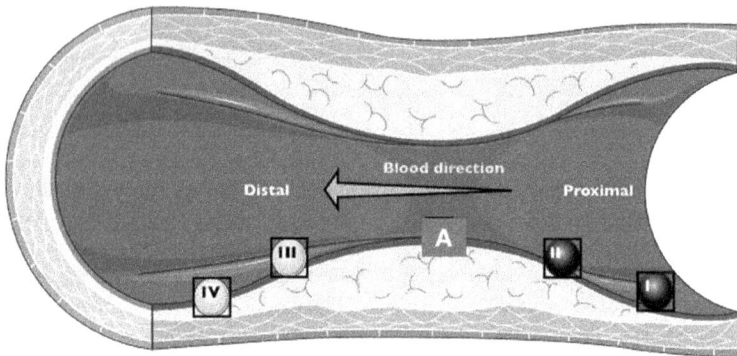

FIGURE 8.5 Hemodynamic simplification of forces acting on atheroma plaque. **I**: Origin; **II**: Up stream; **A**: Apex; **III**: Down stream; **IV**: Terminal. (Inspired by the works of Chen, Kumar, Cicha et http://smart.servier.com.)

Down Stream

Intracerebral arteries were studied with MRI, and minimal WSS was commonly observed at the downstream sections of the stenosis (Chen, 2020). Extracerebral arteries such as carotid arteries were studied on mouse models. Histological evidence showed higher numbers of smooth muscle cells on downstream sections. The downstream region was also characterized by an increased erosion from endothelial cell apoptosis (Cicha, 2011).

Up Stream

In intracerebral stenosis, WSS (expressed in Pascal) on apex and upstream sections were the highest among the five sections studied and shown in Figure 8.5 (Chen, 2020). On coronary stenosis, higher WSS in proximal segments of coronary arteries predicted myocardial infarction. In line with these observations, studies have shown that plaque rupture often occurs in the proximal segments of stenoses. In an analysis from this study, authors found that a 3 Pa increase within the WSS of plaque profile was associated with a 23% increase in myocardial infarction's risk (Kumar, 2018). On mouse histology, about 80% of the ruptured plaques were observed on the upstream site of the stenosis. In this region, neovascularization and hemorrhage were increased as compared with the downstream shoulder of the plaque (Cicha, 2011).

8.1.8 PLAQUE COMPOSITION ASSUMPTION BASED ON WALL SHEAR STRESS VARIATIONS

As it was previously mentioned, the lipid core plays a major role in the plaque vulnerability. With simulated patterns: A shorter lipid core increased both the local and absolute peaks of the shear stress, while also affecting the gradients. Results suggest that shorter lipid cores could be more dangerous than plaques with longer cores, for the same thickness of the fibrous cap. The combined analysis of the stenosis severity and the plaque composition shows that both factors should be simultaneously considered (Alegre-Martínez, 2019).

In another study, the influence of the plaque length was reported as being responsible of a change in the stress distribution for shorter plaques, which makes them more vulnerable than longer plaques. In fact, it appears that shorter plaques were significantly affected by the shear stresses. It has an essential influence on the mechanical response of the vessel wall. As an example, for two plaques having the same stenosis degree (stenosis = 45%), the maximum stress in the fibrous cap is 50% larger for the short plaque than for a large one, and the maximum wall shear stress is increased by 100% (Belzacq, 2012). Mouse carotid analyses permitted comparison of geometric profiles for ruptured and stable plaques. Their histologic analyses showed an increased longitudinal asymmetry of fibrous cap and lipid core thickness in ruptured plaques (Cicha, 2011). In fact, especially in asymmetric plaques, the shear stresses tend to deform the cap by pinching it. This effect was called the pinching effect (Belzacq, 2012). Today, classifications (MRI or histological) exist to differentiate atheroma plaques according to their composition. Table 8.3 introduces the simplification of the common classifications validated for plaque description.

TABLE 8.3
Atheroma Plaque Evolution

	Preatheroma	Lipidic Lesions	Calcified Lesion
Schematic View			
Simplified **Histological** classification	Foam cells infiltration	Lipid core with possible surface defect, hemorrhage, or thrombus	Calcified or fibrotic plaque without lipid core
Simplified **MRI** classification	Intimal thickening with no calcification	Lipid or necrotic core surrounded by fibrous tissue with possible calcification, surface defect, hemorrhage, or thrombus	Calcified or/and fibrotic plaque without lipid core

Source: Inspired by Classification of Human Carotid Atherosclerotic Lesions at http://smart.servier.com. MRI, Magnetic Resonance Imaging.

8.1.9   Plaque Composition Based on Ultrasound Models

The plaque composition can then be precisely described with ultrasound. If stiffness and plaque composition can be identified with ultrasound, we can be more accurate than with a classical technic. In a histological study carried out on mice; their tissues displayed different stiffnesses depending on whether they were non-fibrous or calcified (Loree, 1994). On intracerebral arteries, a study using TOF-MRI explains the finding from a mechanical viewpoint; stenosis percent significantly influences the hemodynamic forces (WSS ratio and pressure ratio) in intra-cerebral arteries. Those leading to an increasing risk of plaque rupture, especially when the percent stenosis was ≥50% (Chen, 2020).

In simulation conditions, Alegre-Martinez et al. used fluid structure interaction (FSI) model and created variations between the plaque composition (lipid core, cap thickness, positive remodeling) and the degree of stenosis (Alegre-Martínez, 2019). Simulations were performed with a finite element method using the COMSOL Multiphysics software. In addition to the first principal stress, modifications of shear stress were shown along the plaque and presented two peaks of pressure, corresponding to the ends of the lipid core according to the authors (Alegre-Martínez, 2019). We can synthesize this rigidity data into one main piece of information. Arteries wall mapping with WSS data would allow us to make a classification of the atherosclerotic plaque profile and help to determine plaque composition. Another clinically relevant information might be the measurement of the pressure drop observed in Table 8.4 and the correlation with lipid core length (Figure 8.6). This critical information might show areas of the plaque sensitive to break. In fact, the pressure drop seems to take a major part of the total mechanical constraints.

8.1.10   Pressure Drop Sites

As shown previously, WSS mapping could provide helpful information. Two possible sites of rupture were identified in a study based on simulation patterns. The first one is related to the global maxima in the stress distribution, and another one is at the beginning and the end of the lipid region (suggested by the authors as being shown by the two pressures drop site) (Alegre-Martínez, 2019).

According to this study, measurements of the WSS along the plaque were done on simulation patterns as shown in Figure 8.7 (Alegre-Martínez, 2019). The constraint levels showed two peaks of pressure drop (**Pd I and II**) on WSS plaque profile. In fact, WSS profile might be largely influenced by the chosen shape of the lipid core length (Table 8.5).

Two information could be useful: firstly, the size of the lipid core could be precise in extrapolation from pressure drop sites. Secondly, different types of profiles can be made, for two plaques of the same degree of stenosis (example: comparison of two stenoses at 60%, with a respective profile; would one of the pressure profiles be more at risk of rupture than the other?). In a study based on MRI of carotids, the local WSS at the stenosis appears to be small compared with the overall loading on plaque. Therefore, the pressure may be the main mechanical trigger for plaque rupture and risk stratification using stress analysis of plaque stability may need to consider the pressure variations (Li, 2009).

TABLE 8.4
Two Groups of Axial Stress Distribution

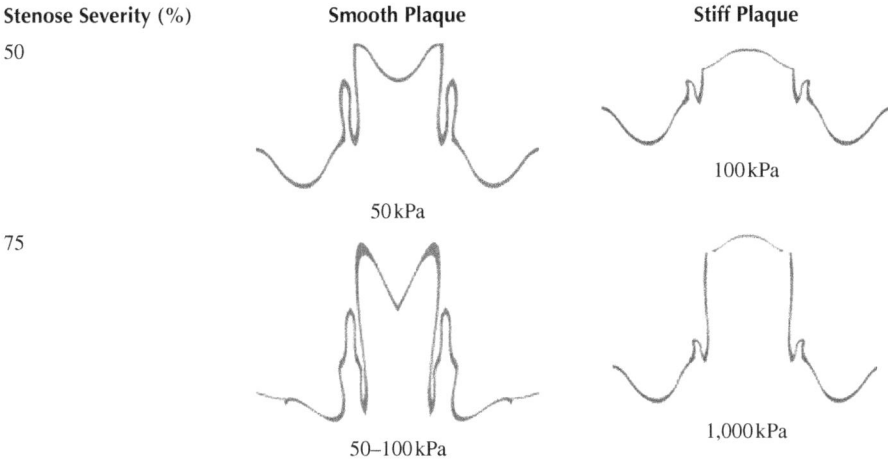

Stenose Severity (%)	Smooth Plaque	Stiff Plaque
50	50 kPa	100 kPa
75	50–100 kPa	1,000 kPa

Source: Inspired by the work of Alegre-Martinez, On the axial distribution of plaque stress: Influence of stenosis severity, lipid core stiffness, lipid core length and fibrous cap stiffness.
Red lines: Higher intensity: maximum stress up and down streams; Blue lines: Lower intensity: maximum stress in the middle of stenosis.
Pa, Pascal.

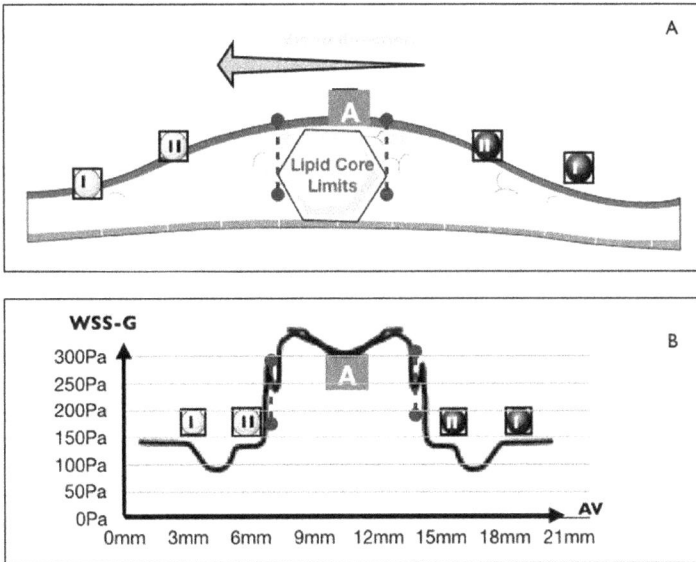

FIGURE 8.6 Representation from a schematically point of view of the ideas comping. Pa, Pascal; mm, millimeters; AV, Axial variations along the plaque; WSS-G, WSS gradient. (From Alegre-Martinez, On the axial distribution of plaque stress.)

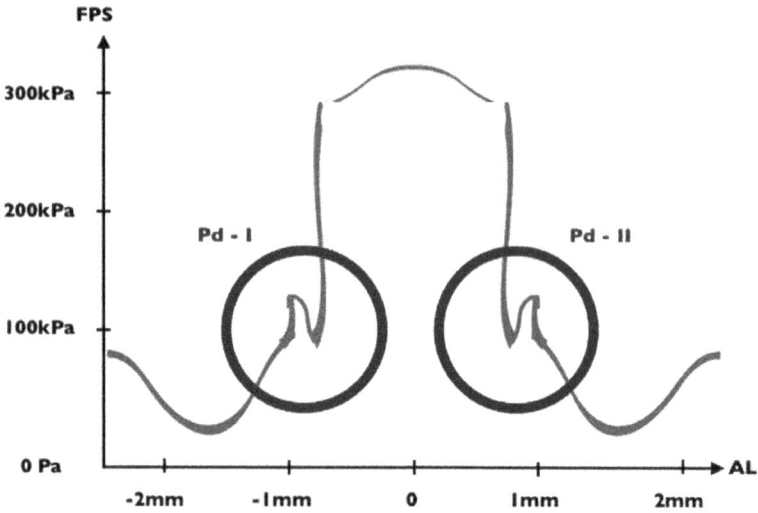

FIGURE 8.7 Stress variation along an atheroma plaque. The + (respect. −) 2 mm site corresponds to the upstream (respect. downstream) side of the stenosis. The 0 mm site corresponds to narrowest lumen site (i.e. the middle point of the plaque). FPS, First Principal Stress; AL, Axial length; PdI and II, Pressure drop sites. (Inspired by the work of Alegre-Martinez, On the axial distribution of plaque stress.)

TABLE 8.5
Simplified Model of Stress Variations

On Intracerebral Arteries	On Carotid Arteries
Simulated patterns showed that the upper half section of the stenosis (between upstream and downstream sections) sustained the majority of total pressure drop, implying that this region carries a significantly larger mechanical load and may be at a higher risk (Chen, 2020). Study based on MRI showed that translesional pressure drop also significantly affects the plaque vulnerability (Li, 2009)	Pressure drop sites were found (upstream and downstream) on histological analyzes of ruptured plaques. In this context, analyses demonstrated that the upstream-ruptured plaque geometry induced a 20% higher pressure drop along the stenosis compared with stable ones (Cicha, 2011)

8.2 INTRODUCING ULTRASOUND VECTOR FLOW IMAGING (UVFI)

As we saw, determining WSS can be very useful. But the measurement of wall shear stress remained challenging in clinical situations. A new ultrasound technic called UVFI allows to access WSS information in clinical situations. Before that, WSS measurement was mainly possible with MRI and Pulse Wave Doppler (PWD). Limitations did exist with these imaging methods. PWD was mainly limited from its angle dependence and MRI was considering a simulation of the blood flow.

Ultrasound Vector Flow Imaging technic was first used to assess the cardiovascular blood flow distribution in an observation plane. This unique, noninvasive technique was derived from the Color Doppler velocity data and generates the velocity fields on 2D images.

8.2.1 WALL DYNAMIC AND CALCULATION OF WALL SHEAR STRESS

In this paragraph, we explain how the ultrasound vector flow imaging technique allows the parietal shear stresses calculation of an artery.

The shear stress is a pressure that is exerted in parallel or tangent to the arterial wall. The shear stress in the case of a Newtonian fluid is given by the following formula (Equation 8.1) expressed in SI unit:

Equation (8.1): Basic equation relating to most of studies exploring shear stress impact.

$$\tau = \mu \frac{\partial v}{\partial r} \tag{8.1}$$

τ: Wall shear stress (WSS) in Pascal (Pa)
μ: Viscosity of the fluid in Pascal seconds (Pa s)
v: Speed in meters per second (m/s)
r: Radius of the vessel in meters (m)
$\frac{\partial v}{\partial r}$: Wall shear rate (WSR) in seconds minus one (s^{-1})

8.2.2 DEFINITION OF THE BLOOD VISCOSITY

Dynamic viscosity is a physical quantity which characterizes the resistance to laminar flow of an incompressible fluid (Mills, 1993). Figure 8.8 shows a simplified illustration of the dynamic viscosity.

The blood is assimilated here to a non-Newtonian fluid. In fact, the viscosity of the blood changes in a non-linear manner with respect to the WSS and the WSR

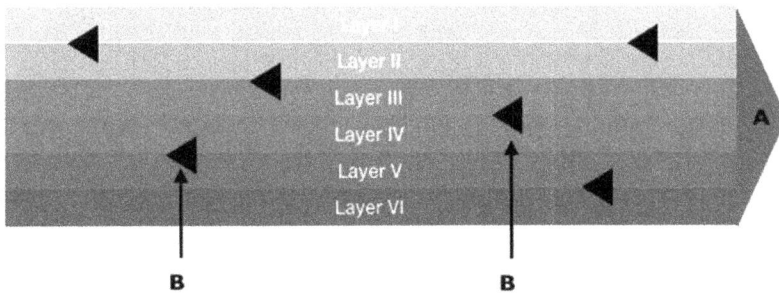

FIGURE 8.8 Illustration from a schematic view's point of the blood viscosity which can be seen as the components' actions (red blood cells, plasma, etc) against blood flow. The presented layers represent the interactions among different components. B: Friction points, resistive forces acting "against" the general motion; A: General motion direction.

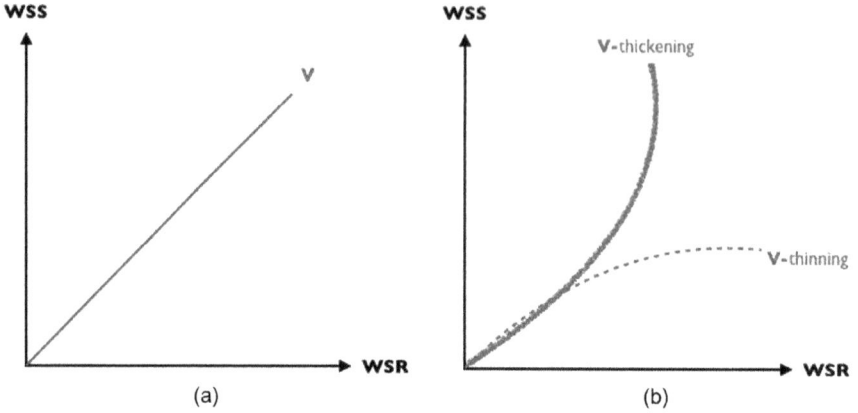

FIGURE 8.9 Comparison between Newtonian fluid (a) and Non-Newtonian fluid (b). (a) Newtonian fluids: Viscosity is evolving **lineraly** with WSS and WSR. (b) Non-Newtonian fluids: Viscosity is evolving **non-lineraly** with WSS and WSR. WSS, Wall shear stress expressed in Pascal; V-thickening, Viscosity compatable to cornstarch mixed with water; V, Viscosity; WSR, Wall shear rate expressed in second^{-1}; V-thinning, Viscosity compatable to ketchup or blood.

(Wall Shear Rate). Blood viscosity is called "thinning" because it is diminishing while WSS increase (Figure 8.9). However, it was found that variation according to different WSR was negligible in clinical situations, and that this value was ~3.5×10^{-3} Pa s (Du, 2020).

8.2.3 FLUID DYNAMIC AND CALCULATION OF VECTORS

Another challenge comes from the fact that the formula was made for a Laminar flow, but the blood in the arteries is frequently turbulent in case of stenosis.

In the case of a laminar fluid, the maximum speed (Vmax) is the speed located at the center of the flow. This makes it possible to define, with the radius of the vessel, the WSR:

Equation (8.2): Basic equation relating to wall shear rate.

$$\text{WSR }(t) = \max_{1 \leq i \leq N} \frac{v_i(t)}{\Delta \; r_i} \tag{8.2}$$

max: maximal speed between all the speed mesured
$1 \leq i \leq N$
v_i: Speed in meters per second (m/s)
r_i: Radius of the vessel in meters (m)
Wall shear rate (WSR) in seconds minus one (s^{-1})

In the context of a non-laminar fluid, several measurements of the velocities (v_N) at different times are carried out by Pulse Wave Doppler technic (PWD). The WSR (Wall Shear Rate) can be calculated according to (Equation 8.2) but still, the use of the PWD induces a dependence on the measurement of the angle (θ). The new Ultrasound Vector Flow Imaging (UVFI=VFM) technic makes WSS possible to obtain without the need of that angle (θ) dependency, by adding the directional information ($\vec{w}$) of the speed vectors (Du, 2020).

8.2.4   VECTOR FLOW IMAGING CUSTOMIZED FOR VESSELS

The conventional UVFI method was applied for the left ventricle, and the flow of interest was in a cavity surrounded by walls. In contrast, on carotids the flow of interest is usually open-ended. The difference in ventricular and vascular geometries affects the boundary conditions, and the VFM (Vector Flow Mapping) algorithm must be modified to accommodate this difference. The continuity equation describes the transport of particles or fluid (Pedlosky, 1987). According to Equation (8.3), we can find that

$$\frac{\partial \rho}{\partial t} + \nabla J = \sigma \qquad (8.3)$$

ρ: volume density in grams per cubic centimeter (g/cm³)
J: stands for the flux ($J = \rho u$)
u: is the velocity field
σ: is the generation of the quantity (cm³/s).
   To apply this equation in our application, i.e. the measurement of hemodynamics, we can consider that the quantity of blood is a conserved quantity, then $\sigma = 0$.
   By assuming that the considered fluid is incompressible then ρ can be considered as constant; Therefore, the continuity equation can be simplified to a volume continuity equation as by Equation (8.4):

$$\nabla J\left(x, y, z\right) = 0 \rightarrow \frac{\partial u\left(x, y, z\right)}{\partial x} + \frac{\partial v\left(x, y, z\right)}{\partial y} + \frac{\partial w\left(x, y, z\right)}{\partial z} = 0 \qquad (8.4)$$

J: flow value
u: steering angle
v: depth velocity
x: transverse direction
y: depth direction
z: elevation direction
   The above equation is the same as the first equation of (Asami, 2019). The authors assumed a planar flow so the motion according to the axis z could be neglected, so the above equation can be simplified as follows (Equation 8.5):

$$\frac{\partial u\left(x, y\right)}{\partial x} + \frac{\partial v\left(x, y\right)}{\partial y} = \partial_x u + \partial_y v = 0 \qquad (8.5)$$

By doing the integration according to the x axis, we can write that (Assi, 2017)

$$u^{\text{ord}}(x,y) = u_a(y) - \int_{x_a}^{x} \partial_y v(x,y)\,dx$$

$$\text{or} \quad u^{\text{inv}}(x,y) = u_p(y) - \int_{x}^{x_p} \partial_y v(x,y)\,dx$$

The aforementioned equation has been considered in ordinary or inverse way (Ord or Inv). In ordinary, we consider the boundary is limited by a or p in the case of inverse (Figure 8.11).

According to the geometric relationship between the linear probe, Doppler steer angle and vessels, some areas might have one, two or no boundaries. To resolve this issue (no boundary condition (Figure 8.10a), a single Doppler beam, called sub-beam, steered at an angle different from the one used for the Color Flow Mapping (CFM) (Figure 8.10b), the new beam, sub-beam, is considered as a new boundary, that generates their Equations (8.4) and (8.5). When multiple calculation pathways exist, the correlation becomes stronger and the final flow velocity is corrected on the basis of the Vector Flow Mapping (VFM) calculated using different boundary conditions.

To choose the optimal vector, many (N) measurements should be done. If we conducted N calculations, then the authors (Assi, 2017) propose the final measurement as the average of the obtained measurements, with the following equation

$$L = \frac{1}{N} \sum_{i=1}^{N} l_{bi}$$

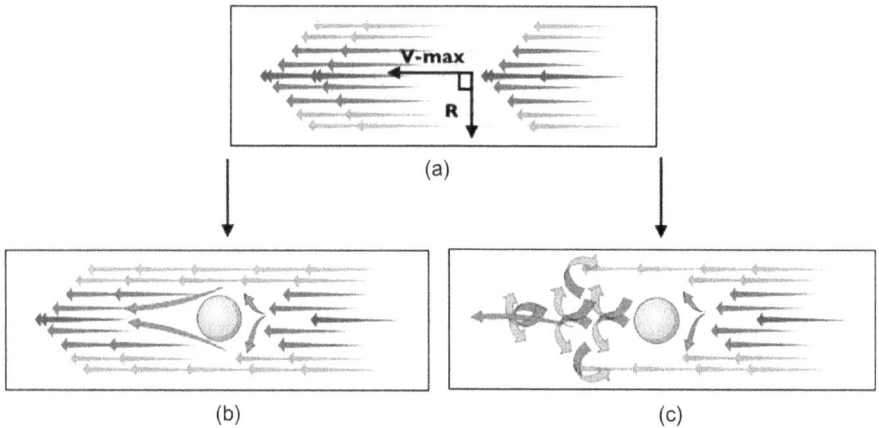

(a)

(b)   (c)

FIGURE 8.10 Different patterns of speed vectors in Laminar and Non-laminar flows. (a) Normal flow in a straight. (b) Laminar flow becoming **laminar** again after passing around a disturbing structure. (c) Flow becoming **turbulent** after passing around a disturbing structure. V-max, Fastest speed; R, Vessel radius.

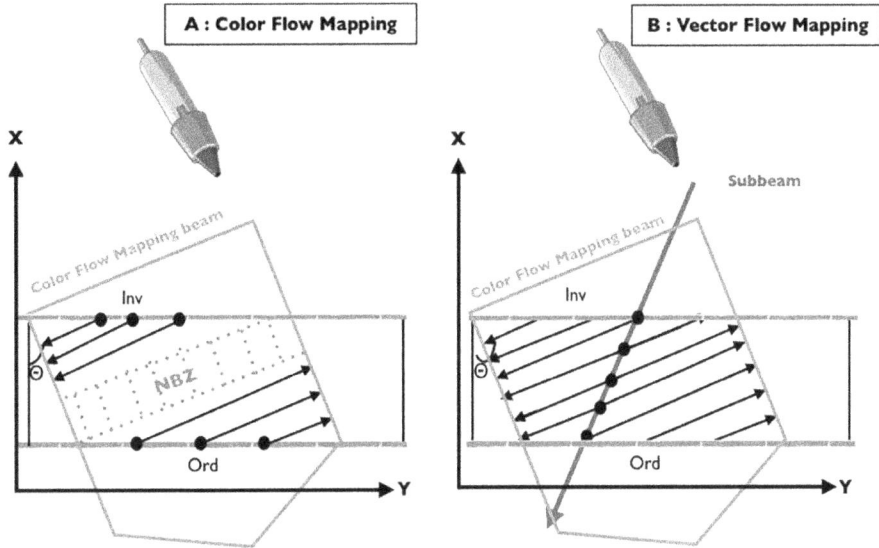

FIGURE 8.11 Comparison between **CFM** and **VFM** technics; **CFM** presents less precision coming from the **NBZ**. CFM, Color Flow Mapping; VFM, Vector Flow Mapping; NBZ, One or No Boundary zone; VW, Vessel wall. (Inspired by the work of Asami et al. (2019).)

To validate their approach, the authors made a Carotid Artery (CA) phantom and they compared their ultrasonic measurements with the results given by an optical particle image velocimetry (PIV). According to them, they achieve a coefficient correlation, $R = 0.95$, between the two sets of results.

These findings have limitations in practical situations because measurements depend on the hypothesis that the flow would be laminar, which is not the case in a carotid bifurcation.

8.2.5   CLINICAL USE OF UVFI DEBUT IN CARDIOLOGY

Ultrasound Vector Flow Imaging, known as UVFI, demonstrates flow dynamics in the heart and vessels in a whole new way evaluating pre and post-surgical ventricular and trans-valvular blood flow, cardiomyopathies along with left ventricular compliance, which are just some of the ways that VFM may help assess complicated hemodynamics and determine surgical strategies. Today, we are able to use VFM technology to easily evaluate vortex. VFM uses 2D cross-sectional images obtained by B mode color Doppler and speckled tracking information in a method to present blood flow as a vector distribution. These vectors indicate velocity and direction with many arrows. Vortex can easily be seen using these vectors and streamlines. Similar vortex patterns are seen when comparing VFM using ultrasound data particle imaging velocimetry (PIV) and 4D-MRI. In addition to vector and streamline display vortex is another parameter that we are able to evaluate.

Today, few studies begin to show VFM tests on peripheric vessels. In carotid arteries, the consequence of a bifurcation in disturbing flow patterns, resulting in a non-uniform distribution of wall shear stress, was shown. Goddi et al. (2017) showed the strong effect of vessel geometries on the flow patterns with Vector Flow technic. In another study, they also showed that VFM is better than CFM to describe complex flow patterns than CFM (Goddi et al., 2018). Fiorina et al. confirmed to be an innovative and intuitive imaging technology to study the flow complexity in the arteriovenous fistulas as well.

8.2.6 Clinical Use of Vascular-UVFI (V-Flow)

In 2020, one of the first clinical studies evaluating ultrasound WSS measurements in pathological situations, directly in contact with the atherosclerotic carotid plaque, showed how much V-Flow can be a powerful tool (Goudot, 2020). V-Flow accuracy was validated with a good correlation ($R^2 = 0.95$, $p < 0.001$) in front of an in-vitro situation, where flows were laminar. Their in-vivo experiment on a prospective series of patients showed feasibility of the realization in clinical situations (Goudot, 2020). If we take their WSS values along carotid bifurcations, the maximum value was found on the plaque peak and was around 3.5 Pa (for an average of 80% stenosis degree). Maximum variation of WSS was found at the same localization of the plaque and was about 1.9 Pa.

VFM allows measurements of the speed and direction of all blood cells flowing through every point of the region of interest in a short moment (Figure 8.12). The flow is represented by many different color vectors. These colors are correlated with the speeds, amplitude and directions that each point of the blood flow takes (the yellow/blue vectors correspond to the low speeds, the orange vectors to the medium speeds and the red vectors to the high speeds (Goddi et al., 2018). Another important point in clinical practice, the interobserver specificity when calculating maximum speeds remains low (Specificity = 56%). In fact, during the follow-up of a patient, the measurements carried out should ideally be identical, which is currently difficult to achieve. These measurements depend not only on the angle used but also on the sampling area chosen. Thanks to VFM, a choice of measurement of the fastest vector speeds can be made. The angle does not define the values of the measured speeds.

Changing the gain will get the fastest vectors and remove the slowest vectors, resulting in a reduced selection of vectors. The operator is therefore more likely to choose the same vector as its namesake. Then, we could imagine that interobserver specificity could be greatly improved and stenosis degree calculation as well.

Here, we can find the main information appearing on the ultrasound machine. The carotid bifurcation is represented there with at the top left the speed scale, at the bottom (**MS**) the set of speeds during an acquisition of 1.5 seconds (ordinate axis in with the speeds in centimeters per second and abscissa the time elapsed in seconds). In the center, the image is displayed first in 2D, and then more precisely the speed vectors are measured in the region of interest (**ROI**). The ROI has to be chosen in the bifurcation area like a classical examination in a Color Mode (as shown in Figure 8.11a). In this same region of interest, a flow can be calculated according to the diameter of the chosen vessels (here the common carotid **CC**). Finally, the shear stresses can

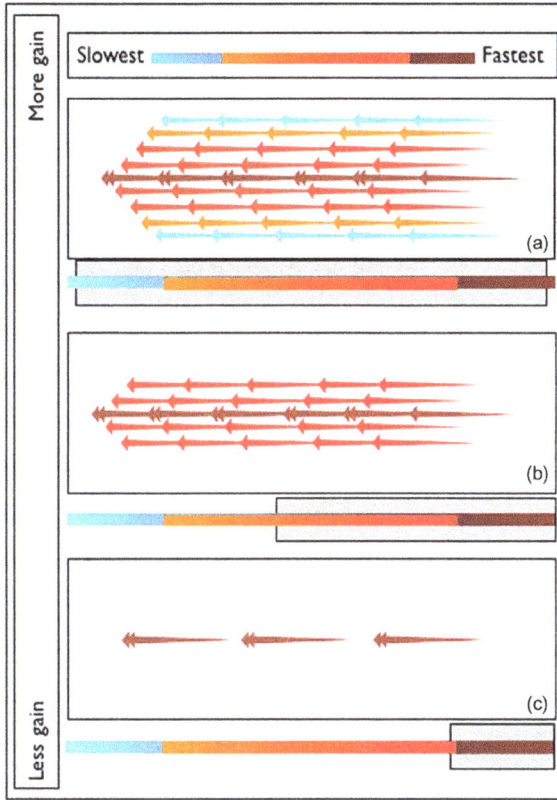

FIGURE 8.12 Schematic view of the blood flow in Vascular-UVFI mode according to a pre-selected speed threshold. The region of Interest (ROI) is the carotid bifurcation. **A**: When the operator can see the whole flow; **B**: When slow vectors are suppressed by a gain modification; **C**: Fastest visible and measurable vectors to determine the optimal speed. VFM, Vector Flow Mapping.

be calculated by selecting the point of interest (**POI**) in contact with the wall of the vessel chosen by the operator (in this case, here on the internal carotid, **IC**). All of these calculus can be done after the acquisition, which means that a simple acquisition of 1.5 seconds by holding the probe stably on the patient's neck can be analyzed immediately afterwards without the risk of moving. In the same way, these calculations are performed automatically without the need for posterior image processing. This explains the great maneuverability of this technique and its promising nature.

8.3 DISCUSSION

8.3.1 PLAQUE COMPOSITION

Despite all, a limitation comes from the graphics, offered by Alegre-Martínez (2019), which are symmetrical. However, it is likely that in vivo these pressures are asymmetrical due to the retrograde flows in vortex existing at the exit of the stenosis.

Consequently, these pressure peaks could just as easily be linked to digital artefacts following data processing and not secondary to a real physical phenomenon. Further studies would be needed to validate this approach.

8.3.2   FLUID DYNAMIC

As shown in Figure 8.13, four types of complex flow patterns were identified by Goddi et al. (2018) on 60 carotid bifurcations (of 30 healthy patients).

No risk stratification was related yet among flow patterns, and studies exploring this assumption can be conducted.

8.3.3   WALL SHEAR STRESS

Hariri et al. (2019) conducted a control case study on 308 patients with carotid stenosis higher than 70% according to NASCET (North America Symptomatic Carotid Endarterectomy Trial). The study suggested a strong relationship (OR = 12.1) between the high Wall Shear Rate ($>8,000\,\text{s}^{-1}$) and the risk of a cerebral ischemic event. It is important to specify the level of risk of these patients, as the strokes occurring in these situations are more due to an embolic event than to hypoperfusion following the reduction in the cerebral flow. In these patients, the gradation of

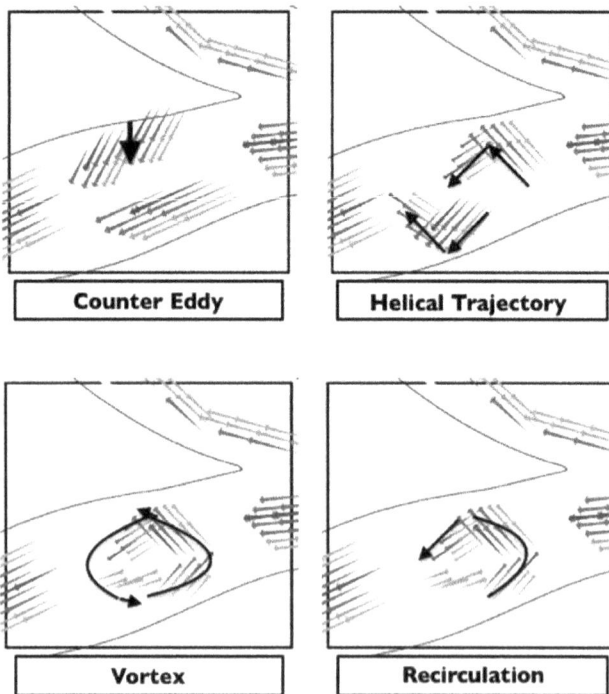

FIGURE 8.13   Fluid dynamic with complex flow patterns identified by Goddi et al. on the carotid bifurcations.

the embolic risk by measuring the parietal constraints could allow better accuracy than the measurement of the degree of stenosis. It would be interesting to study the shear stresses in situations known to be at risk of pathological reduction or elevation. Post-operative monitoring of aneurysms repaired by the endovascular route, graft vascularization and arteriovenous fistulas among others can also be the subject of studies. In a pre-operative graft situation, Casa et al. (2015) used cut-offs of vascular diameter to avoid too low or too high WSR (between 100 and 5,000 s^{-1}). We can even imagine its use in an intraoperative situation to guide the surgeon to the optimal positioning of the endovascular material. According to (Casa et al., 2015), a study of these cut-offs to optimize the geometry of medical devices and thus reduce the risk of thrombosis on contact with them could be interesting.

8.4 CONCLUSION

Stroke is a very common pathology. Advances in the therapeutic field have improved the quality of life of patients as well as reducing mortality. Today, prevention is an active research subject since curative treatments have reached an optimal level of performance. To act at the initial phase of the pathological process, knowledge of these modifications at the cellular level is essential. Precise, in vivo measurement tools would therefore be indisputable in the decision-making process. Consequently, the treatments acting at the microscopic level (according to WSS measurements) can be judged by their effectiveness on a precise and therefore early time scale (provided by UVFI technic). Obviously, the main judgment criterion should focus on improving the quality of life rather than a simple cellular consideration (because this correlation is not always verifiable). Prevention and treatment according to individuals risk obtained with in vivo measurement tools like UVFI would be greatly appreciated. Certain elements remain missing in the diagnostic strategy since, as shown in Figure 8.14; 25% of strokes appear to come from unknown origins.

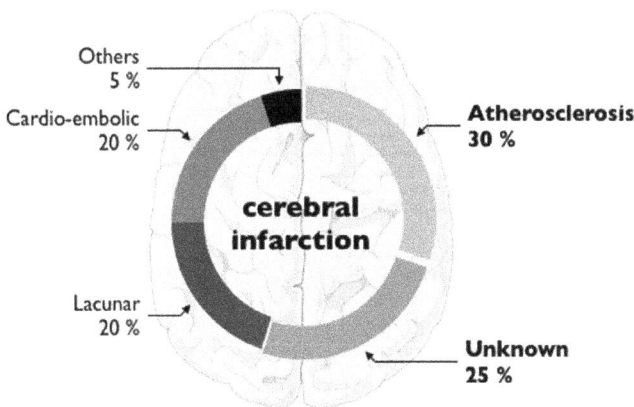

FIGURE 8.14   Cerebral Infarction Etiology according to CEN (Collège des Enseignants de Neurologi), i.e. College of Neurology Teachers.

Today the considered criteria in the grading of risks focus on two types of elements: composition of the plaque and the overall hemodynamic impact (degrees of stenosis). In accordance with the previously given information, prerequisites were proposed by Goddi et al. (2019) for assessing early diagnosis and risk stratification with UVFI technic

- Analyze blood flow patterns to detect flow disturbances and recirculation regions.
- Calculating WSS variations with a series of six measurements, giving maximal and mean WSS values between a time interval.
- Defining risk stratification criteria.

As a conclusion, we propose an improvement of today gradation that could be made with three categories of criteria (Figure 8.15):

- **Solid component:** Composition of the plaque described with WSS variations and echo-structure descriptions.
- **Fluid component:** Stenosis gradation corrected with vector speed precise measurements and flow pattern descriptions.
- **Fluid/solid interaction area:** WSS variations along the plaque and during a follow up of the patient.

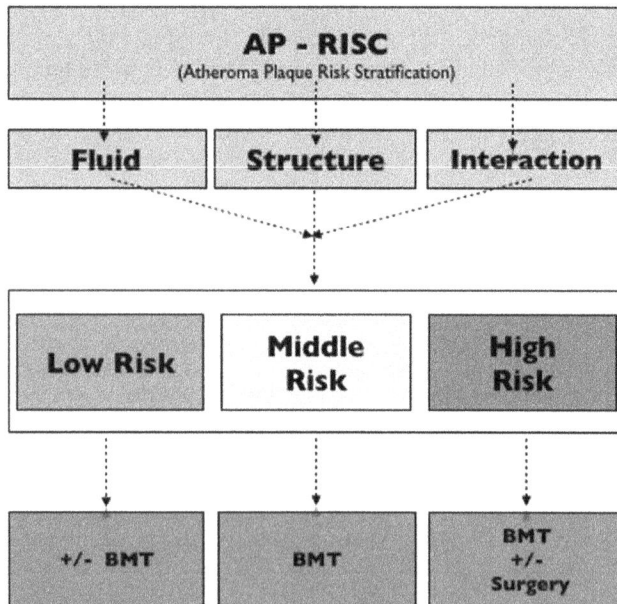

FIGURE 8.15 Illustration of a possible protocol for a risk gradation. **F**: Fluid - Flow as North American Symptomatic Carotid Endarterectomy Trial (**NASCET**) degree corrected with UVFI; **S**: Solid - Plaque Composition; **I**: Interaction - WSS Profile; BMT, Best Medical Treatment.

Studies based on the UVFI technique could provide elements of both etiological and therapeutic response.

REFERENCES

Cesar Alegre-Martínez, Kwing-So Choi, Outi Tammisola, and Donal McNally. 2019. "On the axial distribution of plaque stress: Influence of stenosis severity, lipid core stiffness, lipid core length and fibrous cap stiffness." *Medical Engineering & Physics* 68: 76–84.

Rei Asami, Tomohiko Tanaka, Motochika Shimizu, Yoshinori Seki, Tomohide Nishiyama, Hajime Sakashita, and Takashi Okada. 2019. "Ultrasonic vascular vector flow mapping for 2-D flow estimation." *Ultrasound in Medicine & Biology* 45: 1663–1674.

Kondo Claude Assi, Etienne Gay, Christophe Chnafa, Simon Mendez, Franck Nicoud, Juan F. P. J. Abascal, Pierre Lantelme, François Tournoux, and Damien Garcia. 2017. "Intraventricular vector flow mapping a Doppler-based regularized problem with automatic model selection." *Physics in Medicine & Biology* 62: 7131–7147.

Bassiouny, Hisham S., Yashuhiro Sakaguchi, Susanne A. Mikucki, James F. McKinsey, Giancarlo Piano, Bruce L. Gewertz, and Seymour Glagov. 1997. "Juxtalumenal location of plaque necrosis and neoformation in symptomatic carotid stenosis." *Journal of Vascular Surgery* 26: 585–594.

Kirk W. Beach, Tom Hatsukami, Paul R. Detmer, Jean F. Primozich, Marina S. Ferguson, David Gordon, Charles E. Alpers, David H. Burns, Brett D. Thackray, and Donald Eugene Strandness. 1993. "Carotid artery intraplaque hemorrhage and stenotic velocity." *Stroke* 24: 314–319.

Belzacq, Tristan, Stephane Avril, Emmanuel Leriche, and Alexandre Delache. 2012. "A numerical parametric study of the mechanical action of pulsatile blood flow onto axisymmetric stenosed arteries." *Medical Engineering & Physics* 34(10): 1483–1495.

Burleigh, Mary C., Adrian D. Brigfs, Corinne L. Lendon, Michael J. Davies, Gustav V. R. Born, and Peter D. Richardson. 1992. "Collagen types I and III, collagen content, GAGs and mechanical strength of human atherosclerotic plaque caps: Span-wise variations." *Atherosclerosis* 96: 71–81.

Casa, Lauren D. C., David H. Deaton, and David N. Ku. 2015. "Role of high shear rate in thrombosis." *Journal of Vascular Surgery* 61: 1068–1080.

Chen, Zimo, Haiqiang Qin, Jia Liu, Bokai Wu, Zaiheng Cheng, Yong Jiang, Liping Liu, et al. 2020. "Characteristics of wall shear stress and pressure of intracranial atherosclerosis analyzed by a computational fluid dynamics model: A pilot study." *Frontiers in Neurology* 10: 1372.

Cheng, Caroline, Dennie Tempel, Rien van Haperen, Arjen van der Baan, Frank Grosveld, Mat J. A. P. Daemen, Rob Krams, and Rini de Crom. 2006. "Atherosclerotic lesion size and vulnerability are determined by patterns of fluid shear stress." *Circulation* 113(23): 2744–2753.

Cicha, Wörner, Beronov Urschel, Verhoeven Goppelt-Struebe, and Garlichs Daniel. 2011. "Carotid Plaque Vulnerability: A Positive Feedback Between Hemodynamic and Biochemical Mechanisms." *Stroke* 42(12):3502–3510.

Dodge, J. Theodore, B. Greg Brown, Edward L. Bolson, and Harold T. Dodge. 1992. "Lumen diameter of normal human coronary arteries. Influence of age, sex, anatomic variation, and left ventricular hypertrophy or dilation." *Circulation* 86: 232–246.

Du, Yigang, Alfredo Goddi, Chandra Bortolotto, Yingying Shen, Alex Dell'Era, Fabrizio Calliada, and Lei Zhu. 2020. "Wall shear stress measurements based on ultrasound vector flow imaging: theoretical studies and clinical examples." *Journal of Ultrasound in Medicine* 39(8):1649–1664.

Goddi, Alfredo, Luca Aiani, Yigang Du, Xujin He, Yingying Shen, and Lei Zhu. 2017. "P84 high-frame rate vector flow imaging: Relationship between carotid bifurcation geometry and flow patterns." *Artery Research* 20: 76.

Goddi, A. Alfredo, Chandra Bortolotto, Maria Vittoria Raciti, Ilaria Fiorina, Luca Aiani, Giovanni Magistretti, Andrea Sacchi, Carmine Tinelli, and Fabrizio Calliada. 2018. "High-frame rate vector flow imaging of the carotid bifurcation in healthy adults: Comparison with color Doppler imaging: Vector flow imaging characteristics of the carotid bifurcation." *Journal of Ultrasound in Medicine* 37: 2263–2275.

Goddi, A. Alfredo, Yigang Du, Alex Dell'Era, and Raffaele Novario. 2019. "Evaluation of shear stress in the arterial wall with Hi-Frame rate vector flow." *ECR 2019*, 27 February–3 March, 2019, Austria Center Vienna.

Goudot, Guillaume, Jonathan Poree, Olivier Pedreira, Lina Khider, Pierre Julia, Jean-Marc Alsac, Emeline Laborie, Tristan Mirault, Mickael Tanter, Emmanuel Messas, and Mathieu Pernot. 2019. "Wall shear stress measurement by ultrafast vector flow imaging for atherosclerotic carotid stenosis." *Ultraschall in der Medizin* 42(3): 297–305.

Goudot, Guillaume, Jonathan Poree, Olivier Pedreira, Lina Khider, Pierre Julia, Jean-Marc Alsac, Emeline Laborie, et al. 2020. "Wall shear stress measurement by ultrafast vector flow imaging for atherosclerotic carotid stenosis." *Ultraschall in der Medizin: European Journal of Ultrasound* 42(3): 297–305.

Groen, Gijsen, Ferguson van der Lugt, Yuan Hatsukami, and Wentzel van der Steen. 2008. "High shear stress influences plaque vulnerability." *Netherlands Heart Journal* 16: 280–283.

Haga, Jason H., Yi-Shuan J. Li, and Shu Chien. 2007. "Molecular basis of the effects of mechanical stretch on vascular smooth muscle cells." *Journal of Biomechanics* 40: 947–960.

Hariri, Nizar, Todd Russell, Gregory Kasper, and Fedor Lurie. 2019. "Shear rate is a better marker of symptomatic ischemic cerebrovascular events than velocity or diameter in severe carotid artery stenosis". *Journal of Vascular Surgery* 69(2): 448–452.

Kakkos, Stavros K., Maura B. Griffin, Andrew N. Nicolaides, Efthyvoulos Kyriacou, Michael M. Sabetai, Thomas Tegos, Gregory C. Makris, Dafydd J. Thomas, and George Geroulakos. 2013. "The size of juxtaluminal hypoechoic area in ultrasound images of asymptomatic carotid plaques predicts the occurrence of stroke." *Journal of Vascular Surgery* 57(3): 609–618.

Kumar, Arnav, Elizabeth W. Thompson, Adrien Lefieux, David S. Molony, Emily L. Davis, Nikita Chand, Stephane Fournier, et al. 2018. "High coronary shear stress in patients with coronary artery disease predicts myocardial infarction." *Journal of the American College of Cardiology* 72(16):1926–1935.

Lendon, Corinne L., Gustav Victor Rudolf Born, Michael J. Davies, and Peter D. Richardson. 1992. "Plaque fissure: The link between atherosclerosis and thrombosis." *Nouvelle Revue Francaise d'Hematologie* 34: 27–29.

Li, Zhi-Yong, Valentina Taviani, Tjun Tang, Umar Sadat, Victoria Eleanor L. Young, Andrew J. Patterson, Martin Graves, and Jonathan H. Gillard. 2009. "The mechanical triggers of plaque rupture: Shear stress vs pressure gradient." *The British Institute of Radiology* 82: 39–45.

Loree, Howard M., Alan J. Grodzinsky, Susan Y. Park, Lorna J. Gibson, and Richard T. Lee. 1994. "Static circumferential tangential modulus of human atherosclerotic tissue." *Journal of Biomechanics* 27: 195–204.

Mills, Ian, Tomislav Cvitaš, Klaus Homann, Nikola Kallay, and Kozo Kuchitsu. 1993. *Quantities, Units, and Symbols in Physical Chemistry.* Oxford; Boston, MA; Boca Raton, FL: International Union of Pure and Applied Chemistry.

O'Callaghan, Christopher J. and Bryan Williams. 2000. "Mechanical strain-induced extra-cellular matrix production by human vascular smooth muscle cells: Role of TGF-β." *Hypertension* 36: 319–324.

Pedlosky, Joseph. 1987. *Geophysical Fluid Dynamics,* Second edition. New York: Springer.

Samady, Habib and Arnav Kumar. 2019. "Coupling advanced imaging with computational vascular diagnostics." *JACC: Cardiovascular Imaging* 13(4): 1033–1035.

Sigala, Fragiska, Evangelos Oikonomou, Alexis S. Antonopoulos, George Galyfos, and Dimitris Tousoulis. 2018. "Coronary versus carotid artery plaques. Similarities and differences regarding biomarkers morphology and prognosis." *Current Opinion in Pharmacology* 39: 9–18.

Wang, Liang, Jian Zhu, Habib Samady, David Monoly, Jie Zheng, Xiaoya Guo, Akiko Maehara, et al. 2017. "Effects of residual stress, axial stretch, and circumferential shrinkage on coronary plaque stress and strain calculations: A modeling study using IVUS-based near-idealized geometries." *Journal of Biomechanical Engineering* 139(1): 014501.

Zhang, Bo, Yuqin Ma, and Fang Ding. 2018. "Evaluation of spatial distribution and charac-terization of wall shear stress in carotid sinus based on two-dimensional color Doppler imaging." *BioMedical Engineering OnLine* 17(1):141.

9 A Pre-Screening Technique for Coronary Artery Disease with Multi-Channel Phonocardiography and Electrocardiography

Yue Rong, Matthew Fynn, and Sven Nordholm
Curtin University

9.1 INTRODUCTION

Cardiovascular disease (CVD) including coronary artery disease (CAD) is the leading cause of mortality and morbidity in the world [1], contributing 31% towards all global deaths. Early diagnosis of CAD is important to prevent further development of this disease. Standard methods for diagnosis of CAD such as coronary angiography and myocardial perfusion imaging require specialised equipment and clinical setting. Although these methods are effective in diagnosing CAD, they are highly costly and expose patients to radiation. Table 9.1 shows the performance of different diagnostic methods in Australia in 2012 [2]. The costs were retrieved from http://www9.health.gov.au/mbs in February 2023, with item numbers 11704, 55132, 61324, 57360, and 38310 for the five stress tests in order, respectively.

On the other hand, heart auscultation (the interpretation of heart sounds by a physician) is a cost-effective tool for the pre-screening of CAD. It is well established that partial obstruction of coronary arteries causes disruption of normal, laminar flow and generates flow turbulence and so modifies heart sounds. When the coronary arteries blockage is prevalent, murmur sounds frequently occur. However, auscultation skills are difficult to acquire. Since the heart sound acquired by a stethoscope is often contaminated by various internal (e.g. breathing) and external (e.g. friction) noises, it can be hard for the human auditory system to identify abnormal heart sounds related to CAD. With the aid of computer technology and highly sensitive electronics, digital stethoscopes can be used to detect sounds that are below the human hearing threshold and frequency range. As a result, phonocardiogram (PCG) signal processing

DOI: 10.1201/9781003346678-9

TABLE 9.1

Diagnostic Methods for Coronary Artery Disease Detection

Stress Test	Cost $	Sensit. (%)	Specif. (%)	Advantages	Disadvantages
Electrocar-diography	33	68	70–77	Assessment of exercise capacity Cost effective First line test in absence of contraindications	Lowest sensitivity of all stress tests: risk of false negative test Lower diagnostic accuracy in women
Echocardiography (exercise)	240	80–85	84–86	Assessment of exercise capacity, cardiac structure/function No radiation High specificity	False negatives in single vessel/circumflex territory ischaemia (increased sensitivity with cycle ergometry)
Nuclear perfusion study	653	85–90	70–75	Exercise capacity can be assessed High sensitivity	Radiation False positives due to higher sensitivity/diaphragmatic attenuation
CT coronary angiogram	728	85–99	64–90	High negative predictive value (in low to intermediate risk subjects)	Radiation Functional effect of stenosis not usually assessed, nor exercise capacity
Coronary angiogram	2438	≈ 100	≈ 100	Gold standard	Invasive Radiation Functional effect of stenosis not routinely assessed

combined with computer-aided classification has attracted much interest over the last decade as a low-cost and noninvasive tool for the prescreening of CAD.

In this chapter, we present a multi-channel phonocardiography (PCG) and electrocardiography (ECG) based CAD pre-screening technique, developed by researchers at Curtin University in collaboration with Ticking Heart, a health-tech start-up. With the aid of a background-noise microphone integrated into each digital stethoscope, adaptive signal processing methods are proposed to suppress the environmental noise to improve the integrity of the recorded heartbeat signal. We investigate important factors that can impact the performance of a neural network-based CAD classifier using heart auscultation.

The rest of this chapter is organized as follows. In Section 9.2, a brief overview of the state-of-the-art PCG is provided. An introduction of the multichannel PCG and ECG instrument is presented in Section 9.3. Adaptive PCG noise cancellation algorithm is proposed in Section 9.4 to maintain the integrity of the signal-of-interest. Key factors affecting machine learning-based CAD classification algorithms are discussed in Section 9.5. Conclusions are drawn in Section 9.6.

A list of acronyms used in this chapter is given as follows:

ANN: artificial neural network
BNM: background-noise microphone
CAD: coronary artery disease
CIC: Computing in cardiology
CNN: convolutional neural network
CSA: cardiac sonospectrographic analyzer
CVD: cardiovascular disease
ECG: electrocardiography
FIR: finite impulse response
FL: filter length
HM: heart-sensor microphone
HSMM: hidden semi-Markov model
LTI: linear time-invariant
MLP: multilayer perceptron
MSE: mean-squared error
NLMS: normalized least mean squares
NN: nearest neighbor
PCG: phonocardiography
PSD: power spectral density
SST: synchrosqueezing transform
SVM: support vector machine

9.2   OVERVIEW OF THE STATE-OF-THE-ART PCG

PCG signals are digital recordings of heart sound. They provide a convenient primary diagnostic tool for detecting CAD murmurs [3]. Figure 9.1 displays a PCG signal of a healthy subject, where the regions of systole and diastole, and major heart sounds S1 and S2 are shown.

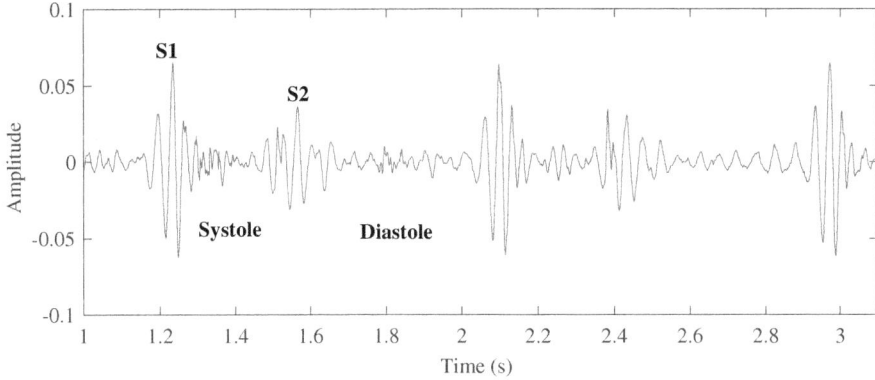

FIGURE 9.1 Phonocardiogram of a healthy subject.

Murmur sounds from the coronary artery blockage affect the basic characteristics of PCG signals, which provides the opportunity to detect these changes using machine learning techniques. Many algorithms have been proposed to classify normal and abnormal heart cycles, aiming at achieving high accuracy, sensitivity, and specificity from their data sets, which are defined below

$$\text{Accuracy} = \frac{T_p + T_n}{T_p + F_p + F_n + T_n} \tag{9.1}$$

$$\text{Sensitivity} = \frac{T_p}{T_p + F_n} \tag{9.2}$$

$$\text{Specificity} = \frac{T_n}{T_n + F_p} \tag{9.3}$$

Here T_p, T_n, F_p, and F_n represent the number of true positives, true negatives, false positives, and false negatives, respectively.

Most PCG-based CAD detection techniques follow a similar set of methodologies, which consists of extracting features from PCG data for the machine learning algorithms to learn. Multiple studies have undertaken time-frequency analysis techniques to extract feature vectors from the heart sound signal. In Ref. [4], an artificial neural network (ANN) model was applied to classify heart sounds using wavelet-based feature extraction. Based on the ANN classification model, a computer-aided diagnosis system was described in Ref. [5] for multiple pathological cases using wavelet decomposition. Support vector machine (SVM) along with wavelet packet decomposition have been applied in Ref. [6] to detect valvular heart sounds as normal or abnormal. In a 2010 study [7], a method was proposed to improve the performance of least-squares SVM to diagnose pathological sounds.

A digital electronic stethoscope named the cardiac sonospectrographic analyzer (CSA) was tested in Ref. [8] for the detection of coronary artery microbruits, where the CSA showed high sensitivity and specificity for the detection of significant early CAD in an outpatient setting. In Ref. [9], nine different types of acoustic features from five overlapping frequency bands were obtained and analysed for the identification of murmur sounds associated with CAD. The result confirms that there is a potential in PCG for the diagnosis of CAD. A dual-input neural network for CAD detection using both ECG and PCG was developed in Ref. [10]. Access to labelled heart sound recordings plays a key role in developing machine learning-based classification algorithms. A list of the currently available open access data sets is provided in a recent paper [11]. Among these data sets, a large public database was created for the 2016 PhysioNet/Computing in Cardiology (CIC) challenge to classify normal and abnormal heart sound recordings [12]. In this database, there are six different data sets available, where each set is composed of heartbeat measurements recorded by different research groups at different institutions. Each recording is labelled as 'normal' or 'abnormal' according to expert diagnosis. A currently largest pediatric heart sound data set is presented in Ref. [11], which contains 215, 780 manually annotated heart sounds. In addition to binary labels, each murmur has been manually annotated by an expert annotator according to its timing, shape, pitch, grading, and quality. The data set [11] is used in the 2022 PhysioNet challenge in detecting abnormal heart function from multi-location PCG recordings of heart sounds [13].

An open access simultaneously recorded PCG and ECG database is presented recently in Ref. [14], where the recording device includes circuitry for three-lead ECG, two digital stethoscope channels for PCG acquisition, and two auxiliary channels to capture the ambient noise. To the best of our knowledge, there is no open access database with multichannel (more than two channels) PCG and ECG recordings. In the next section, we present a multichannel PCG and ECG instrument.

9.3   MULTICHANNEL PCG AND ECG INSTRUMENT

A measurement system has been built by Ticking Heart [15], a health-tech start-up, that incorporates six digital stethoscopes and one three-lead ECG sensor onto a wearable vest that simultaneously measures heartbeat signals and applies machine learning methodologies for pre-screening CAD. The digital stethoscopes are placed in clinically advised positions, see Figure 9.2. In particular, four stethoscopes are located on the left side of the chest to detect sounds from the mitral, aortic, pulmonary, and tricuspid valves. The other two PCG sensors are placed on the right side of the chest to detect sounds from the ascending aorta artery. Compared with systems having only a single stethoscope or two stethoscopes [14], using multiple stethoscopes can improve the performance of classification [16]. Moreover, multichannel PCG and ECG signal processing has the potential to separate the microphone signals with a stronger mapping to the cause of the signal waveform.

As shown in Figure 9.3, each stethoscope has two microphones, one microphone is the heart-sensor microphone (HM) located behind the diaphragm, and the other one is the background-noise microphone (BNM) located at the other end of the stethoscope. The HM acquires the heart signal plus part of the background noise, while the BNM mainly picks up the background noise. Using such two-microphone configuration, the background noise can be reduced from the HM, which contributes

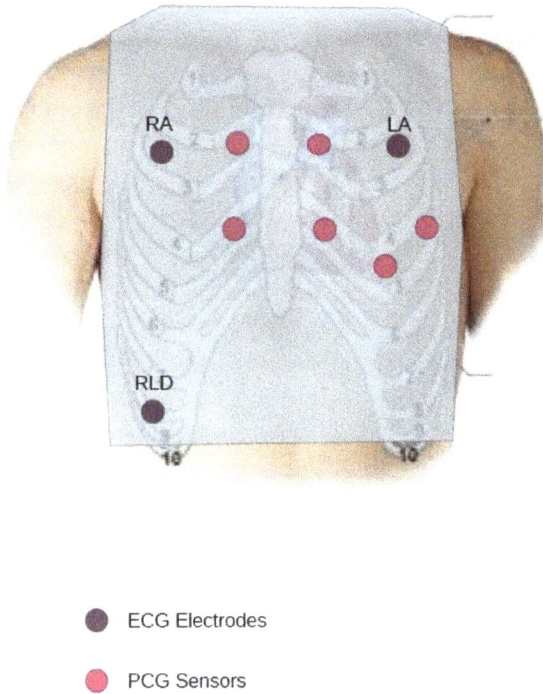

FIGURE 9.2 A vest holding six digital stethoscopes and one three-led ECG sensor (RA, LA and RLD stand for right arm, left arm and right leg drive, respectively)

FIGURE 9.3 Single stethoscope. (a) HM facing upward. (b) BNM facing upward.

to successful diagnosis techniques, as acquired signals from the system are cleaner with higher signal integrity. All signals from the stethoscopes and the ECG sensor are routed to a data collection board as shown in Figure 9.4, for further analog signal conditioning prior to digitalization. Note that Figures 9.3 and 9.4 are not to scale.

An example of the PCG signals measured simultaneously from six digital stethoscopes held by TH's wearable vest is given in Figure 9.5, which shows the waveforms

FIGURE 9.4 Multichannel PCG and ECG data collection board.

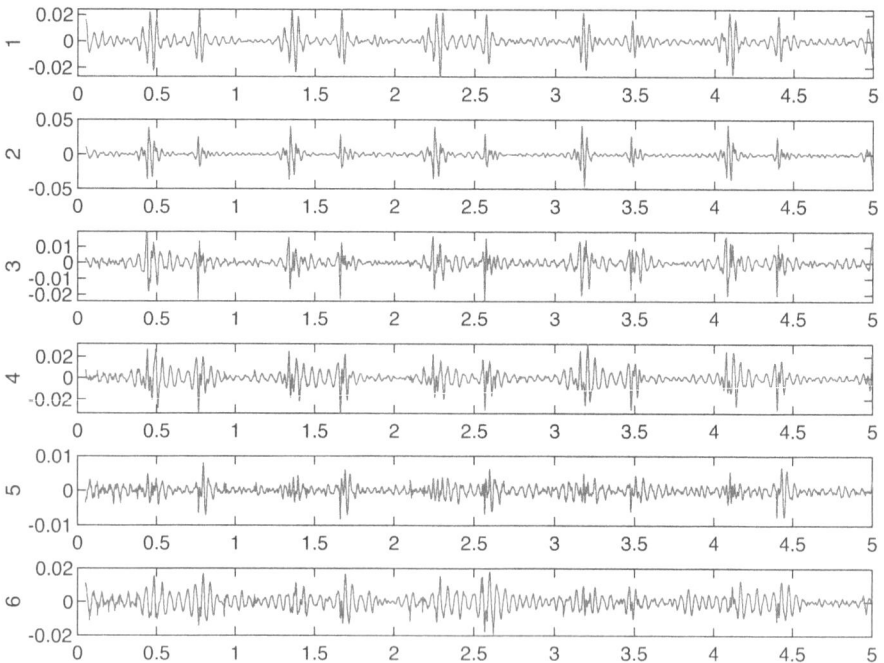

FIGURE 9.5 Waveform of heartbeat signals acquired by the HMs in six digital stethoscopes located in Figure 9.6. The *x*-axis is time in seconds and the *y*-axis represents the amplitude.

of the signals acquired by the HM of each stethoscope sampled at 2 kHz and passed through a high-pass filter with a cutoff frequency at 20 Hz. Thus, the frequency band of signals in Figure 9.5 is from 20 Hz to 1 kHz. The positions of the six stethoscopes on the chest are displayed in Figure 9.6. Note that the six HM signals are synchronized at the signal sample level. We can observe that the signal from Sensor 2 has the

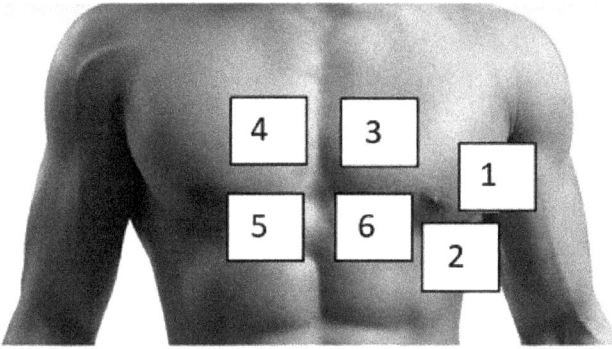

FIGURE 9.6 Positions of the digital stethoscopes for the PCG signals shown in Figure 9.5.

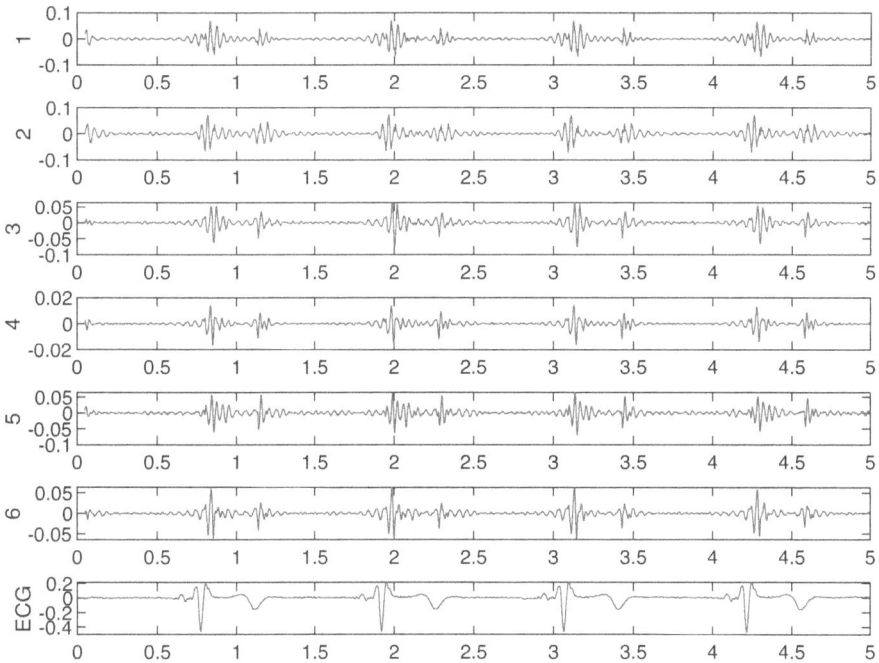

FIGURE 9.7 Waveform of PCG signals and ECG signal recorded simultaneously. The *x*-axis is time in seconds and the *y*-axis represents the amplitude.

highest amplitude, where the pattern of S1 and S2 can be clearly seen, while Sensor 5 yields the weakest signal as it is furthest away from the heart. An advantage of multichannel synchronized PCG signals is that they can provide more information about the heartbeat and murmur sounds, which can be utilized by machine learning algorithms to improve the accuracy of CAD classification.

Waveforms of six PCG signals and one ECG signal recorded simultaneously are shown in Figure 9.7. Synchronization at the signal sample level is achieved through

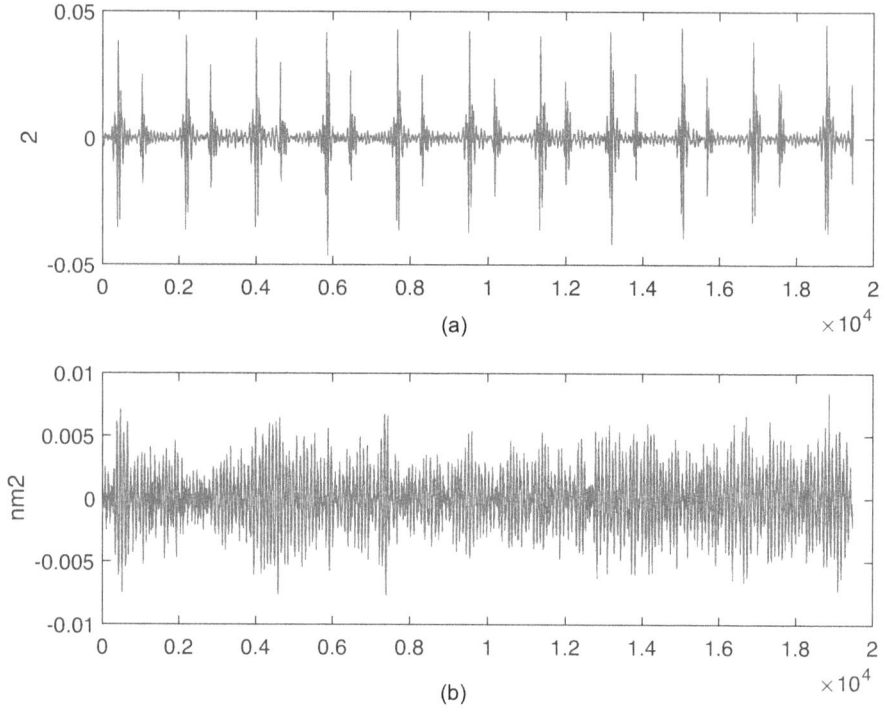

FIGURE 9.8   Waveform of the HM (a) and BNM (b) signals from Sensor 2.

the same multichannel analogue-to-digital converter. We can observe that in one heart cycle, the first peak in ECG (i.e., the R peak) appears slightly ahead of the S1 peak in the PCG signals, while the second peak in ECG coincides with the S2 peak in the PCG signals. In the case of Figure 9.7, the ECG signal appears to be less noisy than PCG signals. Jointly processing the ECG and PCG signals would provide more information on the feature of the heartbeat, which can feed into a classifier to improve the performance of CAD classification.

Figure 9.8 shows the waveform of the signals acquired by the HM and BNM on Sensor 2 in Figure 9.5. We can see that since the signals were recorded in a quiet office environment, the BNM signal (which is mainly external noise) has a much lower amplitude than the HM signal. Moreover, the pattern of S1 and S2 is not visible in the BNM signal, making it possible to cancel the noise in the HM based on the BNM signal without compromising the signal-of-interest, as discussed in detail in the next section. A clinical study is planned in 2023 to evaluate the performance of the vest in real patients.

9.4   NOISE CANCELLATION

As shown in the last section, background noise can couple into the HM and corrupt acoustic heartbeat measurements during heart auscultation. This can decrease heart signal integrity, as the noise cannot be filtered out using conventional frequency-selective filters if it lies within the frequency band of interest (e.g. 10–600 Hz). By using the BNM as a reference for noise, noise cancellation filtering techniques can be applied to attenuate unwanted background noise and restore integrity to the desired signal.

9.4.1 Noise Cancellation Model

A noise cancellation model of the two-microphone stethoscope system is shown in Figure 9.9. The desired heartbeat signal to be measured $d(n)$ is corrupted by various background noise sources $v(n)$ to produce the signal acquired by the HM as $x(n) = d(n) + v_1(n)$. Without any information on $v_1(n)$, it is not possible to remove it from $d(n)$. The BNM measures the noise sources without heartbeat signal $d(n)$, depicted as $v_2(n)$. However, the HM and BNM do not detect the noise in the same way, i.e., $v_1(n) \neq v_2(n)$. This indicates that the desired signal cannot be obtained via a direct subtraction [17]. Instead, a Wiener filter produces an estimate of $v_1(n)$, denoted as $\hat{v}_1(n)$, via the observational measurements of $v_2(n)$. This estimate is subtracted from $x(n)$ to attenuate the background noise. Here, the Wiener-Hopf equation for a finite impulse response (FIR) filter is given by

$$\mathbf{R}_{v_2}\mathbf{w} = \mathbf{r}_{xv_2} \qquad (9.4)$$

where $\mathbf{w}$ is the FIR coefficient vector of the Wiener filter, $\mathbf{R}_{v_2}$ is the autocorrelation matrix of $v_2(n)$, and $\mathbf{r}_{xv_2}$ is the cross-correlation vector of $x(n)$ and $v_2(n)$. The observed signals $x(n)$ and $v_2(n)$ are acoustic signals and their relative time delay is unknown. An extra delay is inserted in $x(n)$ to take into account non-causality [18].

In environments where the background noise is constantly changing, adaptive filters are needed to update filter coefficients $\mathbf{w}$ in Equation (9.4) in real time. An adaptive filter model is shown in Figure 9.10, where the output of the adaptive filter aims to minimize the mean-squared error (MSE) of estimating $v_1(n)$, thus $e(n)$ is the MSE of the desired signal $d(n)$. We use a normalized least mean squares (NLMS) algorithm to update the coefficients of the adaptive filter $\mathbf{w}(i)$. For each sample $x(i)$ from the HM, the error history is given by

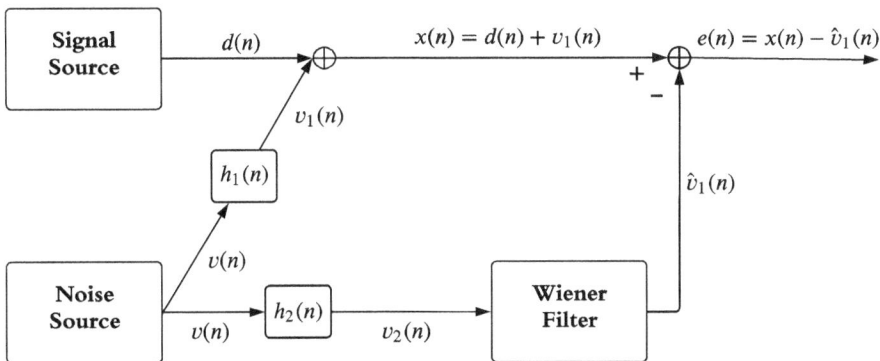

FIGURE 9.9 Noise cancellation model of the two-microphone stethoscope.

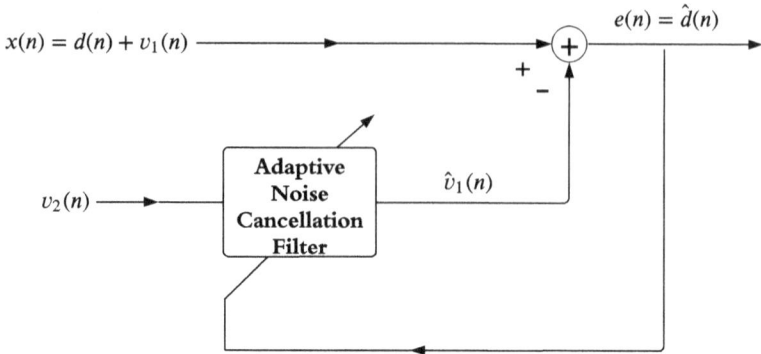

FIGURE 9.10 Adaptive noise canceller.

$$e(i) = x(i) - \mathbf{w}^T(i)\mathbf{v}_2(i) \tag{9.5}$$

where $\mathbf{v}_2(i) = (v_2(i), v_2(i+1), ..., v_2(i+N))^T$, N is the filter length (FL), and $(\cdot)^T$ denotes the matrix transpose. The filter coefficient vector is updated for each sample as

$$\mathbf{w}(i+1) = (1-a)\mathbf{w}(i) + \mu \frac{\mathbf{v}_2(i)e(i)}{|\mathbf{v}_2(i)|^2} \tag{9.6}$$

where α is the leakage coefficient, $|...|$ stands for the vector Euclidean norm, and μ is the step size determining the size at which the filter coefficients are updated. The algorithm specified by Equations (9.5) and (9.6) is a variable leaky NLMS algorithm, which has the potential to significantly outperform the standard LMS algorithm.

9.4.2 COHERENCE FUNCTION

The performance of the NLMS-based method in suppressing the background noise is closely related to the coherence function of signals from the HM and BNM. The coherence between the HM and BNM channels indicates how much noise can be attenuated at particular frequencies [19]. The coherence function of two signals $x(n)$ and $v(n)$ (also referred to as the coherence-squared function) is defined by Ref. [20]

$$\gamma_{vx}^2(f)^2 = \frac{|S_{vx}(f)|^2}{S_{vv}(f)S_{xx}(f)} \tag{9.7}$$

where $S_{vv}(f)$ and $S_{xx}(f)$ are the auto spectral densities and $S_{vx}(f)$ is the cross spectral density. For all frequencies f, Equation (9.7) is bounded by

$$0 \le \gamma_{vx}^2(f) \le 1. \tag{9.8}$$

If we apply (9.7) to a noise-free linear time-invariant (LTI) system governed by $\bar{x}(n) = v(n) * h(n)$, where $h(n)$ is the system impulse response and $*$ denotes the convolution, it can be shown [21] that $\gamma^2(f) = 1$. Let us rearrange Figure 9.9–9.11, where the system of $\hat{h}_1(n)$ represents the concatenated system of $h_2(n)$ and the Wiener filter. It has been shown in Ref. [21] that to minimize the MSE of $e(n)$ in Figure 9.11, there is $\hat{H}_1(f) = H_1(f)$, where $H_1(f)$ and $\hat{H}_1(f)$ are the frequency-domain representation of $h_1(n)$ and $\hat{h}_1(n)$, respectively. Based on this and assuming that $d(n)$ and $v(n)$ are uncorrelated, we can calculate the coherence function in Equation (9.7) as

$$\gamma_{vx}^2(f) = \frac{\left|H_1(f)\right|^2 R_{vv}^2(f)}{R_{vv}\left(\left|H_1(f)\right|^2 R_{vv}(f) + R_{dd}(f)\right)}$$

$$= \frac{1}{1 + \dfrac{R_{dd}(f)}{\left|H_1(f)\right|^2 R_{vv}(f)}} \qquad (9.9)$$

$$= \frac{1}{1 + \dfrac{R_{dd}(f)}{R_{xx}(f) - R_{dd}(f)}}$$

where $R_{vv}(f)$ and $R_{dd}(f)$ are the auto spectral densities of $v(n)$ and $d(n)$, respectively. From Equation (9.9), we have

$$1 - \gamma_{vx}^2(f) = \frac{R_{dd}(f)}{R_{xx}(f)} \qquad (9.10)$$

which establishes the link of the LTI noise cancellation approach to the coherence function.

One of the major problems applying adaptive noise cancellation techniques in real acoustic environments is the low coherence between the noise corrupting the desired signal and the noise measured by the noise sensor. For the scenario depicted in Figure 9.9, a coherence value less than unity indicates that the system relating $x(n)$ and $v_2(n)$ does not fulfill the theoretical assumptions [20], thus the noise at

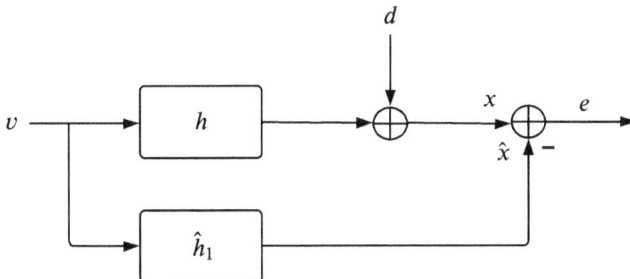

FIGURE 9.11 Linear system model.

those particular frequencies cannot be completely removed. The coherence function assumes an LTI relationship and stationary signals. The ensemble average is used so it furthermore assumes ergodicity. The amount of possible noise cancellation for this scenario as a function of frequency is given below and plotted in Figure 9.12

$$C(f) = \frac{1}{1 - \left| \gamma_{v_2 x}^2 (f) \right|} \tag{9.11}$$

9.4.3 NOISE CANCELLATION PERFORMANCE

Four different sources of background noise were tested on heartbeat measurements including a single 300 Hz tone, multiple tones consisting of 200, 300 and 500 Hz, hospital/clinic noise, and breathing noise. The heartbeat measurements were of the second author and offer no diagnostic insight, thus no ethical approval was required for this research. A FireFace UCX was used with a MATLAB interface to allow simultaneous playback of the background noise through a Fostex 6301B speaker whilst recording from the stethoscope. FireFace UCX is a USB and FireWire audio interface manufactured by RME Audio [22]. The stethoscope was taped to the chest, making sure that the BNM was exposed.

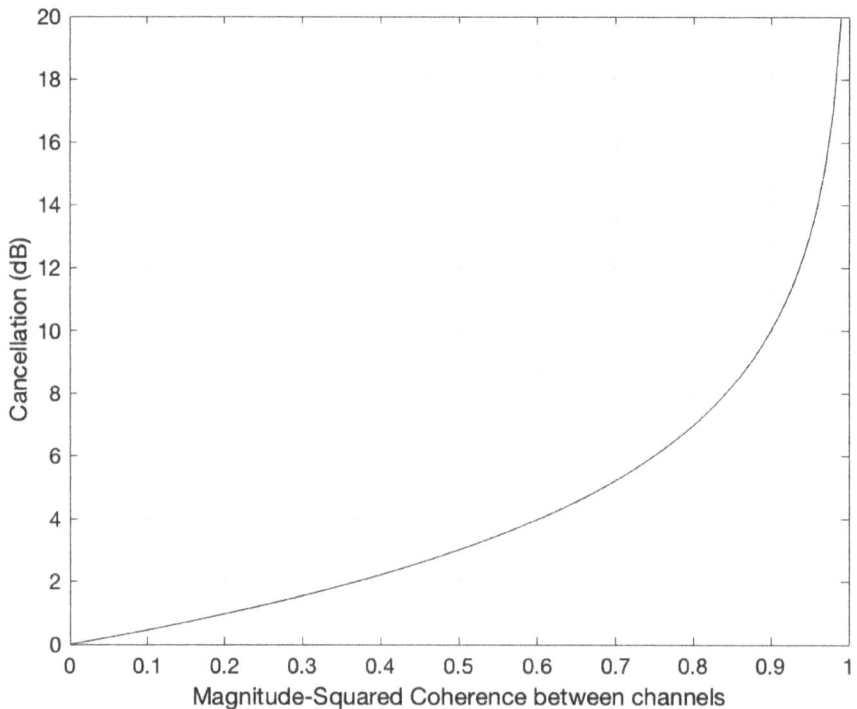

FIGURE 9.12   Adaptive cancellation versus the squared coherence between $v_2(n)$ and $x(n)$.

Heartbeat measurements of 15-second duration were taken for each background noise scenario. The measurements were taken under breath-held conditions, except for the breathing noise scenario, where the speaker was turned off. The FireFace UCX collected data at 44.1 kHz sampling frequency, which was re-sampled down to 2 kHz. The adaptive noise cancellation algorithm in Section 9.4.1 was implemented in MATLAB R2022a and run on an HP laptop.

The leakage coefficient $\alpha=0.001$ was used in Equation (9.6). The filter length N and step size μ were first tuned on the single 300 Hz tone. The combination that achieved the best noise attenuation was then used for the other background noise scenarios after comparing the performance to a conventional LMS algorithm reviewed in Ref. [23]. For each case, the coherence function between the HM and the BNM was plotted via a Welch estimator in MATLAB to indicate expected noise cancellation performance. The Welch estimator applies a 1,024 length Hann window with 512 overlaps calculated across 1,024 samples. Power spectral density (PSD) plots and spectrograms generated through MATLAB allowed visualization of the performance.

When a single 300 Hz tone was played through the speaker, the coherence function between the HM and BNM is shown in Figure 9.13. It can be seen that the coherence function is unity at 300 Hz, indicating that a complete attenuation of the tone is possible using linear techniques.

The NLMS algorithm was tested with $\mu=0.05$ and $N=256$, 512, and 1,024. Figure 9.14 shows the PSD of the filtered HM signals and the unfiltered HM signal. It can be seen that for all three N, the amount of noise attenuation is 27 dB.

Next, we chose $N=512$ for tuning the step size μ of the NLMS algorithm. Values of $\mu=0.01$, 0.05, 0.1, and 0.5 were tested. Figure 9.15 shows the PSD of the filtered

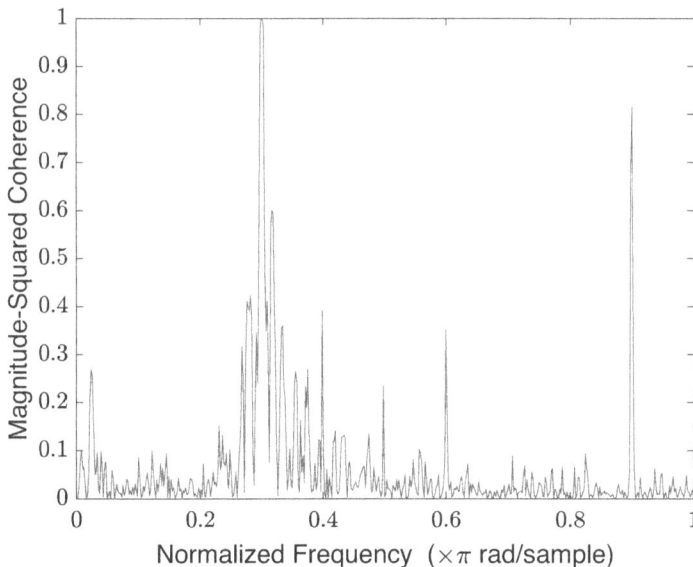

FIGURE 9.13 A 300 Hz tone background noise coherence function.

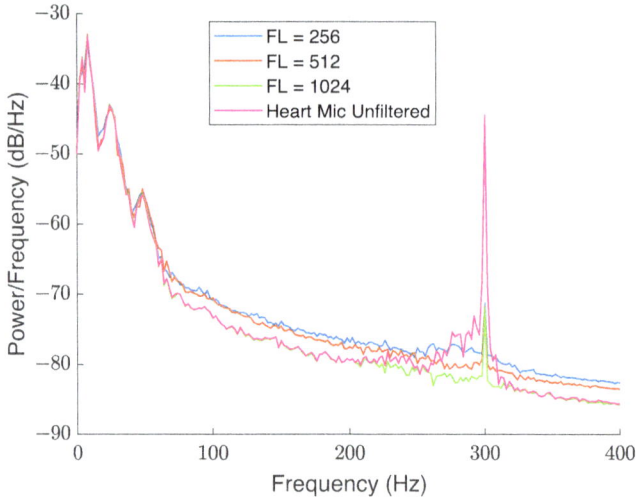

FIGURE 9.14   PSD comparison of various filter lengths with 300 Hz tone background noise and $\mu = 0.05$.

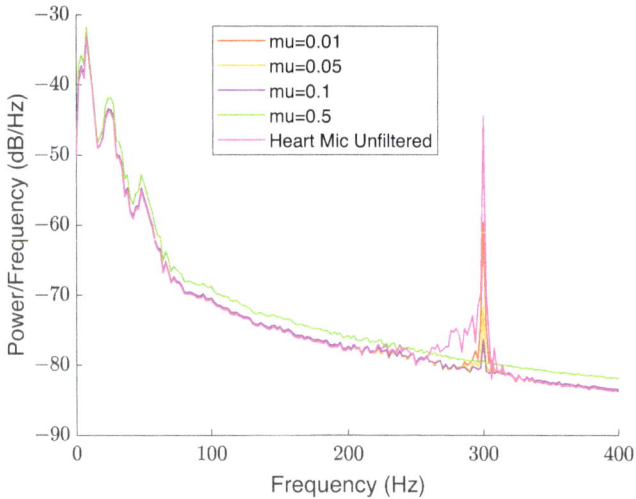

FIGURE 9.15   PSD comparison of various μ with 300 Hz tone background noise and $N = 512$.

HM signal at various μ. It can be observed that the noise attenuation increased as μ increased, where almost complete attenuation (~35 dB) was achieved at $\mu = 0.5$. This agrees with the coherence function which suggests a complete attenuation is possible. A 300 Hz tone is a periodic signal, suggesting why the measured attenuation was high. This may not be the case when other non-periodic noise sources are present.

In the following experiments, we set $N=512$ and $\mu=0.1$. In the second experiment, a multi-tone signal at 200, 300, and 500 Hz was played through the speaker. The coherence between the HM and BNM signals is displayed in Figure 9.16. It can be seen that the 3 tones had unity magnitude-squared coherence, as well as the first harmonic of each tone. This indicates that the first two harmonics of all three tones can be fully attenuated. To test the ability of the algorithm to adapt to different tones, the NLMS algorithm was tested on the scenario where each tone was played for 5 seconds while a heartbeat was recorded. It is shown in Figure 9.17 that a near complete noise attenuation was observed in the PSD comparison of the filtered and unfiltered HM signal. This can be visually observed in the spectrogram shown in Figure 9.18. Here, the BNM, HM and filtered HM signal spectrograms are displayed side-by-side, where the 200, 300, and 500 Hz tones were approximately attenuated by 24.5, 21.4, and 20.3 dB, respectively.

In the third experiment, hospital/clinic background noise was played through the speaker while taking a heartbeat measurement. As this type of noise can be encountered in practical hospital and clinic environment, it is of high interest to attenuate this type of non-stationary noise within the frequency of interest. The magnitude-squared coherence between the HM and BNM signals is shown in Figure 9.19. It can be seen that the coherence function varies from 0.3 to 0.7 (not unity) for the frequency band between 200 and 500 Hz. Thus, linear optimal filtering will not be able to achieve a complete attenuation of the background noise. According to Figure 9.12, around 3–5 dB noise suppression is possible at a magnitude-squared coherence value of 0.5. There is visual evidence of attenuation displayed in the spectrogram comparison in Figure 9.20, where we can see that the noise energy at 500 Hz, and the band between 200 and 300 Hz is slightly suppressed.

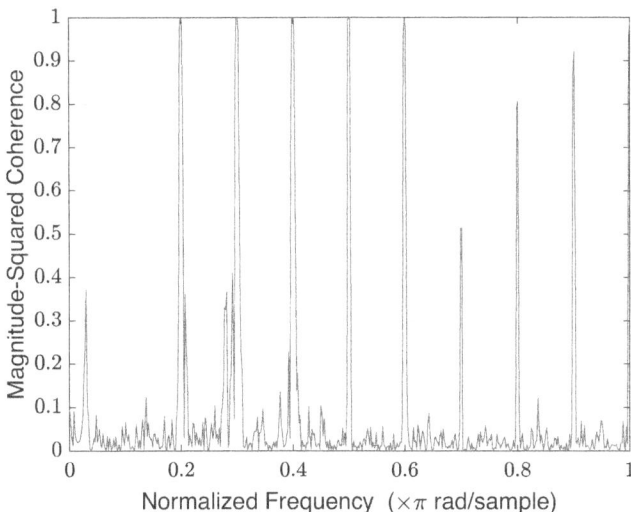

FIGURE 9.16 The 200, 300, and 500 Hz tone background noise coherence function.

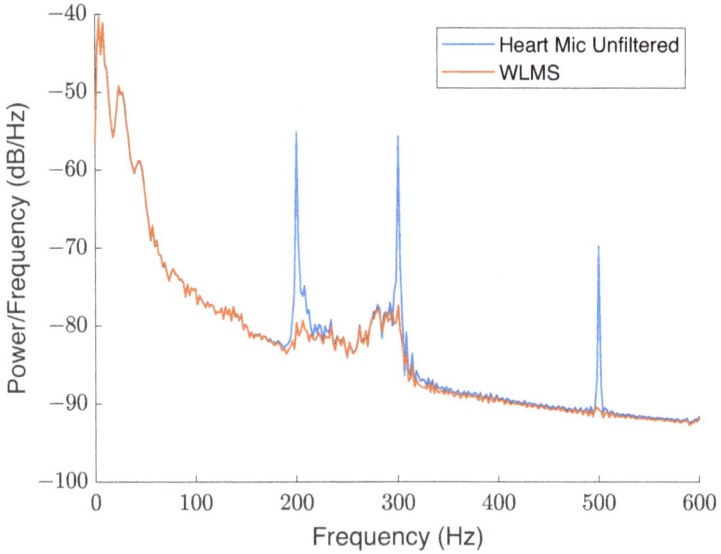

FIGURE 9.17 PSD comparison of BNM signal, filtered and unfiltered HM signal in multi-tone background noise.

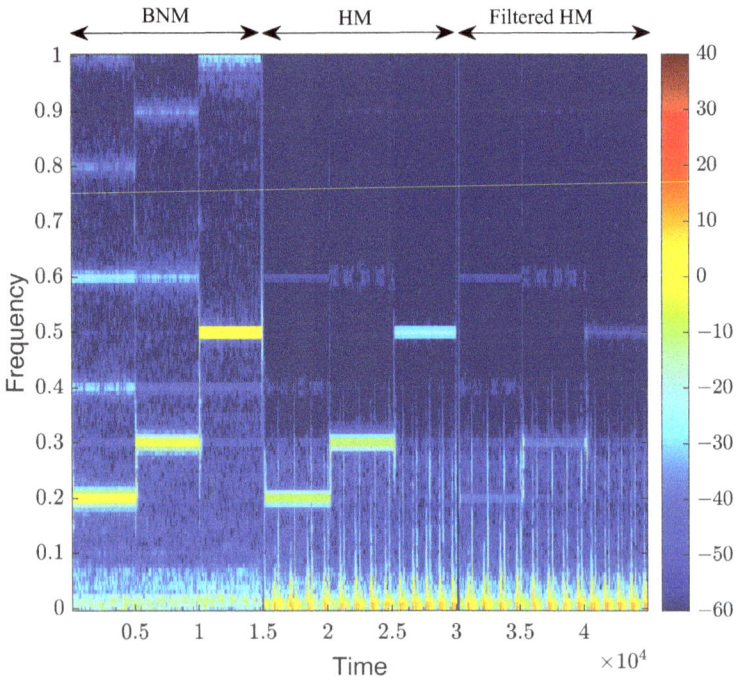

FIGURE 9.18 Spectrogram comparison of filtered and unfiltered HM signal in multi-tone background noise.

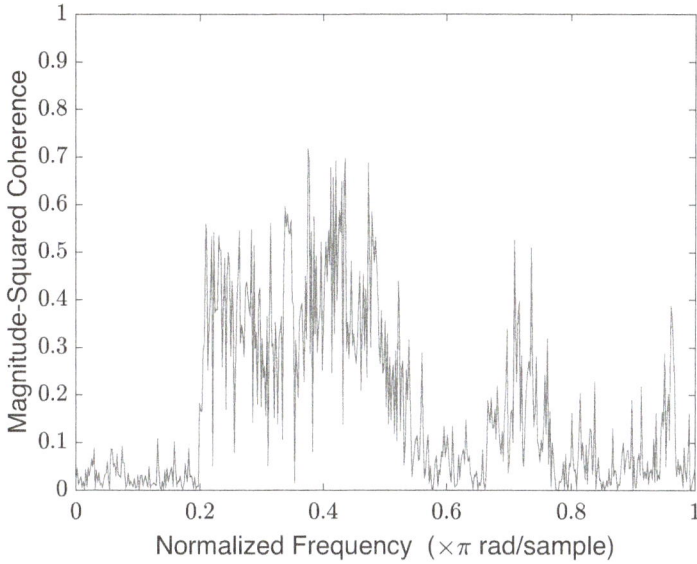

FIGURE 9.19   Hospital/clinic background noise coherence function.

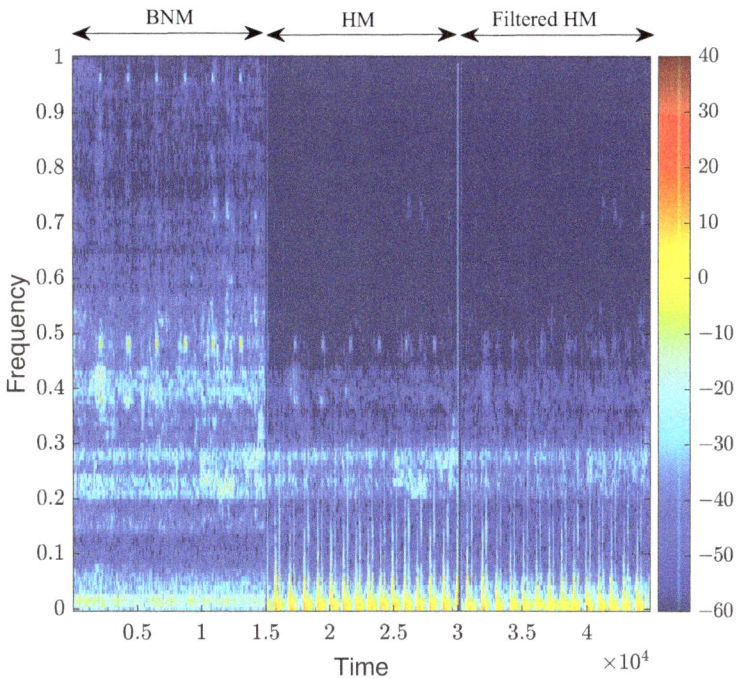

FIGURE 9.20   Spectrogram comparison of filtered and unfiltered HM signal in hospital/clinic background noise.

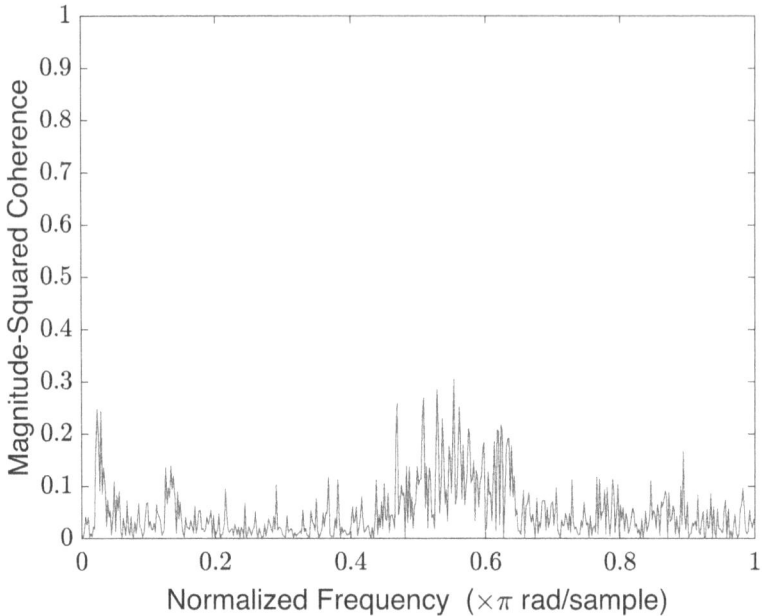

FIGURE 9.21   Breathing noise coherence function.

Many studies in the literature on the detection of CAD require that subjects hold their breath whilst heartbeat measurements are taken [16]. Although this is ideal for CAD detection, extending the stethoscope system to detect lung disease could be performed in unison, where the breathing noise can contain critical information. Thus, it is of great interest to separate the heartbeat sound from the breathing sound, so they can be studied and processed separately.

For the last experiment, the heartbeat signal was measured without holding breath and no extra external background noise sound was played. The magnitude-squared coherence function between the HM and BNM signals is shown in Figure 9.21. It can be seen that the coherence function was <0.3 across the whole frequency band. This suggests that only <1 dB of breathing noise suppression is achievable based on Figure 9.12. This was confirmed in Figure 9.22, where the HM signal before and after filtering appear identical. Non-linear methods may be explored to suppress the breathing noise and enhance the heart signal. These other methods should be carefully designed as to not suppress murmurs that are present in CAD patients, as the diagnostic process will be negatively affected.

9.5   MACHINE LEARNING-BASED CLASSIFICATION

There are many factors which can affect the results of a neural network-based CAD classifier. In this section, we discuss four of these factors: segmentation of heart cycle, the integrity of heartbeat signal, data set size and the neural network structure.

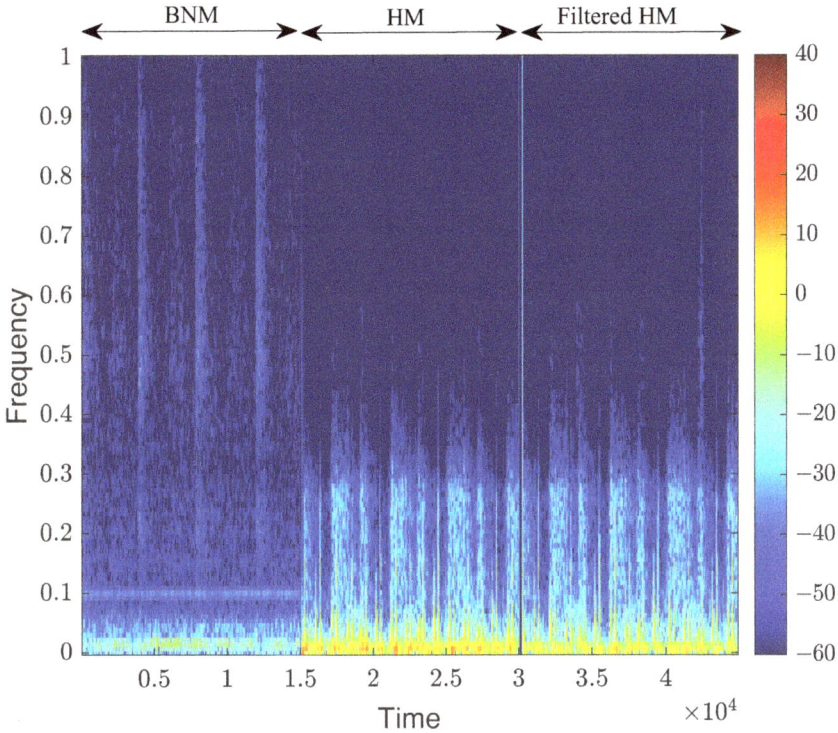

FIGURE 9.22 Spectrogram comparison of filtered and unfiltered HM signal with breathing noise.

We demonstrate that among these factors, the integrity of heartbeat signal has a great impact on the performance of a classifier.

9.5.1 HEART CYCLE SEGMENTATION

Prior to input into a neural network-based classifier, heartbeat signals are usually segmented into heart cycles. Let us take [24] as an example. In Ref. [24], the heart-beat data input to the SVM algorithm is segmented into three epochs, each containing two full heart cycles. Thus, one epoch commences at S1 of the first cycle and finishes at the end of diastole in the second cycle.

The heartbeat signal segmentation can be implemented manually following the methods in Ref. [24]. How- ever, hand-segmenting heartbeat data can be time-consuming and is often not practical for many applications which require automatic data acquisition and pre-processing. In this case, a logistic regression-hidden semi-Markov model (HSMM) based heart sound segmentation algorithm adapted in Ref. [25] can be applied to provide automatic heart cycle segmentation. The segmentation algorithm [25] is a probabilistic model-based approach, which identifies positions that correspond to S1, systole, S2, and diastole. Figure 9.23 displays a hand-segmented heartbeat and a heartbeat segmented via the HSMM-regression algorithm.

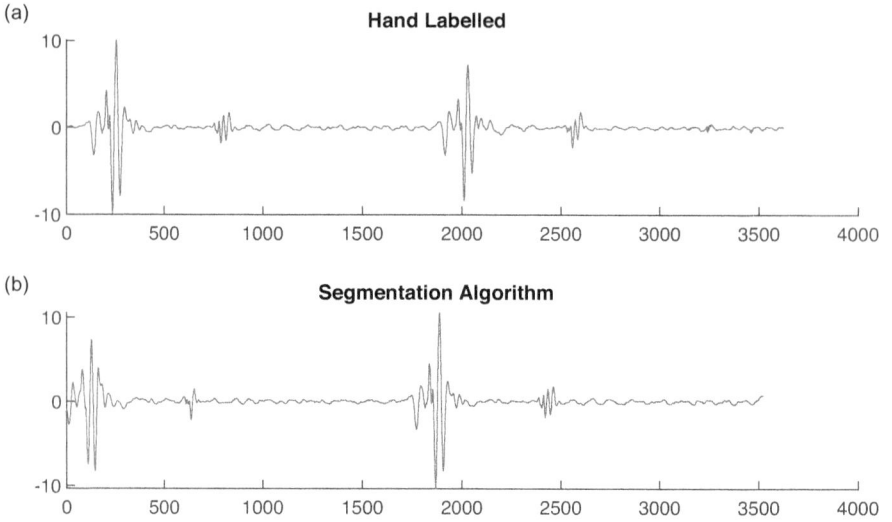

FIGURE 9.23   Hand labelled (a) versus HSMM-regression based segmentation algorithm (b).

It can be seen that the heartbeat cycle starts directly at S1 when using the HSMM-regression algorithm, and there is a small delay shown in Figure 9.23 where the hand-segmentation is based. Ideally, this small shift would not affect the classification results. To test the sensitivity of such shift, an adjustment factor was included in the pre-processing algorithms. This adjustment factor causes a shift in the heartbeat epochs after segmentation using the HSMM-regression algorithm, allowing a slight delay before S1. In particular, 8 different adjustment factors were tested: 0, 0.006, 0.016, ... , 0.066. The adjustment factor is multiplied with the amount of samples in each epoch to obtain the shift value, which is the amount of samples the segmented epoch is delayed by. Figure 9.24 displays the same epoch subjecting to four different

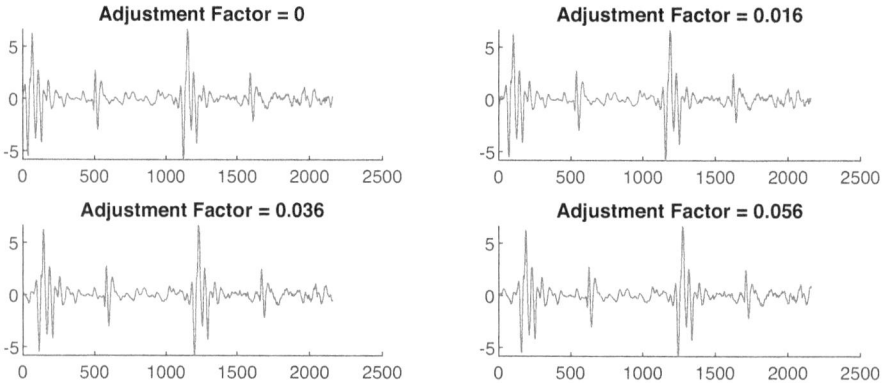

FIGURE 9.24   Epochs with various adjustment factors.

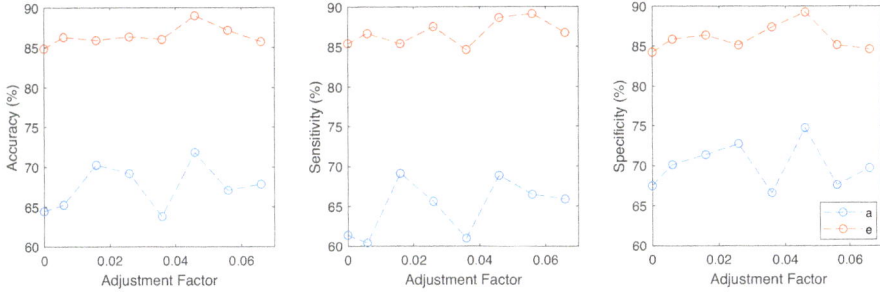

FIGURE 9.25 Accuracy, sensitivity, and specificity versus the adjustment factor for data sets *a* and *e*.

adjustment factors. Note that the higher adjustment factors provide a better representation of the hand-segmented epochs in Figure 9.23, where there is a slight delay before the first S1.

For each adjustment factor, 40 normal and 40 abnormal heartbeats from data set *a* in Ref. [12] were segmented into three epochs using the HSMM-regression algorithm. The k-nearest neighbor (k-NN) SVM algorithm was implemented on the data set of 80 segmented subjects. The algorithm in Ref. [24] including the whole process of performing the synchrosqueezing transform (SST) on each epoch, feature extraction and selection, and training the SVM classifier was repeated each time. The highest accuracy, sensitivity, and specificity were recorded for each test.

The accuracy, sensitivity, and specificity plotted as a function of the adjustment factor are displayed in Figure 9.25. Let us focus on the curves with 'a' at this time. It can be seen that the accuracy varied from 63.8% to 71.8% during the adjustment factor variation test. This 8% difference between the minimum and maximum score occurred when the adjustment was 0.036 and 0.046, respectively. This shows that the SVM algorithm is quite sensitive with respect to pre-processing the heartbeat data. Ideally, the variation should be kept at a minimum as it is impossible to segment each subject's heartbeat in the same way. As it stands, a variation of 8% is too large. Note that when hand-segmenting data set *a*, the accuracy was 68%. Further investigations need to be conducted to determine the nature of this accuracy variation for different adjustment factors, where the behaviour was also apparent in the sensitivity and specificity plots in Figure 9.25. We can investigate utilizing the simultaneous ECG recording (shown in Figure 9.7) for a better defined segmentation. Perhaps segmentation is not necessary, and algorithms can be designed that bypass this pre-processing technique to achieve the best results.

9.5.2 SIGNAL INTEGRITY

The same test in Section 9.5.1 was repeated using 40 normal and 40 abnormal heartbeats from data set *e* in Ref. [12]. Performing the segmentation test on different heartbeat recordings indicates how the performance varies over different environment. We can observe from Figure 9.25 that for all adjustment factors, the accuracy, sensitivity, and specificity for data set *e* were consistently higher than those of data set *a*. As both data sets were acquired in different environment by separate research groups, the amount of noise corrupting the signals was not the same. From the results, we conclude that data set *e* contained less noise in the measurements, due to the accuracy varying between 85% and 90% for each adjustment factor, as opposed to 64% and 71% for data set *a*. This clearly indicates the importance of the data acquisition process. Although both sets of heartbeat signals appeared to be the same visually when filtering out noisy measurements, it is clear that critical features were corrupted in data set *a*, which negatively affected the classification process.

For data set *e*, the maximum accuracy again corresponds to an adjustment factor 0.046. However, the variation was not as large as that in data set *a*, with a minimum accuracy of 84.8% and maximum accuracy of 88.9%. The smaller variation in accuracy can also be explained by the fact that data set *e* was cleaner. Nevertheless, this 4.1% difference still has the ability to degrade the specification goal, despite having cleaner signals.

From the results in Section 9.4, we expect that due to its capability to suppress the background noise, the digital stethoscope with a BNM can obtain measurements with high signal integrity. Due to this reason, it is expected that the accuracy, sensitivity, and specificity of the PCG system in Section 9.2 can be higher than those of data set *e*, fulfilling the specification of exceeding 80%.

9.5.3 DATA SET SIZE

The previous tests consist of 80 subjects in total. In this section, we investigate how the data set size affects the classification accuracy, where it is hypothesized that larger data sets will produce better results. Data sets *a* and *e* were combined to form larger training sets, with 160 and 240 subjects separately trained. Each set had equal numbers of normal and abnormal heartbeats, as well as equal measurements from both data sets. There is not enough clean data to extend the testing beyond 240 subjects. The adjustment factor of 0.046 was used on each data set for heart cycle segmentation, as this factor had the highest performance in the previous test according to Section 9.5.1.

Table 9.2 displays the accuracy, sensitivity, and specificity for different data set sizes made by combining heartbeats from data sets *a* and *e*. It can be seen from Table 9.2 that when increasing the data set size, the accuracy and specificity increased by 1.2% and 3.8%, respectively. However, the sensitivity decreased by 1.4%. More tests with different data sets and sizes need to be investigated to truly understand

---

TABLE 9.2
CAD Classification Results Obtained from Different Data Set Sizes

Data Set Size	Accuracy (%)	Sensitivity (%)	Specificity (%)
160	73.6	72.5	74.7
240	74.8	71.1	78.5

TABLE 9.3
CAD Classification Accuracy of Two Different Neural Networks

Neural Network	Data Set a (%)	Data Set b (%)
SVM	71.8	88.9
CNN	73.8	95.0

the nature between the results and data set size. In saying this, it was encouraging to observe the accuracy and specificity increase.

From the previous tests, the highest accuracy achieved in data sets *a* and *e* was 71.8% and 88.9%, respectively. When joining these data sets together the accuracy was not at the midpoint, instead achieving just 73.6%. Thus, the lower quality data negatively affected the entire data set. This again indicates the importance of signal integrity. Even if most of the data is of high quality, a certain number of poorer measurements arising from body movements or unexpected noise have the potential to disproportionately compromise the results.

9.5.4 Neural Network Structure

Lastly, we study the impact of the neural network structure on the performance of a CAD classifier by comparing the SVM-based classifier [24] and the convolutional neural network (CNN) driven classifier in Ref. [26], which contains two convolution layers followed by a multilayer perceptron (MLP) network. An adjustment factor of 0.046 was adopted for the SVM approach, which has the highest accuracy. No adjustment was used for the CNN-based classifier, i.e., the HSMM-regression-based segmentation algorithm was directly applied. The accuracy results of both approaches are shown in Table 9.3. We can see that for both data sets *a* and *e*, the CNN-based neural network achieved higher accuracy. This is mainly because it has a larger number of layers and parameters than the SVM to adapt to the features of the heartbeat signals, rendering a higher classification accuracy. However, we note that a deeper neural network in general needs a larger amount of labelled training data. Interestingly, we also observe that the improvement in data set *e* is 6%, while only 2% for data set *a*. This suggests that by using deep neural network to cleaner data set, one can achieve a higher accuracy.

9.6 CONCLUSIONS

In this chapter, we discussed a CAD pre-screening technique using PCG. A multichannel PCG and ECG measurement device was presented. For the topic of adaptive noise cancellation in PCG signals, we showed in this chapter that the coherence function can be used to predict the performance of an NLMS-based method in suppressing the background noise. By applying existing machine learning-based CAD classification algorithms to heart sound recordings in the open database of the 2016 PhysioNet/Computing in Cardiology Challenge, we demonstrated that the integrity of the heart sound signal has a significant impact on the accuracy of classification.

Another discovery we showed in the chapter is that the identification of the correct starting point of heartbeat cycle affects the classification results, which raises an interesting question if segmentation of the PCG signal is absolutely necessary in machine learning-based classification methods.

REFERENCES

1. G. A. Roth et al., "Global burden of cardiovascular diseases and risk factors, 1990-2019: Update from the GBD 2019 study," *Journal of the American College of Cardiology*, vol. 76, no. 25, pp. 2982–3021, 2020.
2. A. McLellan and D. Prior, "Cardiac stress testing stress electrocardiography and stress echocardiography," *Australian Family Physician*, vol. 41, no. 3, pp. 119–122, 2012. [Online]. Available: https://www.racgp.org.au/afp/2012/march/cardiac-stress-testing/.
3. A. K. Kumar and G. Saha, "Interpretation of heart sound signal through automated artifact-free segmentation," *Heart Research Open Journal*, vol. 2, no. 1, pp. 25–34, 2015.
4. I. Cathers, "Neural network assisted cardiac auscultation," *Artificial Intelligence in Medicine*, vol. 7, no. 1, pp. 53–66, 1995.
5. T. R. Reed, N. E. Reed, and P. Fritzson, "Heart sound analysis for symptom detection and computer-aided diagnosis," *Simulation Modelling Practice and Theory*, vol. 12, no. 2, pp. 129–146, 2004.
6. S. Choi, "Detection of valvular heart disorders using wavelet packet decomposition and support vector machine," *Expert Systems with Applications*, vol. 35, no. 4, pp. 1679–1687, 2008.
7. S. Ari, K. Hembram, and G. Saha, *"Detection of cardiac abnormality from PCG signal using LMS based least square SVM classifier,"* *Expert Systems with Applications*, vol. 37, no. 12, pp. 8019–8026, 2010.
8. A. N. Makaryus, J. N. Makaryus, A. Figgatt, D. Mulholland, H. Kushner, J. L. Semmlow, J. Mieres, and A. J. Taylor, "Utility of an advanced digital electronic stethoscope in the diagnosis of coronary artery disease compared with coronary computed tomographic angiography," *The American Journal of Cardiology*, vol. 111, no. 6, pp. 786–792, 2013.
9. S. Schmidt, C. Holst-Hansen, J. Hansen, E. Toft, and J. Struijk, *"Acoustic features for the identification of coronary artery disease,"* IEEE Transactions on Biomedical Engineering, vol. 62, no. 11, pp. 2611–2619, 2015.
10. H. Li, X. Wang, C. Liu, Y. Wang, P. Li, H. Tang, L. Yao, and H. Zhang, "Dual-input neural network integrating feature extraction and deep learning for coronary artery disease detection using electrocardiogram and phonocardiogram," *IEEE Access*, vol. 7, pp. 146457–146469, 2019.
11. J. Oliveira et al., "The CirCor DigiScope dataset: From murmur detection to murmur classification," *IEEE Journal of Biomedical and Health Informatics*, vol. 26, no. 6, pp. 2524–2535, 2022.
12. G. D. Clifford, C. Liu, B. Moody, D. Springer, I. Silva, Q. Li, and R. G. Mark, "Classification of normal/abnormal heart sound recordings: The PhysioNet/Computing in cardiology challenge 2016," In *Proceedings of Computing in Cardiology Conference*, Vancouver, BC, Canada, Septemper 11–14, 2016.
13. George B. Moody PhysioNet Challenge, https://moody-challenge.physionet.org/2022/.
14. A. Kazemnejad, P. Gordany, and R. Sameni, "An open-access simultaneous electrocardiogram and phonocardiogram database," bioRxiv preprint. [Online]. Available: https://doi.org/10.1101/2021.05.17.444563.
15. Ticking Heart Pty Ltd, https://www.tickingheart.com.

16. P. Samanta, A. Pathak, K. Mandana, and G. Saha, "Classification of coronary artery diseased and normal subjects using multi-channel phonocardiogram signal," *Biocybernetics and Biomedical Engineering*, vol. 39, no. 2, pp. 426–443, 2019.

17. D. Della Giustina, M. Riva, F. Belloni, and M. Malcangi, "Embedding a multichannel environmental noise cancellation algorithm into an electronic stethoscope," *International Journal of Circuits, Systems and Signal Processing*, vol. 5, no. 2, pp. 184–191, 2011.

18. B. Widrow, J. R. Glover, J. M. McCool, J. Kaunitz, C. S. Williams, R. H. Hearn, J. R. Zeidler, E. Dong, and R. C. Goodlin, "Adaptive noise cancelling: Principles and applications," *Proceedings of the IEEE*, vol. 63, no. 12, pp. 1692–1716, 1975.

19. J. A. Zhang, N. Murata, Y. Maeno, P. N. Samarasinghe, T. D. Abhayapala, and Y. Mitsufuji, "Coherence-based performance analysis on noise reduction in multichannel active noise control systems," *The Journal of the Acoustical Society of America*, vol. 148, no. 3, pp. 1519–1528, 2020.

20. J. S. Bendat and A. G. Piersol, *Random Data: Analysis and Measurement Procedures*. John Wiley & Sons: Hoboken, NJ, 2011.

21. M. Fynn, S. Nordholm, and Y. Rong, "Coherence function and adaptive noise cancellation performance of an acoustic sensor system for use in detecting coronary artery disease," *MDPI Sensors*, vol. 22, no. 17, 6591, 2022.

22. Fireface UCX, https://www.rme-audio.de/fireface-ucx.html.

23. S. Dixit and D. Nagaria, "LMS adaptive filters for noise cancellation: A review". *International Journal of Electrical and Computer Engineering*, vol. 7, 2520, 2017.

24. A. Pathak, P. Samanta, K. Mandana, and G. Saha, "Detection of coronary artery atherosclerotic disease using novel features from synchrosqueezing transform of phonocardiogram," *Biomedical Signal Processing and Control*, vol. 62, 102055, 2020.

25. D. B. Springer, L. Tarassenko, and G. D. Clifford, "Logistic regression-HSMM-based heart sound segmentation," *IEEE Transactions on Biomedical Engineering*, vol. 63, no. 4, pp. 822–832, 2015.

26. C. Potes, S. Parvaneh, A. Rahman, and B. Conroy, "Ensemble of feature-based and deep learning-based classifiers for detection of abnormal heart sounds," In *Proceedings of Computing in Cardiology Conference*, Vancouver, BC, Canada, Septemper 11–14, 2016.

10 Exploring the Feasibility of Estimating the Carotid-to-Femoral Pulse Wave Velocity Using Machine Learning Algorithms

Mohamed A. Bahloul
Alfaisal University

Juan M. Vargas
King Abdullah University of Science and Technology

Zehor Belkhatir
University of Southampton

Taous-Meriem Laleg-Kirati
King Abdullah University of Science and Technology
National Institute for Research in Digital
Science and Technology (INRIA)

10.1 INTRODUCTION

Assessment of cardiovascular disease (CVD) risk typically includes evaluating different biomarkers, including arterial stiffness (AS), which is considered a high-risk marker that affects heart and vascular physiology [1]. Over the years, various techniques for evaluating AS have been explored, with pulse wave velocity (PWV) being recognized as an efficient tool to assess vascular stiffness [2,3]. Carotid to femoral PWV (cf-PWV) is known as the 'Gold standard' measurement of AS and has been shown to correlate strongly with vascular aging, hypertension severity, and atherosclerosis [4,5]. The non-invasive measurement process of cf-PWV is considered intrusive and operator-dependent, requiring expertise and prone to inaccuracies and error. Further discussion about the precision of cf-PWV measurements, different

DOI: 10.1201/9781003346678-10

modalities used in the clinical settings, and the main challenges and strengths of each technique can be found in Ref. [6]. Clinicians urge the need for an easy-to-use, non-invasive, and smart cf-PWV measurement tool that can be implemented in daily clinical routine [7].

Over the last decade, the world of artificial intelligence and machine learning (ML) has seen exponential maturation [8] and has been widely adopted in different biomedical applications, namely as a prediction tool for CVDs and to assist in their assessment and prevention [9–11]. Indeed, several studies have introduced and demonstrated the feasibility of ML in the analysis of the onset of certain disorders and diseases and in constructing an insightful relationship between symptoms and underlying medical conditions [12,13]. ML offers new ways and techniques to analyze and explore the different types of health data, which are usually considered to be big data and diverse due to variety of major resources and modalities, as well as structured or non-structured and dynamic. Various ML algorithms and neural network-based architectures have opened up novel guidelines for designing unique paradigms and predictive procedures for clinical services. In accordance with this path, over the last 5 years, some research groups have explored the application of ML to predict cf-PWV. For instance, the authors in Ref. [7] have developed an artificial intelligence tool to estimate cf-PWV using *intrinsic frequency* features based on uncalibrated carotid pressure waveform [14] combined with typical clinical variables. Bahloul et al. [15–17] introduced the use of a multi-layer perceptron-based cf-PWV estimation using timing and magnitude of the fiducial points-based features, which are extracted from the photoplethysmogram (PPG) signal and its first, second, and third time-derivative. Another investigation by Weiwei Jin et al. [18] has proposed two ML pipelines, namely the *Gaussian process regression* and *recurrent neural network* to estimate cf-PWV from the radial blood pressure waveform measured by applanation tonometry. The two proposed ML pipelines were fed by key features generated from the timing and magnitude of the fiducial points as well as heart rate. This chapter presents a conceptual framework and comparative study of applying well-known ML-based regression models to estimate cf-PWV from a single peripheral arterial pulse wave, namely the blood pressure or the PPG signal collected at the digital or radial, or brachial site. Figure 10.1 illustrates the main step of the adopted

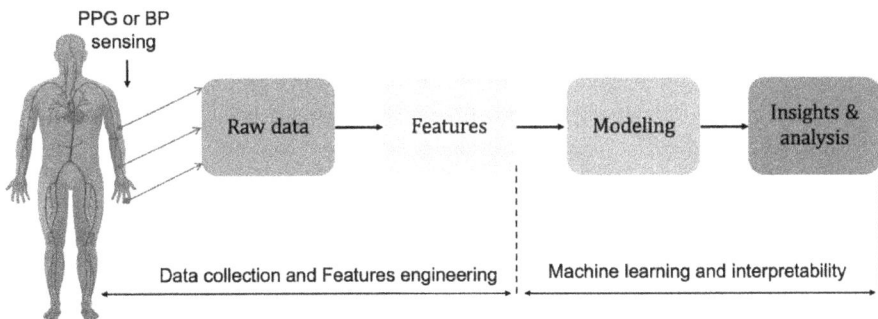

FIGURE 10.1   Framework illustration of machine learning-based regression models to estimate cf-PWV.

framework. In addition, the main contributions of this study can be summarized in the following points:

- **Features engineering:** In this work, we investigate the use of different domain knowledge to extract useful features from the time series data of blood pressure or PPG signals. We adopt the following three different types of features:
  1. **Time domain-based features:** Timing and magnitude of the fiducial points-based features extracted from the pulse wave signal and its first, second, and third time-derivative [15],
  2. **Frequency domain-based features:** Intrinsic Frequency method [14], has been applied to extract features from the pulse wave signal.
  3. **Semi-classical signal analysis (SCSA)-based features:** Features extracted using SCSA tools, see Ref. [19,20] and references therein.
- **Machine learning modeling:** Seven machine learning regression models have been investigated in this study. Each model has been trained and tested using the above features extracted from blood pressure (BP) and PPG signals from three different arterial sites: digital, radial, and brachial arteries. Each algorithm has been tested in noise-free and noisy cases (various noise levels have been tested as well).
- **Insight and analysis:** A comparative discussion has been conducted analyzing the performance versus complexity for each model and features-based domain of knowledge.

10.2 PULSE WAVE FEATURE EXTRACTION AND SIGNAL PROCESSING TECHNIQUES

This section presents the main signal processing-based techniques to extract features from the peripheral pulse waveforms, namely the blood pressure and PPG signals. The adopted methods use as input the time series data of BP or PPG and generate features from different domain knowledge nature: time domain, frequency domain, combined frequency and time domains, and the discrete spectrum of the Schrödinger operator domain based on the semi-classical signal processing method.

10.2.1 TIME DOMAIN-BASED FEATURES

The temporal features were extracted from the peripheral pulse wave (BP or PPG measured at the digital, radial, or brachial level) (ω) and its derivative, namely the first time-derivative (ω'), second-time-derivative ω'' and third-derivative ω'''. ω denotes the time-varying blood pressure or photoplethysmography signal. The key features are calculated based on the timing and magnitude of the fiducial points as well as the heart rate detected on the pulse wave and its derivative. Figure 10.2 shows an example of a PPG waveform simulated at the level of the radial artery and its derivatives. The red circles on these curves mark the fiducial points:

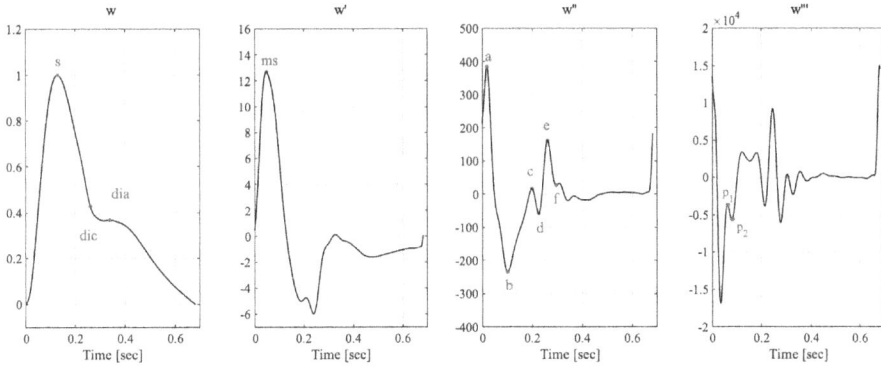

FIGURE 10.2 Detection of fiducial points on the PPG signal (ω) and its first derivative (ω'), second derivative (ω'') and third derivative (ω''').

- Three fiducial points on the original pulse wave, ω that are: (s) the systolic peak, (dic) the dicrotic notch and (dia) the diastolic peak.
- One fiducial point, (ms), on the first time-derivative (ω') of the pulse wave that corresponds to the maximum upslope [29].
- Five fiducial points $(a, b, c, d,$ and $e)$ on the second time-derivative (ω'') of the pulse wave [30].
- Two fiducial points $(p_1$ and $p_2)$ from the third derivative (ω''') of the pulse wave that denote the early and late systolic components [31].

To process the pulse wave signal and calculate its derivative in this work, we use the well-known and publicly available *Pulse Wave Analysis Algorithm*.[1] The detailed instruction and the time-derivative-based method adopted in this algorithm can be found in Refs. [29,32]. Table 10.1 shows the evaluated feature based on the detected fiducial points and the corresponding formula. Many studies have revealed the physiological insight and interpretability of each extracted feature that can be reviewed in the corresponding reference listed in Table 10.1 along with the key feature.

10.2.2 FREQUENCY DOMAIN-BASED FEATURES

10.2.2.1 Intrinsic Frequency (IF) Method

The arterial pulse wave consists of systolic and diastolic phases that represent the opening and closing of the aortic valve. The frequency analysis of the pulse wave provides important information about the physiological condition of the cardiac system. The *Intrinsic frequency* (IF) method is a technique used to analyze the pulse wave, which involves two dominant frequency components before and after the dicrotic notch. This method can be applied to un-calibrated signals like the PPG collected by a smartphone-integrated optical sensor [7,14,33]. For a pulse wave signal, the IF technique can be formulated as follows:

TABLE 10.1

Features Detected from an Example of PPG Pulse Waves

Pulse	Features	Formula
	ΔT, [21]	$t(\text{dia}) - t(\text{s})$
PPG, ω	CT, [22]	$t(\text{s})$
	prop_s, [23]	$t(\text{s})/T$
	t_{sys}, [24]	$t(\text{dic})$
	t_{dia}, [24]	$T - t(\text{dic})$
	t_{ratio}, [24] $\text{prop}_{\Delta T}$, [24]	$t(\text{s})/t(\text{dic})$
	$t_{p1-\text{dia}}$, [24]	$(t(\text{dia}) - t(\text{s}))/T$
	$t_{p2-\text{dia}}$,[24]	$t(\text{dia}) - t(p_1)$
	IPR, [24]	$t(\text{dia}) - t(p_2)$
	AI, [25]	$60/T$
	RI, [21]	$(\omega(p_2) - \omega(p_1))/\omega(\text{s})$
	RI_{p1}, [24]	$\omega(\text{dia})/\omega(\text{s})$
	RI_{p2},[24]	$\omega(\text{dia})/\omega(p_1)$
	Ratio_{p2-p1}, [24]	$\omega(\text{dia})/\omega(p_2)$
	A1, [24]	$\omega(p_2)/\omega(p_1)$
	A2, [24]	Area from pulse foot to dicrotic notch
	IPA, [24]	Area from dicrotic notch to pulse end
		A2/A1
PPG', ω'	ms, [22]	$\omega''(\text{ms})/\omega''(p_1)$
	b/a, [25]	$\omega''(b)/\omega''(a)$
PPG", ω''	c/a, [25]	$\omega''(c)/\omega''(a)$
	d/a, [25]	$\omega''(d)/a''(a)$
	e/a, [25]	$\omega''(e)/\omega''(a)$
	AGI, [25]	$\left(\omega''(b) - \omega''(c) - \omega''(d) - \omega''(e)\right)/\omega''(a)$
	AGI_{int}, [26]	$\left(\omega''(b) - \omega''(e)\right)/\omega''(a)$
	AGI_{mod}, [27]	$\left(\omega''(b) - \omega''(c) - \omega''(d)\right)/\omega''(a)$
	t_{b-c}, [24]	$t(c) - t(b)$
	t_{b-d}, [24]	$t(d) - t(b)$
	Slope_{b-c}, [24]	d/dt of straight line between b and c,
	Slope_{b-d}, [24]	normalized by a
		d/dt of straight line between b and d,
		normalized by a
Combined (ω'' and ω)	IPAID, [24]	$(\text{A2/A1}) + d/a$
	k, [28]	$\omega''(\text{s})/\left(\omega(\text{s}) - \omega(\text{ms})\right)/\omega(\text{s})$
		$\omega''(\text{s})/(\omega(\text{s}) - \omega(\text{ms}))/\omega(\text{s})$

Definitions: t, time since pulse onset (beginning of systolic upslope); ω, ω', ω'', ω''', PPG signal and derivatives; T, duration of the cardiac cycle (s).

$$S\left(a_i, b_i, \bar{p}, \omega_i, t\right) = \left(a_1 \cos(\omega_1 t) + b_1 \sin(\omega_1 t) + \bar{p}\right) 1_{[0, T_0)}(t)$$
$$+ \left(a_2 \cos(\omega_2 t) + b_2 \sin(\omega_2 t) + \bar{p}\right) 1_{[T_0, T)}(t),$$

(10.1)

where a_1, b_1, a_2, and b_2 are the envelopes of the IF model fit. ω_1 and ω_2 correspond to the Intrinsic Frequencies of the pulse wave, and $\bar{p}$ denotes the mean value of the time signal for the period $[0, T)$. An arterial pulse wave is usually periodic and continues in time. Accordingly, a continuity condition at T_0 (the time of the dicrotic notch) and periodicity at T (the duration of the cardiac cycle) is ensured. The indicator $1_{[x,y)}(t)$ function is defined as:

$$1_{[x,y)}(t) = \begin{cases} 1 & x \le t < y \\ 0 & \text{else} \end{cases}. \tag{10.2}$$

The goal of the IF model (10.1) is to extract a fit, called Intrinsic Mode Function (IMF), that carries most of the energy (information) from a pulse waveform $s(t)$ in one period. The latter is performed by solving the following optimization problem.

$$\min_{a_i, b_i, w_i, \bar{p}} \left\| s(t) - S\left(a_i, b_i, \bar{p}, \omega_i, t\right) \right\|_2^2 \tag{10.3}$$

$$S.t. \begin{cases} a_1 \cos\left(\omega_1 T_0\right) + b_1 \sin\left(\omega_1 T_0\right) = a_2 \cos\left(\omega_2 T_0\right) + b_1 \sin\left(\omega_2 T_0\right) \\ a_1 = a_2 \cos\left(\omega_2 T_0\right) + b_1 \sin\left(\omega_2 T_0\right) \end{cases} \tag{10.4}$$

Here, $\| \ \|_2$ is the $L_{2\text{-norm}}$ defined on $[0, T)$. One assumption in this optimization is that the extracted IMF is continuous at the dicrotic notch time T_0 (the first condition in Equation (10.4)). The other assumption is that the extracted IMF is periodic (the second condition in Equation (10.4)). The details of the method of the solution of the optimization problem mentioned by Equations (10.3) and (10.4) can be found in Ref. [14].

10.2.2.2 Intrinsic Frequency-Based Features

Different parameters-based features can be extracted from the solutions of Equations (10.3) and (10.4). These parameters are both of mathematical and physiological importance. For instance, ω_1 and ω_2 can be normalized with respect to the systolic and diastolic periods, T_0 and $(T - T_0)$, respectively. Also, ω_1 and ω_2 can be normalized with respect to the whole cardiac cycle T. The used set of features is defined as

$$F_{\text{IF}} = \left\{ \omega_1, \omega_2, \bar{\omega}_1 = \omega_1 T_0, \ \bar{\omega}_2 = \omega_2 (T - T_0), \ \omega_{1c} = \omega_1 \sqrt{T_0}, \right.$$

$$\omega_{2c} = \omega_2 T^2, \ \omega_{1n} = \omega_1 T, \ \omega_{2n} = \omega_2 T, \ \rho = \frac{s(T_0) - \min\left(s(t)\right)}{\max\left(s(t)\right) - \min\left(s(t)\right)}, \tag{10.5}$$

$$\left. E_r = \frac{\sqrt{a_1^2 + b_1^2}}{\sqrt{a_2^2 + b_2^2}}, \ C_r = \frac{\bar{p} - \min\left(s(t)\right)}{\max\left(s(t)\right) - \min\left(s(t)\right)}, \ T, \ \frac{1}{T - T_0}, \ \frac{1}{T_0} \right\}.$$

10.2.3 SEMI CLASSICAL SIGNAL ANALYSIS-BASED FEATURES

10.2.3.1 Semi Classical Signal Analysis Background

The Semi Classical Signal Analysis (SCSA) is a signal-dependent adaptive decomposition method, inspired by the *Solitons* propagation model of the blood pressure waveform [34–36]. The SCSA is well adapted for representing, denoising, and preprocessing of pulse-shaped signals [37,38]. It has also been used for signal characterization, where SCSA parameters have been used as signals' features for feeding machine learning algorithms [19]. One of the main advantages of SCSA is that it can be used as an instructive features extraction generator while denoising the explored signal [39]. The main idea of the SCSA is to decompose the signal into the set of the squared eigenfunctions of the Schrödinger operator.

The generalized SCSA representation is defined by:

$$V_{n,h,\gamma}(z) = \frac{h^n}{L_{n,\gamma}^{cl}} \sum_{m=1}^{M_h} \left[\left(-\kappa_{mh} \right)^\gamma \psi_{mh}^2(z) \right]^{\frac{1}{\gamma + \frac{n}{2}}}, \tag{10.6}$$

where n is the signal dimension ($n = 1$ for a signal and $n = 2$ for an image), $z \in \mathbb{R}^n$, $\gamma \in \mathbb{R}_+$, $h \in \mathbb{R}_+^*$. The latter is known as the semi-classical parameter. κ_{mh} are the negative eigenvalues ($\kappa_{1h} > \cdots > \kappa_{M_h}$), such as and $\{\psi_{1h}, \psi_{2h}, \ldots, \psi_{M_h}\}$ correspond to their associated L^2-normalized eigenfunctions ($m = 1, \ldots, M_h$ the number of eigenvalues), and $L_{n,\gamma}^{cl}$ is the suitable semi-classical constant defined as:

$$L_{n,\gamma}^{cl} = \frac{1}{\left(2\sqrt{\pi} \right)^n} \frac{\Gamma(\gamma + 1)}{\Gamma\left(\gamma + 1 + \frac{n}{2} \right)}, \tag{10.7}$$

where Γ in the Gamma function. This SCSA representation is based on the n-dimensional semi-classical Schrödinger operator associated with a potential $V_{n,h,\gamma}$:

$$H_{n,h}\left(V_{n,h,\gamma} \right)\psi = -h^2 \Delta \psi - V_{n,h,\gamma}\psi, \tag{10.8}$$

where Δ_n is the n-dimensional laplacian operator, defined as:

$$\Delta = \sum_{i=1}^n \frac{\partial^2}{\partial x_i^2}, \tag{10.9}$$

where x_i is the Cartesian coordinates.

Here, we focus on 1D-SCSA to extract features from peripheral pulse waves (BP or PPG). The 1D-SCSA, ($n = 1$ in Equation (10.6)), is defined as follows [40]:

Let $S(t)$ be a real positive signal to be analyzed, the 1D-SCSA representation $S_h(t)$ is defined as:

$$S_h(t) = 4h \sum_{m=1}^{M_h} k_{mh} \psi_{mh}^2(t), \quad t \in \mathbb{R}, \tag{10.10}$$

10.2.3.2   Features Based on 1D-SCSA

The SCSA method provides spectral parameters corresponding to the Schrödinger operator extracted from the original signal i.e., the eigenfunctions and their corresponding eigenvalues. These spectral parameters (eigenvalues) have been used to extract features from the 1D-SCSA providing morphological and physiological details [40,41]. In the following, we define the explored 1D-SCSA features. We refer to M_h to the number of eigenvalues used to reconstruct the pulse wave signal. The first features calculated were the three first invariants consisting in some momentum of the negative eigenvalues, defined as:

$$\left\{ \begin{aligned} \text{INV}_1 &= 4h \sum_{m=1}^{M_h} \kappa_{mh} \\ \text{INV}_2 &= \frac{16h}{3} \sum_{m=1}^{M_h} \kappa_{mh}^3 \\ \text{INV}_3 &= \frac{256h}{7} \sum_{m=1}^{M_h} \kappa_{mh}^7 \end{aligned} \right. \tag{10.11}$$

The above invariant parameters have been used by Refs. [42,43], showing relevant performance in predicting arterial BP using PPG signal.

Furthermore, based on the properties of the eigenfunctions described by Refs. [40,41] where the first eigenvalues have fewer oscillations and approximate the general profile of the signal (or images), we use the first three eigenvalue parameters as follows:

$$\left\{ \begin{aligned} E_i &= \kappa_{ih} & i &= [1,2,3] \\ E_i^2 &= \kappa_{ih}^2 & i &= [1,2,3] \end{aligned} \right. \tag{10.12}$$

To capture some features related to the eigenfunctions, we calculated the ratio between the eigenvalue of the first oscillatory eigenfunction and h (Equation 10.13) and the ratio between the median of the eigenvalues related to the eigenfunctions with an oscillatory behavior and h (Equation 10.14, [44]), as follows:

$$R_h = \frac{\kappa_{1h}}{h} \tag{10.13}$$

$$MR_h = \frac{\text{Median}(\kappa_{mh})}{h} \tag{10.14}$$

Finally, we use the number of eigenvalues M_h since these values give valuable information about the shape of the signal [44] and some mathematical moments such as the mean, the median, and the standard derivation of the negative eigenvalues of the signal due to the localization properties of the SCSA that provides shape information [44], helping to identify changes in signal morphology.

The result features matrix were:

$$F_{\text{SCSA}} = \left\{ \text{INV}_1, \text{INV}_2, \text{INV}_3, E_1, E_2, E_3, E_1^2, E_2^2, \right.$$
$$\left. E_3^2, R_h, MR_h, \text{Mean}(\kappa_{mh}), \text{Median}(\kappa_{mh}), \text{std}(\kappa_{mh}) \right\}$$
(10.15)

10.3   MACHINE LEARNING-BASED REGRESSION MODELS

Different machine-learning regression models have been investigated in this study. In the following, we briefly define each method.

1. **Multiple Linear Regression (LR)** attempts to model the relationship between two or more features and a response by fitting a linear equation to observed data.
2. **Gaussian Regression (GR)** is a nonparametric, Bayesian approach to regression that is making waves in the area of machine learning. GPR has several benefits, including working well on small datasets and having the ability to provide uncertainty measurements on the predictions.
3. **Random Forest Regression (RF)** is a supervised learning algorithm that uses an ensemble learning method for regression. The ensemble learning method is a technique that combines predictions from multiple machine learning algorithms to make a more accurate prediction than a single model.
4. **Gradient Boosting (GB)** is a machine learning technique used in regression. This model gives a prediction in the form of an ensemble of weak prediction models, which are typically decision trees. Boosting is a method of converting weak learners into strong learners. In boosting, each new tree is a fit on a modified version of the original data set. The main idea of this algorithm is to improve upon the predictions of the first tree.
5. **Support Vector Regression (SVR)** is a non-parametric supervised machine learning algorithm where the objective is to define a margin of error in a plane or hyperplane where the values can rely on the prediction. This error margin will change depending on the complexity of the model, which could be linear, radial, or polynomial.
6. **Multilayer Perceptron (MLP)** is a widely used and robust supervised machine learning algorithm based on a feedforward artificial neural network where each node, apart from the input nodes, has a nonlinear activation function that provides the non-linearity of the model, helping to obtain very complex regression.
7. **Long Short-Term Memory (LSTM)** cells are an RNN based on deep learning architecture that is suited for time series data. These models can learn long-term sequences due to their set of different gates with the capability to add or remove information by regulating the flow of information into and out of the cell in long sequences. This allows one to forget or remember a set of key features of the data to predict continuous values.

10.4   MATERIAL AND METHODS

In this study, an in-silico hemodynamic database was used to validate the proposed approach since real hemodynamic data was not available. The publicly available database[2] contains simulated pulse waves at different arterial locations and has been used in various pre-clinical assessment studies and hemodynamic analyzer algorithms. The study used photoplethysmogram and blood pressure waveforms from the radial, brachial, and digital arteries to estimate cf-PWV and regression models were trained and tested using these features.

In this study, the MLP and LSTM regression models were tested and fed with the samples of once cardiac cycle pulse wave (BP and PPG) in addition to the above-prescribed features. To test the robustness of the approach against noise, different intensities of high-frequency *Gaussian* white noise were generated and added to the original pulse wave waveforms. Each algorithm was tested in noise-free and noisy cases, with noise levels of 2%, 5%, and 10%. The features of all the subjects were divided into two sets: the training set (70% of the total dataset) and the testing set (30% of the total dataset). The training dataset-based features were then fed to the regression algorithms. A randomized search technique for hyper-parameter tuning was applied in the training phase. In addition, a fivefold cross-validation method was used to avoid over-fitting during the training and show the model's generalization capability. The estimation performance is measured by the mean absolute percentage error (MAPE) of the testing set as follows:

$$\text{MAPE}[\%] = \frac{1}{N}\sum_{n=1}^{N}\frac{\left|\text{cf-PWV}_{\text{real}}^{n} - \text{cf-PWV}_{\text{predicted}}^{n}\right|}{\text{cf-PWV}_{\text{predicted}}^{n}}, \qquad (10.16)$$

where N is the size of the training set. cf-PWV$_{\text{real}}^{n}$ and cf-PWV$_{\text{real}}^{n}$ are the real and estimated cf-PWV, respectively.

In addition, for ease of visualization of the goodness of the proposed model, we assessed the sum of the squared differences (SSE), R-square (R^2) which is the square of the correlation and, the root mean square error (RMSE) between the real values and the predicted response values, as follows:

$$\left\{ \begin{array}{l} \text{SSE} = \sum_{n=1}^{N}\left(\text{cf-PWV}_{\text{real}}^{n} - \text{cf-PWV}_{\text{predicted}}^{n}\right)^2 \\[2mm] R^2 = 1 - \dfrac{\text{SSE}}{\sum_{n=1}^{N}\left(\text{cf-PWV}_{\text{real}}^{n} - \mu\left(\text{cf-PWV}_{\text{real}}\right)\right)^2}, \\[2mm] \text{RMSE} = \sqrt{\dfrac{\text{SSE}}{N}}, \end{array} \right. \qquad (10.17)$$

where μ is a function that evaluates the mean of cf-PWV$_{\text{real}}$ over N subjects.

10.5 RESULTS AND DISCUSSION

Various machine learning pipelines were utilized in this study, using features extracted from peripheral pulse waveforms, namely blood BP and PPG signals collected at the digital, radial, and brachial arteries. The explored ML models varied in complexity, and different modalities were investigated to generate informative input features while reducing the dimensionality of the time series peripheral PWs (BP and PPG). These modalities included feature extraction based on the timings and magnitudes of fiducial points identified on the waveform and its first, second, and third derivatives (referred to as FP), features generated using the discrete spectrum of the Schrodinger operator (SCSA), features generated using the IF method (IF), and the IF-based features added to some routine clinical variables such as age, heart rate (Hr), brachial systolic blood pressure (SBP_b), brachial diastolic blood pressure (DBP_b), brachial pulse blood pressure (PP_b), and brachial mean blood pressure (MBP_b) (referred to as IF_C).

10.5.1 MLP-Based cf-PWV Estimation Using PPG Signal

This section focuses specifically on the MLP architecture fed with FP-based peripheral PPG signal.

In the following, we detail the different steps adopted to generate the input features for the developed model.

> **Step 1:** Detection of the fiducial points from the PPG and its derivatives signals (first derivative: PPG′, second derivative: PPG″, and third derivative: PPG‴). Subsequently, generating features based on FP techniques. In addition to these features, some routine clinical variables such as age, heart rate (Hr), the brachial systolic blood pressure (SBP_b), brachial diastolic blood pressure (DBP_b), brachial pulse blood pressure (PP_b) and brachial mean blood pressure (MBP_b) were also used as extra features.
>
> **Step 2:** Feature selection is a crucial step in machine learning model development as not all features may impact the output variable estimation, and irrelevant features can reduce performance and increase complexity. We used the filter method to select the most relevant features for cf-PWV estimation based on Pearson correlation with the output variable. Only features with correlations above 0.5 (absolute value) were selected, resulting in 18 features from the initial 52 (All features).
>
> **Step 3**: Regression models setting; The extracted and selected features are fed to different Multi-layer Perceptron (MLP) or Neural Network (NN) models. The NN models have two hidden layers with a '*ReLu*' activation function for each hidden layer. The used model uses the '*lbfgs*' solver, which is an optimizer in the family of quasi-Newton methods. The optimal L_2 penalty parameter '*alpha*' is set to 0.001. We have examined different architectures in this work by varying the number of neurons per hidden layer from 1 to 1,000. For this regression experiment, half of the randomly shuffled data is used for training, and the remaining half is used for testing.

Figure 10.3 shows the estimation performance evaluated by the mean absolute percentage error (MAPE) (top row) and R-square (R^2) (Bottom row) versus the number of neurons per hidden layer. These performances have been evaluated in three different arterial distal sites, mainly the digital, radial, and brachial arteries. Besides, the neural network has been fed either with all the features listed in Table 10.1 along with routine clinical variables such as age, heart rate (Hr), the brachial systolic blood pressure (SBP$_b$), brachial diastolic blood pressure (DBP$_b$), brachial pulse blood pressure (PP$_b$), and brachial mean blood pressure (MBP$_b$) or with selected features as described in step 3 in the feature extraction from PPG signal subsection. It is clear from these results that although using only selected features, the number of inputs has been reduced to 19 from 52 features, the estimation performances are better. Overall, in both cases (all features or selected features) using two hidden layers and independently of the number of neurons per layer, the prediction performances are acceptable. The maximum achieved mean absolute percentage error is around 4%, and the square of the correlation is generally greater than 0.9. Using only selected features, we could reach good estimation performances with an R^2 around 0.95 and MAPE less than 2.22% based on features extracted from PPG at the brachial artery level, an R^2 around 0.98 and MAPE less than 1.71% based on features extracted from PPG at the radial artery level and R^2 around 0.97 and MAPE less than 1.88% based on features extracted from PPG at the digital artery level.

Table 10.2 reports the estimation performance-based criterion, {MAPE, SSE, R^2, and RMSE}, of the proposed model, using all the features based on PPG signals

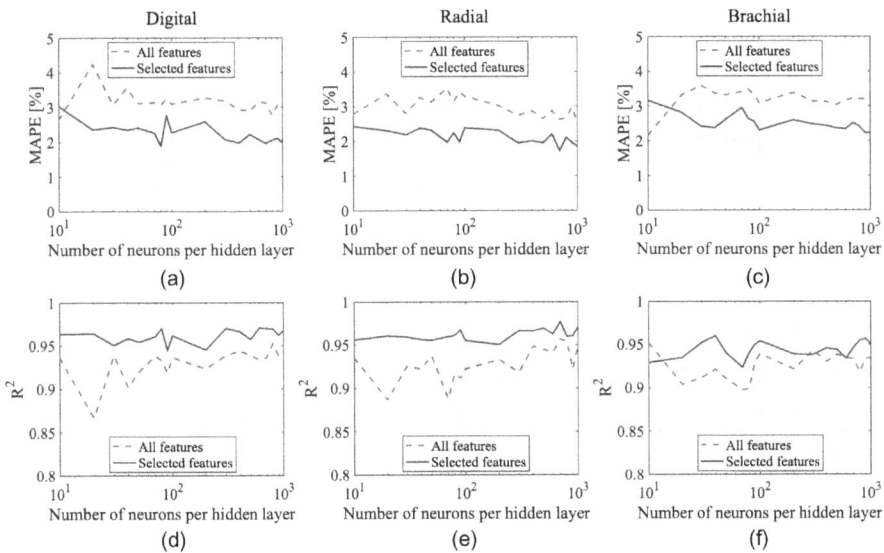

FIGURE 10.3 Estimation performance evaluated by the mean absolute percentage error (MAPE) (a–c) and R-square (R^2) which is the square of the correlation between the real values and the predicted response values (d–f). In this simulation, we varied the number of neurons per hidden layer and evaluated the performance of estimating the carotid-femoral PWV at the digital, radial, and brachial sites.

TABLE 10.2

Estimation Performance Is Evaluated as R-Square (R^2), Which Is the Square of the Correlation and the Root Mean Square Error (RMSE) between the Real Values and the Predicted Response Values

Arterial Site	N_h	N_n	R^2	RMSE
Brachial	2	100	0.94	0.50
Radial	2	100	0.96	0.40
Digital	2	100	0.96	0.39

N_h and N_n correspond to the number of hidden layers and neurons per layer, respectively.

collected at different measurement sites of the arterial network: the digital, radial, and brachial arteries. Overall, the model performs properly for the three sites and shows acceptable capabilities using only two hidden layers with 100 neurons. The best goodness of prediction, on the whole, validation dataset, was obtained at the level of the radial artery with a MAPE value equal to 1.94%. Besides, this criterion was marginally larger for the digital and brachial sites, but it does not exceed 2.5%. In Figure 10.4, we show, in the first row, the $\left(\text{cf-PWV}_{predicted}\right)$ versus $\left(\text{cf-PWV}_{real}\right)$ of the validation dataset for the three arterial sites. The square of the correlation, R^2, between the tested data and the estimated response is around 0.96 for the radial and digital site and marginally smaller for the brachial site. The second row of Figure 10.4

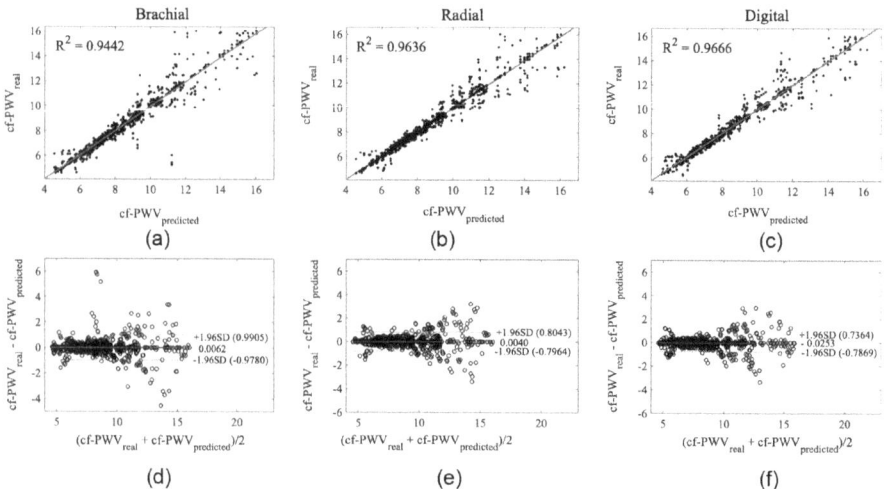

FIGURE 10.4 cf–PWV estimated performance on the testing set using features-based PPG waveforms and their derivatives collected at the level of brachial, radial, and digital arteries. The panel (a–c) represents the plots of the predicted cf–PWV versus the real tested cf–PWV. The panel (d–f) shows the cf–PWV Bland-Altman plots along with the limits of agreement and the mean difference. The unit of all the axes is (m/s). We can observe from these plots that the more estimation error corresponds to the larger cf–PWV (more than 12 m/s).

shows Bland-Altman plots along with the mean difference value and the limits of agreements of the prediction results for each PPG-based measurement site. The mean difference is minimal in all cases. For instance, the mean difference value in the case of collecting PPG from the radial site is equal to 0.004, and the limits agreement is approximately equal to $(+0.8043)$ and (-0.7964). From these plots, it is clear that the proposed model is performing better for the prediction of cf-PWV less than 12 m/s, and above this value, the error starts to increase.

The MLP has shown promising results in estimating the central AS index, cf-PWV, using PPG signals collected at the level of distal arteries. A potential advantage of this approach is the ability to estimate cf-PWV non-invasively and inexpensively using an ideal ambulatory device, such as PPG. Furthermore, PPG technology is widely available in commercial fitness devices, smartphones, and tablets, making it accessible to a large population without specialized training or guidance.

10.5.2 COMPARATIVE ANALYSIS

For each regression method presented in Section 10.3 and for each noise level, the estimation performances were evaluated by $(RMSE)$ and (R^2) and reported in tables in the Appendix. The estimation's performances have been evaluated in three different arterial distal sites, mainly the digital, radial, and brachial arteries. In Section 10.6, the results are based on features extracted from the BP pulse wave, whereas Section 10.6 presents the regression's results based on features extracted from the PPG pulse wave. Overall, the obtained results showed the potential of the proposed features and algorithms in estimating cf-PWV and assessing AS non-invasively. However, it is clear that there is a noticeable variation in the estimation performance between the investigated ML algorithms and with respect to the level of noise as well as the nature of the used features. For ease of visualization, Figure 10.5 summarizes the estimation performances (RMSE and R^2) of the proposed machine learning models using different types of features based on noise-free BP signal collected at the finger level. The observed results demonstrate that using BP signals, the minimum R^2 is obtained in the case of IF-based features using LR and GR ML models with an RMSE equal to 0.91 and 0.92, respectively.

Apart from IF-based features, all the other techniques lead to an R^2 larger than 0.90 regardless of the complexity of the applied ML model. This result is very important and reveals the potential of estimating cf-PWV from a distal BP signal. Figure 10.6 shows the estimation performances (RMSE and R^2) of the proposed machine learning models using different types of features based on noise-free PPG signal collected at the finger level. It is clear from this figure and in comparison with the results shown in Figure 10.5 that using the PPG signal, the estimation performances have decreased. For instance, applying the simplest ML algorithm, LR, and utilizing SCSA-based features, R^2 have dropped from 0.95 to 0.8 when using the PPG signal. This drop is more apparent when using IF-based IF$_C$-based features and applying the LR method. However, it is clear that FP-based and SCSA-based features present a good performance when processing PPG signals.

The study includes performance evaluations of various machine learning models, and the results are presented in the appendix tables. The models utilize different

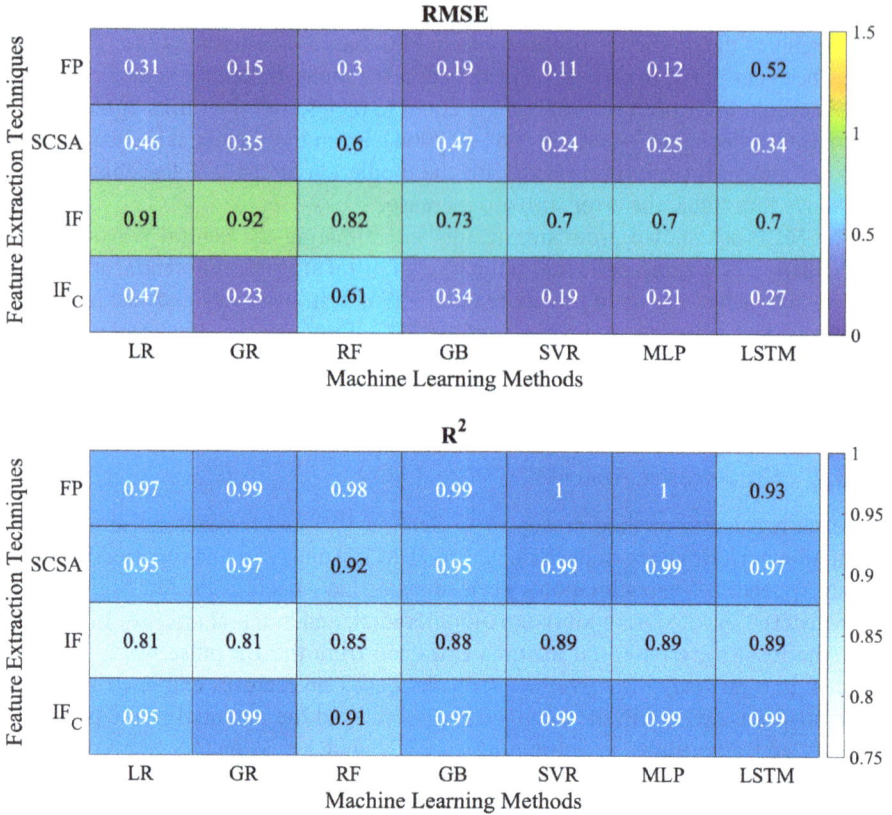

FIGURE 10.5 Estimation performances of the proposed machine learning models using a different type of features based on noise-free BP signal collected at the finger level.

types of features extracted from either blood pressure (BP) or photoplethysmography (PPG) signals obtained from the radial artery site. These estimations of performance are demonstrated for a range of noise levels, from 0% to 10%.

The study has found that the PPG pulse wave-based features are more sensitive to noise than BP pulse wave-based features, regardless of the feature extraction technique used. The results indicate that features based on BP signals using frequency-domain analysis (FP) exhibit high noise resistance. In contrast, those based on PPG or BP signals using time-domain analysis (IF) are significantly impacted by noise.

As the level of noise increases, the coefficient of determination (R^2) decreases, and the root-mean-square error (RMSE) increases. Moreover, the study finds that the inclusion of standard clinical routine variables, such as systolic blood pressure (SBP), diastolic blood pressure (DBP), mean blood pressure (MBP), and heart rate (HR), measured at the brachial level, into IF-based features (creating IF$_C$) enhances the results and mitigates the impact of noise.

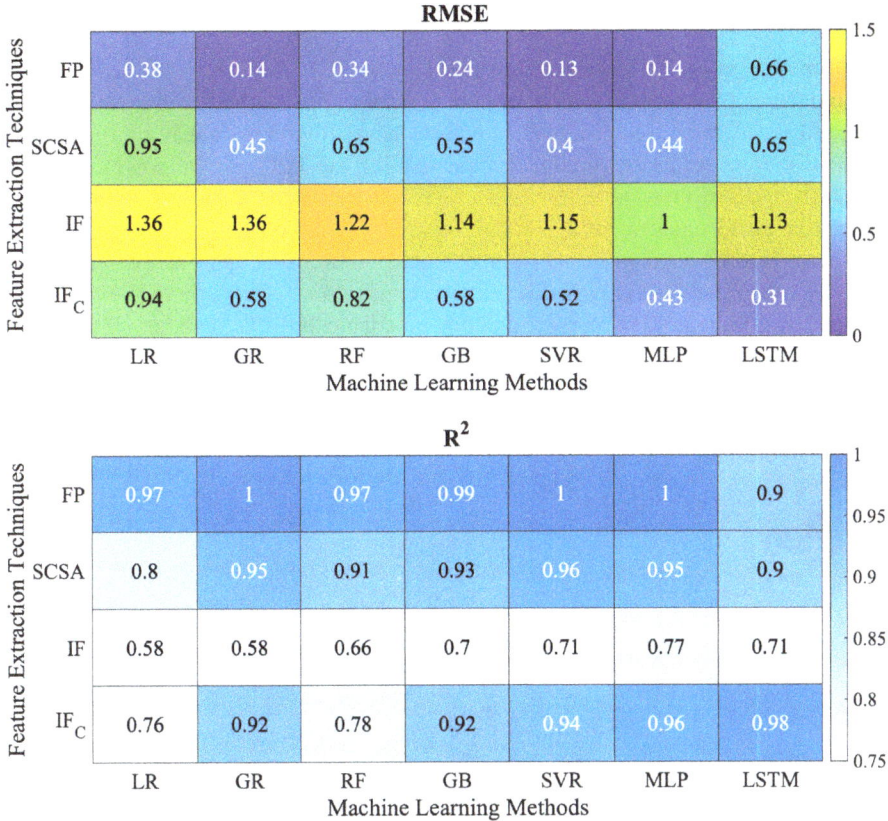

FIGURE 10.6 Estimation performances of the proposed machine learning models using a different type of features based on noise-free PPG signal collected at the finger level.

The study's finding highlights the critical role of these standard clinical variables in improving the performance of machine learning models, especially when dealing with noisy signals obtained from PPG or BP signals.

The use of LSTM models to estimate cf-PWV was also evaluated in this study. To this end, the model was trained using samples of a single cardiac cycle pulse wave, either BP or PPG, as input features. Interestingly, the results indicate that using these signal samples as features enhances the accuracy of cf-PWV estimation, likely due to their ability to capture all relevant physiological information. In addition, LSTM models have been shown to excel at time series prediction, making them a fitting choice for this application. LSTM models are considered the most robust approach to learning from sequential data, making them a promising tool for the analysis of time series data in clinical settings. Overall, the results suggest that LSTM models trained on BP and PPG data hold great promise for estimating cf-PWV non-invasively and with high accuracy, thus providing a valuable tool for the diagnosis and prognosis of AS-related CVDs.

10.6   CONCLUSION

In conclusion, given the high prevalence and risks of CVDs worldwide, the diagnosis and prognosis of AS have become of utmost importance. Although the clinical gold standard measure of AS, cf-PWV, is non-invasive; it remains costly and intrusive. This work proposes an artificial technique that utilizes PPG and BP waveforms collected from distal arteries to estimate cf-PWV. Our results demonstrate promising estimation performances, although limitations such as the use of a virtual database and noise impact need to be addressed. In the future, this approach could potentially be integrated into personal healthcare applications. The code of the proposed method is publicly available online and downloadable: https://github.com/Bahloulm/cf-PWV-estimation.

Appendix

A.1 CF-PWV ESTIMATION RESULTS USING PERIPHERAL BP SIGNAL

A.1.1 Noise-Free Case

TABLE 10.3
Estimation Performance (Fiducial Points-Based Features)

	LR		GR		RF		GB		SVR		MLP		LSTM	
	R^2	RMSE	R^2	RMSE	R^2	RMSE	R^2	RMSE	R^2	RMSE	R^2	RMSE	R^2	RMSE
Finger	0.97	0.31	0.99	0.15	0.98	0.30	0.99	0.19	1.0	0.11	1.0	0.12	0.93	0.52
Radial	0.97	0.30	1.0	0.12	0.98	0.26	0.99	0.17	1.0	0.11	1.0	0.12	0.93	0.53
Brachial	0.97	0.19	0.97	0.19	0.97	0.34	0.99	0.19	0.99	0.10	0.99	0.10	0.93	0.52

TABLE 10.4
Estimation Performance (SCSA-Based Features)

	LR		GR		RF		GB		SVR		MLP		LSTM	
	R^2	RMSE	R^2	RMSE	R^2	RMSE	R^2	RMSE	R^2	RMSE	R^2	RMSE	R^2	RMSE
Finger	0.95	0.46	0.97	0.35	0.92	0.60	0.95	0.47	0.99	0.24	0.99	0.25	0.97	0.34
Radial	0.85	0.80	0.95	0.48	0.88	0.75	0.95	0.47	0.96	0.42	0.97	0.35	0.97	0.34
Brachial	0.86	0.80	0.95	0.48	0.66	1.25	0.87	0.86	0.97	0.38	0.98	0.30	0.68	1.18

TABLE 10.5
Estimation Performance (IF-Based Features)

	LR		GR		RF		GB		SVR		MLP		LSTM	
	R^2	RMSE	R^2	RMSE	R^2	RMSE	R^2	RMSE	R^2	RMSE	R^2	RMSE	R^2	RMSE
Finger	0.81	0.91	0.81	0.92	0.85	0.82	0.88	0.73	0.89	0.70	0.89	0.70	0.89	0.70
Radial	0.81	0.91	0.81	0.91	0.85	0.82	0.87	0.75	0.89	0.70	0.89	0.69	0.86	0.78
Brachial	0.57	1.37	0.57	1.37	0.68	1.19	0.72	1.10	0.76	1.04	0.78	0.97	0.69	1.18

TABLE 10.6
Estimation Performance (IF-Based Features and Clinical Routine-Based Features)

	LR		GR		RF		GB		SVR		MLP		LSTM	
	R^2	RMSE	R^2	RMSE	R^2	RMSE	R^2	RMSE	R^2	RMSE	R^2	RMSE	R^2	RMSE
Finger	0.95	0.47	0.99	0.23	0.91	0.61	0.97	0.34	0.99	0.19	0.99	0.21	0.99	0.27
Radial	0.95	0.48	0.99	0.22	0.92	0.60	0.97	0.33	0.99	0.19	0.99	0.21	0.99	0.20
Brachial	0.88	0.73	0.95	0.47	0.88	0.75	0.95	0.49	0.97	0.34	0.97	0.37	0.95	0.47

TABLE 10.7
Estimation Performance (Pulse Wave's Samples-Based Features)

	MLP		LSTM	
	R^2	RMSE	R^2	RMSE
Finger	0.99	0.17	0.97	0.34
Radial	0.99	0.18	0.98	0.29
Brachial	0.99	0.23	0.97	0.40

A.1.2 Noise = 2%

TABLE 10.8
Estimation Performance (Fiducial Points-Based Features)

	LR		GR		RF		GB		SVR		MLP		LSTM	
	R^2	RMSE	R^2	RMSE	R^2	RMSE	R^2	RMSE	R^2	RMSE	R^2	RMSE	R^2	RMSE
Finger	0.97	0.33	0.99	0.15	0.96	0.37	0.99	0.22	1.0	0.11	0.99	0.13	0.93	0.50
Radial	0.97	0.33	1.00	0.10	0.97	0.34	0.99	0.19	1.0	0.11	1.0	0.13	0.93	0.53
Brachial	0.96	0.19	0.98	0.14	0.95	0.24	0.98	0.16	0.99	0.12	0.99	0.11	0.84	0.41

TABLE 10.9
Estimation Performance (SCSA-Based Features)

	LR		GR		RF		GB		SVR		MLP		LSTM	
	R^2	RMSE	R^2	RMSE	R^2	RMSE	R^2	RMSE	R^2	RMSE	R^2	RMSE	R^2	RMSE
Finger	0.76	1.01	0.93	0.58	0.83	0.89	0.90	0.65	0.94	0.52	0.96	0.40	0.85	0.82
Radial	0.92	0.60	0.97	0.37	0.84	0.82	0.94	0.52	0.98	0.28	0.98	0.26	0.90	0.66
Brachial	0.45	1.54	0.81	0.92	0.48	1.51	0.57	1.37	0.85	0.81	0.85	0.81	0.48	1.51

TABLE 10.10
Estimation Performance (IF-Based Features)

	LR		GR		RF		GB		SVR		MLP		LSTM	
	R^2	RMSE	R^2	RMSE	R^2	RMSE	R^2	RMSE	R^2	RMSE	R^2	RMSE	R^2	RMSE
Finger	0.79	0.95	0.66	1.35	0.82	0.88	0.84	0.85	0.84	0.83	0.84	0.83	0.82	0.94
Radial	0.79	0.95	0.62	1.49	0.82	0.88	0.84	0.85	0.85	0.82	0.85	0.81	0.84	0.85
Brachial	0.53	1.43	0.51	1.74	0.65	1.23	0.75	1.05	0.77	1.02	0.81	0.91	0.76	1.02

TABLE 10.11
Estimation Performance (IF-Based Features and Clinical Routine-Based Features)

	LR		GR		RF		GB		SVR		MLP		LSTM	
	R^2	RMSE	R^2	RMSE	R^2	RMSE	R^2	RMSE	R^2	RMSE	R^2	RMSE	R^2	RMSE
Finger	0.95	0.51	0.98	0.27	0.91	0.63	0.97	0.37	0.99	0.22	0.99	0.25	0.99	0.26
Radial	0.94	0.52	0.99	0.26	0.91	0.63	0.97	0.38	0.99	0.22	0.99	0.25	0.98	0.30
Brachial	0.83	0.86	0.94	0.53	0.85	0.85	0.91	0.61	0.96	0.44	0.95	0.48	0.89	0.68

TABLE 10.12
Estimation Performance (Pulse Wave's Samples-Based Features)

	MLP		LSTM	
	R^2	RMSE	R^2	RMSE
Finger	1.0	0.14	0.98	0.28
Radial	1.0	0.14	0.98	0.32
Brachial	0.99	0.18	0.89	0.74

A.1.3 Noise = 5%

TABLE 10.13
Estimation Performance (Fiducial Points-Based Features)

	LR		GR		RF		GB		SVR		MLP		LSTM	
	R^2	RMSE	R^2	RMSE	R^2	RMSE	R^2	RMSE	R^2	RMSE	R^2	RMSE	R^2	RMSE
Finger	0.97	0.33	0.99	0.15	0.96	0.37	0.99	0.22	1.0	0.11	0.99	0.14	0.93	0.50
Radial	0.97	0.33	1.0	0.10	0.97	0.34	0.99	0.19	1.0	0.10	0.99	0.14	0.93	0.51
Brachial	0.96	0.20	0.99	0.12	0.95	0.26	0.98	0.17	0.99	0.11	0.99	0.12	0.84	0.43

TABLE 10.14
Estimation Performance (SCSA-Based Features)

	LR		GR		RF		GB		SVR		MLP		LSTM	
	R^2	RMSE	R^2	RMSE	R^2	RMSE	R^2	RMSE	R^2	RMSE	R^2	RMSE	R^2	RMSE
Finger	0.77	1.01	0.90	0.66	0.73	1.09	0.83	0.88	0.85	0.84	0.96	0.44	0.65	1.25
Radial	0.69	1.17	0.90	0.67	0.81	0.94	0.89	0.70	0.94	0.49	0.95	0.48	0.70	1.15
Brachial	0.62	0.80	0.85	0.81	0.68	1.19	0.76	1.02	0.84	0.83	0.92	0.59	0.60	1.32

TABLE 10.15

Estimation Performance (IF-Based Features)

	LR		GR		RF		GB		SVR		MLP		LSTM	
	R^2	RMSE	R^2	RMSE	R^2	RMSE	R^2	RMSE	R^2	RMSE	R^2	RMSE	R^2	RMSE
Finger	0.76	1.03	0.61	1.50	0.79	0.96	0.81	0.91	0.82	0.88	0.83	0.88	0.77	1.06
Radial	0.76	1.03	0.61	1.53	0.79	0.96	0.81	0.92	0.82	0.89	0.82	0.88	0.77	1.03
Brachial	0.49	1.49	0.49	1.84	0.64	1.26	0.73	1.09	0.74	1.08	0.80	0.93	0.73	1.10

TABLE 10.16

Estimation Performance (IF-Based Features and Clinical Routine-Based Features)

	LR		GR		RF		GB		SVR		MLP		LSTM	
	R^2	RMSE	R^2	RMSE	R^2	RMSE	R^2	RMSE	R^2	RMSE	R^2	RMSE	R^2	RMSE
Finger	0.92	0.59	0.97	0.36	0.90	0.65	0.96	0.43	0.98	0.26	0.98	0.31	0.97	0.37
Radial	0.92	0.59	0.98	0.33	0.90	0.65	0.96	0.43	0.98	0.26	0.98	0.31	0.97	0.38
Brachial	0.82	0.88	0.93	0.55	0.85	0.85	0.91	0.64	0.94	0.50	0.93	0.56	0.92	0.60

TABLE 10.17

Estimation Performance (Pulse Wave's Samples-Based Features)

	MLP		LSTM	
	R^2	RMSE	R^2	RMSE
Finger	1.0	0.15	0.98	0.34
Radial	0.99	0.17	0.97	0.34
Brachial	0.99	0.17	0.97	0.40

A.1.4 Noise = 10%

TABLE 10.18
Estimation Performance (Fiducial Points-Based Features)

	LR		GR		RF		GB		SVR		MLP		LSTM	
	R^2	RMSE	R^2	RMSE	R^2	RMSE	R^2	RMSE	R^2	RMSE	R^2	RMSE	R^2	RMSE
Finger	0.97	0.33	0.99	0.15	0.96	0.37	0.99	0.22	1.0	0.11	0.99	0.14	0.93	0.50
Radial	0.97	0.33	1.0	0.10	0.97	0.34	0.99	0.19	1.0	0.10	1.0	0.13	0.93	0.51
Brachial	0.96	0.20	0.99	0.12	0.95	0.25	0.98	0.16	0.99	0.09	0.99	0.10	0.85	0.40

TABLE 10.19
Estimation Performance (SCSA-Based Features)

	LR		GR		RF		GB		SVR		MLP		LSTM	
	R^2	RMSE	R^2	RMSE	R^2	RMSE	R^2	RMSE	R^2	RMSE	R^2	RMSE	R^2	RMSE
Finger	0.85	0.81	0.95	0.48	0.83	0.87	0.91	0.62	0.96	0.40	0.96	0.43	0.82	0.89
Radial	0.69	1.16	0.90	0.66	0.79	0.99	0.88	0.73	0.94	0.50	0.94	0.53	0.79	0.96
Brachial	0.83	0.87	0.93	0.57	0.79	0.96	0.89	0.70	0.94	0.52	0.94	0.52	0.79	0.95

TABLE 10.20
Estimation Performance (IF-Based Features)

	LR		GR		RF		GB		SVR		MLP		LSTM	
	R^2	RMSE	R^2	RMSE	R^2	RMSE	R^2	RMSE	R^2	RMSE	R^2	RMSE	R^2	RMSE
Finger	0.70	1.14	0.58	1.55	0.75	1.05	0.78	0.98	0.78	0.98	0.80	0.94	0.75	1.09
Radial	0.71	1.14	0.54	1.66	0.75	1.05	0.76	1.02	0.77	1.00	0.78	0.97	0.75	1.12
Brachial	0.45	1.55	0.54	1.63	0.64	1.26	0.72	1.10	0.71	1.13	0.79	0.95	0.72	1.12

TABLE 10.21
Estimation Performance (IF-Based Features and Clinical Routine-Based Features)

	LR		GR		RF		GB		SVR		MLP		LSTM	
	R^2	RMSE	R^2	RMSE	R^2	RMSE	R^2	RMSE	R^2	RMSE	R^2	RMSE	R^2	RMSE
Finger	0.89	0.69	0.96	0.42	0.89	0.69	0.94	0.51	0.97	0.37	0.97	0.37	0.95	0.48
Radial	0.89	0.70	0.96	0.41	0.89	0.70	0.94	0.52	0.97	0.37	0.96	0.40	0.94	0.51
Brachial	0.82	0.90	0.92	0.59	0.85	0.84	0.90	0.65	0.94	0.52	0.91	0.62	0.87	0.77

TABLE 10.22
Estimation Performance (Pulse Wave's Samples-Based Features)

	MLP		LSTM	
	R^2	RMSE	R^2	RMSE
Finger	0.99	0.21	0.98	0.29
Radial	0.99	0.20	0.99	0.25
Brachial	0.99	0.24	0.97	0.47

A.2   CF-PWV ESTIMATION RESULTS USING PERIPHERAL PPG SIGNAL

A.2.1   Noise-Free

TABLE 10.23
Estimation Performance (Fiducial Points-Based Features)

	LR		GR		RF		GB		SVR		MLP		LSTM	
	R^2	RMSE	R^2	RMSE	R^2	RMSE	R^2	RMSE	R^2	RMSE	R^2	RMSE	R^2	RMSE
Finger	0.97	0.38	1.0	0.14	0.97	0.34	0.99	0.24	1.0	0.13	1.0	0.14	0.90	0.66
Radial	0.92	0.60	0.99	0.22	0.92	0.62	0.96	0.41	0.99	0.20	0.99	0.26	0.91	0.64
Brachial	0.97	0.34	0.99	0.15	0.96	0.39	0.99	0.23	1.0	0.13	1.0	0.14	0.92	0.59

TABLE 10.24
Estimation Performance (SCSA-Based Features)

	LR		GR		RF		GB		SVR		MLP		LSTM	
	R^2	RMSE	R^2	RMSE	R^2	RMSE	R^2	RMSE	R^2	RMSE	R^2	RMSE	R^2	RMSE
Finger	0.80	0.95	0.95	0.45	0.91	0.65	0.93	0.55	0.96	0.40	0.95	0.44	0.90	0.65
Radial	0.78	0.99	0.88	0.72	0.83	0.86	0.87	0.75	0.92	0.61	0.90	0.65	0.89	0.71
Brachial	0.81	0.85	0.95	0.48	0.90	0.62	0.93	0.53	0.97	0.39	0.95	0.43	0.92	0.60

TABLE 10.25
Estimation Performance (IF-Based Features)

	LR		GR		RF		GB		SVR		MLP		LSTM	
	R^2	RMSE	R^2	RMSE	R^2	RMSE	R^2	RMSE	R^2	RMSE	R^2	RMSE	R^2	RMSE
Finger	0.58	1.36	0.58	1.36	0.66	1.22	0.70	1.14	0.71	1.15	0.77	1.00	0.71	1.13
Radial	0.62	1.29	0.53	1.71	0.71	1.12	0.77	1.00	0.77	1.02	0.78	0.97	0.76	1.04
Brachial	0.64	1.25	0.64	1.25	0.71	1.12	0.77	1.00	0.79	0.97	0.83	0.85	0.71	1.15

TABLE 10.26
Estimation Performance (IF-Based Features and Clinical Routine-Based Features)

	LR		GR		RF		GB		SVR		MLP		LSTM	
	R^2	RMSE	R^2	RMSE	R^2	RMSE	R^2	RMSE	R^2	RMSE	R^2	RMSE	R^2	RMSE
Finger	0.76	0.94	0.92	0.58	0.78	0.82	0.92	0.58	0.94	0.52	0.96	0.43	0.98	0.31
Radial	0.81	0.94	0.87	0.80	0.85	0.84	0.89	0.70	0.91	0.66	0.90	0.68	0.91	0.65
Brachial	0.88	0.73	0.98	0.32	0.92	0.61	0.96	0.43	0.99	0.25	0.99	0.25	0.97	0.34

TABLE 10.27
Estimation Performance (Pulse Wave's Samples-Based Features)

	MLP		LSTM	
	R^2	RMSE	R^2	RMSE
Finger	0.98	0.28	0.98	0.30
Radial	0.98	0.29	0.99	0.19
Brachial	0.99	0.24	0.99	0.21

A.2.2 NOISE = 2%

TABLE 10.28
Estimation Performance (Fiducial Points-Based Features)

	LR		GR		RF		GB		SVR		MLP		LSTM	
	R^2	RMSE	R^2	RMSE	R^2	RMSE	R^2	RMSE	R^2	RMSE	R^2	RMSE	R^2	RMSE
Finger	0.88	0.74	0.99	0.17	0.95	0.49	0.97	0.35	0.99	0.19	0.99	0.23	0.90	0.67
Radial	0.91	0.64	0.99	0.21	0.94	0.51	0.97	0.35	0.99	0.21	0.98	0.29	0.91	0.67
Brachial	0.87	0.76	0.99	0.23	0.94	0.54	0.97	0.37	0.99	0.22	0.98	0.29	0.92	0.61

TABLE 10.29
Estimation Performance (SCSA-Based Features)

	LR		GR		RF		GB		SVR		MLP		LSTM	
	R^2	RMSE	R^2	RMSE	R^2	RMSE	R^2	RMSE	R^2	RMSE	R^2	RMSE	R^2	RMSE
Finger	0.78	0.97	0.95	0.48	0.90	0.68	0.93	0.54	0.96	0.43	0.95	0.46	0.91	0.64
Radial	0.79	0.96	0.94	0.50	0.88	0.72	0.92	0.59	0.97	0.39	0.96	0.44	0.90	0.66
Brachial	0.82	0.89	0.96	0.44	0.90	0.65	0.94	0.50	0.97	0.38	0.96	0.42	0.91	0.62

TABLE 10.30
Estimation Performance (IF-Based Features)

	LR		GR		RF		GB		SVR		MLP		LSTM	
	R^2	RMSE	R^2	RMSE	R^2	RMSE	R^2	RMSE	R^2	RMSE	R^2	RMSE	R^2	RMSE
Finger	0.58	1.35	0.41	2.17	0.71	1.12	0.78	0.98	0.81	0.92	0.84	0.84	0.78	1.01
Radial	0.62	1.29	0.65	1.39	0.72	1.10	0.83	0.87	0.78	0.99	0.81	0.92	0.80	0.96
Brachial	0.64	1.26	0.61	1.51	0.72	1.10	0.79	0.95	0.84	0.86	0.86	0.77	0.83	0.90

TABLE 10.31
Estimation Performance (IF-Based Features and Clinical Routine-Based Features)

	LR		GR		RF		GB		SVR		MLP		LSTM	
	R^2	RMSE	R^2	RMSE	R^2	RMSE	R^2	RMSE	R^2	RMSE	R^2	RMSE	R^2	RMSE
Finger	0.88	0.71	0.99	0.21	0.90	0.69	0.96	0.44	0.99	0.20	0.99	0.22	0.98	0.28
Radial	0.84	0.84	0.96	0.44	0.85	0.81	0.92	0.59	0.92	0.59	0.96	0.40	0.93	0.57
Brachial	0.88	0.73	0.98	0.31	0.90	0.70	0.95	0.45	0.98	0.26	0.98	0.30	0.96	0.41

TABLE 10.32
Estimation Performance (Pulse Wave's Samples-Based Features)

	MLP		LSTM	
	R^2	RMSE	R^2	RMSE
Finger	0.99	0.17	0.98	0.33
Radial	0.99	0.24	0.90	0.68
Brachial	0.99	0.19	0.92	0.59

A.2.3 Noise = 5%

TABLE 10.33

Estimation Performance (Fiducial Points-Based Features)

	LR		GR		RF		GB		SVR		MLP		LSTM	
	R^2	RMSE	R^2	RMSE	R^2	RMSE	R^2	RMSE	R^2	RMSE	R^2	RMSE	R^2	RMSE
Finger	0.83	0.86	0.99	0.20	0.94	0.55	0.97	0.38	0.99	0.22	0.99	0.24	0.90	0.69
Radial	0.88	0.74	0.98	0.33	0.90	0.66	0.95	0.49	0.99	0.26	0.98	0.32	0.90	0.68
Brachial	0.83	0.87	0.98	0.27	0.93	0.55	0.99	0.42	0.98	0.26	0.98	0.32	0.91	0.63

TABLE 10.34

Estimation Performance (SCSA-Based Features)

	LR		GR		RF		GB		SVR		MLP		LSTM	
	R^2	RMSE	R^2	RMSE	R^2	RMSE	R^2	RMSE	R^2	RMSE	R^2	RMSE	R^2	RMSE
Finger	0.78	0.97	0.95	0.45	0.89	0.70	0.94	0.51	0.96	0.41	0.95	0.46	0.92	0.58
Radial	0.82	0.89	0.95	0.45	0.89	0.70	0.93	0.53	0.97	0.37	0.96	0.43	0.92	0.62
Brachial	0.82	0.88	0.97	0.37	0.91	0.65	0.95	0.47	0.97	0.34	0.97	0.39	0.92	0.60

TABLE 10.35
Estimation Performance (IF-Based Features)

	LR		GR		RF		GB		SVR		MLP		LSTM	
	R^2	RMSE	R^2	RMSE	R^2	RMSE	R^2	RMSE	R^2	RMSE	R^2	RMSE	R^2	RMSE
Finger	0.57	1.37	0.55	1.71	0.71	1.13	0.78	0.98	0.81	0.92	0.85	0.81	0.80	0.95
Radial	0.61	1.30	0.56	1.73	0.72	1.10	0.82	0.89	0.79	0.97	0.80	0.94	0.80	0.96
Brachial	0.63	1.27	0.58	1.55	0.72	1.11	0.78	0.97	0.83	0.89	0.86	0.78	0.83	0.89

TABLE 10.36
Estimation Performance (IF-Based Features and Clinical Routine-Based Features)

	LR		GR		RF		GB		SVR		MLP		LSTM	
	R^2	RMSE	R^2	RMSE	R^2	RMSE	R^2	RMSE	R^2	RMSE	R^2	RMSE	R^2	RMSE
Finger	0.89	0.71	0.98	0.30	0.90	0.70	0.96	0.44	0.99	0.19	0.98	0.30	0.98	0.29
Radial	0.84	0.84	0.96	0.45	0.86	0.81	0.92	0.59	0.96	0.40	0.94	0.50	0.93	0.57
Brachial	0.88	0.73	0.98	0.31	0.90	0.70	0.95	0.45	0.98	0.26	0.97	0.34	0.98	0.34

TABLE 10.37
Estimation Performance (Pulse Wave's Samples-Based Features)

	MLP		LSTM	
	R^2	RMSE	R^2	RMSE
Finger	0.99	0.19	0.98	0.31
Radial	0.99	0.21	0.86	0.78
Brachial	0.99	0.16	0.91	0.62

A.2.4 Noise = 10%

TABLE 10.38
Estimation Performance (Fiducial Points-Based Features)

	LR		GR		RF		GB		SVR		MLP		LSTM	
	R^2	RMSE	R^2	RMSE	R^2	RMSE	R^2	RMSE	R^2	RMSE	R^2	RMSE	R^2	RMSE
Finger	0.76	1.02	0.98	0.32	0.92	0.63	0.96	0.42	0.98	0.27	0.98	0.28	0.88	0.72
Radial	0.78	1.0	0.97	0.36	0.89	0.72	0.95	0.49	0.98	0.28	0.97	0.36	0.88	0.76
Brachial	0.78	0.97	0.98	0.33	0.92	0.62	0.96	0.43	0.98	0.29	0.97	0.34	0.90	0.68

TABLE 10.39
Estimation Performance (SCSA-Based Features)

	LR		GR		RF		GB		SVR		MLP		LSTM	
	R^2	RMSE	R^2	RMSE	R^2	RMSE	R^2	RMSE	R^2	RMSE	R^2	RMSE	R^2	RMSE
Finger	0.80	0.93	0.97	0.36	0.90	0.67	0.95	0.48	0.97	0.34	0.97	0.39	0.93	0.56
Radial	0.83	0.87	0.96	0.40	0.89	0.69	0.94	0.52	0.97	0.34	0.96	0.40	0.90	0.65
Brachial	0.83	0.85	0.97	0.35	0.91	0.62	0.95	0.45	0.98	0.31	0.97	0.35	0.94	0.52

TABLE 10.40
Estimation Performance (IF-Based Features)

	LR		GR		RF		GB		SVR		MLP		LSTM	
	R^2	RMSE	R^2	RMSE	R^2	RMSE	R^2	RMSE	R^2	RMSE	R^2	RMSE	R^2	RMSE
Finger	0.58	1.35	0.53	1.68	0.70	1.15	0.75	1.05	0.79	0.99	0.81	0.92	0.73	1.09
Radial	0.60	1.32	0.58	1.67	0.71	1.14	0.79	0.95	0.79	0.97	0.82	0.88	0.80	0.95
Brachial	0.62	1.28	0.60	1.52	0.72	1.11	0.76	1.02	0.81	0.94	0.84	0.85	0.82	0.89

TABLE 10.41
Estimation Performance (IF-Based Features and Clinical Routine-Based Features)

	LR		GR		RF		GB		SVR		MLP		LSTM	
	R^2	RMSE	R^2	RMSE	R^2	RMSE	R^2	RMSE	R^2	RMSE	R^2	RMSE	R^2	RMSE
Finger	0.88	0.72	0.99	0.25	0.89	0.71	0.95	0.45	0.99	0.23	0.98	0.27	0.97	0.35
Radial	0.84	0.84	0.96	0.45	0.85	0.81	0.91	0.61	0.96	0.44	0.95	0.47	0.92	0.61
Brachial	0.87	0.75	0.98	0.31	0.89	0.70	0.95	0.46	0.98	0.30	0.98	0.33	0.96	0.43

TABLE 10.42
Estimation Performance (Pulse Wave's Samples-Based Features)

	MLP		LSTM	
	R^2	RMSE	R^2	RMSE
Finger	0.99	0.16	0.98	0.31
Radial	0.99	0.20	0.90	0.65
Brachial	0.99	0.15	0.90	0.67

NOTES

1 https://peterhcharlton.github.io/pulse-analyse/index.html.
2 https://peterhcharlton.github.io/pwdb/index.html.

REFERENCES

1. J.L. Gade, C.J. Thore, B. Sonesson, and J. Stålhand, In vivo parameter identification in arteries considering multiple levels of smooth muscle activity, *Biomechanics and Modeling in Mechanobiology* 20(4) (2021), pp. 1547–1559.
2. T.T. van Sloten, M.T. Schram, K. van den Hurk, J.M. Dekker, G. Nijpels, R.M. Henry, and C.D. Stehouwer, Local stiffness of the carotid and femoral artery is associated with incident cardiovascular events and all-cause mortality: The Hoorn study, *Journal of the American College of Cardiology* 63 (2014), pp. 1739–1747.
3. M.A. Bahloul and T.M.L. Kirati, Fractional-order model representations of apparent vascular compliance as an alternative in the analysis of arterial stiffness: An in-silico study, *Physiological Measurement* 42 (2021), p. 045008.
4. C.U. Choi, E.B. Park, S.Y. Suh, J.W. Kim, E.J. Kim, S.W. Rha, H.S. Seo, D.J. Oh, and C.G. Park, Impact of aortic stiffness on cardiovascular disease in patients with chest pain: Assessment with direct intra-arterial measurement, *American Journal of Hypertension* 20 (2007), pp. 1163–1169.
5. T.M. Laleg and M.A. Bahloul, Mathematical biomarker for arterial viscoelasticity assessment. US Patent App. 17/284,072 (2021).
6. M.W. Rajzer, W. Wojciechowska, M. Klocek, I. Palka, M. Brzozowska-Kiszka, and K. Kawecka-Jaszcz, Comparison of aortic pulse wave velocity measured by three techniques: Complior, sphygmocor and arteriograph, *Journal of Hypertension* 26 (2008), pp. 2001–2007.
7. P. Tavallali, M. Razavi, and N.M. Pahlevan, Artificial intelligence estimation of carotidfemoral pulse wave velocity using carotid waveform, *Scientific Reports* 8 (2018), pp. 1–12.
8. J. Qiu, Q. Wu, G. Ding, Y. Xu, and S. Feng, A survey of machine learning for big data processing, *EURASIP Journal on Advances in Signal Processing* 2016 (2016), pp. 1–16.
9. C. Krittanawong, H.U.H. Virk, S. Bangalore, Z. Wang, K.W. Johnson, R. Pinotti, H. Zhang, S. Kaplin, B. Narasimhan, T. Kitai, et al., Machine learning prediction in cardiovascular diseases: A meta-analysis, *Scientific Reports* 10 (2020), pp. 1–11.
10. A. Magbool, M.A. Bahloul, T. Ballal, T.Y. Al-Naffouri, and T.M. Laleg-Kirati, Aortic blood pressure estimation: A hybrid machine-learning and cross-relation approach, *Biomedical Signal Processing and Control* 68 (2021), p. 102762.
11. A. Magbool, M.A. Bahloul, T. Ballal, T.Y. Al-Naffouri, and T.M. Laleg-Kirati, Combining machine learning and blind estimation for central aortic blood pressure reconstruction, In *2021 43rd Annual International Conference of the IEEE Engineering in Medicine & Biology Society (EMBC)*. IEEE (2021), pp. 5512–5517, Mexico.
12. M.A. Myszczynska, P.N. Ojamies, A.M. Lacoste, D. Neil, A. Saffari, R. Mead, G.M. Hautbergue, J.D. Holbrook, and L. Ferraiuolo, Applications of machine learning to diagnosis and treatment of neurodegenerative diseases, *Nature Reviews Neurology* 16 (2020), pp. 440–456.
13. M. Bahloul, Z. Belkhatir, Y. Aboelkassem, and M.T. Laleg-Kirati, Physics-based modeling and data-driven algorithm for prediction and diagnosis of atherosclerosis, *Biophysical Journal* 121 (2022), pp. 419a–420a.
14. P. Tavallali, T.Y. Hou, D.G. Rinderknecht, and N.M. Pahlevan, On the convergence and accuracy of the cardiovascular intrinsic frequency method, *Royal Society Open Science* 2 (2015), p. 150475.

15. M.A. Bahloul, A. Chahid, and T.M. Laleg-Kirati, A multilayer perceptron-based carotid-to-femoral pulse wave velocity estimation using PPG signal, In *2021 IEEE EMBS International Conference on Biomedical and Health Informatics (BHI)*. IEEE (2021), pp. 1–6, Athens.

16. M.A. Bahloul, T.M. Laleg-Kirati, et al., Spectrogram image-based machine learning model for carotid-to-femoral pulse wave velocity estimation using PPG signal, In *2022 IEEE-EMBS International Conference on Biomedical and Health Informatics (BHI)*. IEEE (2022), pp. 01–04, Ioannina.

17. J.M.V. Garcia, M.A. Bahloul, and T.M. Laleg-Kirati, A multiple linear regression model for carotid-to-femoral pulse wave velocity estimation based on schrodinger spectrum characterization, *In 2022 44th Annual International Conference of the IEEE Engineering in Medicine & Biology Society (EMBC). IEEE (2022)*, pp. 143–147, Glasgow.

18. W. Jin, P. Chowienczyk, and J. Alastruey, Estimating pulse wave velocity from the radial pressure wave using machine learning algorithms, *PLoS One* 16 (2021), p. e0245026.

19. P. Li and T.M. Laleg-Kirati, Central blood pressure estimation from distal ppg measurement using semiclassical signal analysis features, *IEEE Access* 9 (2021), pp. 44963–44973.

20. J.M. Vargas Garcia, M.A. Bahloul, and T.M. Laleg, A learning-based image processing approach for pulse wave velocity estimation using spectrogram from peripheral pulse wave signals: An in-silico study, *Frontiers in Physiology* 14 (2023), p. 189.

21. P.J. Chowienczyk, R.P. Kelly, H. MacCallum, S.C. Millasseau, T.L. Andersson, R.G. Gosling, J.M. Ritter, and E.E. Änggård, Photoplethysmographic assessment of pulse wave reflection: Blunted response to endothelium-dependent beta2-adrenergic vasodilation in type II diabetes mellitus, *Journal of the American College of Cardiology* 34 (1999), pp. 2007–2014.

22. S.R. Alty, S.C. Millasseau, P. Chowienczyc, and A. Jakobsson, Cardiovascular disease prediction using support vector machines, In *2003 46th Midwest Symposium on Circuits and Systems*, vol. 1. IEEE (2003), pp. 376–379, Cairo.

23. H.T. Wu, C.C. Liu, P.H. Lin, H.M. Chung, M.C. Liu, H.K. Yip, A.B. Liu, and C.K. Sun, Novel application of parameters in waveform contour analysis for assessing arterial stiffness in aged and atherosclerotic subjects, *Atherosclerosis* 213 (2010), pp. 173–177.

24. J.M. Ahn, New aging index using signal features of both photoplethysmograms and acceleration plethysmograms, *Healthcare Informatics Research* 23 (2017), pp. 53–59.

25. K. Takazawa, N. Tanaka, M. Fujita, O. Matsuoka, T. Saiki, M. Aikawa, S. Tamura, and C. Ibukiyama, Assessment of vasoactive agents and vascular aging by the second derivative of photoplethysmogram waveform, *Hypertension* 32 (1998), pp. 365–370.

26. H.J. Baek, J.S. Kim, Y.S. Kim, H.B. Lee, and K.S. Park, Second derivative of photoplethysmography for estimating vascular aging, In *2007 6th International Special Topic Conference on Information Technology Applications in Biomedicine*. IEEE (2007), pp. 70–72, Tokyo.

27. T. Ushiroyama, Y. Kajimoto, K. Sakuma, and M. Ueki, Assessment of chilly sensation in Japanese women with laser Doppler fluxmetry and acceleration plethysmogram with respect to peripheral circulation, *Bulletin of the Osaka Medical College* 51 (2005), pp. 76–84.

28. C.C. Wei, Developing an effective arterial stiffness monitoring system using the spring constant method and photoplethysmography, *IEEE Transactions on Biomedical Engineering* 60 (2012), pp. 151–154.

29. P.H. Charlton, P. Celka, B. Farukh, P. Chowienczyk, and J. Alastruey, Assessing mental stress from the photoplethysmogram: a numerical study, *Physiological Measurement* 39 (2018), p. 054001.

30. M. Elgendi, Standard terminologies for photoplethysmogram signals, *Current Cardiology Reviews* 8 (2012), pp. 215–219.
31. C.S. Hayward and R.P. Kelly, Gender-related differences in the central arterial pressure waveform, *Journal of the American College of Cardiology* 30 (1997), pp. 1863–1871.
32. P.H. Charlton, J. Mariscal Harana, S. Vennin, Y. Li, P. Chowienczyk, and J. Alastruey, Modeling arterial pulse waves in healthy aging: a database for in silico evaluation of hemodynamics and pulse wave indexes, *American Journal of Physiology-Heart and Circulatory Physiology* 317 (2019), pp. H1062–H1085.
33. N.M. Pahlevan, P. Tavallali, D.G. Rinderknecht, D. Petrasek, R.V. Matthews, T.Y. Hou, and M. Gharib, Intrinsic frequency for a systems approach to haemodynamic waveform analysis with clinical applications, *Journal of the Royal Society Interface* 11 (2014), p. 20140617.
34. T.M. Laleg-Kirati, E. Crèpeau, and M. Sorine, Semi-classical signal analysis, *Mathematics of Control, Signals, and Systems* 25 (2013), pp. 37–61.
35. T.M. Laleg, E. Crèpeau, and M. Sorine, Separation of arterial pressure into a nonlinear superposition of solitary waves and a windkessel flow, *Biomedical Signal Processing and Control* 2 (2007), pp. 163–170.
36. E. Crèpeau and M. Sorine, A reduced model of pulsatile flow in an arterial compartment, *Chaos, Solitons & Fractals* 34 (2007), pp. 594–605.
37. T.M. Laleg-Kirati, C. Mèdigue, Y. Papelier, F. Cottin, and A. Van de Louw, Validation of a semi-classical signal analysis method for stroke volume variation assessment: A comparison with the PiCCO technique, *Annals of Biomedical Engineering* 38 (2010), pp. 3618–3629.
38. A. Chahid, H. Serrai, E. Achten, and T.M. Laleg-Kirati, A new ROI-based performance evaluation method for image denoising using the Squared Eigenfunctions of the Schrödinger operator, In *2018 40th Annual International Conference of the IEEE Engineering in Medicine and Biology Society (EMBC)*. IEEE (2018), pp. 5579–5582, Honolulu, HI.
39. A. Chahid, Pre-processing and feature extraction methods for smart biomedical signal monitoring: Algorithms and applications, Ph.D. Dissertaion, King Abdullah University of Science and Technology (2020).
40. T.M. Laleg-Kirati, E. Crèpeau, and M. Sorine, Semi-classical signal analysis, *Mathematics of Control, Signals, and Systems* 25 (2013), pp. 37–61.
41. P. Li and T. Laleg-Kirati, Signal denoising based on the Schrödinger operator's Eigenspectrum and a curvature constraint, *IET Signal Processing* 15 (2021), pp. 195–206.
42. P. Li and T.M. Laleg-Kirati, Central blood pressure estimation from distal PPG measurement using semiclassical signal analysis features, *IEEE Access* 9 (2021), pp. 44963–44973.
43. T.M. Laleg-Kirati, C. Mèdigue, Y. Papelier, F. Cottin, and A. Van De Louw, Validation of a semi-classical signal analysis method for stroke volume variation assessment: A comparison with the PiCCO technique, *Annals of Biomedical Engineering* 38 (2010), pp. 3618–3629.
44. P. Li, E. Piliouras, V. Poghosyan, M. AlHameed, and T.M. Laleg-Kirati, Automatic detection of epileptiform EEG discharges based on the Semi-Classical Signal Analysis (SCSA) method, In *2021 43rd Annual International Conference of the IEEE Engineering in Medicine & Biology Society (EMBC)*. IEEE (2021), pp. 928–931, Mexico.

11 DVT Diagnosis Based on HOS and Scattering Operators

Thibaud Berthomier
ENSTA-Bretagne
CHU de la Cavale Blanche

Ali Mansour
ENSTA-Bretagne

*Luc Bressollette, Clément Hoffman,
and Dominique Mottier*
CHRU Cavale Blanche

11.1 INTRODUCTION

Our project, a collaborative project between an engineering school (ENSTA Bretagne) and a Medical Research Group (EA GETBO 3878), focuses on the Venous Thrombo-Embolic Disease (VTE). A thrombus (also called blood clot) is the result of blood coagulation which is a natural process to prevent bleeding. There are mainly three components in a thrombus: platelets, red blood cells and a mesh of fibrins. Three physio-pathological mechanisms can contribute, isolated or combined, to the development of a DVT: venous stasis, endothelial injury and hypercoagulability.

The ability to predict the risk of pulmonary embolism (PE) would also be a major advance. The French cardiology association estimates that PE affects around 100,000 cases annually with 10,000–20,000 fatal cases just in France. It should be also noted that PE is the third cause of vascular death after myocardial infarction and stroke. etc. This multifactorial disease that is related to advanced age, immobility, surgery or obesity) is an important public health issue. From the medical point of view, some phlebitides may cause PE. Another risk of phlebitides is related to the recurrence aspects of such disease; A thrombophlebitis is likely to be recurrent, and an actual PE may be the alarm of a future and more dangerous one. The occurrence of phlebitis can be multifactorial combining genetic and acquired factors. Our study aims to analyze the characteristics of the clot in order to:

a. Predict the risk of PE and analyze the clot;
b. Evoke a pathophysiological mechanism at the origin of thrombosis;

DOI: 10.1201/9781003346678-11

c. Date the clot;

d. Detect a possible underlying cancer

e. Assess the impact of anticoagulants on the clot.

To reach our goals, we are collecting ultrasonography (echogenicity) and elastography (stiffness) of human thrombus. Our approaches to characterize the thrombus structure with ultrasound images are described. To extract the features from these ultrasound images, we propose a hybrid approach based on scattering operators and high-order statistics. The obtained features are analyzed using several classification technics to find the main cause of the DVT or the presence of PE. Experimental results are presented and discussed.

11.2   DEEP VENOUS THROMBUS

Normally, a thrombus should be developed as a result of an injured blood vessel to limit bleeding [36]. This phenomenon, which transforms the blood from a liquid state to a more or less solid, is called blood clotting. A thrombus is mainly made up of platelets (small parts of the blood) and fibrinogens (soluble proteins made by the liver). During the coagulation, the chain of chemical reactions, involving various substrates and enzymes plasma, transforms fibrinogen into an insoluble protein, fibrin, which forms the thrombus. The thrombus therefore becomes increasingly more solid over time. Once the vessel is repaired, the thrombus is dissolved in order to restore blood flow. This vital process, called fibrinolysis, removes fibrin waste in order to clean the vessel.

Numerous physiopathological mechanisms can disrupt the blood circulation and cause the formation or non-dissolution of a thrombus. Deep venous thrombosis (DVT) corresponds to the partial or total occlusion of a deep vein (as opposed to a so-called superficial surface vein). It most often begins in the veins of the calf and extends gradually toward the veins of the thigh. Venous thrombosis, associated with inflammation of the veins, is more commonly known as phlebitis. According to the triad described by Virchow in 1956 [3], three types of phenomena, associated or not, favoring the occurrence of thrombosis:

1. **Venous stasis**: Arteries carry oxygenated blood from the heart to various organs through rhythmic contractions of the heart muscle. Veins return the blood from organs to the heart [36]. Venous stasis, *i.e.* the slowing down of the venous flow, promotes the formation of a thrombus, especially in the lower limbs. A study [47] showed that the risk of thrombosis during a flight lasting more than 5 hours is estimated between 2 and 4 per 10,000 passengers. Likewise, heart failure and varicose veins of the lower limbs are possible factors to generate venous stasis.

2. **Endothelial injury**: Veins have valve systems to prevent backflow of blood and have walls much finer and more elastic than arteries. They are therefore more sensitive to surgery, trauma and aging which can deteriorate them and therefore limit their effectiveness. Moreover, in case of infection, inflammation of the veins occurs and therefore venous returns can cause phlebitis. Phlebographic studies [47] showed a 10%–20% prevalence of new venous

thrombosis within 4–5 weeks after discharge from the hospital without prophylaxis (*i.e.* without measures to prevent the onset of phlebitis). The incidence of venous thrombosis increases exponentially with age: the risk doubles every 10 years from the age of 40 [47,76].

3. **Hypercoagulable state**: Thrombophilia brings together all acquired or genetic mechanisms responsible for a strengthening of the physical coagulation or an anticoagulation deficit. C and S proteins, synthesized by the liver, work together to inhibit blood clotting. Deficits in proteins C or S, or malfunctions of one of the two proteins can lead to excessive activation of coagulation and therefore cause venous thrombosis. For example, a pregnancy (in the third trimester) may reduce levels of these two proteins and, conversely, an estrogen-progestogen therapy may increase their levels: the risk of thrombosis would become two to five times greater [47] compared to a person not pregnant or under any hormonal treatment. Genetic abnormalities (factor V Leiden, FII gene G20210A) or specific cases (obesity, liver disease, kidney disease, serious infections, and cancers [22]) may decrease their production or increase their consumption and therefore provoke a thrombosis [36]. Some studies show an incidence of cigarette smoking on venous thrombosis [10,26] (Figure 11.1).

The PE or ThromboEmbolic Venous Disease (TEVD) occurs when the thrombus gets stuck in a pulmonary artery. Such an event may not show any symptom or it can present a sidelight, dyspnea, respiratory distress, coughing up blood and, in the worst case, a sudden death.

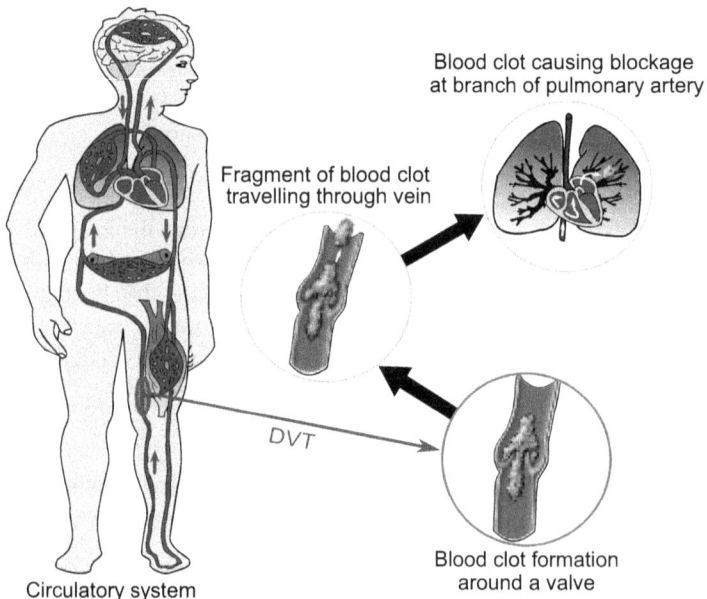

FIGURE 11.1   DVT's major complication.

11.3 ACOUSTIC IMAGING FOR THE CHARACTERIZATION OF HUMAN TISSUES

Ultrasounds propagate in an elastic medium of a density ρ (in kg/m^3) according to two modes of vibration defined according to the compression modulus K (in Pa) and the shearing modulus μ (in Pa) [29] (Figure 11.2):

4. A **longitudinal wave**, or a compression wave, travels parallel to the direction of the ultrasound propagation at a speed: $c_l = \sqrt{\dfrac{3K + 4\mu}{3\rho}}$

5. A **transverse wave**, or a shear wave, moves perpendicular to the ultrasound propagation at a speed: $c_t = \sqrt{\dfrac{\mu}{\rho}}$.

The velocity of these waves is directly connected to the elastic modulus E (or Young's modulus). In soft biological tissues [6], the compressional velocity is much higher ($c_l \approx 10\,\text{m/s}$), so Young's modulus can be approximated by: $E \approx 3\rho c_s^2$. The human body is a complex acoustic environment made up of numerous sub-environments with specific characteristics. The human tissues are therefore non-linear media. In such environments, the sound velocity is no longer constant. Positive peaks of the wave, due to tissue compression, have a faster propagation speed than the negative peaks, where there is a tissue relaxation. This results in a wave distortion that changes a pure original sinusoidal shape to a sawtooth shape and enriches the spectrum with harmonic frequencies.

Ultrasound imaging is based on the determination of the amplitude and the delay of an acoustic signal reflected by the interfaces of a medium to represent its structure. The ultrasound probe operates in two distinct phases: transmission and reception. The delay reflects the depth of the interface (tissue, organ, vein, etc.) and

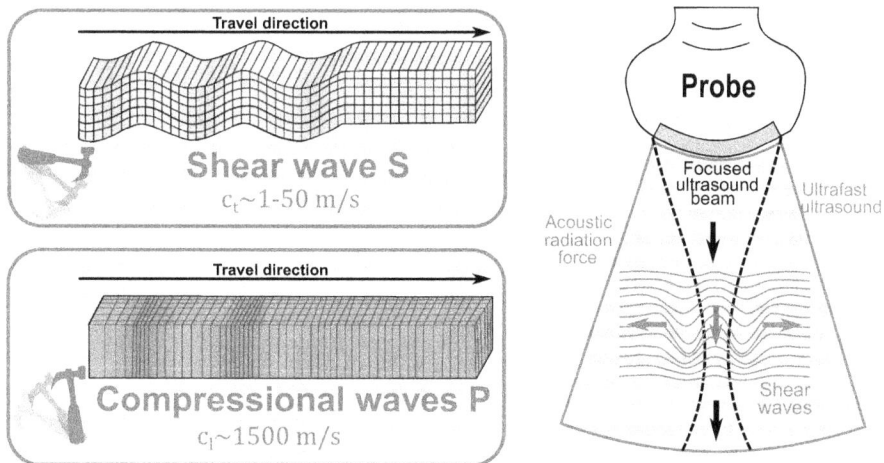

FIGURE 11.2 Ultrasounds propagation.

the amplitude of its echogenicity. **Mode A** (Amplitude) is the oldest and simplest mode. It uses the emission of an ultrasonic pulse and the reception of its echo along a single line of propagation. Using this mode, we may precisely measure the delay and therefore the distances between echogenic structures, but it can't be used to identify them. The most used mode in medical imaging is the **Mode B** (Brightness), which is a two-dimensional image formed by the juxtaposition of a large number of lines each corresponding to an ultrasound in mode A. The gray level, called brightness, represents the amplitude of the echoes (Figures 11.3 and 11.4).

When an ultrasonic wave with a fundamental frequency f_0 passes through a linear and homogenous tissue, its propagation speed becomes constant. The reflected signal is used in order to build ultrasound image and this imaging mode is called the **fundamental mode**. Human tissues are non-linear media and the ultrasound wave is distorted during its propagation. The spectrum of the reflected signal contains some harmonics. In the fundamental mode, the system performs a band-pass filtering to keep only the echoes at the fundamental frequency and eliminate other frequencies which are considered to be noise. The **harmonic mode**, *i.e.* **Tissue Harmonic Imaging** (THI), removes the fundamental component of the echo and treat only the harmonic components. In practice, the amplitude of the harmonics decreases sharply. Therefore, harmonic systems are only interested in echoes of the second harmonic (*i.e.* $2f_0$) (Figure 11.5).

In differential tissue harmonic imaging (DTHI) modes developed by Canon [42], the probe transmits simultaneously two pulses with different frequencies f_1 and f_2 $(f_1 < f_2 < 3f_1)$. The probe receptor picks up a signal composed of fundamental frequencies f_1 and f_2, their sum $f_1 + f_2$, their difference $f_1 - f_2$ and the harmonics $2f_1$ and $2f_2$ (other harmonics being neglected). The fundamental components are eliminated by an impulse subtraction technique (similar to pulse reversal [13]). The frequencies $f_1 + f_2$, $2f_2$ and other harmonics are filtered due to the bandwidth of the probe. Then, only the frequencies $f_2 - f_1$ and $2f_1$ are processed.

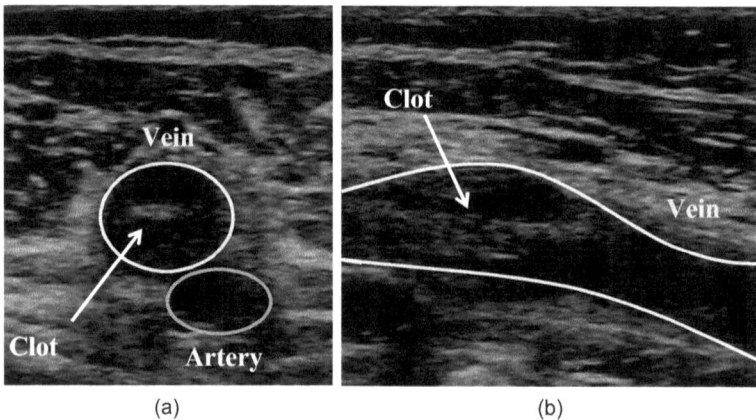

(a)   (b)

FIGURE 11.3   Transverse (a) and longitudinal (b) ultrasonographies of a blood clot in the femoral vein. (Images obtained with Canon Aplio.)

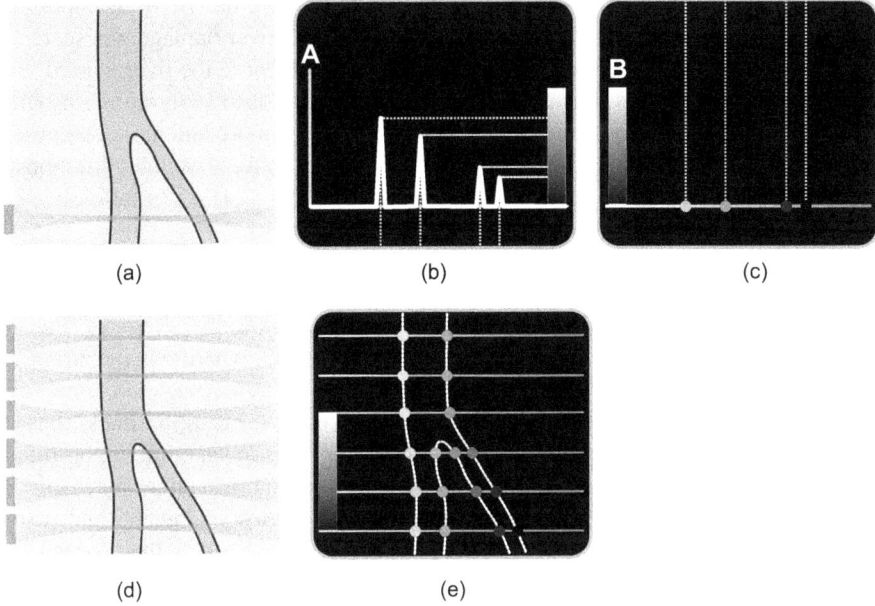

FIGURE 11.4 Mono dimensional mode A and biomedical images using mode B. (a) Exploring one-line, (b) mode A, (c) mode B, (d) exploring multi-lines, and (e) mode B.

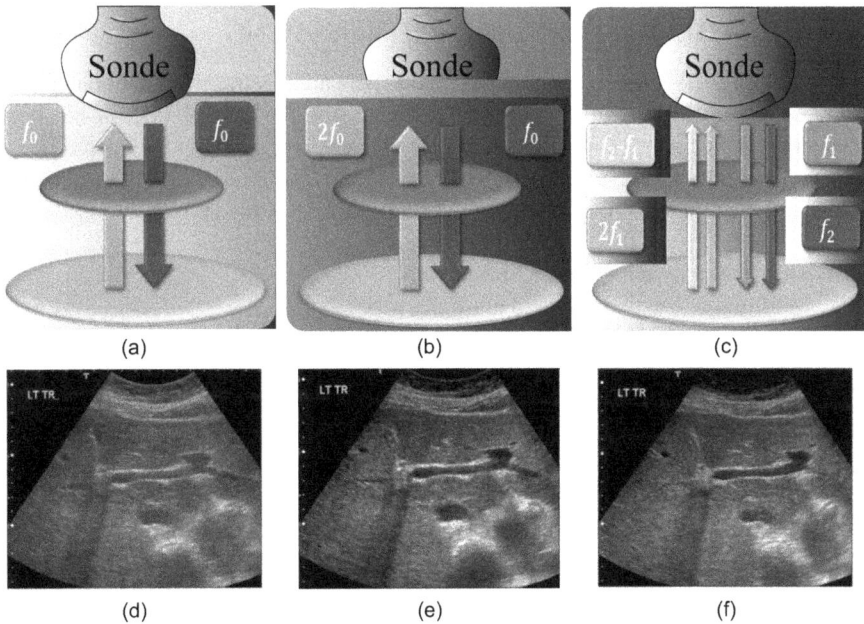

FIGURE 11.5 Three main modes of echography. (a) Fundamental mode, (b) Tissue Harmonic Imaging (THI) mode, (c) Differential Tissue Harmonic Imaging (DTHI) mode, (d) fundamental mode, (e) THI mode, and (f) DTHI mode.

Historically, tissue hardness, assessed by palpation [19], has been an important measure for diagnose liver fibrosis. Indeed, for a chronic liver damage, the scarring replaces the damaged liver cells with a fibrous scar. The more the liver is hard, the more fibrosis is important. The main objective of our study is to analyze the elasticity thrombus which normally hardens with age (because it is more and more composed of fibrin). However, the role of elastometry in the diagnosis of venous thrombosis remains to be defined. The major methods of elastography [24] are as follows:

1. **Quasi-static methods**: When an external static stress, σ, is applied to the surface of a solid, a deformation occurs depending on the material hardness E (Young's modulus in Pa). For a elastic deformation, the stress, σ (Pa), follows a linear law, called the Hooke's law: $\sigma = \epsilon E$, where ϵ is the strain (dimensionless).
2. **Dynamic or pulse methods**: They do not directly measure hardness, but the speed propagation of shear waves. The latter is as much more important than the middle is hard. The system generates an impulse either mechanical or ultrasound to generate shear waves to a region of interest and, in parallel, emits an ultrasound (at a higher frequency) to follow the propagation of shear waves in tissues. The processing of the echoes (by complex intercorrelation methods) allows to create a hardness or speed map (called elastography) and to estimate the hardness (or speed) average of a user-specified area (called elastometry).

11.4   DATA ACQUISITION AND PREPROCESSING

In the case of a DVT suspicion, the medical expert locates the blood clot head using compression ultrasonography. During the diagnosis, the patient lies on his back. Then, the physician switches to the Shear Wave Elastography (SWE) mode and selects a region of interest (ROI). In the shear wave elastography (SWE) mode, three display modes are available on Canon's system (Figure 11.6 and 11.7):

1. Shear velocity named speed mode (m/s);
2. Modulus of elasticity named elasticity mode (kPa);
3. Propagation mode.

To reduce human error (due to *e.g.* probe pressure on the skin, localization), two experts make 10 measures for every patient. Then, a second check-up should be made 3 months later. In most cases, the treatment dissolves the blood clot but new measurements are performed again. The new system limits operator errors and makes elastography images more stable. The data are exported from Aplio in a format named **Digital Imaging and COmmunication in Medicine** (DICOM). A DICOM file contains a certain number of attributes (metadata such as name, age, *etc.*) and also a special attribute containing the image pixel data (ultrasound, X-rays, **magnetic resonance angiography** (MRA), **computed tomography angiography** (CTA), *etc.*).

Ultrasounds are mainly limited by their low contrast and the presence of noise (speckle). The aim of the preprocessing methods is to improve the contrast of images,

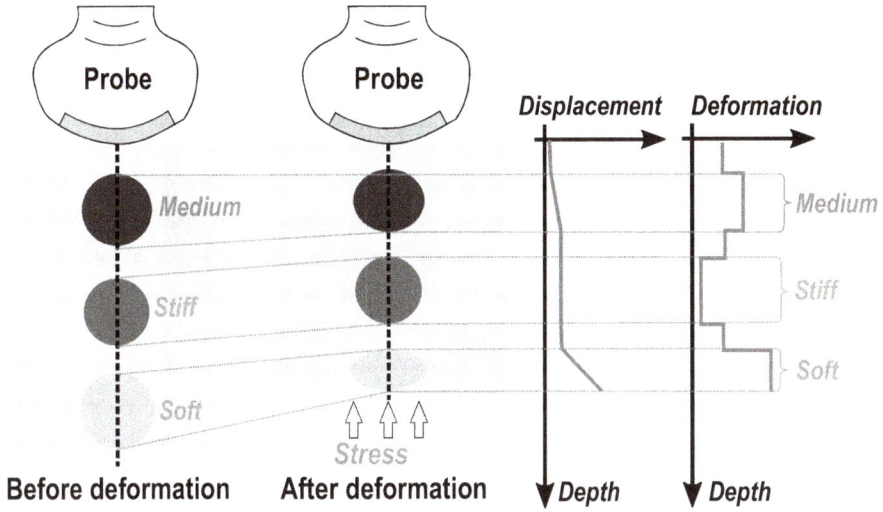

FIGURE 11.6 Elastic deformation of soft tissues with and without external stress.

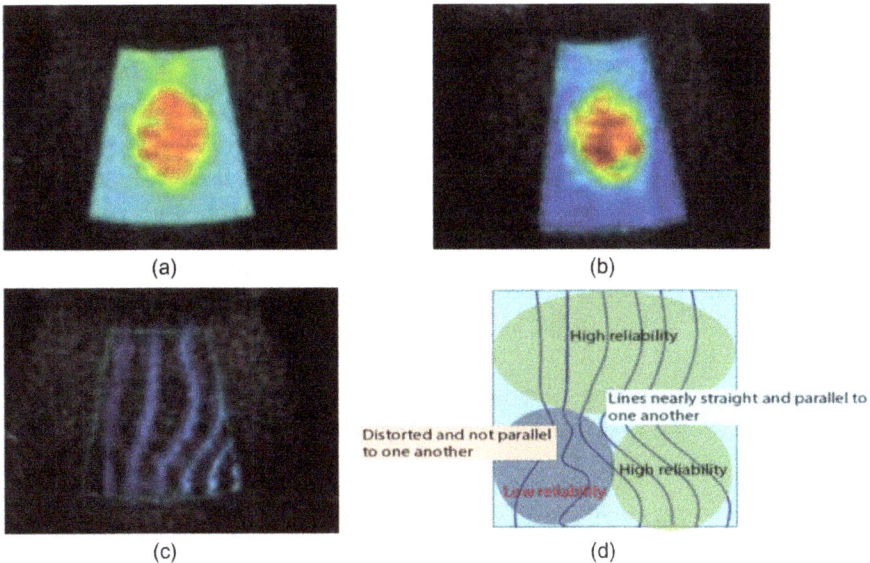

FIGURE 11.7 Three display modes: shear wave velocity (speed mode), modulus of elasticity (elasticity mode) and propagation mode [8]. (a) Speed mode (m/s), (b) elasticity mode (kPa), (c) propagation mode, and (d) reliability of the propagation display.

in particular in the regions of interest by reducing noise without losing useful information for the characterization of the targeted tissue. We can distinguish four main types of approaches to reduce the speckle (Figure 11.8):

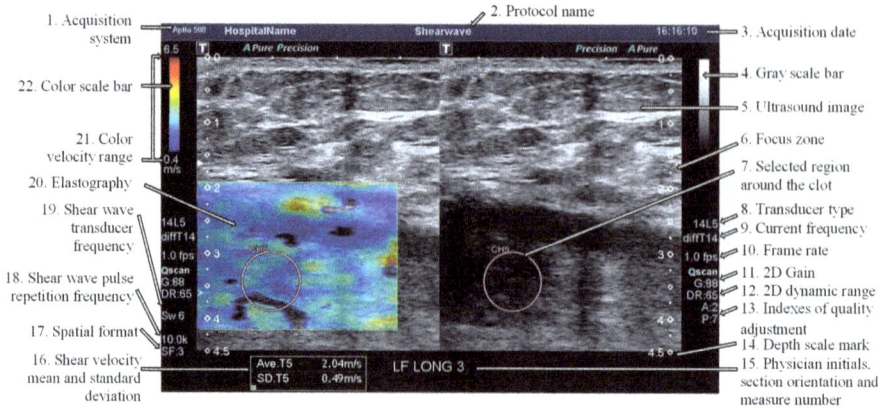

FIGURE 11.8   Canon elastography and ultrasound images; with metadata displayed on the border of the image.

1. **Equalization Histogram (EH)**: Ordinary Histogram Equalization (HE) or Contrast-Limited Adaptive Histogram Equalization (CLAHE) [86]: For improving the local contrast and enhancing the definitions of edges in each region of an image, HE can be applied over each region with a transformation function derived from their neighbor pixels. This method is called Adaptive Histogram Equalization (AHE) and a major drawback concerning the overamplifying small amounts of noise in largely homogeneous regions. This problem can be reduced by limiting contrast enhancement specifically in homogenous areas characterized by a high peak.
2. **Filtering [51]**: Linear filters (Average, RIF, Moving Average), nonlinear filters (Adaptive median, Maximum A Posteriori, anisotropy, geometric filters, *etc.*).
3. **Wavelets**: Wavelet shrinkage [11], Bayesian Approaches [1], Filtered Version [70].
4. **Image compounding**: Averaging of ultrasounds acquired at different frequencies and acquisition angles (Figure 11.9).

11.5   STATE OF ART AND BACKGROUNDS OF DVT ANALYSIS

The standard examination for DVT is **venography** (also called **phlebography** or ascending phlebography). This examination is rarely performed (because of the use of a contrast medium, exposure to X-rays, and its cost) and is replaced by ultrasound (compression test). Currently, to study the properties viscoelastic of a thrombus, elastography images are often taken in parallel with ultrasound. We can mainly distinguish two major approaches based on Echography or Elastography.

A. **Echography**: The authors of [17] present six thrombus characterization methods to conduct a neural network classification (supervised learning):

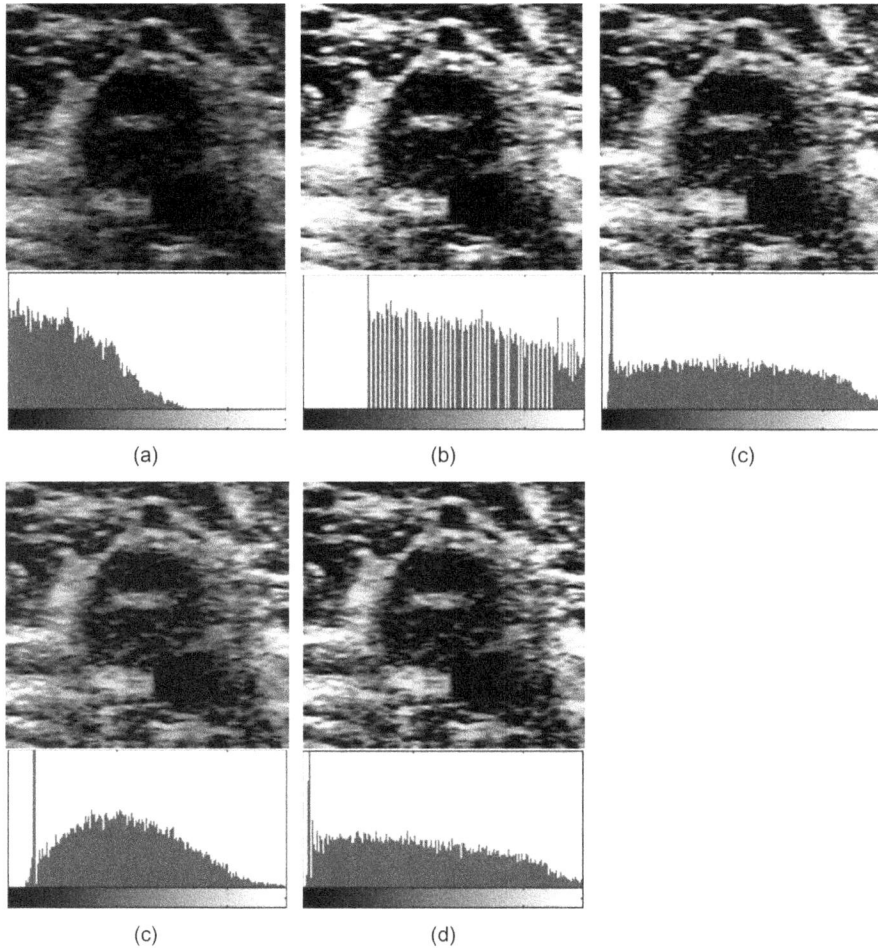

FIGURE 11.9 Ultrasonography contrast-enhanced images. (a) Ultrasonography, (b) HE, (c) CLAHE – uniform, (d) CLAHE – Rayleigh, and (e) CLAHE – exponential.

a. **Histogram**: ultrasound images are represented by their histogram (256 values).
b. **Sliding window**: a region of interest (30×30 pixels) is extracted from each ultrasound and then is proceeded by a sliding average (3×3 pixels). A vector containing 100 values is constructed from the normalized mean values of each image.
c. **Mean and variance**: each image is represented by its mean and its variance.
d. **Threshold**: the descriptor vectors are composed of a set of 0 and 1 resulting from a thresholding of ultrasound images.
e. **Wavelets**: the descriptors come from the decomposition into biorthogonal wavelets of all images.

   f. **Co-occurrence matrices**: after calculating the 16 co-occurrence matrices (for four directions and four different distances), each image is represented by energy, entropy, inertia and the homogeneity of these 16 matrices.

B. **Elastography**: Static elastography imaging shows the possibility to quantify the age of thrombosis. Nevertheless, despite its promising results obtained in vitro or in vivo on animal thrombi, this technique has significant limitations: quantitative, operator-dependent, and limited sensitivity (classification as recent or very old thrombi). To overcome these limitations, new quantitative approaches based on dynamic elastography have been proposed in recent years. Schmitt [74] reviewed existing techniques on dynamic elastography and presented its own method called Shear Wave Induced Resonance Elastography (SWIRE).

11.6 SEGMENTATION AND FEATURE EXTRACTION

In the processing chain, **segmentation** is an essential step coming just after data acquisition and pre-processing (see Section 11.4). It consists in delimiting a region of interest (lesion or targeted tissue). This step can be manually performed by the operator making the acquisition. However, to segment the thrombus only from ultrasound, several segmentation techniques can be considered, such as (Table 11.1):

Regarding **Feature Extraction**, the objective is to describe the preprocessed and segmented data by relevant features. In practice, the raw data (such as the data of thrombi) are immersed in very intertwined surfaces. The feature extraction makes

---

TABLE 11.1
Segmentation Techniques

Approaches	Description	Advantages	Drawbacks
Histogram thresholding [34,38]	After smoothing the images, we set one or more thresholds to separate healthy from damaged tissues	Easy and fast implementation	Threshold setting
Active contours (Snake [40])	After defining a rough outline, displacement of the contour according to the energy card until it best matches the object outline	Simple and independent of the object's form	Time-consuming related to the number of required iterations
Pixel classification [14,44,48,82]	Segmentation seen as a classification problem, one can use spectral classification, neural networks, etc.	Automatic extraction	How to find the class number
Markov field [56,63,71]	Modeling of ultrasonic textures by Markovian fields leading to local distribution laws	Precession and efficiency	Time and computing efforts

TABLE 11.2
Feature Extraction Techniques

Approaches	Examples
Statistical	Statistical Feature Matrix SFM [85], Pixel correlation matrix [35], ratio of variance or correlation coefficients [38], cooccurrence matrix [23,60,80], first and second order statistics [39,64,80], Block Difference of Inverse Probabilities (BDIP) [35]
Spectral	Discrete Fourier Transform (DFT) [80], Discrete Cosine Transform (DCT), relative frequency of contour elements [23], ratio of high frequencies to lower ones [38]
Texture related morphological	Energy, entropy, local homogeneity [80], wavelets analysis [59,60], contrast [38], dissimilarity [38]
	Contour form Ref. [12], echogenicity [81], protuberance and hallows [12], perimeter and area of damaged tissue [12]
Distribution models	K distribution, Rayleigh distribution of Nagami [77,78], Power-Law Shot Noise model (PLSN) [52]

it possible to "untwist" these surfaces (**manifold untangling**) so that the classes are more easily separable. In the literature, there is a wide variety of approaches using the image itself, its statistics, its energy, its texture or geometrical properties (Table 11.2).

11.7 FEATURE OR DIMENSIONALITY REDUCTION

Feature or dimensionality reduction methods can be applied as pre or post-processing to the feature extraction step. In both cases, the goal is to reduce the size of space while retaining relevant information. The difficulty of this task is to identify the relevant information and reduce redundancy. We can choose, for example, to select the frequency components with the highest energy, or to project the data into a new reduced representation space while retaining the most energetic information, or to reduce the space while preserving the distances between the images (Figure 11.10).

1. **Frequency approaches**: A simple solution to compress a signal (or an image) is to reduce it in the Fourier's domain and then perform a reverse transformation. The **Fourier Transform** (FT) or the **Discrete Fourier Transform** (DFT) breaks down the spectrum of the signal into elementary frequencies. Let $s(t)$ be an assumed continuous time signal finite energy:

$$\|s(t)\|^2 = \int_{-\infty}^{+\infty} |s(t)|^2 \, dt < \infty. \text{ The spectrum of } s(t) \text{ can be obtained by applying}$$

a FT [65]: $S(f) = \int_{-\infty}^{+\infty} s(t)e^{-i2\pi ft} dt = \langle s(t), e^{i2\pi ft} \rangle$, where $i = \sqrt{-1}$ stands for the

complex number, $a(t), b(t)$ is the inner product. For a discrete signal (i.e. discrete-time signal) $s(n)$, the Discrete Fourier Transform is defined over N

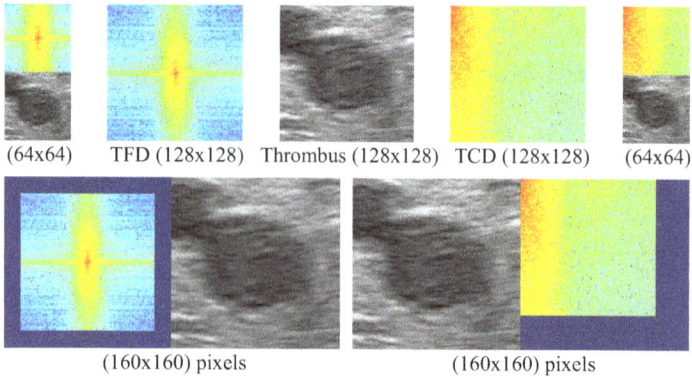

(64x64)   TFD (128x128)   Thrombus (128x128)   TCD (128x128)   (64x64)

(160x160) pixels   (160x160) pixels

FIGURE 11.10   Data compression, shrinkage and enlargement of image size using DFT and DCT.

available samples: $S(k) = \sum_{n=0}^{N-1} s(n) e^{-i2\pi\frac{nk}{N}}$. By selecting the high energy parts of the spectrum and cancelling the nonsignificant parts and evaluating an inverse Fourier transform, one can reduce the dimensionality of raw data (or images). Similar to DFT, the **Discrete Cosine Transform** (DCT) projects the signal on the orthogonal real basis of cosines. We should mention that DCT has been widely applied for data compression in digital audio (as MP3), digital video (MPEG), or images (JPEG).

2. **Principal Component Analysis (PCA)**: can reduce the dimension of the representation space for the data (or descriptors) by distorting reality as little as possible [37]. It transforms correlated variables into new uncorrelated ones. The new variables, called principal components (or axes), can be reduced by choosing the most energetic ones in the sense of variance. The retaining axes generate a new space of reduced size. PCA is therefore a geometric approach (representation of data in a new space according to directions of maximum inertia) and statistics (search for independent axes best explaining the variability of the data).

3. **t-distributed Stochastic Neighbor Embedding (t-SNE)**: In practice, the data (raw or at the output of the preprocessing stage) are very large (number of pixels for images). Their descriptors, although small in size, are often greater than three which makes the visualization of the data surface complicated. To project these data from large to smaller dimensions, for which the visualization will be possible and more intuitive, one may select the first two or three PCA components. Another popular method, **t-distributed Stochastic Neighbor Embedding (t-SNE)**, has been developed by Maaten and Hinton [54]. t-SNE chooses a low-dimensional space by respecting the distances, especially for near points in the starting space. From a mathematical point of view [54], the similarity p_{ij} between two data (x_i, x_j) in the initial space is modeled by:

$$p_{ij} = \frac{\exp\left(-\frac{\|x_i - x_j\|^2}{2\sigma^2}\right)}{\sum_{k \neq l} \exp\left(-\frac{\|x_k - x_l\|^2}{2\sigma^2}\right)}$$

where $P = (p_{ij})$ is the $D \times D$ similarity matrix, D is the basis dimension, σ is the dispersion parameter, $\|\cdot\|$ stands for the Euclidian Distance. Using a student distribution (t which is representing the t in the name of t-SNE), one can evaluate the similarity in the projection space as follows:

$$q_{ij} = \frac{\left(1 + \|y_i - y_j\|^2\right)^{-1}}{\sum_{k \neq l} \left(1 + \|y_k - y_l\|^2\right)^{-1}}$$

where $Q = (q_{ij})$ is the $D \times D$ similarity matrix in the final space. The final (projection) space is built based on the minimization of Kullback-Leibler Divergence $KL(P//Q)$. To reach this goal, one should minimize the following loss function (Figure 11.11):

$$C = KL(P//Q) = \sum_{i,j} p_{ij} \log\left(\frac{p_{ij}}{q_{ij}}\right)$$

By comparing the representation of the data before and after the feature extraction step, the t-SNE algorithm makes it possible to visualize the untangling manifold (see Figure 11.12) and to give an indication of class separability. However, if obtained points belong to the same cluster, they may be coming from different clusters in a higher dimension space, a powerful classification algorithm may separate these points into different clusters.

4. **Multi-Dimensional Scaling (MDS) [83]**: Similar to t-SNE, Multi-Dimensional Scaling (MDS) reduces the feature dimension by preserving the distances among the data points.

5. **ISOMAP [25]**: ISOMAP is a nonlinear dimensionality reduction. ISOMAP presents the data as a graph and reduces the dimension by maintaining the distances among the various nodes of that graph. The algorithm estimates

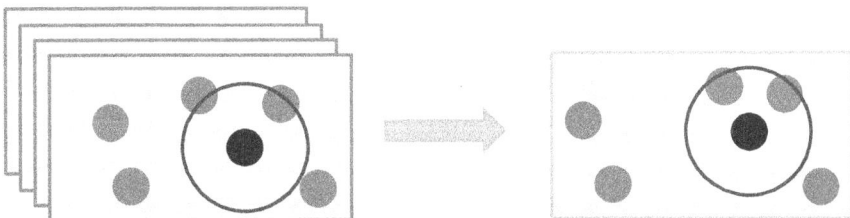

FIGURE 11.11 The concept of the t-distributed stochastic neighbor embedding method.

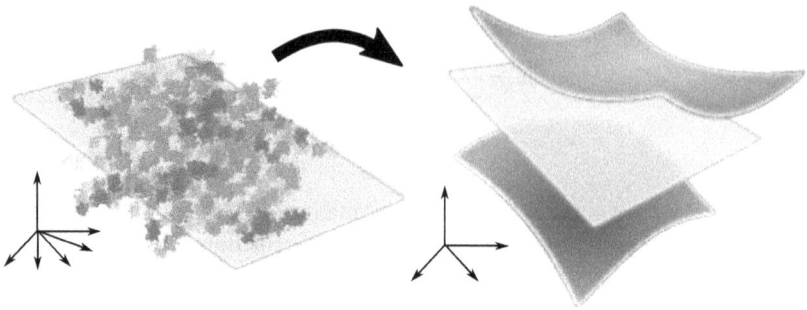

FIGURE 11.12   Manifold untangling.

the intrinsic geometry of a data manifold based on each data point's
neighbors.

6. **Kernel Principal Component Analysis (Kernel–PCA) [75]**: Using a projection kernel (by projecting the data into a space of higher dimension and
performing linear classification tasks into that new space, the linear decision
boundaries obtained in this space may correspond to complex non-linear
boundaries in the original space), one can indirectly reduce the dimension
by performing complex nonlinear projections.

7. **Linear Discriminant Analysis (LDA), Normal Discriminant Analysis
(NDA), or discriminant function analysis [25]**: The LDA is a generalization of Fisher's linear discriminant method. This method is similar to
PCA in which the principal axes (or components) will be replaced by the
discriminant axes in order to define a new hyper projection plane.

8. **Locally Linear Embedding (LLE) [72]**   measures the linear relations
among points of the original data and their closest neighbors in order to find
a smaller dimension representation which preserves the local relations.

9. **Wavelets**: In a similar way to DCT, Wavlets can be used to reduce the data
dimension. Such technics have been used in data compression and image
compression (such as JEPG 2000).

11.8   CLASSIFICATION

Identifying groups of similar data in a large data set has become an important issue
in data analysis. Nowadays, automatic classification (data clustering), which consists
of grouping together the most similar data (cluster), is applied in many fields such as
computer vision, biomedical, finance, insurance or mass distribution. As the notions
of similarity and group can be explained in many ways, many automatic classification
methods have been proposed. The most widely used classifier, due to its simplicity
and efficiency, is the k-means algorithm (k-means). However, to obtain good performance, this algorithm requires that the subsets be relatively distinct and elliptical in
shape. To overcome this constraint on the form of subsets, other partitioning methods have been developed such as the hierarchical, density-based or model-based.

TABLE 11.3
Major Supervised Classification Methods

Supervised Approaches	Examples	Description
Linear classifier	Linear discriminant analysis [78,80], Affine scattering space selection [7]	Construction a representation of a small dimension that preserves the data distances by linearly combining original variables
Probabilistic approaches	k-Nearest Neighbors (kNN), Bayesian classifier, Bayesian network [39]	Using a priori probability to define the different classes
Decision tree	Classification and Regression Trees (CART) [50], Inductive of Decision Trees (IDT) [67]	Hierarchical clustering according to established learning rules
Neural networks	Artificial Neural Network (ANN) [12,81], Deep Neural Network (DNN) [15,39,48,79]	Learning on huge data to estimate the best neural network parameters
Statistical approaches	Support Vector Machines (SVM) [35,69], Independent Component Analysis (ICA) [4,16,45]	Data are represented into classes based on some examples (support points), these classes form the hyperplanes used to separate the test data

One of the most common unsupervised classifications is Spectral Clustering [31] which is based on the use of the spectrum (eigenvalues and eigenvectors) of the similarity matrix. In the literature, many techniques make it possible to obtain a measure of similarity between two objects. The choice of this measure depends essentially on the application and the type of data (numeric, binary, text, images, voices, *etc.*) (Table 11.3).

The preprocessing stage, the automatic segmentation and the feature extraction stages are definitely essential steps of our project; However, the classification is considered as the ultimate step of our study. Indeed, we hope through our project to assist physicians with their diagnose and decision-making duties. Before addressing our strategy to extract new features and how to classify the patient conditions into some predefined classes for further diagnose and treatment, we present in the following two tables the major classification supervised or non-supervised approaches (Table 11.4).

As an important part of our project, hereinafter, we present two classical supervised approaches, and we discuss the potential of deep learning approaches. Then, we describe the concepts of popular unsupervised algorithms.

11.8.1 SUPERVISED CLASSIFICATION

Supervised classification consists of learning a model based on predefined classes and examples, and then comparing the new data to each of the learned models to deduce their class.

TABLE 11.4

Main Unsupervised Classification Approaches

Supervised Approaches	Examples	Description
Clustering by point sets	k-means clustering algorithm [32,44], Minimum Distance, Bayes Quadratic, and Voting k-NN Classifiers [39], Constrained k-means [32]	Estimation of cluster centers and grouping the points according to their minimal distance to the centers
Hierarchical clustering	Waikato Environment for Knowledge Analysis (WEKA) [30], Robust Clustering using linKs (ROCK) [28], Cobweb [21]	Hierarchical decomposition of a set of points based on predefined criteria
Density clustering	Density-Based Spatial Clustering of Applications with Noise (DBSCAN) [20], Ordering Points to Identify the Clustering Structure (OPTICS) [2]	Grouping closet points i.e. that two points are in the same cluster if their distance is less than a threshold
Model clustering	Expectation Maximization (EM) [44]	Clustering according to a mathematical model
Neural networks	Self-Organizing Maps (SOM) [41], Boltzmann's Machin, sparse coding	Self-organized network, minimization of criteria

11.8.1.1 Affine Scattering Space Selection (ASSS)

According to Ref. [7], PCA can also be used to generate affine models of each class. In the beginning, the database should be split into two subsets: Learning and Testing. When learning, PCA (or another algorithm such as linear discriminant analysis or kernel PCA) is used to select the affine model of the different classes: each class is represented by a center, the average of its members, and by a certain number of new uncorrelated variables with high variance (deduced from the original features of PCA). The number of the new variables must be defined (Figure 11.13).

The descriptor classification may start just after the modulation of all classes. Each vector is projected into the reduced space of each class (*i.e.* the affine models) and is associated with the closest class (*i.e.* whose distance between the projected vector and the center of the model is the smallest one). From a mathematical point of view [7], if we denote S_I the vector descriptor of a data item, I, of the test base, then I is associated with a class k according to:

$$k(I) = \arg\min d_{Euc}\left(S_I - P_{A_{d,k}}(S_I)\right) \tag{11.1}$$

Here K is the total number of classes, d_{Euc} is the total number of classes, $P_{A_{d,k}}(S_I)$ is the projection of S_I is the projection of the kth class and has d dimension. The dimension of the affine space d should be optimized [7].

11.8.1.2 Support Vector Machine

Support vector machine (SVM) is a generalization of linear classifiers like the ASSS: they are able to perform powerful binary nonlinear classifications [46].

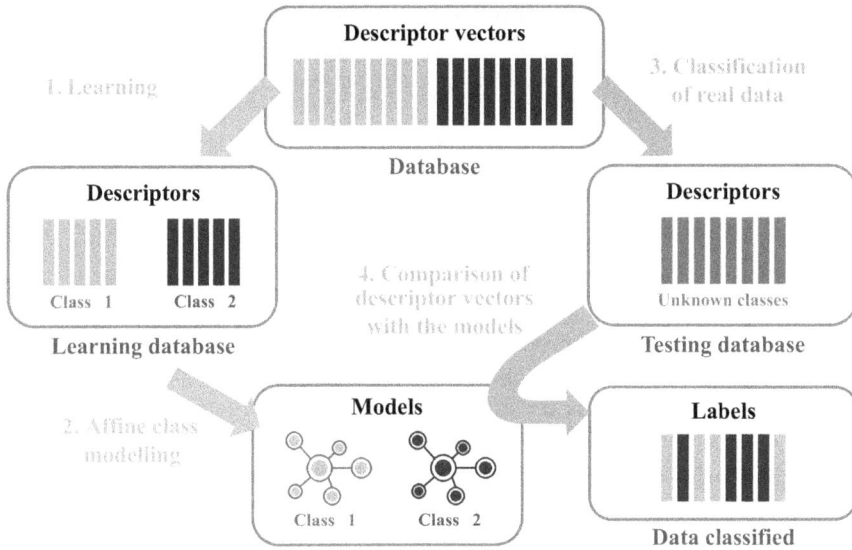

FIGURE 11.13   The concept of Affine Scattering Space Selection (ASSS).

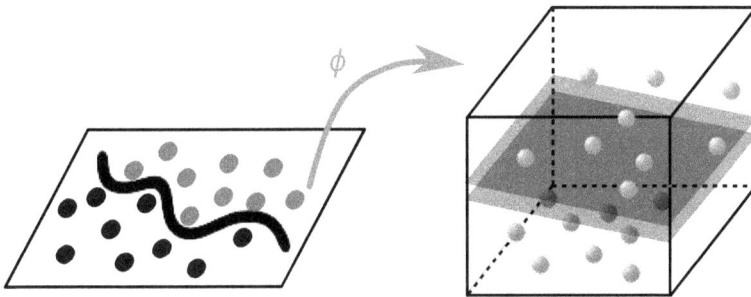

FIGURE 11.14   The concept of the support vector machine.

Firstly, proposed in the 1990s [25], they replaced many applications of neural network systems which require much more expensive training time and data. SVM consists in building a maximum width path between two observation groups (*i.e.* two classes) and to use this path as a decision boundary for the data to be classified. Compared to linear classifiers, SVM replaces a problem of complex origin with a simplified one at the cost of an increase in the dimensions (Figure 11.14).

The main idea of SVM is to apply a nonlinear transformation ϕ to transform the learning data $x_i \in R^p$ of dimension p in a new space, $\phi(R^p)$, of higher dimension. In this new space, the classes are more easily separable. SVM doesn't require explicit knowledge of ϕ: the idea is to calculate a linear separation band of maximum width by involving only the distances and the scalar products between the points to be separated (their explicit coordinates are not necessary). Let $\phi(x_i)$ denote the images of the point, x_i, of the original space, then we should only know the result of the scalar product, $K(x_i, x_j) = \langle \phi(x_i), \phi(x_j) \rangle$, between two considered points. The kernel function, $K()$, allows the PCA to classify nonlinear data. The most common kernels [25] are:

Linear: $K(x_i, x_j) = x_i^T x_j$ **Polynomial:** $K(x_i, x_j) = (\gamma x_i^T x_j + r)^d$

Radial Gaussian: $K(x_i, x_j) = \exp(-\gamma \|x_i - x_j\|^2)$ **Sigmoidal**: $K(x_i, x_j) =$ $\tanh(\gamma x_i^T x_j + r)$

where x^T stands for the transpose of x and $\|x\|$ is the Euclidean norm for x. An SVM classifier adjusts the widest possible margin (represented by the two dotted lines in Figure 11.15). The margin is the distance between the hyperplane separating two classes and the observations closest to this hyperplane. These observations are called support vectors, hence the name of "support vector machine". SVMs are sensitive to the scale difference of variables (this issue can be solved using normalized variables) and outliers: data in transformed space must be linearly separable. To avoid the latter problem, it is possible to allow a certain number of points to be inside the class, or even on the wrong side. This is referred to as a "soft margin" classification as opposed to "hard margin".

11.8.1.3 Deep Learning

For several years, technologies based on artificial intelligence have invaded gradually our daily lives [46]: voice recognition (Siri, Cortana or Google now), facial identification, machine translation, video games, autonomous robots, and automatic driving. The concept of Neural Networks (NN) is not, however, new since the first attempts date from the 1960s. Twenty years later, the first back-propagation error algorithm was a success for RNs. However, the computing power required to train

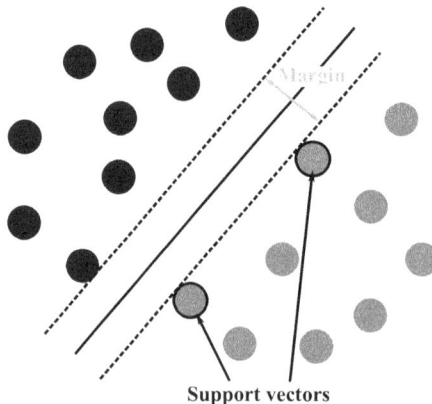

FIGURE 11.15 SVM creates the path corresponding to the greatest margin between the two classes.

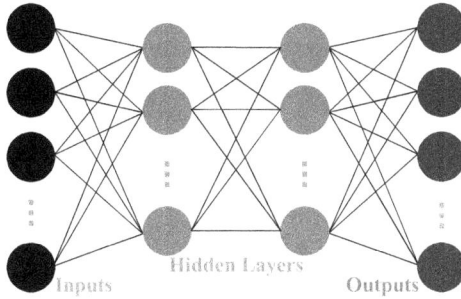

FIGURE 11.16 A neural network with two hidden layers.

such networks and the arrival of alternative methods such as SVMs have contributed to idle NNs [27].

The principle of NNs is inspired from the mammal visual cortex [33]. In a simple version, each neuron of the network has an activation level between 0 and 1. The interconnection diagram between neurons defines the architecture of the network. A classic architecture consists in organizing neurons in successive layers with interconnections limited to adjacent layers as shown in Figure 11.16. A classic example is the automatic classification of handwritten numbers. The activation levels of the input layer correspond to the levels of gray of the pixels (or descriptors) of the input image. The output layer indicates the probability of this image to be one of the ten digits. Network training consists of finding the neuron weights of hidden layers such as the output layer accurately classifies the images from the learning base.

The emergence of new unsupervised learning models of **deep networks (Deep Belief Nets)** combined with exponential growth in computing power of machines (CPU, GPU) has attracted the interest of the scientific community [73]. The first notable result is to be credited to Microsoft's Neural Network which won the voice recognition challenge. The second very important success took place during the ImageNet challenge (one million labeled images divided into thousand categories) by the **AlexNet** network [43]. In many areas, the arrival of **convolutional neural networks**, as AlexNet, made a spectacular leap in performance compared to other references.

Since 2012, many neural network architectures [27] such as:

- *ConvNet* [84] developed by Oxford University, see Figure 11.17;
- *Caffe*, *Caffe2* and *DeCAF* first developed by Berkeley Vision and Learning Center (BVLC);
- *Theano* (Python), developed by the University of Montreal;
- *TensorFlow* originally designed by Google;
- *Keras* library, created by Google and based on *TensorFlow* and *Theano*.

The different stages of *ConvNet* presented in Figure 11.17 [57] are:

- **Convolution stage**: This layer processes the input data with a set of convolutional filters to extract certain features (descriptors). The first layer can extract simple features, such as brightness or edges, while the deep layers

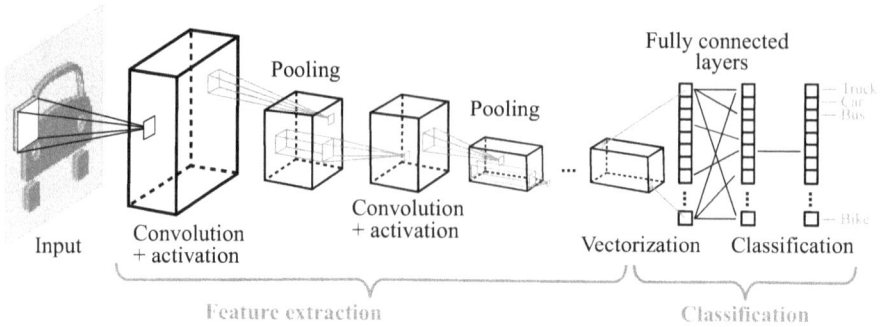

FIGURE 11.17   Convolutive Neurol Networks (CNN).

will bring out more complex characteristics which will uniquely define the object. Unlike traditional methods, convolution filters are learned by the network during the training phase: they are first initialized and then updated by a gradient back-propagation. These filters are considered as the weights of the convolution layer.

- **Activation stage**: The activation layer generally follows a convolution layer and corresponds to applying a nonlinear function to the descriptors extracted in the previous layer. This non-linearity helps to solve classification problems with complex boundaries. A popular activation function is the ReLU (**Rectified Linear Units**) which replaces all negative values with zeros. This layer speeds up and optimizes the learning because only activated neurons are transferred to the next layer.

- **Pooling stage**: It is an important step to reduce the spatial size of descriptor maps (*i.e.* the outputs of the first two stages, convolution-activation layers) through a stride which adds a partial invariance in translation and increases the internal stability of the representations. Among the most used pooling operators, there is the max function, the mean or the L_p norm (*e.g.* Euclidean norm if p equals 2). This layer improves the network efficiency by reducing the parameter number of the network and limiting the risk of overfitting.

- **Fully connected network stage**: This layer receives a vector from the vectorization layer (flatten). By linear combinations, it produces an output vector that indicates the probability for the input image to belong to each class. To calculate these probabilities, the fully connected layer multiplies each input element by a weight, makes the sum, and then apply an activation function (softmax for example). The matrix of weights is updated during the network training phase.

A convolutional neural network can have several dozen or even hundreds of layers that learn from each other to identify different characteristics of an image. For a given application, there are generally two alternatives for learning the network filters:

start from scratch or use a pre-trained model. In the first case, the network architect must define the number of layers and filters, as well as other adjustable parameters. It must also have a large amount of data, sometimes millions of samples. The other alternative is to use a pre-trained model to automatically extract features from a new set of data. This method, called transfer learning, is a practical solution to apply a deep learning network without having a very large dataset or performing very consuming computational efforts related to learning phases. This type of alternative has given good performance in medical imaging [49].

11.8.2 Unsupervised Classification

Unsupervised classification seeks to group data into different clusters (classes) without a prior knowledge. It is often necessary to transform the data representation space (dimension reduction and/or feature extraction). In our project, we mainly used three approaches to classify our thrombi images: two partitioning algorithms and a method that transforms space before applying a partitioning algorithm.

11.8.2.1 K-means

The k-means [32] is the most popular clustering algorithm for its simplicity and efficiency. Its principle, illustrated in Figure 11.18, is as follows:

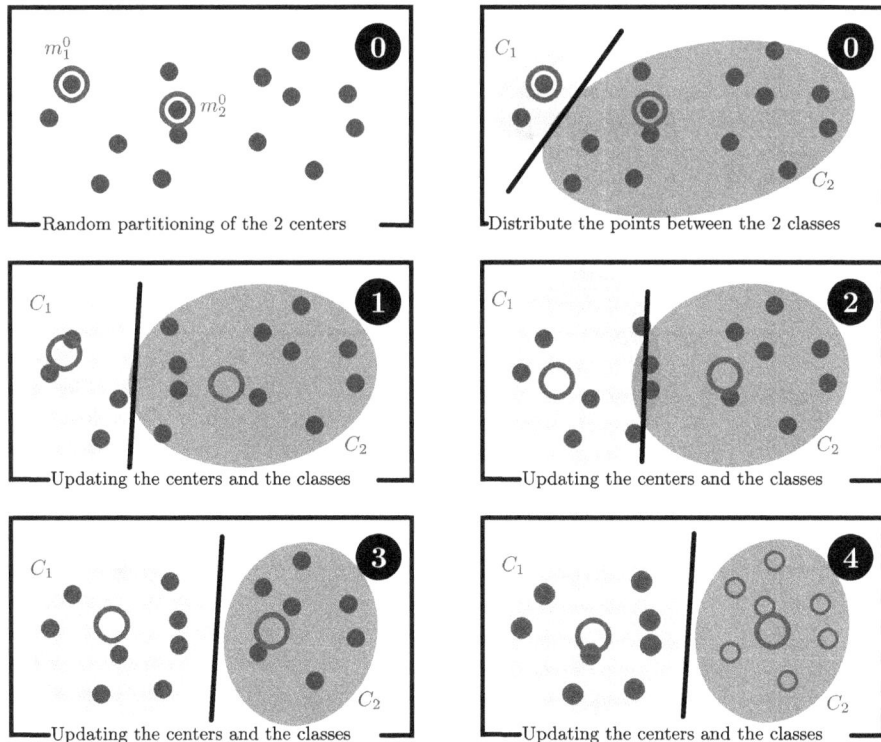

FIGURE 11.18 Main steps of k-means.

1. Random partitioning of points into k subsets (clusters) and calculation of centroids (*i.e.* center) of each cluster;
2. As long as the centroids are not stabilized (*i.e.* do not converge toward a local minimum):
   a. Calculation of the distance between each point and each centroid;
   b. Assignment of each point to the nearest cluster (*i.e.* with the smallest distance to its center);
   c. Update the centroids of each cluster.

In practice, the algorithm stops when two successive iterations lead to the same partition or when a certain criterion is met. The two most common criteria are the maximum number of iterations and the minimum number of points to change classes. The main drawbacks of k-means are the predefined number of classes, and its sensitivity to initial points and outliers. The algorithm is only suitable for hyper-elliptical clusters. Therefore, it is often preferable to transform the data representation space with techniques such as PCA [61], spectral clustering [14] or the scattering operator.

11.8.2.2 Density-Based Spatial Clustering of Applications with Noise (DBSCAN)

Density-Based Spatial Clustering of Applications with Noise (DBSCAN) is a density-based algorithm [20] that identifies in the representation space the high-density areas (clusters) surrounded by low-density areas, see Figure 11.19. Considering a set of given points, the points closest to each other (*i.e.* dense) will be grouped together.

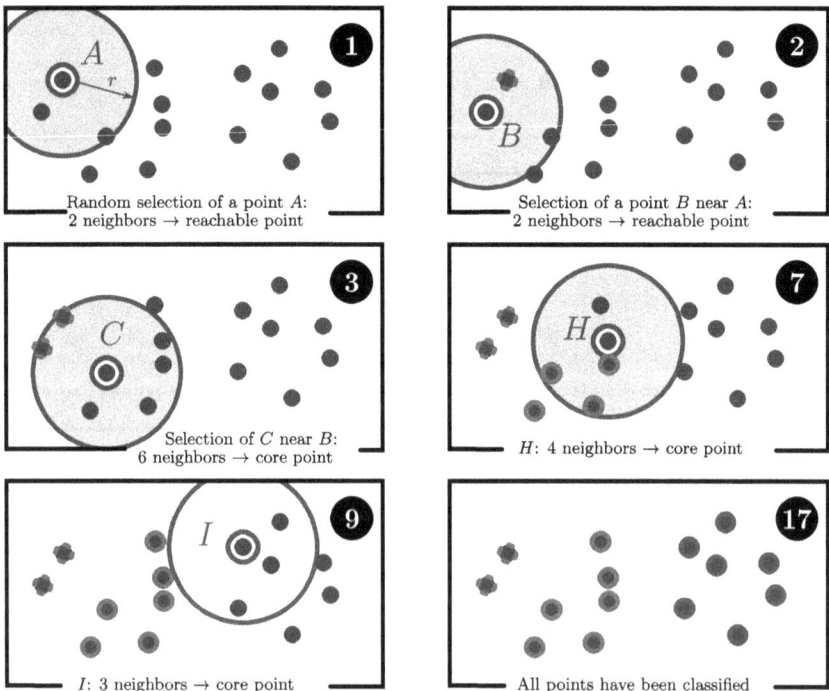

FIGURE 11.19   Concept of DBSCAN.

In practice, the algorithm begins by randomly selecting a point A and searches for all its neighbors. The first key parameter to define is therefore the maximum radial distance r to define the neighborhood. The second is the minimum number of points needed to create a cluster. Depending on this number, there will be three scenarios:

1. Point A has no neighbor and is therefore considered to be an outlier;
2. Point A has enough neighbors so it belongs to a cluster and is named a core point;
3. Point A does not have enough neighbors, so it is called reachable point and it will be assigned to the closet cluster (if its neighbors are outliers, it will also be classified as outlier).

If the point A has neighbors, the algorithm selects one of them and looks at its neighborhood. Thus, for each added point, we expand the cluster zone. The larger this zone (sphere of radius r) becomes the more likely the cluster will grow. The notion of neighborhood (or density) is therefore a key point. Unlike k-means, DBSCAN does not require a predefined number of classes and has no constraint on the cluster shape. It is worth to be highlighted that the choice of the radius r becomes important. For example, if we have a class D (where the data is very dense) and another class E (where its data are well spaced), the choice of a small r can result in splitting the class E into subclasses or outliers, while a large radius r may generate a single cluster. This drawback can be overcome by using the **OPTICS** algorithm (**Ordering points to identify the clustering structure**) which represents data in the form of a graph where each node is attached to the nodes closest to it.

11.8.2.3   Spectral Clustering

To overcome the limitations of k-means (hyperelliptic classes) and DBSCAN (constant neighborhood radius), it is often preferable to transform the space of representation. To group together the most similar data, the Spectral Clustering [31] generates the new space from the spectrum (eigenvalues and eigenvectors) of the similarity matrix. In the literature, many techniques are proposed as similarity measurements depending on applications and the type of data (real or binary data, texts, images). In Ref. [14], they successfully applied spectral clustering on MRIs images of prostates. In our project, we implemented and used the algorithms published in Ref. [31].

11.8.2.3.1   Main Idea

The main idea of Spectral Clustering is to represent all the data as a weighted graph and then cut this graph at the level of the weakest links. A graph is defined by: V is the set of all the nodes (each node represents a data), E contains all the links between two points $\left(v_i, v_j\right)$ and W corresponds to the symmetric neighborhood matrix. The coefficient W_{ij} takes as value the weight of the link between the nodes v_i and v_j. In practice, for a given set of nodes, there are several ways to define a graph:

- Graph of ϵ-neighbors: Node links less ϵ are only considered;
- Graph of k-nearest neighbors: each node is connected to the k-nearest neighbors;
- Fully connected graph: all nodes are interconnected.

TABLE 11.5
Laplacian, Degree and Similarity Matrices

L	1	2	3	4	5	6	7	8	9
1	1.5	-0.7	-0.7	-0.1	0	0	0	0	0
2	-0.7	1.5	-0.8	0	0	0	0	0	0
3	-0.7	-0.8	2.5	0	0	-0.2	-0.6	-0.1	-0.4
4	-0.1	0	0	1.6	-0.7	-0.8	0	0	0
5	0	0	0	-0.7	1.5	-0.6	0	-0.2	0
6	0	0	-0.2	-0.8	-0.6	2.4	-0.3	-0.5	0
7	0	0	-0.6	0	0	-0.3	2.4	-0.5	-0.9
8	0	0	-0.1	0	-0.2	-0.5	-0.6	2	-0.6
9	0	0	-0.4	0	0	0	-0.9	-0.6	1.9

(a) Laplacian $L = D - W$

D	1	2	3	4	5	6	7	8	9
1	2.5	0	0	0	0	0	0	0	0
2	0	2.5	0	0	0	0	0	0	0
3	0	0	3.5	0	0	0	0	0	0
4	0	0	0	2.6	0	0	0	0	0
5	0	0	0	0	2.5	0	0	0	0
6	0	0	0	0	0	3.4	0	0	0
7	0	0	0	0	0	0	3.4	0	0
8	0	0	0	0	0	0	0	3	0
9	0	0	0	0	0	0	0	0	2.9

(b) Degree matrix D

W	1	2	3	4	5	6	7	8	9
1	1	0.7	0.7	0.1	0	0	0	0	0
2	0.7	1	0.8	0	0	0	0	0	0
3	0.7	0.8	1	0	0	0.2	0.6	0.1	0.4
4	0.1	0	0	1	0.7	0.8	0	0	0
5	0	0	0	0.7	1	0.6	0	0.2	0
6	0	0	0.2	0.8	0.6	1	0.3	0.5	0
7	0	0	0.6	0	0	0.3	1	0.5	0.9
8	0	0	0.1	0	0.2	0.5	0.6	1	0.6
9	0	0	0.4	0	0	0	0.9	0.6	1

(c) Similarity matrix W

In our simulations, no a priori knowledge on the links between thrombi images is considered. Therefore, we use fully connected graphs. The similarity and distance play a key role in the performance of spectral clustering methods. Various similarity measures can be found in the literature, in our work, we used the most common ones: **Gaussian similarity**: $W(I,J) = \exp\left(-\dfrac{\|I-J\|^2}{2\sigma^2} \right)$ & **Euclidean distance**:

$D_{\text{Euc}} = \sqrt{\sum_{i=1}^{N}(I_i - J_i)^2}$. Here, I and J stand for two images, N is the pixel number, σ

is the dispersion parameter. The similarity matrix W has coefficients between 0 and 1 (Table 11.5).

After evaluating the similarity matrix, one should calculate the connection degree of each node with all other nodes of the graph. The degree matrix D is a diagonal matrix defined by its coefficient $\left(D_i = \sum_{j=1}^{n} W_{ij} \right)$ of a node i. Once the graph is established, then the second step of the Spectral Clustering consists in cutting this graph into a certain number of groups k. The intuitive idea is to cut it at the weakest connections. Let X and Y be two sets, the interconnection Cut (*i.e.* the connection between the two subsets): $\text{Cut}(X,Y) = \sum_{i \in X, j \in Y} W_{ij}$. By considering the three sets, A, B and C introduced in Figure 11.20, we can evaluate their interconnections Cut as follows: $\text{Cut}(A,B) = \sum_{i \in A, j \in B} W_{ij} = 0.3$, $\text{Cut}(A,C) = \sum_{i \in A, j \in C} W_{ij} = 1.1$,

$\text{Cut}(B,C) = \sum_{i \in B, j \in C} W_{ij} = 1.0$, $\text{Cut}\left(A,\bar{A}\right) = \text{Cut}\left(A,\{B,C\}\right) = \text{Cut}(A,B) + \text{Cut}(A,C) = 1.4$,

$\text{Cut}\left(B,\bar{B}\right) = \text{Cut}\left(B,\{A,C\}\right) = \text{Cut}(B,A) + \text{Cut}(B,C) = 1.3$, and $\text{Cut}\left(C,\bar{C}\right) = \text{Cut}\left(C,\{A,B\}\right) = \text{Cut}(A,C) + \text{Cut}(B,C) = 2.1$.

Let us consider the subset X formed by the points $\{1,4,8\}$ and its complement $\bar{X} = \{2,3,5,6,7,9\}$, we obtain an interconnection of 4.9 much greater than those found previously. To partition the graph into three classes, we must therefore find the three subsets which minimize the interconnection function (*i.e.* find A, B and C in the example): $J_{k\,\text{Cut}}(X_1,...,X_k) = \sum_{i=1}^{k} \text{Cut}\left(X_i, \bar{X_i}\right) = J_{k\text{Cut}}(u_k) = \sum_{i=1}^{k} u_i^T L u_i$; With $u_i = 1$ if

$i \in X_k$ and 0 else, the Laplacian $L = D - W$, X^T is the transpose of X, using the example presented in Figure 11.20, we can find that: $u_A = [111000000]^T$; $u_B = [000111000]^T$;

$u_C = [000000111]^T$; $J_{kCut}(A,B,C) = \sum_{i \in \{A,B,C\}}^{k} u_i^T L u_i$.

The solution to minimize $J_{kCut}(u_k)$ is in the space of the L eigenvectors associated with the k smallest none zero eigenvalues. Figure 11.20 shows the projection of the graph shown in Figure 11.21b in the plane formed by the second and the third eigenvectors (the first being associated with 0). After the projection, one applies a

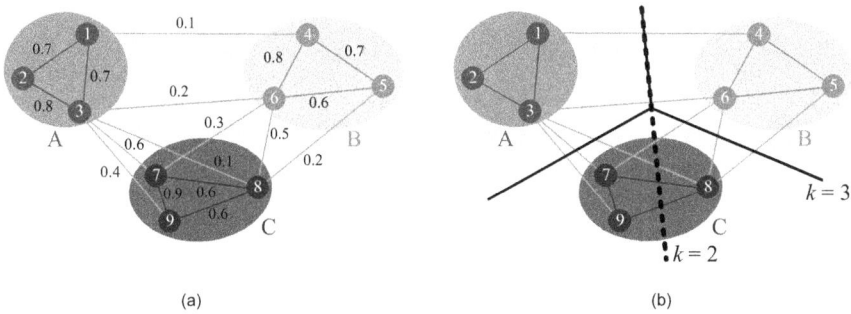

(a)   (b)

FIGURE 11.20 Example of weighted and cut graph into two or three clusters.

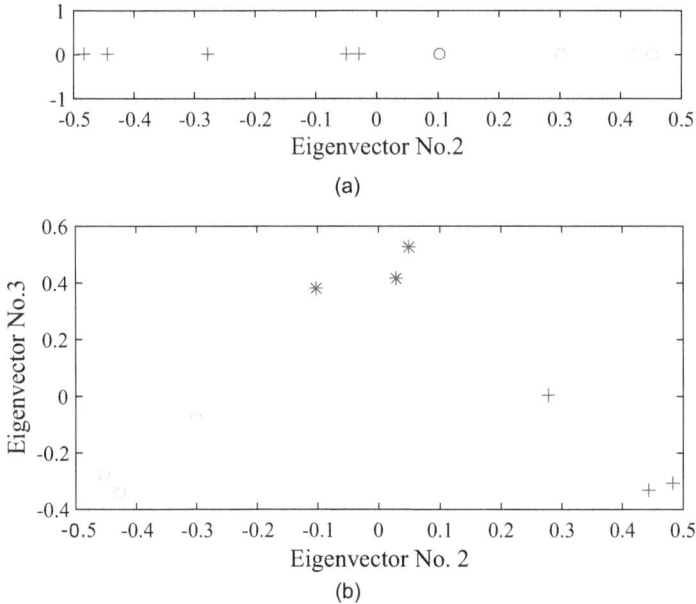

(a)

(b)

FIGURE 11.21 The projection of the example presented in Figure 20. The points of A (resp. B and C) are presented in blue (resp. in green and red). (a) Projection according to the second eigenvector. (b) Projection according to second and third eigenvectors.

classification algorithm, as k-means, to group these points into k classes in this simplified space. Figure 11.21a shows the projection according to the second eigenvector in the case of partitioning into two classes. The results given in Figure 11.21b show two and three classes. We can thus note that the choice of the class number is also an important parameter.

11.8.2.3.2   Selected Algorithms

To cluster the data into k classes, one should minimize $J_{kCut}(u_k)$. To reach this goal, many algorithms can be found in the literature. Ng, Jordan and Weiss (NJW) [62] developed an efficient algorithm in six steps:

1. Build the graph and calculate the similarity matrix W as mentioned previously;
2. Calculate the normalized Laplacian $L = D^{-1/2}(D-W)D^{-1/2}$
3. Evaluate the matrix U containing the eigenvectors of L associated with the smallest none zero eigenvalues;
4. Normalize every raw of U and form the following matrix

$$T = \left(t_{ij} = \frac{u_{ij}}{\sqrt{\sum_{k=1}^{n} u_{ik}^2}} \right);$$

5. Classify the elements of every raw of T into k sub-sets using any classification algorithm;
6. Label every node of the graph by the result obtained in the previous step.

Among various feature extraction approaches, we have considered those based on wavelets or statistics because they are concerned with the texture of the image and its distribution, respectively. Morphological Descriptors are not well adapted to our problem and there is not, to date, a mathematical model capable of translating the structure and/or evolution of a thrombus. To observe if our data can be classified and if the obtained clusters can be linked to real pathologies, we first consider a visualizing method (t-SNE), a scattering space selection supervised method (ASSS) and three unsupervised algorithms (k-means, DBSCAN and spectral clustering).

11.9   MULTIRESOLUTION APPROACHES TO CHARACTERIZE THROMBUS

The Scattering Operator, introduced by Mallat [7], can improve the automatic classification of signals (or images). Considering for example a cloud of points representing two different but non-dissociable classes of images, our objective becomes to find a function Φ to change the representation space, such that the two classes become distinguishable.

The scattering operator convolves the input images with scaled and oriented wavelets. Every wavelet is obtained from the mother wavelet by a rotation θ and a scaling factor j:

$$\psi_\lambda(u) = 2^{-2j}\psi\left(2^{-j}r_\theta^{-1}u\right)$$

where $\lambda = 2^j r_\theta$, $u = (u_1, u_2)^T$ is the spatial position vector, $r_\theta = \begin{pmatrix} \cos\theta & -\sin\theta \\ \sin\theta & \cos\theta \end{pmatrix}$ is

the rotation matrix. We used the Morlet's wavelet [7] (Figure 11.22).

Wavelets may fit better natural signals and are stable to small deformations. However, they are affected by translation. To become translation invariance, we use the modulus and apply a low-pass filter (the father wavelet). The modulus can shift the high-frequency information toward lower frequencies; then a low pass filter extracts the high-frequency information. The first Scattering Operator coefficient for an image x is given by: a. Where * stands for the convolution product, the approximation coefficient $S[\phi]x$ is invariant with translation and it doesn't contain the low pass information (Figure 11.23).

The low pass information can be obtained by calculating the wavelet coefficients $U[\lambda_1]x = |x * \phi_J|$ where $\lambda_1 = 2^{j_1} r_{\theta_1} \in \Lambda_J = \left\{\lambda = 2^j r_\theta, 0 \le j \le J, \theta = \dfrac{l\pi}{L}, \text{with } 0 \le l \le L\right\}$. The wavelet coefficients $U[\lambda_1]x$ are stable for small deformation but variant with

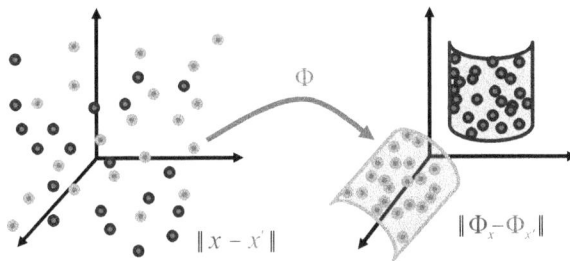

FIGURE 11.22   Space transformation by the function Φ.

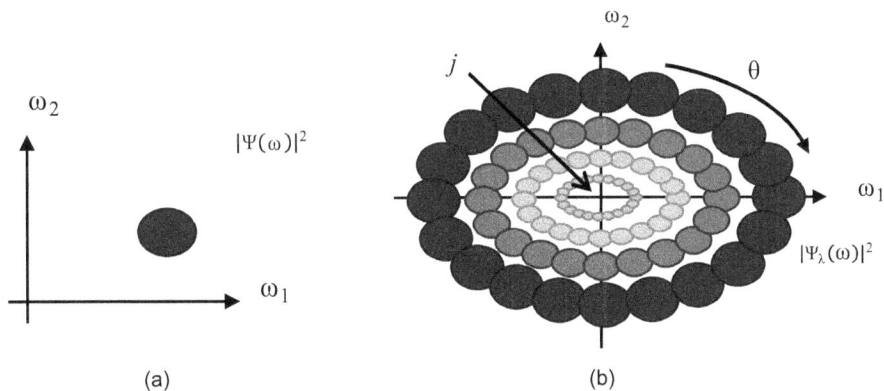

FIGURE 11.23   The mother wavelet on the left side and other wavelets with rotation and scaling.

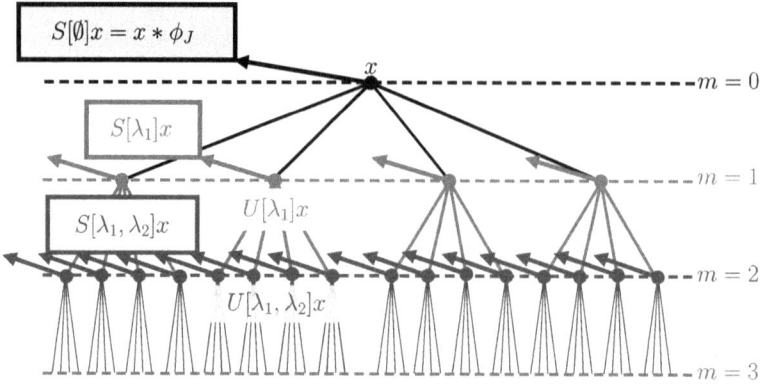

FIGURE 11.24 The scattering operator coefficients with three levels of decomposition.

translation. To obtain the invariance property, we must apply a filter to all the wavelet coefficient: $S[\lambda_1]x = U[\lambda_1]x * \phi_J = |x * \psi_{\lambda_1}| * \phi_J$. The obtained coefficients $S[\lambda_1]x$ are the first-order scattering operator coefficients. By repeating the same operation, we can obtain the higher order scattering operator coefficients: $U[p]x = U[\lambda_m]...U[\lambda_2]U[\lambda_1]x = \left\| |x * \psi_{\lambda_1}| * \psi_{\lambda_2}|...\psi_{\lambda_m} \right\| * \phi_J$.

The Scattering Operator can be seen as a network of wavelet filters. All obtained coefficients for different orders of decomposition form a stable representation to deformations, invariant in translation and therefore effective for classification. Scattering operator also preserves the energy of the original image. Indeed, the energy of the scattering coefficients decreases exponentially in function of the order of decomposition m. After four orders of decomposition, the coefficients are sufficient to recover over 99% of the image energy [7].

To make the Scattering Operator coefficients invariant to rotation, we must first evaluate all the coefficients, $S[\phi]x$, $U[\lambda_1]x$, as mentioned before. In addition to the spatial filter ϕ_J and for each fixed scale j_1, the obtained coefficients for each rotation θ_1 should be averaged. The coefficients of the first order with invariance to rotation are defined by: $S[j_1]x = U[j_1,\theta_1]x \circledast \phi_J$; where $\phi_J(u,\theta) = \bar{\phi}(\theta)\phi_J(u)$, $\circledast$ is the convolution product according to the rotation (Figure 11.24).

11.10 STATISTICAL APPROACHES TO CHARACTERIZE THROMBUS

Random signals can be partially or completely described through their probability distributions. Conventional signal processing techniques are based on a simplified description using moments of order one or two (mean, variance, standard deviation, or correlation) [68]. However, higher-than-two order statistics provide other useful information. These statistics, developed primarily for astronomy, seismic prospecting and communications, can be used to resolve certain problems not accessible to order two such as blind source separation [66]. Nowadays, they have been applied in many application fields [58]: geophysics, speech [55], sonar, radar and more recently image processing [53] or biomedical [5].

11.10.1 Spatial Moments and Cumulants

In this section, the definitions and the properties of moments and cumulants are briefly introduced. For further details, see [68] and therein cited references. The **moment of the** q^{th} **order** is defined by $\mathcal{M}_q[I](i,j) = E\big[I(m,n)I(m+i_1,n+j_1)...I(m+i_{q-1},n+j_{q-1})\big]$; where I is the considered image and $I(m,n)$ represent the pixels of I, $i=(i_1,...,i_{q-1})$ and $j=(j_1,...,j_{q-1})$ denote the delays, E stands for the expectation, the moment of the qth order can be obtained by the qth derivative of the first characteristic function. The first characteristic function can be considered as the "Fourier" Transform of the probability density function. The second characteristic function is the logarithm of the first characteristic function. Using the successive derivative of the second characteristic function, one can define the qth **order cumulant**: $\mathcal{C}_q[I](i,j) = \mathcal{C}\big[I(m,n)I(m+i_1,n+j_1)...I(m+i_{q-1},n+j_{q-1})\big]$. We should mention that the qth order cumulant can be calculated as a polynomial function of all moments of order q and lower (as proposed by Leonov and Shiryayev). The qth moment of a $M \times N$ size image I can be estimated by: $\hat{\mathcal{M}}_q[I](i,j) =$

$$\frac{1}{MN}\sum_{m=0}^{M-1}\sum_{n=0}^{N-1}I(m,n)I(m+i_1,n+j_1)...I(m+i_{q-1},n+j_{q-1})$$

The **autocorrelation** is the second-order cumulant, for a zero mean image I, see Figure 11.25:

$$\hat{\mathcal{C}}_2[I](i,j) = \hat{\mathcal{M}}_2[I](i,j) = \frac{1}{MN}\sum_{m=0}^{M-1}\sum_{n=0}^{N-1}I(m,n)I(m+i_1,n+j_1)$$

The **cross-correlation** between two images I and J can be considered as a similarity measurement:

$$\hat{\mathcal{C}}_2[I,J](i,j) = \hat{\mathcal{M}}_2[I,J](i,j) = \frac{1}{MN}\sum_{m=0}^{M-1}\sum_{n=0}^{N-1}I(m,n)J(m+i_1,n+j_1)$$

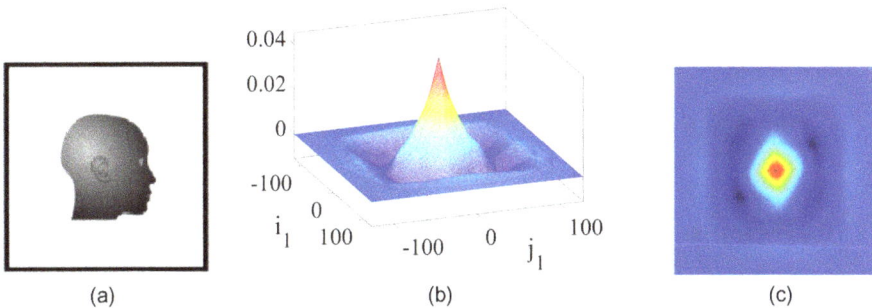

FIGURE 11.25 Autocorrelation of an image I. (a) Image I, (b) 3D autocorrelation, (c) 2D autocorrelation.

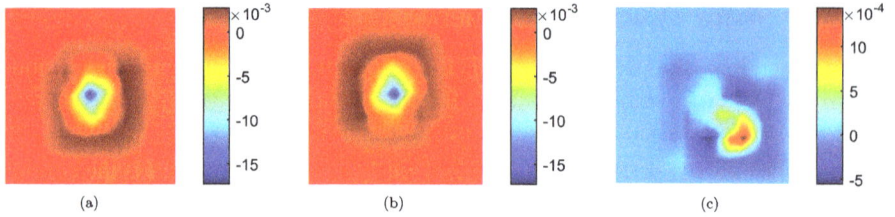

FIGURE 11.26   Bicorrelation of an image I presented in Figure 11.25. (a) $\hat{C}_3[I](i,j,i,j)$, (b) $\hat{C}_3[I](0,0,i,j)$, and (c) $\hat{C}_3[I](M/2,N/2,i,j)$.

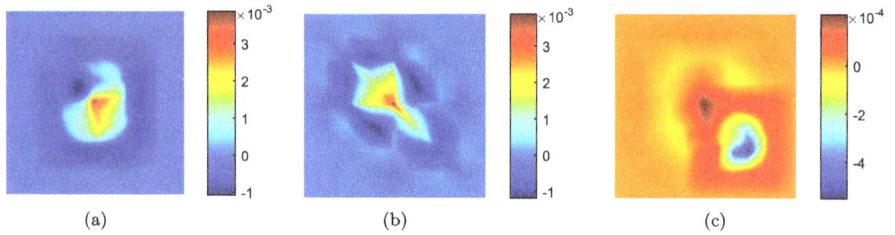

FIGURE 11.27   Triple-correlation of an image I presented in Figure 11.25. (a) $\hat{C}_4[I](i,j,i,j,i,j)$, (b) $\hat{C}_4[I](0,0,i,0,j,0)$, and (c) $\hat{C}_4[I](0,0,M/2,N/2,i,j)$.

The **bicorrelation** is the third-order cumulant for a zero mean image I, it can be estimated by fourth order Tensor (Figures 11.26 and 11.27):

$$\hat{C}_3[I](i,j) = \frac{1}{MN}\sum_{m=0}^{M-1}\sum_{n=0}^{N-1} I(m,n)I(m+i_1,n+j_1)I(m+i_2,n+j_2)$$

For zero mean images, the **Triple-correlation** is a fourth-order cumulant given by:

$$\hat{C}_4[I](i_1,i_2,i_3,i_4,j_1,j_2,j_3,j_4) = \hat{\mathcal{M}}_4[I](i_1,i_2,i_3,i_4,j_1,j_2,j_3,j_4)$$
$$- \hat{\mathcal{M}}_2[I](i_1,j_1)\,\mathrm{hat}\mathcal{M}_2[I](i_2-i_3,j_2-j_3)$$
$$- \hat{\mathcal{M}}_2[I](i_2,j_2)\,\mathrm{hat}\mathcal{M}_2[I](i_1-i_3,j_1-j_3)$$
$$- \hat{\mathcal{M}}_2[I](i_3,j_3)\,\mathrm{hat}\mathcal{M}_2[I](i_1-i_2,j_1-j_2)$$

11.10.2   FEATURE EXTRACTION

Multi-correlation of order higher than 2 can only be presented by tensors. The higher the order of the multi-correlation becomes, the higher the order of the tensor will be, and therefore the tensor size increases rapidly. For an image of 64×64 pixels, its bicorrelation will include more than 260 million values. Therefore, it will be not impossible to calculate all the values. In practical situations, one can downsample the images or reduce their sizes. However, even by reducing the image to 16×16 pixels, the bicorrelation would still contain almost a million points. Another way consists of calculating values only along certain privileged axes. Cardoso et al. propose to reduce the fourth-order cumulant tensor by a matrix generated from the eigenvectors of the tensor (Joint Approximation Diagonalization of Eigenmatrices [9]). The authors of [18] also present an algorithm to simultaneously diagonalize the tensor into slices of order three. Likewise, Comon proposes to diagonalize the third-order tensor by maximizing its trace [16]. These approaches make it possible to perform a tensor contraction (*e.g.* to lower the tensor order as similar to a projection). To characterize the thrombi images, we propose to calculate multicorrelations along one (or more) privileged axis(es) on the plane made up of the last two dimensions of the tensor, *i.e.*: three horizontal axes, three vertical axes and two diagonals. These axes are shown in Figure 11.28. For example, we can obtain from these axes:

- Autocorrelation axis 1: $\hat{C}_2[I](\tau_1) = \hat{C}_2[I](\tau, \tau) =$

$$\frac{1}{MN} \sum_{m=0}^{M-1} \sum_{n=0}^{N-1} I(m,n) I(m+\tau, n+\tau)$$

- Autocorrelation axis 2: $\hat{C}_2[I](\tau_2) = \hat{C}_2[I](\tau', \tau') = \hat{C}_2[r_{\pi/2}(I)](\tau, \tau)$; where $r_{\pi/2}$ represents 90deg rotation.
- Bicorrelation axis 3: $\hat{C}_3[I](\tau_3) = \hat{C}_3[I](0,0,0,\tau) =$

$$\frac{1}{MN} \sum_{m=0}^{M-1} \sum_{n=0}^{N-1} I^2(m,n) I(m,n+\tau)$$

- Bicorrelation axis 6:

$$\hat{C}_3[I](\tau_6) = \left(\hat{C}_3[I](\tau_3), \left|\hat{C}_3[I](\tau_4)\right.\right)^T = \left(\hat{C}_3[I](0,0,0,\tau), \left|\hat{C}_3[I](0,0,\tau,0)\right.\right)^T$$

- Bicorrelation axis 8:

$$\hat{C}_2[I](\tau_8) = \left(\hat{C}_2[I]\left(-\frac{M}{2}, \tau\right), \hat{C}_2[I](0,\tau), \left|\hat{C}_2[I]\left(\frac{M}{2}, \tau\right)\right.\right)^T$$

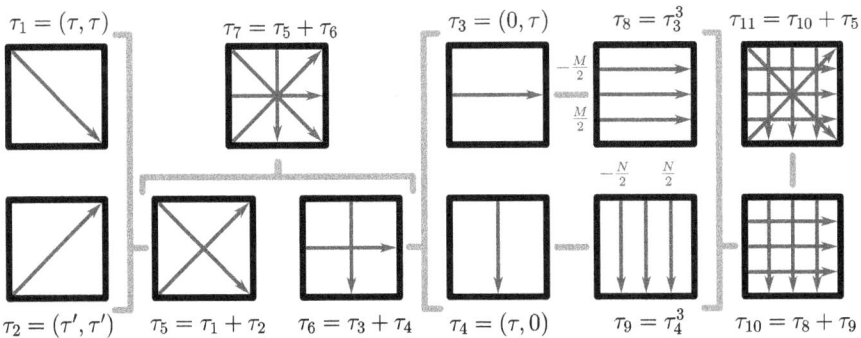

FIGURE 11.28 Multi-correlation projected along various axes.

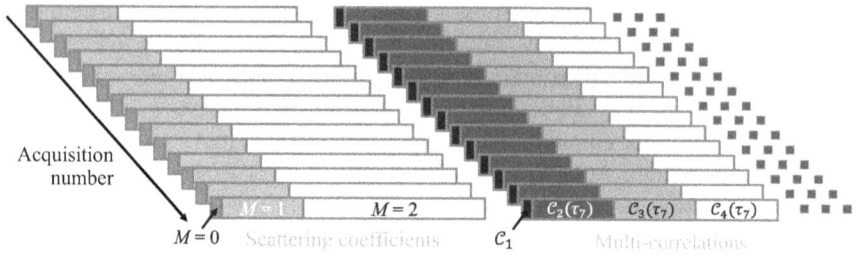

FIGURE 11.29 Our features are based on a combination of Scattering coefficients and Multi-correlations.

In our study, multi-correlations with up to eleven combinations of axes have been used as features of our thrombi images. Another way to process our data consists of using the dimension reduction methods presented in Section 11.7. The advantage of choosing preselected axes is that all images have the same set of features; Unlike the dimensionality reduction which can project the images in different reduced spaces (Figure 11.29).

11.11   EXPERIMENTAL RESULTS

The validation and optimization of feature extraction approaches (Scattering Operator and High Order Statistics) are difficult to be directly performed on the basis of ultrasound and elastography images for two main reasons: a lack of labelled data, and reliable classified and processed data. Parametric optimization may require relatively big classes well identified and containing about hundreds or thousands of data. Therefore, we optimized our parameters based on a reference database: KTH-TIPS (Kungliga Tekniska Högskolan - Textured under varying Illumination, Pose and Scale). After having validated our approaches on a reference database and obtained promising results. The objective of our simulations is to classify obtained images with respect to Deep Vein Thrombosis (DVT). Three classes are considered:

- **Blood stasis** due to prolonged immobilization or reduced mobility;
- **Alteration of the venous walls** due to a surgery or aging veins;
- **Thrombophilia** is a state of hypercoagulability of the blood due to genetic predisposition, hormones or cancer.

It is hard to clearly define our classes. For example, cancer may cause blood stasis and hypercoagulability. In addition, some patients present with idiopathic DVT which means that the main trigger cannot be identified from the patient's medical history. Potentially, these patients can therefore suffer from undeclared cancer which has favored the development of the thrombus [22]. DVTs due to immobilization can be distinguished from other DVT because the thrombi are often older (about 15 days) and therefore more echogenic. Finally, a clustering according to the presence of PE can be considered as a positive result. The main complexity remains that a diagnosed DVT without PE can provoke (without the anticoagulant treatment) eventually

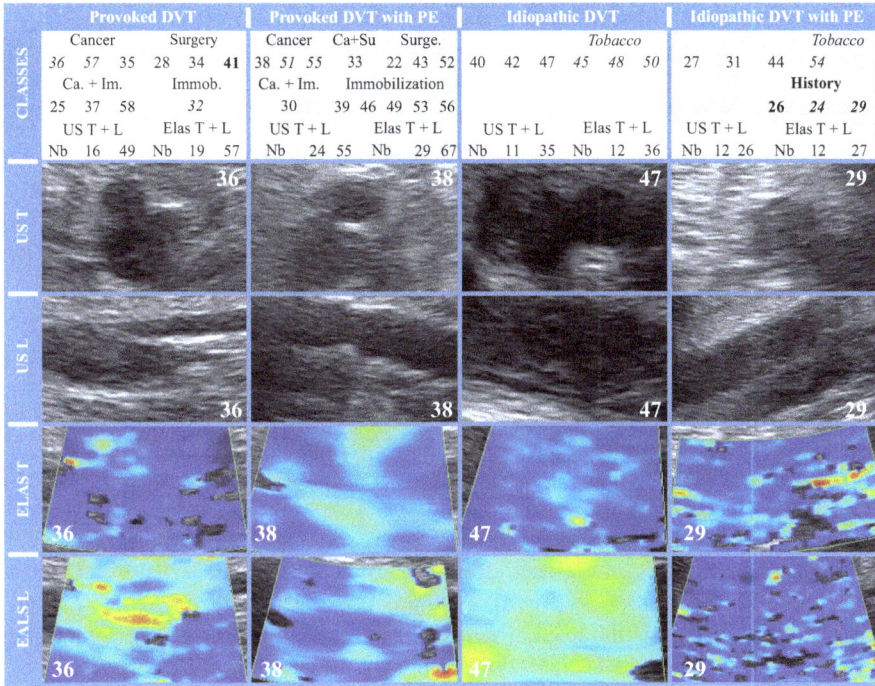

FIGURE 11.30 Thrombi base organized into four classes according to induced DVT or not (idiopathic DVT) and the presence or not of PE: Patients' IDs are sorted according to triggering factors; Bold IDs indicate previous history of DVT; Italic IDs mean smoking; The image numbers by type (ultrasound and elastography) or section (transverse and longitudinal) are provided.

a thromboembolic case. Our project tries to classify the thrombi according to their triggers. From Figure 11.30, we distinguish 5 classes:

1. Surgery: patients 22, 28, 34, 41, 43 and 52;
2. Genetics (ie, for this basis, with family history): patients 24, 26 and 29;
3. Cancer: patients 25, 30, 33, 35, 36, 37, 38, 51, 55, 57 and 58;
4. Idiopathic: patients 27, 31, 40, 42, 44, 45, 47, 48, 50 and 54;
5. Immobilization: patients 32, 39, 46, 49, 53 and 56.

We can already note that some patients could belong to two different classes since, for example, four of the base patients reported DVT following immobilization and in the presence of cancer. For the simulations, these patients are assigned to the Cancer class because we believe cancer may have more influence on the structure thrombus than immobilization (or surgery for patient 33).

11.11.1 IDENTIFICATION OF THE TRIGGERING FACTOR

Here, we aimed to correlate the structure of the thrombus and their generating processus. As mentioned above, we split the dataset according to four triggers (surgery,

TABLE 11.6

ASSS Confusion Matrix (The Percentage Is the Average over Ten Partitions) According to Venous Thrombi Triggers: Surgery (Su.), Genetics (Ge.), Cancer (Ca.), Idiopathic (Id.) & Immobilization (Im.)

C_{ASSS}	Su.	Ge.	Ca.	Id.	Im.
Su.	0.0	0.0	33.3	66.7	0.0
Ge.	0.0	0.0	80.0	20.0	0.0
Ca.	0.0	0.0	58.0	42.0	0.0
Id.	1.7	1.7	21.7	75.0	0.0
Im.	0.0	0.0	16.7	76.7	6.7

(a) Elastography - $S18$ - $SO_{log} + C_2$ (42.2%)

C_{ASSS}	Su.	Ge.	Ca.	Id.	Im.
Su.	0.0	0.0	26.7	73.3	0.0
Ge.	0.0	0.0	20.0	80.0	0.0
Ca.	0.0	0.0	42.0	58.0	0.0
Id.	0.0	0.0	25.0	75.0	0.0
Im.	0.0	0.0	50.0	50.0	0.0

(b) Ultrasound - $S02$ - $SO_{log} + C_2$ (38.8%)

genetics, cancer and immobilization) and the idiopathic DVT (no trigger). The classification results are shown according to five classes mentioned in Table 11.6. Table 11.6a shows a good classification rates by dedicating 50% of thrombi from each class to learning process (for 10 random partitions from each class). For example, the Surgery class is modeled based on the data of 3 patients and the remaining data are used for the test. The rows of Table 11.6a correspond to the simulation settings (image type, standardization of thumbnails by acquisition, standardization and merger of descriptors images by acquisition) and the columns indicate the descriptor vectors used (images vectorized, scattering SO coefficients, C multicorrelations and combined descriptors).

According to Table 11.7a, every simulation gives relatively low rates (32% on average) which fluctuate between 26% and 42%. The maximum is obtained for simulation S18 (elastography with fusion of descriptors by acquisition) with the combined descriptors ($SO_{log} + C_2$). The Confusion Matrix corresponding to this maximum rate (Table 11.6a) shows that the majority of elastographies are assigned to the Cancer and Idiopathic classes. These results could indicate that thrombi of cancerous and genetic origin may have a different structure from idiopathic thrombi, due to immobilization or surgery. However, the confusion matrix of the same simulation applied to ultrasound scans (Table 11.6b) shows a different distribution of thrombi in the two same classes.

According to supervised classification results, a classification in two classes can certainly be more suitable (than in five classes). Looking at Table 11.7b, we can verify this statement on k-means results (in five classes) which are similar to those of the ASSS (average percentage of 33%), since they oscillate between 24% and 43%. By focusing on the confusion matrix of two best classification rates (Table 11.8), we observe different results from those of the supervised classification. In both cases, we observe significant confusion between the Surgery and Cancer classes. In elastography (Simulation S22), the thrombi due to immobilization are rather well classified (60%), whereas in ultrasound (S01), they are distributed among all clusters.

On the other hand, Table 11.7c indicates that the spectral classification is slightly more efficient, to classify thrombi elastographies, than the two previously

TABLE 11.7
Thrombi Classification According to Triggers

Simulations / Descriptors:				Images	SO	SO$_{log}$	c_2	c_3	c_4	$c_{1,2}$	$c_{2,4}$	$c_{1,4}$	SO+ c_2	SO+ $c_{1,2}$	SO+ $c_{2,4}$	SO+ $c_{1,4}$	SO$_{log}$+ c_2	SO$_{log}$+ $c_{1,2}$	SO$_{log}$+ $c_{2,4}$	SO$_{log}$+ $c_{1,4}$
T	Img	Descrip.	N°																	
Ultrasound			S01		32	31	35	31	31	27	31	32	32	31	34	33	33	38	26	30	30
		F	S02		30	31	36	31	32	32	38	33	33	35	36	29	29	30	39	32	32
	N		S03		32	32	34	28	32	35	30	34	34	28	27	32	33	36	35	31	32
	N	F	S04		30	31	34	26	37	30	29	34	34	28	28	33	33	32	31	34	33
N			S05		31	28	34	29	31	27	29	33	32	29	31	32	32	35	34	29	29
N		F	S06		35	31	34	30	30	32	31	34	34	30	29	32	32	34	36	28	29
N	N		S07		31	32	32	28	30	35	30	34	34	28	27	32	33	35	34	31	31
N	N	F	S08		35	30	34	26	37	30	28	33	33	27	28	33	33	32	31	34	34
Ultrasound HE			S09		32	31	35	31	31	27	31	32	32	31	34	33	33	38	26	30	30
		F	S10		30	31	36	31	32	32	38	33	33	35	36	29	29	39	39	32	32
	N		S11		32	32	34	28	30	35	30	34	34	28	27	32	33	36	35	31	32
	N	F	S12		30	31	34	26	27	30	29	34	34	28	28	33	33	32	31	34	33
N			S13		31	28	34	29	31	27	29	33	32	29	31	32	32	35	34	29	29
N		F	S14		35	31	34	30	30	32	31	34	34	30	29	32	32	34	36	28	29
N	N		S15		31	32	32	28	30	35	30	34	34	28	27	32	33	35	34	31	31
N	N	F	S16		35	30	34	26	37	30	28	33	33	27	28	33	33	32	31	34	34
Elastography			S17		37	31	35	33	35	32	36	32	32	36	33	30	30	38	37	37	37
		F	S18		37	32	32	32	35	29	33	29	30	29	34	29	30	42	33	34	34
	N		S19		37	28	32	32	35	33	33	30	33	51	34	28	29	24	35	29	29
	N	F	S20		37	28	29	32	33	27	32	51	32	31	33	32	32	34	34	33	33
N			S21		30	35	31	33	27	32	33	32	32	35	35	33	33	38	34	36	35
N		F	S22		34	35	27	34	34	30	34	31	31	36	34	32	32	32	31	28	28
N	N		S23		30	28	32	32	33	32	34	31	34	31	34	28	29	33	34	32	30
N	N	F	S24		24	28	29	32	35	30	32	31	33	33	32	32	33	34	35	33	32

(a) ASSS with 50% of thrombi used for learning stage (total standardization by type).

Simulations / Descripteurs :				Images	SO	SO$_{log}$	c_2	c_3	c_4	$c_{1,2}$	$c_{2,4}$	$c_{1,4}$	SO+ c_2	SO+ $c_{1,2}$	SO+ $c_{2,4}$	SO+ $c_{1,4}$	SO$_{log}$+ c_2	SO$_{log}$+ $c_{1,2}$	SO$_{log}$+ $c_{2,4}$	SO$_{log}$+ $c_{1,4}$
T	Img	Descrip.	N°																	
Echographie			S01		29	32	39	34	27	32	28	32	35	28	32	25	29	38	42	41	40
		F	S02		35	37	30	32	30	33	34	31	32	34	36	30	32	29	35	34	34
	N		S03		31	27	35	35	40	35	31	31	32	32	26	37	32	36	34	33	34
	N	F	S04		36	30	31	30	35	34	31	34	33	29	26	32	30	29	31	32	33
N			S05		32	28	29	37	29	31	24	34	32	33	28	32	33	32	32	32	31
N		F	S06		31	33	36	37	30	32	35	32	32	34	35	33	32	37	35	35	31
N	N		S07		36	28	35	35	38	36	37	31	32	28	26	32	32	36	37	32	32
N	N	F	S08		31	31	32	30	35	36	32	32	32	28	31	28	28	31	30	33	36
Echographie EH			S09		29	32	39	34	28	31	28	32	33	28	32	25	29	38	42	41	40
		F	S10		35	37	30	32	30	33	34	31	32	34	36	30	30	29	35	34	34
	N		S11		31	27	35	35	40	34	31	31	32	26	27	31	33	36	34	33	33
	N	F	S12		34	31	32	32	35	34	36	33	35	31	32	34	32	32	32	32	34
N			S13		37	27	28	36	27	31	28	33	32	35	27	32	32	32	31	31	31
N		F	S14		31	33	35	37	32	33	37	30	31	36	36	33	33	35	34	33	31
N	N		S15		36	28	35	35	39	34	37	31	32	28	26	32	32	36	37	32	32
N	N	F	S16		31	31	32	30	32	35	32	32	32	28	31	28	28	31	30	33	36
Elastographie			S17		26	31	43	38	36	41	36	36	41	39	34	32	32	42	41	43	43
		F	S18		34	33	42	36	34	32	35	38	36	35	31	33	35	40	39	40	43
	N		S19		26	34	28	27	28	33	25	36	40	31	33	29	37	34	37	38	34
	N	F	S20		36	34	32	35	35	34	29	35	36	32	37	33	32	29	31	34	32
N			S21		35	27	30	31	34	38	37	37	35	32	27	28	29	41	38	31	31
N		F	S22		29	38	33	28	37	37	29	40	40	31	29	35	37	43	35	31	31
N	N		S23		35	35	30	36	34	36	30	38	37	33	34	35	34	41	39	39	38
N	N	F	S24		29	38	37	28	38	34	30	35	34	33	35	38	37	33	36	32	36

(b) k-means (total standardization by type).

Simulations / Descriptors:				Images	SO	SO$_{log}$	c_2	c_3	c_4	$c_{1,2}$	$c_{2,4}$	$c_{1,4}$	SO+ c_2	SO+ $c_{1,2}$	SO+ $c_{2,4}$	SO+ $c_{1,4}$	SO$_{log}$+ c_2	SO$_{log}$+ $c_{1,2}$	SO$_{log}$+ $c_{2,4}$	SO$_{log}$+ $c_{1,4}$
T	Img	Descrip.	N°																	
Ultrasound			S01		25	35	35	28	21	26	26	24	27	38	30	29	25	35	32	27	26
		F	S02		32	33	33	28	31	33	33	28	27	29	25	31	34	36	36	35	35
	N		S03		26	25	35	33	40	31	28	28	29	29	28	32	32	36	35	14	36
	N	F	S04		32	35	23	31	34	35	29	28	28	30	29	32	30	32	32	32	29
N			S05		26	28	26	32	19	26	29	26	24	26	27	24	24	32	32	27	27
N		F	S06		28	30	35	28	30	34	33	32	28	31	27	27	27	33	31	28	26
N	N		S07		25	27	34	33	41	35	29	26	27	30	28	28	30	36	36	33	35
N	N	F	S08		27	35	25	33	31	36	31	28	28	31	30	30	27	31	32	28	32
Ultrasound HE			S09		25	35	35	28	20	33	26	24	27	38	30	29	25	35	32	27	26
		F	S10		35	35	33	28	31	34	33	33	28	27	26	29	31	34	36	36	35
	N		S11		26	25	35	33	41	36	28	28	29	30	29	32	34	36	36	16	36
	N	F	S12		33	36	25	32	32	33	31	31	29	30	32	32	31	30	33	28	31
N			S13		26	30	26	28	21	26	29	22	23	28	27	24	25	31	32	27	27
N		F	S14		28	29	33	28	29	34	31	29	29	28	30	26	27	32	32	26	27
N	N		S15		25	27	34	33	40	35	29	26	27	30	28	28	30	36	36	33	35
N	N	F	S16		27	35	25	33	31	36	31	28	28	31	30	30	27	31	32	28	32
Elastography			S17		36	37	31	37	38	33	38	26	34	38	43	37	39	48	47	47	48
		F	S18		35	33	37	29	32	34	23	36	35	27	29	37	37	36	37	35	39
	N		S19		33	34	39	29	28	37	30	37	38	32	53	45	48	33	24	38	39
	N	F	S20		32	40	33	31	30	31	35	36	37	32	33	27	28	31	31	39	40
N			S21		28	38	29	31	31	31	30	32	33	39	36	31	41	48	46	44	44
N		F	S22		30	43	36	36	34	33	31	33	37	39	33	36	40	40	43	36	39
N	N		S23		28	34	38	38	31	36	31	42	42	36	35	43	42	31	34	42	39
N	N	F	S24		29	41	31	30	33	27	29	38	38	34	34	35	33	32	30	40	37

(c) Spectral clustering with k-means (total standardization by order).

TABLE 11.8

Confusion Matrix (Percentage over Ten Tests) by k-Means According to Triggers of Venous Thrombi

C_{kmn}	Su.	Ge.	Ca.	Id.	Im.
Su.	3.3	1.7	61.7	23.3	10.0
Ge.	6.7	60.0	30.0	0.0	3.3
Ca.	2.0	2.0	49.0	32.0	15.0
Id.	1.5	23.1	13.8	42.3	19.2
Im.	0.0	0.0	15.7	24.3	60.0

(a) Elastography - $S22$ - $SO_{log} + C_2$ (42.6%)

C_{kmn}	Su.	Ge.	Ca.	Id.	Im.
Su.	58.3	0.0	13.3	20.0	8.3
Ge.	0.0	33.3	50.0	16.7	0.0
Ca.	29.0	0.0	48.0	3.0	20.0
Id.	26.9	6.2	19.2	39.2	8.5
Im.	32.0	10.0	14.0	20.0	24.0

(b) Ultrasound - $S01$ - $SO_{log} + C_{1,2}$ (42.2%)

TABLE 11.9

Confusion Matrix According to Triggers of Venous Thrombi

C_{SC}	Su.	Ge.	Ca.	Id.	Im.
Su.	28.3	11.7	41.7	16.7	1.7
Ge.	16.7	56.7	26.7	0.0	0.0
Ca.	29.0	8.0	39.0	11.0	13.0
Id.	9.2	7.7	2.3	65.4	15.4
Im.	0.0	0.0	10.0	50.0	40.0

(a) Elastography - $S17$ - $SO_{log} + C_2$ (47.7%)

C_{SC}	Su.	Ge.	Ca.	Id.	Im.
Su.	33.3	6.7	13.3	26.7	20.0
Ge.	0.0	33.3	60.0	0.0	6.7
Ca.	20.0	10.0	40.0	20.0	10.0
Id.	36.9	9.2	0.0	46.2	7.7
Im.	20.0	20.0	20.0	0.0	40.0

(b) Ultrasound - $S07$ - C_3 (40.5%)

mentioned algorithms. The combination of multiresolution descriptors and statistics makes it possible to achieve 48% good classification. The corresponding confusion matrix, displayed in Table 11.9a, shows that four of the five classes are moderately classified: the majority of idiopathic DVTs are grouped together, as DVTs with a genetic history. In ultrasounds, the classification rates are a little lower: the maximum rate is 40%. The Confusion Matrix corresponding is given by Table 11.9b and shows that each class is well classified with at best 46%. We can nevertheless note that 60% of TVP associated with cancer are grouped together with DVTs with a genetic history.

Finally, the results presented in this section indicate that the classification into five classes, defined in relation to triggering factors, seems complicated. ASSS Supervised classification suggests rather a classification in two classes and the unsupervised classification in three or four classes (there is at least one cluster that does not have many points). The relatively low (around 40%) classification rates may indicate that:

- The triggering factor does not have a visible impact on the structure of the thrombus;
- The possible impact of the trigger is not visible on ultrasound and/or elastography;

- The used features and/or the classifiers are not suitable for thrombi data: the boundaries between the classes may be nonlinear, which would explain the difficulty of modeling classes by ASSS;
- Variations in acquisition parameters (such as gain) alter classification results;
- The age of the thrombi (not necessarily known) is not considered for the classification: two thrombi caused by the same triggering factor but at different ages (one week apart for example) will probably have different structures;
- The five chosen classes are not sufficiently well defined (we have notably seen that some thrombi could indeed be classified into different classes).

However, whatever the reason(s) for these relatively low rates, these results should be put into perspective in the sense that our current base is still small (around 30 patients). Therefore, we redefined our classes and simplified them to improve the classification performance.

11.11.2   Triggering Factor of DVT and Presence of PE

We considered DVTs according to two criteria: identification of the main triggering factor (Provoked/idiopathic DVT) and the complication of DVT in PE. The classification results, in function of the simulation parameters and the choice of descriptors, are given in Table 11.10.

Supervised classification rates vary between 25% and 45% for an average rate of 35% (see Table 11.10a). The modeling in the four desired classes seems still complicated for ASSS. We can note that the rates obtained from the elastographies are once again higher than those obtained from ultrasounds. Moreover, by analyzing the confusion matrices of the two simulations (see Table 11.11), with maximum rates in elastography (45% for $S22 - SO_{log} + C_{1,2}$) and in ultrasound (41.7% for $S06 - C_4$), we observe that these rates are misleading and that, in reality, the vast majority of thrombi are classified as Idiopathic with PE. SMA fails to correctly reach desired classes.

In unsupervised mode, the k-means algorithm shows (see Table 11.10b) similar performances (35% on average with rates between 25% and 46%). Contrary to previous results, the simulations carried out on the ultrasounds give better rates of good classification than those performed on elastographies. By studying two simulations in particular ($S17 - SO_{log} + C_2$ and $S07 - SO_{log} + C_2$), their confusion matrices, shown in Table 11.12, show more interesting results than those presented for the ASSS. The simulation S07 (on ultrasounds) reaches 73.4% for trigged thromboses ($73.4 = (57.8 + 55.7 + 13.3 + 20.0) / 2$) and 57.25% of idiopathic thromboses. On one hand, there is significant confusion among classes with or without PE. On the other hand, with elastographies, simulation S17 shows a good percentage of thrombosis idiopathic classification (without PE) but the other three classes are still mixed.

With spectral clustering, the performances obtained again appear to be superior to those of the other two classification methods. We obtained 48% correct classification on elastography and up to 49% on ultrasounds. The simulations detailed in Table 11.13 show that more than 60% of triggered and idiopathic thromboses are

TABLE 11.10
Classification of Thrombi with Respect to Triggers (Known Triggering Factor or Idiopathic) of DVT and the Presence of PE

Simulations / Descriptors:			Images	SO	SO$_{log}$	c_2	c_3	c_4	$c_{1,2}$	$c_{2,4}$	$c_{1,4}$	SO+ c_2	SO+ $c_{1,2}$	SO+ $c_{2,4}$	SO+ $c_{1,4}$	SO$_{log}$+ c_2	SO$_{log}$+ $c_{1,2}$	SO$_{log}$+ $c_{2,4}$	SO$_{log}$+ $c_{1,4}$
T	Img	Descrip.	N°																

(a) ASSE with 50% of thrombi used for learning stage (total standardization by type).

(b) k-means (total standardization by type).

(c) Spectral clustering with k-means (total standardization by order).

correctly separated. However, the identification of the presence of PE is still confusing for these two simulations.

The four classes defined and presented in Tables 11.11–11.13 seem to be still relatively complicated regardless of the descriptors and the classification algorithms. However, the analysis of the confusion matrices of the best simulations (for classifier and image type) is encouraging since we have seen that DVTs caused and idiopathic

---

TABLE 11.11

ASSS Confusion Matrix as a Function of DVT Triggering Factors and Presence of PE: Provoked (Prov.), Provoked with PE (Pr. PE), Idiopathic (Idiop.) and Idiopathic with PE (Id. PE)

C_{ASSS}	Prov.	Pr. PE	Idiop.	Id. PE
Prov.	0.0	0.0	2.5	97.5
Pr. PE	0.0	0.0	26.7	73.3
Idiop.	0.0	0.0	30.0	70.0
Id. PE	0.0	0.0	1.4	98.6

(a) Elastography - $S20$ - $SO_{log} + C_{1,2}$ (45.0%)

C_{ASSS}	Prov.	Pr. PE	Idiop.	Id. PE
Prov.	30.0	0.0	7.5	62.5
Pr. PE	3.3	0.0	10.0	86.7
Idiop.	0.0	10.0	15.0	75.0
Id. PE	3.3	0.0	8.3	88.3

(b) Ultrasound - $S06$ - C_4 (41.7%)

TABLE 11.12

Confusion Matrix by k-Means with Respect to Triggers of DVT and the Presence of PE

C_{kmn}	Prov.	Pr. PE	Idiop.	Id. PE
Prov.	8.9	0.0	10.0	81.1
Pr. PE	2.9	41.4	15.7	40.0
Idiop.	2.5	0.0	70.0	27.5
Id. PE	5.3	0.0	43.3	51.3

(a) Elastography - $S17$ - $SO_{log} + C_2$ (43.6%)

C_{kmn}	Prov.	Pr. PE	Idiop.	Id. PE
Prov.	57.8	13.3	0.0	28.9
Pr. PE	55.7	20.0	14.3	10.0
Idiop.	23.8	32.5	17.5	26.3
Id. PE	18.5	10.8	1.5	69.2

(b) Ultrasound - $S07$ - $SO_{log} + C_2$ (45.9%)

TABLE 11.13

Confusion Matrix by Spectral Clustering with Respect to Triggers of DVT and the Presence of PE

C_{SC}	Prov.	Pr. PE	Idiop.	Id. PE
Prov.	66.7	15.6	17.8	0.0
Pr. PE	20.0	37.1	14.3	28.6
Idiop.	26.3	7.5	40.0	26.3
Id. PE	10.7	20.0	24.0	45.3

(a) Elastography - $S23$ - $SO_{log} + C_{2..4}$ (47.7%)

C_{SC}	Prov.	Pr. PE	Idiop.	Id. PE
Prov.	61.1	27.8	0.0	11.1
Pr. PE	50.0	35.7	14.3	0.0
Idiop.	25.0	25.0	25.0	25.0
Id. PE	7.7	23.1	7.7	61.5

(b) Ultrasound - $S07$ - $SO_{log} + C_2$ (48.7%)

are fairly well separated and that the confusion was rather for the detection of the presence of PE. To support this observation, we conducted further simulations by only considering two classes: provoked or idiopathic.

11.11.3 TRIGGERING FACTOR OF DVT

In this subsection, the objective is to dissociate provoked from idiopathic DVT. We are looking to identify whether having an identifiable trigger factor (surgery, immobilization, cancer, *etc.*) gives the thrombus a different structure than if it was formed without an apparent reason. Part of our classification results (with these two classes) is given in Table 11.14.

According to Table 11.15a, ASSS couldn't correctly classify the data into the two desired classes. Indeed, classification rates vary between 44% and 60%. The confusion matrices, corresponding to the two best simulations by elastography and ultrasound, show that the majority of thrombi are labeled idiopathic.

However, the two unsupervised classification techniques (the k-means in Table 11.14b and the spectral clustering in Table 11.14c) provide much better results, especially in ultrasound. Indeed, by focusing on the two simulations (*cf.* Tables 11.16a and 11.17a), we note that the majority of thrombi are grouped together in the same cluster (hence about 65% of good classification). But, in ultrasound, the confusion matrices of the two best simulations of each algorithm (Tables 11.16b and 11.17b) show that ~70% of thrombi are well partitioned, which is quite satisfactory.

By analyzing Tables 11.14b and 11.14c, we find that the best rates are obtained with the multiresolution descriptors (with the logarithm for the k-means, without logarithm for CS), multicorrelations of order two (C_2 or $C_{1,2}$ for CS) and/or with combined descriptors. Multicorrelations of orders three and four do not seem relevant to discriminate between thrombosis caused and idiopathic thrombosis by ultrasound. At the level of simulation parameters, the standardization of multiresolution descriptors by acquisition seems to improve the unsupervised classification performance. Statistical descriptors appear to be more efficient if they are merged and if our images are standardized by acquisition.

11.11.4 PRESENCE OF PE

In the previous subsection, we succeeded in dissociating the thromboses caused by idiopathic thrombosis by ultrasound with ~70% success. Hereinafter, we are only interested in the presence, or absence, of PE associated with DVT. The classification results are given in Table 11.19.

Table 11.19a shows that ASSS gives more satisfactory than in the simulations presented in previous sub-sections. Based on Statistical and/or combined descriptors, we were able to detect the presence of PE with an accuracy of approximately 65% (ultrasound and elastography). The confusion matrices of simulations S24 and S05 (*cf.* Table 11.18) confirm these results since their diagonals show rates close to 65%.

For unsupervised classification (k-means in Table 11.19b or spectral clustering in Table 11.19c), the echogenicity of the thrombus seems more likely to allow the detection of the presence of PE (up to 76% good rating for ultrasounds and up to 63%

TABLE 11.14
Thrombi Classification According to (Known or Unknown) Triggers

Simulations / Descriptors:				Images	SO	SO_{log}	c_2	c_3	c_4	$c_{1,2}$	$c_{2,4}$	$c_{1,4}$	$SO+$ c_2	$SO+$ $c_{1,2}$	$SO+$ $c_{2,4}$	$SO+$ $c_{1,4}$	$SO_{log}+$ c_2	$SO_{log}+$ $c_{1,2}$	$SO_{log}+$ $c_{2,4}$	$SO_{log}+$ $c_{1,4}$
T	Img	Descrip.	N°																	
Ultrasound			S01	55	56	50	58	50	54	57	47	48	58	59	43	45	52	48	50	50
		F	S02	57	55	52	57	56	57	37	55	53	58	57	51	51	52	54	50	50
	N		S03	55	58	51	53	57	57	54	52	53	53	54	53	52	51	51	54	54
	N	F	S04	57	52	53	57	57	57	56	56	56	54	54	55	56	51	51	54	55
	N		S05	55	57	48	53	50	54	54	47	47	55	58	46	46	50	49	51	51
	N	F	S06	56	58	51	58	57	56	57	54	55	57	56	51	50	52	51	49	49
	N	N	S07	55	58	52	53	52	57	54	52	53	53	54	53	52	51	51	54	54
	N	N F	S08	56	52	53	57	56	56	57	56	56	54	54	55	56	51	51	54	55
Ultrasound HE			S09	55	56	50	58	50	54	57	47	48	58	59	45	45	52	48	50	50
		F	S10	57	55	52	57	56	57	57	55	53	58	57	51	51	52	54	50	50
	N		S11	55	58	51	53	52	57	54	52	53	53	54	53	52	51	51	54	54
	N	F	S12	57	52	53	57	57	57	56	57	56	54	54	55	56	51	51	54	55
	N		S13	55	57	48	53	50	54	54	47	47	55	58	46	46	50	49	51	51
	N	F	S14	56	56	51	58	57	57	58	57	54	55	57	56	51	50	52	51	49
	N	N	S15	55	58	52	53	52	57	54	52	53	53	54	53	52	51	51	54	54
	N	N F	S16	56	52	53	53	56	56	57	56	56	54	54	55	56	51	51	54	55
Elastography			S17	59	59	58	60	47	44	58	45	45	59	57	48	48	57	58	58	58
		F	S18	58	59	58	58	57	58	57	55	54	57	58	54	54	58	57	56	56
	N		S19	59	59	58	55	54	54	54	58	58	57	57	57	58	58	58	58	58
	N	F	S20	58	58	58	58	58	58	57	58	58	58	58	57	57	58	58	58	58
	N		S21	57	58	58	54	56	44	57	46	46	56	58	52	53	58	58	58	58
	N	F	S22	58	59	57	58	57	58	57	56	56	58	58	57	57	58	58	56	56
	N	N	S23	57	56	58	58	55	54	52	58	58	57	57	57	57	58	58	58	59
	N	N F	S24	58	58	58	58	58	58	58	58	58	58	58	58	58	58	57	58	58

(a) ASSS with 50% of thrombi used for learning stage (total standardization by type).

Simulations / Descriptors:				Images	SO	SO_{log}	c_2	c_3	c_4	$c_{1,2}$	$c_{2,4}$	$c_{1,4}$	$SO+$ c_2	$SO+$ $c_{1,2}$	$SO+$ $c_{2,4}$	$SO+$ $c_{1,4}$	$SO_{log}+$ c_2	$SO_{log}+$ $c_{1,2}$	$SO_{log}+$ $c_{2,4}$	$SO_{log}+$ $c_{1,4}$
T	Img	Descrip.	N°																	
Ultrasound			S01	57	57	56	56	62	58	56	57	57	56	59	58	58	54	51	57	57
		F	S02	59	58	55	57	62	60	56	58	60	57	58	58	57	64	54	56	55
	N		S03	57	60	68	44	61	46	48	45	44	56	54	43	43	57	55	48	47
	N	F	S04	58	60	64	56	58	49	57	46	46	67	67	56	55	63	64	49	49
	N		S05	53	54	58	56	61	57	55	58	55	55	57	58	58	61	62	60	61
	N	F	S06	47	50	70	58	62	60	61	59	59	60	58	60	61	68	63	68	68
	N	N	S07	51	60	69	44	62	46	50	45	45	54	54	43	43	55	54	47	47
	N	N F	S08	47	57	64	58	58	50	57	46	46	62	65	56	56	63	63	49	49
Ultrasound HE			S09	57	57	56	56	60	58	56	57	57	56	59	58	58	54	51	57	57
		F	S10	59	58	55	57	62	59	56	58	60	57	58	58	57	64	54	56	55
	N		S11	57	60	68	44	61	46	48	45	44	56	54	43	43	55	55	47	47
	N	F	S12	58	56	64	56	58	50	57	46	46	59	62	55	56	62	63	49	49
	N		S13	52	54	58	55	61	58	56	58	57	56	57	58	58	61	62	61	59
	N	F	S14	47	51	70	61	62	59	57	60	59	59	57	59	60	68	68	69	68
	N	N	S15	51	60	69	44	61	46	50	45	45	54	54	43	43	55	54	47	47
	N	N F	S16	47	57	64	58	58	50	57	46	46	62	65	56	56	63	63	49	49
Elastography			S17	57	59	59	62	51	49	62	59	59	67	56	59	59	51	57	59	60
		F	S18	58	56	52	63	47	59	63	59	59	62	57	58	58	55	52	52	55
	N		S19	57	53	57	56	49	54	57	53	55	53	51	56	56	57	54	56	56
	N	F	S20	56	54	58	62	49	54	62	55	54	60	62	54	53	54	54	55	55
	N		S21	53	53	53	67	54	59	64	59	59	67	64	51	51	61	52	55	55
	N	F	S22	55	57	52	55	50	56	58	59	50	54	52	54	54	51	53	55	56
	N	N	S23	53	52	58	58	52	56	57	51	51	53	49	51	51	57	54	51	51
	N	N F	S24	55	53	57	60	55	54	61	50	53	60	61	51	49	54	53	52	53

(b) k-means (total standardization by type).

Simulations / Descriptors:				Images	SO	SO_{log}	c_2	c_3	c_4	$c_{1,2}$	$c_{2,4}$	$c_{1,4}$	$SO+$ c_2	$SO+$ $c_{1,2}$	$SO+$ $c_{2,4}$	$SO+$ $c_{1,4}$	$SO_{log}+$ c_2	$SO_{log}+$ $c_{1,2}$	$SO_{log}+$ $c_{2,4}$	$SO_{log}+$ $c_{1,4}$
T	Img	Descrip.	N°																	
Ultrasound			S01	57	49	56	50	50	44	48	42	42	51	49	61	62	56	56	51	51
		F	S02	59	50	53	52	53	36	55	54	57	51	51	56	55	56	57	58	58
	N		S03	57	53	69	46	62	54	46	51	51	55	53	50	51	63	64	51	51
	N	F	S04	59	53	59	59	63	52	59	50	50	63	61	50	49	61	64	49	50
	N		S05	49	52	57	56	50	45	57	45	42	59	59	38	38	57	57	59	60
	N	F	S06	55	64	70	72	54	56	70	65	65	69	47	63	68	68	68	71	72
	N	N	S07	49	53	69	46	62	54	46	51	51	55	53	50	50	63	63	51	51
	N	N F	S08	53	56	58	59	61	51	59	50	50	62	63	49	49	64	59	50	50
Ultrasound HE			S09	57	49	56	50	50	44	48	42	42	51	49	61	62	56	56	51	51
		F	S10	59	50	53	52	53	36	55	54	57	51	51	56	55	56	57	56	56
	N		S11	57	53	69	46	62	54	46	51	51	55	42	51	49	63	58	51	51
	N	F	S12	59	53	58	59	63	52	59	51	50	63	62	50	50	61	64	50	49
	N		S13	49	52	57	55	50	44	57	45	42	59	59	38	38	57	57	60	60
	N	F	S14	53	63	70	71	54	56	70	65	65	69	57	63	66	68	68	72	72
	N	N	S15	49	56	69	46	62	54	46	51	51	55	53	50	50	63	63	51	51
	N	N F	S16	53	56	58	59	61	51	59	50	50	62	63	49	49	64	59	50	50
Elastography			S17	50	51	58	51	52	59	49	55	52	51	54	53	52	57	57	57	56
		F	S18	48	48	54	51	47	61	55	58	51	50	53	56	54	59	51	51	51
	N		S19	50	56	58	56	57	49	58	51	51	54	50	51	51	59	57	49	50
	N	F	S20	49	54	51	56	51	51	50	53	52	60	60	53	52	55	55	52	52
	N		S21	52	49	49	66	51	59	60	65	64	57	56	56	56	56	54	54	52
	N	F	S22	49	51	49	51	48	62	51	55	55	49	49	53	59	50	50	49	50
	N	N	S23	52	56	58	58	62	49	55	49	49	52	49	51	51	57	56	51	51
	N	N F	S24	48	53	59	58	54	51	61	53	52	60	61	51	52	54	55	52	52

(c) Spectral clustering with k-means (total standardization by order).

TABLE 11.15

Confusion Matrix by ASSS with Known Triggers and Idiopathic DVT

C_{ASSS}	Provoked	Idiopathic
Provoked	33.8	66.3
Idiopathic	20.9	79.1

(a) Elastography - $S17$ - C_2 (60.0%)

C_{ASSS}	Provoked	Idiopathic
Provoked	17.5	82.5
Idiopathic	7.0	93.0

(b) Ultrasound - $S01$ - SO $+ C_{1,2}$ (59.4%)

TABLE 11.16

Confusion Matrix by k-Means with Known Triggers and Idiopathic DVT

C_{kmn}	Provoked	Idiopathic
Provoked	20.6	79.4
Idiopathic	0.0	100.0

(a) Elastography - $S17$ - SO $+ C_2$ (67.4%)

C_{kmn}	Provoked	Idiopathic
Provoked	63.8	36.3
Idiopathic	25.7	74.3

(b) Ultrasound - $S06$ - SO_{log} (69.7%)

TABLE 11.17

Confusion Matrix by Spectral Clustering with Known Triggers and Idiopathic DVT

C_{SC}	Provoked	Idiopathic
Provoked	46.3	53.8
Idiopathic	20.9	79.1

(a) Elastography - $S21$ - C_2 (65.6%)

C_{SC}	Provoked	Idiopathic
Provoked	75.0	25.0
Idiopathic	31.0	69.0

(b) Ultrasound - $S06$ - $SO_{log} + C_{1..4}$ (71.6%)

TABLE 11.18

Confusion Matrix by ASSS as a Function of the Presence of PE

C_{ASSS}	No PE	With PE
No PE	67.5	32.5
With PE	34.5	65.5

(a) Elastography - $S24$ - SO $+ C_{2..4}$ (66.3%)

C_{ASSS}	No PE	With PE
No PE	63.8	36.3
With PE	36.0	64.0

(b) Ultrasound - $S05$ - $C_{1..4}$ (63.8%)

for elastographies). The confusion matrices in Tables 11.20 and 11.21 indicate that the elastographies allow better grouping of thrombi not associated with PE (more than 70%) and that ultrasounds are more effective at detecting the presence of PE (over 85%). The spectral clustering of second-order multicorrelations ($S14 - C_{1,2}$) shows

TABLE 11.19
Thrombi Classification According to the Presence of PE

Simulations / Descriptors:				Images	SO	SO_{log}	C_2	C_3	C_4	$C_{1,2}$	$C_{2,4}$	$C_{1,4}$	$SO+ / C_2$	$SO+ / C_{1,2}$	$SO+ / C_{2,4}$	$SO+ / C_{1,4}$	$SO_{log}+ / C_2$	$SO_{log}+ / C_{1,2}$	$SO_{log}+ / C_{2,4}$	$SO_{log}+ / C_{1,4}$
T	Img	Descrip.	N°																	
Ultrasound			S01																	
		F	S02																	
	N		S03																	
	N	F	S04																	
	N		S05																	
	N	F	S06																	
	N	N	S07																	
	N	N F	S08																	
Ultrasound HE			S09																	
		F	S10																	
	N		S11																	
	N	F	S12																	
	N		S13																	
	N	F	S14																	
	N	N	S15																	
	N	N F	S16																	
Elastography			S17																	
		F	S18																	
	N		S19																	
	N	F	S20																	
	N		S21																	
	N	F	S22																	
	N	N	S23																	
	N	N F	S24																	

(a) ASSS with 50% of thrombi used for learning stage (total standardization by type).

Simulations / Descriptors:				Images	SO	SO_{log}	C_2	C_3	C_4	$C_{1,2}$	$C_{2,4}$	$C_{1,4}$	$SO+ / C_2$	$SO+ / C_{1,2}$	$SO+ / C_{2,4}$	$SO+ / C_{1,4}$	$SO_{log}+ / C_2$	$SO_{log}+ / C_{1,2}$	$SO_{log}+ / C_{2,4}$	$SO_{log}+ / C_{1,4}$
T	Img	Descrip.	N°																	
Ultrasound			S01	54	54	44	54	46	55	49	54	54	55	51	53	53	52	49	49	49
		F	S02	52	53	51	46	58	51	49	52	51	46	52	53	53	61	51	53	53
	N		S03	54	74	64	49	58	54	51	57	57	68	68	57	57	58	58	56	55
	N	F	S04	53	69	64	51	61	56	52	54	54	71	71	52	53	64	66	51	51
	N		S05	53	54	64	55	47	54	55	53	53	55	54	53	52	66	65	62	62
	N	F	S06	55	44	62	54	57	51	60	51	52	56	58	51	50	68	65	62	62
	N	N	S07	53	74	63	49	57	54	51	57	57	68	68	57	57	58	59	55	55
	N	N F	S08	53	69	65	50	62	55	53	54	54	67	70	52	52	64	65	51	51
Ultrasound HE			S09	54	54	44	54	48	55	49	54	54	55	51	53	53	52	49	49	49
		F	S10	52	53	51	46	57	51	49	52	51	46	52	53	53	61	51	53	53
	N		S11	54	74	64	49	58	54	51	57	57	68	68	57	57	58	58	55	55
	N	F	S12	53	68	65	51	61	56	51	54	54	65	69	53	52	64	66	51	51
	N		S13	52	54	64	55	47	53	55	53	54	55	54	52	53	66	65	63	61
	N	F	S14	55	48	62	55	59	51	52	51	52	55	58	52	51	67	65	64	62
	N	N	S15	53	74	63	49	58	54	51	57	57	68	68	57	57	58	59	55	55
	N	N F	S16	53	69	65	50	61	56	53	54	54	67	70	52	52	64	65	51	51
Elastography			S17	47	49	50	51	54	56	56	56	55	51	51	49	50	51	52	50	49
		F	S18	54	56	49	50	53	50	51	56	56	57	55	55	54	50	49	48	48
	N		S19	47	50	51	57	49	56	56	55	55	58	59	56	56	62	62	56	55
	N	F	S20	54	55	62	57	54	55	69	53	53	54	54	51	51	48	46	51	51
	N		S21	50	49	45	46	50	56	54	56	55	46	54	53	52	50	49	49	48
	N	F	S22	45	50	49	55	50	56	52	56	56	54	52	53	55	53	53	49	48
	N	N	S23	49	59	52	52	49	56	55	56	56	57	59	56	56	61	62	56	56
	N	N F	S24	45	57	63	58	47	55	55	55	54	56	56	55	55	46	48	54	54

(b) k-means (total standardization by type).

Simulations / Descriptors:				Images	SO	SO_{log}	C_2	C_3	C_4	$C_{1,2}$	$C_{2,4}$	$C_{1,4}$	$SO+ / C_2$	$SO+ / C_{1,2}$	$SO+ / C_{2,4}$	$SO+ / C_{1,4}$	$SO_{log}+ / C_2$	$SO_{log}+ / C_{1,2}$	$SO_{log}+ / C_{2,4}$	$SO_{log}+ / C_{1,4}$
T	Img	Descrip.	N°																	
Ultrasound			S01	54	51	59	47	53	50	45	50	50	48	46	45	45	49	50	49	49
		F	S02	51	49	51	49	61	52	46	51	54	49	49	49	47	50	49	50	50
	N		S03	54	68	64	57	57	54	57	50	59	65	64	58	59	58	59	59	59
	N	F	S04	51	69	62	55	54	54	55	53	52	69	64	53	53	64	67	52	52
	N		S05	53	64	61	55	53	50	53	50	50	62	62	41	41	65	65	59	59
	N	F	S06	53	66	62	74	60	52	75	57	57	71	54	71	74	65	65	66	66
	N	N	S07	54	69	63	57	57	54	57	50	59	65	63	58	58	58	59	59	59
	N	N F	S08	53	68	61	53	54	54	54	58	53	68	70	52	53	64	66	52	53
Ultrasound HE			S09	54	51	59	47	53	51	45	50	50	48	46	45	45	49	50	49	49
		F	S10	51	49	51	49	59	52	46	51	54	49	49	49	47	50	49	50	50
	N		S11	54	68	64	57	57	54	57	50	59	65	64	59	58	58	59	59	59
	N	F	S12	53	69	62	55	54	54	56	53	52	69	66	53	52	64	66	53	52
	N		S13	54	65	62	56	52	50	56	50	50	62	62	41	41	65	65	59	59
	N	F	S14	54	66	60	74	60	52	76	57	57	72	54	73	74	65	65	66	66
	N	N	S15	54	69	63	57	57	54	57	59	59	65	65	58	58	58	59	59	59
	N	N F	S16	53	68	61	53	54	54	54	58	58	68	70	52	53	64	66	52	53
Elastography			S17	53	46	50	49	51	44	54	46	49	51	51	50	50	51	51	51	51
		F	S18	54	46	51	49	50	46	47	51	54	52	49	54	51	50	48	49	48
	N		S19	53	49	47	52	51	55	55	56	56	56	57	56	56	62	58	59	59
	N	F	S20	54	58	59	58	52	55	59	60	57	56	54	60	59	45	45	59	57
	N		S21	55	48	49	50	50	44	53	53	52	54	54	60	60	53	51	51	50
	N	F	S22	60	54	54	52	55	54	57	58	58	53	54	55	56	49	49	49	49
	N	N	S23	55	49	48	48	51	55	57	59	59	55	53	59	59	63	59	59	59
	N	N F	S24	60	58	60	58	51	55	54	58	58	55	55	59	59	45	45	59	59

(c) CSpectral clustering with k-means (total standardization by order).

TABLE 11.20

Confusion Matrix by k-Means as a Function of the Presence of PE

C_{kmn}	No PE	With PE
No PE	76.5	23.5
With PE	48.2	51.8

(a) Elastography - $S24$ - SO_{log} (62.6%)

C_{kmn}	No PE	With PE
No PE	54.7	45.3
With PE	10.0	90.0

(b) Ultrasound - $S07$ - SO (73.8%)

TABLE 11.21

Confusion Matrix by Spectral Clustering as a Function of the PE Presence

C_{sc}	No PE	With PE
No PE	71.8	28.2
With PE	43.6	56.4

(a) Elastography - $S23$ - $SO_{log} + C_2$ (63.1%)

C_{sc}	No PE	With PE
No PE	65.9	34.1
With PE	15.0	85.0

(b) Ultrasound - $S14$ - $C_{1,2}$ (76.2%)

promising performances with 76% good ranking. Indeed, SO shows better results when combined with k-means. On the other hand, these simulations as well as those of previous sub-sections suggest that the multicorrelations (of order two in particular) are more useful to the clustering spectral than k-means.

11.11.4.1 Summary and Outcomes

This section focuses on the classification of thrombi images. At first, we tried to identify the triggers of deep vein thrombosis (surgery, genetic factor, cancer, immobilization and idiopathic) by ultrasound and elastography. However, obtained results were not really satisfactory (especially by ASSS) and therefore this type of classification is achieved. Besides, we don't have enough data in every class of considered classes (seven on average). In addition, certain thromboses associated with cancers (and therefore labeled Cancer) present a second risk factor (surgery or immobilization) and could therefore have been labeled in another class.

To avoid any confusion related to multifactorial thrombosis, we decided to reduce the class number to two: triggered thromboses (*i.e.* with at least one known risk factor) and idiopathic thrombosis. These two groups were then separated into two subsets based on the occurrence of PE. The obtained classification of four classes gave better results (especially in unsupervised). However, the results were not yet completely satisfactory (less than 50% of correct classification) but the confusion matrices of certain simulations were interesting, as: the confusion was mainly about the presence or absence of PE but the triggered and idiopathic thromboses appeared to be fairly well separated. This observation was confirmed in latter simulations where the data was classified into two classes: triggered thrombosis or idiopathic. The simulations on the ultrasounds reached 71% of good classification. The best performances were obtained with the logarithm of the scattering coefficients, with multicorrelations of order two, or with the combination of these two types of descriptors. In future work, it will be interesting to check this result on a larger database. On the other hand, the simulations on the elastographies showed greater confusion between the two classes.

Later on, we were interested in separating the data according to two classes depending on the occurrence of PE. The results turned out to be very encouraging since favorable results on ultrasound resulted in up to 76% good classification. Best performances were obtained on multiresolution descriptors (without logarithm) with k-means, on statistical descriptors of order two with spectral clustering or on the combination of the two types of descriptors (regardless of the classification algorithm). The results on elastography are also encouraging (up to 66%), especially in supervised, but remain below those obtained on ultrasounds.

In other simulations, not presented in this chapter, we found that the classification of thromboses in relation to their association with cancer gave weak performance. This result can be expected as the presence of cancer does not have a great influence on echogenicity or the hardness of the thrombus. Similarly, studying the impact of smoking on the structure of the thrombus (classification into two classes depending on whether the patients are smokers) did not prove to be much more efficient although some favorable simulations gave almost 64% good classification. We run many simulations to identify the impact of ultrasound acquisition parameters on the classification results. The results of these simulations have shown that, most of the time, the change in ultrasound mode (fundamental or harmonic) and the gain adjustment do not have an excessive impact on the classification results. To push further our study and measure more precisely this impact, we can consider building a database in which the acquisitions of each patient would be collected for different predefined sets of parameters.

11.12 CONCLUSION

Our project aims to identify and characterize a thrombus venous by acoustic imaging. Inappropriate formation of a thrombus in one or more veins of the deep blood network (as opposed to superficial) leads to the development of a deep vein thrombosis (DVT), better known as phlebitis. DVT may cause a thrombus fragment (or all of it) to detach from the venous wall and move up toward the heart by being carried along by the blood circulation to finally lodge in a pulmonary artery. This complication, called pulmonary embolism, is the third leading cause of death in vascular diseases after myocardial infarction and stroke. Three physiopathological mechanisms intervene, in isolation or in combination, in the DVT creation: venous stasis, venous wall alteration and thrombophilia (*i.e.* increased coagulation or a deficit of anticoagulation). Among the most frequent risk factors, cancer, a surgery act (anesthesia, bleeding, placement of a catheter), immobilization or pregnancy can trigger one or more of these three mechanisms and lead to thrombosis. However, it is common for thrombosis to occur for no known reason; it is then called idiopathic. Our project concerns the analysis of the structure of human venous thrombi in order to predict the PE risk and/or to identify the origin of the thrombosis. It is known that a thrombus contains red blood cells and platelets in a fibrin network whose stability and microstructure are dependent on all of the biological processes that result in its formation. The challenge of our project is therefore to observe (by acoustic imaging) if the structure of the thrombus depends on the generating mechanism and following the PE occurrence.

To reach our goal, the Echo-Doppler and Vascular Medicine Unit, with a Canon Aplio system, allowed us to collect ultrasound and elastography of venous thrombi. The Echo-Doppler and Vascular Medicine Unit developed an arm articulated to limit intra- and inter-operator variability (the pressure force of the probe on the skin may change the elastography). However, we noticed that the intra-operator variability (*i.e.* same operator, same patient, same area examined) of the elastographies obtained with the linear probe was still quite large. Using a convex probe instead of a linear one helped stabilize elastography acquisitions. After preparing the thrombi data, we propose two distinct approaches to extract descriptors from images: Scattering Operator (SO) or Higher Order Statistics (HOS). SO is a powerful tool to extract multiresolution features suitable for discriminating between textures. Many image processing techniques (such as principal component analysis) are based on statistics of order two. However, it has been shown that the use of statistics greater than two improves and extends existing solutions, or even solves problems not accessible before. These two approaches also have the advantage of being complementary since they do not use the same source of information.

Database Images have been represented by multiresolution descriptors (obtained by SO) and HOS. Then, we considered four classification algorithms: Affine Scattering Space selection in supervised mode, k-means, DBSCAN and Spectral clustering in unsupervised mode. As our database is relatively small, we couldn't use some supervised classification techniques such as SVMs or neural networks. After obtaining conclusive results on our reference database, we conducted simulations based on the thrombi to find the main triggers; but we observed great confusion among the proposed classes. However, this result is not illogical because some patients have several risk factors (for example cancer and immobilization). In addition, if immobilization thrombosis is usually due to stasis venous (and possibly thrombophilia), cancers or surgery can trigger several mechanisms responsible for the coagulation. Therefore, we decided to classify thrombi depending on: the provoked or idiopathic nature of the thrombosis, and the occurrence of a PE. Classification in four classes are better than the one based on the triggers but not yet sufficiently satisfactory. On the other hand, considering the classification of these two parameters separately we got much more interesting results. In elastography, the classification performance reached 66% correct classification of thrombosis according to the PE presence. In ultrasound, simulations separate correctly triggered and idiopathic thrombosis up to 70% and detect the PE presence up to 76%.

In summary, in our project, we adapted a set of signal and image processing tools to medical data: ultrasound and elastography. We have devoted ourselves to every step of the image processing chain with particular attention on the feature extraction step. Our processing channel was first validated on synthetic images of textures and then on medical data. To characterize the structure of venous thrombi and identify whether the latter differs depending on the triggers and the presence (or absence) of PE, we have set up an experimental protocol to analyze the performance of classification for different settings of the processing chain (descriptors, considering the vicinity of the thrombus, *etc.*). The classification results obtained are promising, in particular for the PE prediction from the thrombi images. Therefore, our pioneer works open up many perspectives both medical, application and in image processing.

REFERENCES

1. A. Achim, A. Bezerianos, and P. Tsakalides. Novel Bayesian multiscale method for speckle removal in medical ultrasound images. *IEEE Transactions on Medical Imaging*, 20(8): 772–783, August 2001.
2. M. Ankerst, M.M. Breunig, H.P. Kriegel, and J. Sander. OPTICS: Ordering points to identify the clustering structure. In *International Conference on Management of Data*, pp. 49–60, Philadelphia, PA, June 1999.
3. C.N. Bagot and R. Arya. Virchow and his triad: A question of attribution. *British Journal of Haematology*, 143(2): 180–190, September 2008.
4. A.K. Barros, A. Mansour, and N. Ohnishi. Removing artifacts from electrocardiographic signals using independent components analysis. *Neurocomputing*, 22(1-3): 173–186, November 1998.
5. K. Bensafia, A. Mansour, and S. Haddab. Blind source subspace separation and classification of ECG signals. In *Conférence Internationale en Automatique & Traitement de Signal*, Sousse, Tunisia, March 2017.
6. E. Bournay Bouchereau. Analyse d'images par transformées en ondelettes. Application aux images sismiques. PhD thesis, Université Joseph-Fourier, Grenoble, France, March 1997.
7. J. Bruna and S. Mallat. Invariant scattering convolution networks. *IEEE Transactions on Pattern Analysis and Machine Intelligence*, 35(8): 1872–1886, August 2013.
8. R. Cadene, T. Robert, N. Thome, and M. Cord. M2CAI workflow challenge: Convolutional neural networks with time smoothing and hidden Markov model for video frames classification. *Computing Research Repository*, arXiv:1610.05541, December 2016.
9. J.F. Cardoso and A. Souloumiac. Blind beamforming for non-Gaussian signals. *IEE Proceedings F: Radar and Signal Processing*, 140(6): 362–370, December 1993.
10. P. Carruzzo, M. Méan, A. Limacher, D. Aujesky, J. Cornuz, and C. Clair. Association between smoking and recurrence of venous thromboembolism and bleeding in elderly patients with past acute venous thromboembolism. *Thrombosis Research*, 138: 74–79, February 2016.
11. S.G. Chang, B. Yu, and M. Vetterli. Spatially adaptive wavelet thresholding with context modeling for image denoising. *IEEE Transactions on Image Processing*, 9(9): 1522–1531, September 2000.
12. C.M. Chen, Y.H. Chou, K.C. Han, G.S. Hung, C.M. Tiu, H.J. Chiou, and S.Y. Chiou. Breast lesions on sonograms: Computer-aided diagnosis with nearly setting-independent features and artificial neural networks. *Radiology*, 226(2): 504–514, February 2003.
13. S. Chiou, F. Forsberg, T.B. Fox, and L. Needleman. Comparing differential tissue harmonic imaging with tissue harmonic and fundamental gray scale imaging of the liver. *Journal of Ultrasound in Medicine*, 26(11): 1557–1563, November 2007.
14. P. Chuzel, A. Mansour, J. Ognard, J. Gentric, L. Bressollette, D. Hamad, and N. Betrouni. Automatic clustering for MRI images, application on perfusion MRI of brain. In *International Conference on Frontiers of Signal Processing*, pp. 63–66, Warsaw, Poland, October 2016.

15. D. Ciresan, A. Giusti, L.M. Gambardella, and J. Schmidhuber. Deep neural networks segment neuronal membranes in electron microscopy images. In F. Pereira, C.J.C. Burges, L. Bottou, and K.Q. Weinberger, editors, *Neural Information Processing Systems (NIPS 2012)*, pp. 2843–2851. Lake Tahoe, Nevada, December 2012.

16. P. Comon and C. Jutten. *Handbook of Blind Source Separation: Independent Component Analysis and Applications.* Elsevier Science: Amsterdam, Netherlands, February 2010.

17. A. Dahabiah, J. Puentes, B. Guias, L. Bressollette, and B. Solaiman. Comparative neural network based venous thrombosis echogenicity and echostructure characterization using ultrasound images. In *Information and Communication Technologies*, vol. 1, pp. 992–997, Damascus, Syria, April 2006.

18. L. De Lathauwer, B. De Moor, and J. Vandewalle. Independent component analysis and (simultaneous) third-order tensor diagonalization. *IEEE Transactions on Signal Processing*, 49(10): 2262–2271, October 2001.

19. Société Française de Médecine Vasculaire, Collège des enseignants de médecine Vasculaire, and Collège Français de Pathologie Vasculaire. *Traité de médecine vasculaire. Tome 2: Maladies veineuses, lymphatiques et microcirculatoires, thérapeutique.* Elsevier Health Sciences: Amsterdam, Netherlands, February 2012.

20. M. Ester, H.P. Kriegel, J. Sander, and X. Xu. Density-based spatial clustering of applications with noise. In *International Conference on Knowledge Discovery and Data Mining*, pp. 226–231, Portland, OR, August 1996.

21. D.H. Fisher. Knowledge acquisition via incremental conceptual clustering. *Machine Learning*, 2(2): 139–172, July 1987.

22. J.P. Galanaud, A.C. Arnoult, M.A. Sevestre, C. Genty, M. Bonaldi, A. Guyard, P. Giordana, O. Pichot, M. Colonna, I. Quéré, and J.L. Bosson. Impact of anatomical location of lower limb venous thrombus on the risk of subsequent cancer. *Thrombosis and Haemostasis*, 112(06): 1129–1136, December 2014.

23. B.S. Garra, B.H. Krasner, S.C. Horii, S. Ascher, S.K. Mun, and R.K. Zeman. Improving the distinction between benign and malignant breast lesions: The value of sonographic texture analysis. *Ultrasonic Imaging*, 15(4): 267–285, October 1993.

24. J.L. Gennisson, T. Deffieux, M. Fink, and M. Tanter. Ultrasound elastography: Principles and techniques. *Diagnostic and Interventional Imaging*, 94(5): 487–495, May 2013.

25. A. Géron. *Machine Learning avec Scikit-Learn: Mise en oeuvre et cas concrets.* Dunod: Paris, France, August 2017.

26. B.A. Golomb, V.T. Chan, J.O. Denenberg, S. Koperski, and M.H. Criqui. Risk marker associations with venous thrombotic events: A cross-sectional analysis. *British Medical Journal*, 4(3): e003208, March 2014.

27. I. Goodfellow, Y. Bengio, and A. Courville. Deep Learning. MIT Press: Cambridge, MA, November 2016.

28. S. Guha, R. Rastogi, and K. Shim. Rock: A robust clustering algorithm for categorical attributes. In *International Conference on Data Engineering*, pp. 512–521, Sydney, Australia, March 1999.

29. M.E. Hachemi, S. Callé, and J.P. Remeniéras. Utilisation des ondes de cisaillement ultrasonores pour l'imagerie d'élasticité des tissus biologiques. *Traitement du signal et cancérologie*, 23(3-4-NS): 247–258, December 2006.

30. M. Hall, E. Frank, G. Holmes, B. Pfahringer, P. Reutemann, and I.H. Witten. The WEKA data mining software: An update. *ACM SIGKDD Explorations Newsletter*, 11(1): 10–18, June 2009.

31. D. Hamad and P. Biela. Introduction to spectral clustering. In *International Conference on Information and Communication Technologies: From Theory to Applications*, pp. 1–6, Damascus, Syria, April 2008.

32. J.A. Hartigan and M.A. Wong. Algorithm AS 136: A k-means clustering algorithm. *Journal of the Royal Statistical Society*, 28(1): 100–108, January 1979.
33. J. Herault and C. Jutten. *Réseaux neuronaux et traitement du signal*. Hermes Science: Germany, July 1994.
34. K. Horsch, M.L. Giger, L.A. Venta, and C.J. Vyborny. Automatic segmentation of breast lesions on ultrasound. *Medical Physics*, 28(8): 1652–1659, August 2001.
35. Y.L. Huang, K.L. Wang, and D.R. Chen. Diagnosis of breast tumors with ultrasonic texture analysis using support vector machines. *Neural Computing & Applications*, 15(2): 164–169, April 2006.
36. F. Jobin. *La thrombose*. Presses de l'Université Laval: Quebec City, Canada, February 1995.
37. I.T. Jolliffe. *Principal Component Analysis*. Springer: Berlin, Germany, April 2002.
38. S. Joo, Y.S. Yang, W.K. Moon, and H.C. Kim. Computer-aided diagnosis of solid breast nodules: Use of an artificial neural network based on multiple sonographic features. *IEEE Transactions on Medical Imaging*, 23(10): 1292–1300, October 2004.
39. Y.M. Kadah, A.A. Farag, J.M. Zurada, A.M. Badawi, and A.B.M. Youssef. Classification algorithms for quantitative tissue characterization of diffuse liver disease from ultrasound images. *IEEE Transactions on Medical Imaging*, 15(4): 466–478, August 1996.
40. M. Kass, A. Witkin, and D. Terzopoulos. Snakes: Active contour models. *International Journal of Computer Vision*, 1(4): 321–331, January 1988.
41. T. Kohonen. The self-organizing map. *Proceedings of the IEEE*, 78(9): 1464–1480, September 1990.
42. C. Kollmann. New sonographic techniques for harmonic imaging-underlying physical principles. *European Journal of Radiology*, 64(2): 164–172, November 2007.
43. A. Krizhevsky, I. Sutskever, and G.E. Hinton. Imagenet classification with deep convolutional neural networks. In *Neural Information Processing Systems*, pp. 1097–1105, Lake Tahoe, NV, December 2012.
44. T.H. Lee, M.F.A. Fauzi, and R. Komiya. Segmentation of ct brain images using k-means and em clustering. In *International Conference on Computer Graphics, Imaging and Visualisation*, pp. 339–344. Penang, Malaysia, August 2008.
45. W.L. Lee, T. Tan, T. Falkmer, and Y.H. Leung. Single-trial event-related potential extraction through one-unit ica-with-reference. *Journal of Neural Engineering*, 13(6): 066010, October 2016.
46. P. Lemberger, M. Batty, M. Morel, and J.L. Raffaëlli. *Big Data et Machine Learning: Les concepts et les outils de la data science*. Dunod: Paris, France, October 2016.
47. P. Léger, D. Barcat, C. Boccalon, J. Guilloux, and H. Boccalon. Thromboses veineuses des membres inférieurs et de la veine cave inférieure. *EMC - Cardiologie-Angéiologie*, 1(1): 80–96, February 2004.
48. P. Liskowski and K. Krawiec. Segmenting retinal blood vessels with deep neural networks. *IEEE Transactions on Medical Imaging*, 35(11): 2369–2380, November 2016.
49. G. Litjens, T. Kooi, B.E. Bejnordi, A.A.A. Setio, F. Ciompi, M. Ghafoorian, J. Van Der Laak, B. Van Ginneken, and C.I. Sánchez. A survey on deep learning in medical image analysis. *Medical Image Analysis*, 42: 60–88, December 2017.
50. W.Y. Loh. Classification and regression trees. *Data Mining and Knowledge Discovery*, 1(1): 14–23, January 2011.
51. C.P. Loizou, C.S. Pattichis, C.I. Christodoulou, R.S.H. Istepanian, M. Pantziaris, and A. Nicolaides. Comparative evaluation of despeckle filtering in ultrasound imaging of the carotid artery. *IEEE Transactions on Ultrasonics, Ferroelectrics and Frequency Control*, 52(10): 1653–1669, December 2005.
52. S.B. Lowen and M.C. Teich. Power-law shot noise. *IEEE Transactions on Information Theory*, 36(6): 1302–1318, November 1990.

53. J. Lu, G. Wang, and P. Moulin. Image set classification using holistic multiple order statistics features and localized multi-kernel metric learning. In *International Conference on Computer Vision*, pp. 329–336, Sydney, New South Wales, Australia, December 2013.
54. L. Maaten and G. Hinton. Visualizing data using t-SNE. *Journal of Machine Learning Research*, 9: 2579–2605, November 2008.
55. A. Martin and A. Mansour. High order statistic estimators for speech processing. In *International Conference on ITS Telecommunications*, pp. 295–300, Brest, France, June 2005.
56. M. Martın-Fernández and C. Alberola-Lopez. An approach for contour detection of human kidneys from ultrasound images using Markov random fields and active contours. *Medical Image Analysis*, 9(1): 1–23, February 2005.
57. MathWorks. Le deep learning à matlab. https://fr.mathworks.com/discovery/deep-learning.html#withmatlab. [Online: accessed on July 17, 2018].
58. J.M. Mendel. Tutorial on higher-order statistics (spectra) in signal processing and system theory: Theoretical results and some applications. *Proceedings of the IEEE*, 79(3): 278–305, March 1991.
59. K.V. Mogatadakala, K.D. Donohue, C.W. Piccoli, and F. Forsberg. Detection of breast lesion regions in ultrasound images using wavelets and order statistics. *Medical Physics*, 33(4): 840–849, March 2006.
60. C. Molder, H. Thomas, and A. Quinquis. Classification des sédiments marins par analyse de texture. In *Congré Reconnaissance des Formes et Intelligence Artificielle*, Angers, France, January 2002.
61. A. Nasser, D. Hamad, and C. Nasr. K-means clustering algorithm in projected spaces. In *International Conference on Information Fusion*, pp. 1–6, Florence, Italy, July 2006.
62. A.Y. Ng, M.I. Jordan, and Y. Weiss. On spectral clustering: Analysis and an algorithm. In *Proceedings of the Advances in Neural Information Processing Systems*, pp. 849–856, Cambridge, MA, December 2002.
63. F. Nicolas, A. Arnold-Bos, I. Quidu, and B. Zerr. Markov-based approaches for ternary change detection between two high resolution synthetic aperture sonar tracks. In *OCEANS*, pp. 1–9, Aberdeen, United Kingdom, June 2017.
64. D. Paoliello, T. Tan, and A. Mansour. Classification of electroencephalogram signals for human motor actions. In *World Congress of Biomechanics*, pp. 1374–1377, Singapore, August 2010.
65. B. Pesquet-Popescu and J. Pesquet. *Ondelettes et Applications*. Edition Techniques Ingénieur, Paris, August 2001.
66. C.G. Puntonet, A. Mansour, C. Bauer, and E. Lang. Separation of sources using simulated annealing and competitive learning. *Neurocomputing*, 49(1–4): 39–60, December 2002.
67. J.R. Quinlan. Induction of decision trees. *Machine Learning*, 1(1): 81–106, March 1986.
68. A. Quinquis, A. Mansour, and E. Radoi. *Signaux et systèmes en questions: signaux, filtrage et décision*. Lavoisier, Hermes, Paris, March 2019.
69. E. Radoi, F. Totir, A. Quinquis, and L. Anton. Superresolution imagery based SVM classification of radar targets. In *European Conference on Synthetic Aperture Radar*, pp. 1–4, Dresde, Germany, May 2006.
70. Y. Rangsanseri and W. Prasongsook. Speckle reduction using wiener filtering in wavelet domain. In *International Conference on Neural Information Processing*, pp. 792–795, Singapore, November 2002.
71. K. Rousseeuw, E. P. Caillault, A. Lefebvre, and D. Hamad. Hybrid hidden Markov model for marine environment monitoring. *IEEE Journal of Selected Topics in Applied Earth Observations and Remote Sensing*, 8(1): 204–213, January 2015.

72. S.T. Roweis and L.K. Saul. Nonlinear dimensionality reduction by locally linear embedding. *Science*, 290(5500): 2323–2326, December 2000.
73. J. Schmidhuber. Deep learning in neural networks: An overview. *Neural Networks*, 61: 85–117, January 2015.
74. C. Schmitt. L'élastographie ultrasonore dynamique vasculaire: une nouvelle modalité d'imagerie non-invasive pour la caractérisation mécanique de la thrombose veineuse. PhD thesis, Université de Montréal, Montréal, Canada, April 2011.
75. B. Schölkopf, A. Smola, and K.R. Müller. Kernel principal component analysis. In *International Conference on Artificial Neural Networks*, pp. 583–588, Lausanne, Switzerland, October 1997.
76. E. Sellier, J. Labarere, M.A. Sevestre, J. Belmin, H. Thiel, P. Couturier, and J.L. Bosson. Risk factors for deep vein thrombosis in older patients: A multicenter study with systematic compression ultrasonography in postacute care facilities in France. *Journal of the American Geriatrics Society*, 56(2): 224–230, February 2008.
77. P.M. Shankar. A general statistical model for ultrasonic backscattering from tissues. *IEEE Transactions on Ultrasonics, Ferroelectrics, and Frequency Control*, 47(3): 727–736, May 2000.
78. P.M. Shankar, V.A. Dumane, T. George, C.W. Piccoli, J.M. Reid, F. Forsberg, and B.B. Goldberg. Classification of breast masses in ultrasonic B scans using Nakagami and K distributions. *Physics in Medicine and Biology*, 48(14): 2229–2240, July 2003.
79. H. Shin, H.R. Roth, M. Gao, L. Lu, Z. Xu, I. Nogues, J. Yao, D. Mollura, and R.M. Summers. Deep convolutional neural networks for computer-aided detection: CNN architectures, dataset characteristics and transfer learning. *IEEE Transactions on Medical Imaging*, 35(5): 1285–1298, May 2016.
80. M. Singh, S. Singh, and S. Gupta. An information fusion based method for liver classification using texture analysis of ultrasound images. *Information Fusion*, 19: 91–96, September 2014.
81. J.H. Song, S.S. Venkatesh, E.F. Conant, T.W. Cary, P.H. Arger, and C.M. Sehgal. Artificial neural network to aid differentiation of malignant and benign breast masses by ultrasound imaging. In *Proceedings Volume 5750, Medical Imaging 2005: Ultrasonic Imaging and Signal Processing,* vol. 5750, pp. 148–152, San Diego, CA, February 2005.
82. G. Tartare, D. Hamad, M. Azahaf, P. Puech, and N. Betrouni. Spectral clustering applied for dynamic contrast-enhanced MR analysis of time-intensity curves. *Computerized Medical Imaging and Graphics*, 38(8): 702–713, December 2014.
83. J.B. Tenenbaum, V. De Silva, and J.C. Langford. A global geometric framework for nonlinear dimensionality reduction. *Science*, 290(5500): 2319–2323, December 2000.
84. A. Vedaldi and K. Lenc. Matconvnet: Convolutional neural networks for matlab. In *ACM Multimedia Conference*, pp. 689–692, Brisbane, Australia, October 2015.
85. C.M. Wu and Y.C. Chen. Statistical feature matrix for texture analysis. *CVGIP: Graphical Models and Image Processing*, 54(5): 407–419, September 1992.
86. K. Zuiderveld. Contrast limited adaptive histogram equalization. In P.S. Heckbert, editor *Graphics Gems IV*, pp. 474–485. Academic Press, San Diego, CA, May 1994.

12 Non-Invasive AI-Assisted Techniques for 3D Printing of the Heart via Image Analysis
Current State, Challenges, and Future Directions

Najmeh Fayyazifar
Curtin University
University of Western Australia
Fiona Stanley Hospital

Najmeh Samadiani
CSIRO Manufacturing

Mohammed Bennamoun
University of Western Australia

Girish Dwivedi
University of Western Australia
Fiona Stanley Hospital

Andrew Maiorana
Curtin University
Fiona Stanley Hospital

12.1 INTRODUCTION TO 3D PRINTING

The use of additive manufacturing (AM) has gained popularity as a rapid prototyping method for creating *three-dimensional* (3D) objects. It is also referred to as 3D printing, which involves adding layers of materials one on top of another in order to create dimensionally accurate parts. This layer-by-layer process is controlled by computer-aided design (CAD) software, which enables the creation of customized intricate products in one step without the limitations of traditional methods such as

DOI: 10.1201/9781003346678-12

casting and subtractive technology (Nasiri and Khosravani 2021). Freedom in design reduces the weight and number of components comprised in a part and consequently minimizes the assembly task required for building a part. In addition, AM is capable of printing multiple sizes of any design and has shortened the building times for fabricating small-volume parts. It has also facilitated the identification and replacement of critical or obsolete components in a timely manner (DebRoy et al. 2018). With its numerous advantages over conventional manufacturing, AM is increasingly being used by both the industrial sector and academics to produce high-performance objects for various applications in the medical, aerospace, automotive, and energy industries. In general, an AM process includes three main phases, as shown in Figure 12.1: design, manufacturing, and quality control.

In the *design phase*, a component with the desired shape and properties is designed using CAD systems and converted to a Standard Triangle Language (STL) file format to be read by 3D printers. *In the manufacturing phase* (step II), the machine loads the STL file and sets up the process parameters. The 3D object is then printed using a layer-by-layer printing process, and then it is removed from the build plate and some post-processing is applied, if necessary. In the *quality control phase*, the produced object is inspected to confirm if all desired requirements are met (Abdulhameed et al. 2019). If some defects exist in the produced object, the manufacturing process should be repeated, and new process parameters should be initialized to prevent anomaly creation.

Biomedical and medical products like implants are mostly customized and constructed according to the patient's body parts; thus, it is essential to provide more precise part specifications, which are required for 3D design. To this end, the ordinary design phase is extended to cover more sub-tasks. Volumetric image acquisition is the first step, which assists the designers in characterizing the precise geometry of 3D models as well as determining tissue properties. They could also lead to choosing the pure material for creating the part. Computed tomography (CT) scans

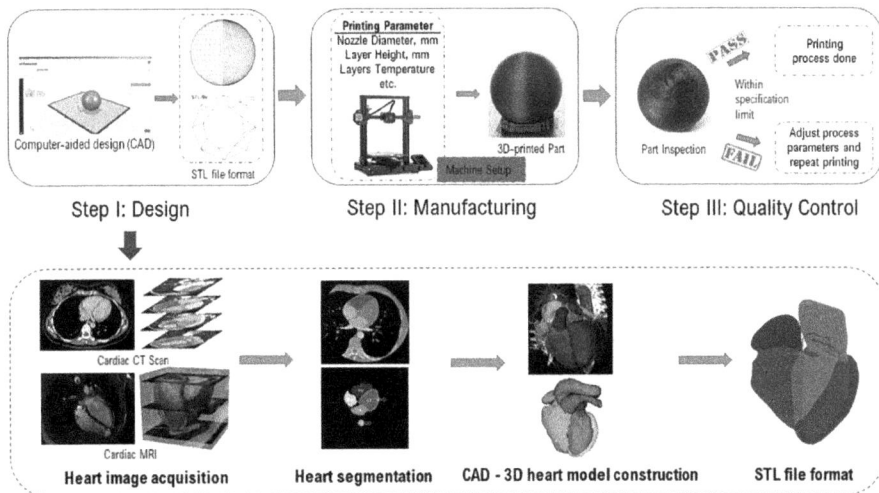

FIGURE 12.1 The first row shows the general AM phases for printing a 3D product. The second row details the design phase for the 3D-printed cardiac process.

and magnetic resonance imaging (MRI) are the most common 3D image acquisition techniques used for 3D printing. These imaging techniques produce volumetric digital imaging and communication in medicine (DICOM) images. The next stage in CAD model design and thereby, STL file generation is image segmentation which identifies regions of interest (ROI) in the medical image (Wang, Qian, et al. 2021). The segmentation step can be manually performed by physicians, which is time-consuming and labor-intensive, or cutting-edge artificial intelligence (AI) techniques could be used to automate the segmentation process.

Three kinds of materials, solid, liquid, and powder, can be used in different AM machines. The American Society for Testing Materials (ASTM) has classified the AM processes into seven categories: (i) vat photopolymerization, (ii) powder bed fusion, (iii) direct energy deposition, (iv) binder jetting, (v) material jetting, (vi) sheet lamination, and (vii) material extrusion (Astm 2010). Vat photopolymerization uses ultraviolet light to construct chains between molecules of photopolymers i.e., liquid radiation-curable resins, and turn them into solid (Gibson, Rosen, and Stucker 2015). Stereolithography (SLA), continuous digital light processing (CDLP), and digital light processing (DLP) are three photopolymerization technologies (Pagac et al. 2021) that are utilized in different applications such as hearing aids (Artec3D 2022), cell scaffold (Zheng et al. 2019), and cardiovascular tissue structures (Yang et al. 2021; Mahmud et al. 2021). Powder bed fusion (PBF) techniques, including selective laser sintering (SLS), direct metal laser sintering (DMLS), and selective laser melting (SLM) processes, apply an energy source e.g., laser and electron beam, to powder particles to fuse them in a layer-by-layer process. One of PBF applications is printing metallic medical devices, such as dental parts, trauma medical implants (e.g., screws and intramedullary rods), and orthopaedic medical products (Fina, Gaisford, and Basit 2018). Direct energy deposition (DED) technique like PBF melts the material, using a heat source but includes a coaxial feed of wire or powder (Gushchina et al. 2022). It is widely utilized in architecture, marine, and aerospace but is very limited in medicine (Pantermehl et al. 2021).

Material extrusion, known as fused deposition modeling (FDM) or fused filament fabrication (FFF), is an AM technique in which a spool of materials is driven through a heated nozzle and selectively deposited layer by layer to build 3D products (Zhuo et al. 2021). It applies to pharmaceutics, dentistry, and medicine, by fabricating implants using metals, polymers, and reinforced composites (Salmi 2021). Binder jetting, called powder bed inkjet printing, consists of a powder bed similar to the PBF process, but instead of a heat source, a binder solution or dispersion is settled to distribute small liquid droplets between powder layers helping to fuse the chosen areas (Rahman et al. 2020). This technique is used in the personalized medication industry (Kozakiewicz et al. 2021) and in making colorful cardiovascular structures for illustration (Wang, Qian, et al. 2021). Material jetting, another AM process, utilizes the building materials in lieu of liquid. Two multi-jet printing (MJP) and PolyJet modelings are grouped in this category (Piłczyńska 2022). 3D products are created by bonding thin sheets of materials in a sheet lamination process, which is used only for making medical models or phantoms (Salmi 2021).

Rapid prototyping is a costly manufacturing technique, but it is increasingly being used in medicine and cardiology due to its ability to quickly create patient-specific products. Among the materials that have been studied for 3D medical printing, titanium and related alloys have shown excellent properties for use in the human body. Therefore, three AM processes – vat photopolymerization, powder bed fusion, and binder jetting – are commonly used in the construction of 3D medical and cardiology objects. However, there is still room to further examine the 3D printing process and the feasibility of using new materials in medicine and cardiology. Artificial intelligence (AI) techniques, as a cutting-edge method for learning and improving existing approaches, can be used in cardiac 3D printing to address the challenges in AM processes. In this chapter, we will provide more details about the existing challenges in heart 3D printing and the role of AI in resolving them.

12.2 CURRENT STATE OF 3D PRINTING AND CHAPTER CONTRIBUTIONS

As mentioned in Section 12.1, various AM or 3D printing processes are utilized in different industries from aerospace, automotive, architecture, and energy to medical and biomedical. Even during the COVID-19 pandemic period, 3D printing solutions were utilized to address the shortage of nasopharyngeal swabs, and personal protective equipment, such as face masks and shields (Sandhu et al. 2022). All these applications are great examples to demonstrate the benefits of AM techniques in the manufacturing of various 3D objects, which is derived through their unique capability of customization and building complex objects.

Although 3D printing techniques and materials continue to develop at a rapid pace, some challenges prevent their wide usage across various domains. Creating objects with different anomalies/defects results in low accuracy of AM, which may cause serious problems in mission-critical scenarios. For instance, in medicine, an implant with inaccuracies in its dimensions could not be well fitted in the patient's body and would cause negative consequences for the patient. The unknown relationships between different factors affect the printing process. Dozens of process parameters are one of those reasons that may create anomalies in the final part. These parameters require proper initialization at the start of the printing process and should be accordingly adjusted during the printing process to prevent flaws.

The other constraining factor in the widespread usage of 3D printers relates to long manufacturing times that could be reduced if the design phase was improved. In medicine, the segmentation challenges add an extra burden to the design step. As stated before, automatic segmentation techniques are used to extract the ROI from the volumetric 3D images, captured by medical devices. Selecting an accurate segmentation method to segment ROI neatly and flawlessly plays an important role in the next steps, including creating 3D mesh and its fabrication. This chapter aims to highlight the role of AI in reducing the mentioned existing challenges of using AM in medicine, more particularly in cardiac applications. This is the main difference compared to existing review papers (Wang, Qian, et al. 2021; Uccheddu et al. 2018; Sun et al. 2019). To investigate these AI techniques used for enhancing cardiac 3D printing, we chose articles published from 2017 to 2022. We have listed our contributions as follows:

- Reviewing the current literature to explain the cutting-edge applications of 3D printing in medicine and cardiac analysis.
- Presenting the latest deep learning-based segmentation techniques and demonstrating their applications to 3D printing in the medical field.
- Analyzing how AI can assist AM in advancing the design approaches and in-situ monitoring in medical applications.

The chapter is organized as follows: applications of 3D printing in cardiology and cardiac image acquisition techniques used for 3D printing are reviewed in Section 12.3. Followed by in Section 12.4, a detailed description of AI segmentation techniques for medical image processing is introduced. The role of AI in improving the "design" and "process monitoring", for producing accurate cardiac 3D models are then discussed in Section 12.5. Lastly, the unresolved challenges in cardiac 3D printing and future research directions to alleviate them are summarized in Sections 12.6 and 12.7, respectively.

12.3   3-D PRINTING IN CARDIOLOGY

12.3.1   CARDIAC 3-D MODELS APPLICATIONS

3D printed models have been successfully used in a wide range of medical applications such as reconstructive breast surgery (Di Rosa 2022), skull implant, ear (Lee et al. 2017), bone tissues (Haleem et al. 2020), and dental implants (Balamurugan and Selvakumar 2021). We discuss a detailed explanation of 3D printing applications in cardiology in this chapter.

12.3.1.1   Surgical Simulation, Surgical Planning, and Education

Cardiac 3D printed models are useful tools for surgical simulation and surgical planning. Surgery simulation is beneficial for both novice and experienced surgeons. In vitro simulated surgeries provide novice cardiovascular practitioners with the opportunity to practice the surgical procedure on patient-specific 3D models (Yamada et al. 2017). In addition, cardiac 3D models assist medical students in understanding the relationship between cardiac images and in vivo cardiac anatomy, which will enhance the cardiovascular intervention success rate (Wu et al. 2021). Experienced surgeons can utilize 3D models in complex cases where more than one surgical procedure is applicable, and in extreme instances, some of which are controversial. In these cases, cardiac 3D printed models can be used to analyze the anatomical characteristics of patients, simulate surgery, asses the surgical outcomes, and select the most efficient treatment based on the patient's cardiac anatomy (Hussein et al. 2020). Furthermore, these patient-specific 3D models can assist cardiovascular physicians and surgeons in careful surgical planning by predicting potential complications such as annular rupture, paravalvular leak, the requirement for a pacemaker, and coronary artery occlusion (Sardari Nia et al. 2017).

12.3.1.2   Patient Education and Doctor-Patient Communication

3D printed models of the heart can be used to visualize the cardiac structure and help physicians to explain conditions to patients. This can assist doctor-patient communication and reduce patients' perioperative anxiety and pain (Hui 2017). It has also been shown that the interaction of patients with 3D printed models during clinical visits can improve the patient's understanding of treatment options, patient engagement, and stratification (Wang, Qian, et al. 2021).

12.3.1.3   Structural Cardiac Disease Intervention

Structural heart disease includes a range of different cardiac conditions such as congenital heart disease, hypertrophic cardiomyopathy, and valvular cardiac conditions. Cardiac 3D modeling has applications in the repair and replacement of mitral valves, aortic valves, and pulmonic valves (Kuk et al. 2017). Application of 3D printing in mitral valve intervention includes Mitral Valve Perforation Repair (Little et al. 2016), Transcatheter Mitral Valve Replacement (TMVR) (Wang, Eng, et al. 2016), and Mitral Valve Annuloplasty (Dankowski et al. 2014). 3D-printed cardiac models have also been used in Aortic Valve Replacement Resternotomy (Schmauss et al. 2015) and Transcatheter Aortic Valve Implantation (TAVI) (Schmauss et al. 2012). To assess the stability of implanted pulmonic valve in patients with irregularly shaped right ventricular, cardiac 3D models have been employed and proved effective in positioning a metal stent inside a catheter-delivered valve (Phillips et al. 2016).

In addition to the above-mentioned applications of cardiac 3D models, they enhanced the quality of treatment in other cardiac conditions. For example, (Giannopoulos et al. 2015) guided spatial navigation of occlude device during surgery. Furthermore, 3D models were successfully employed in cases where open heart procedures were required. This includes improving the accuracy of implanting the Ventricular Assist Device (Farooqi et al. 2016) and locating a primary cardiac tumor in the right ventricle (Schmauss, Gerber, and Sodian 2013).

Overall, 3D printed models have a wide range of applications in the medical domain, and more specifically in cardiology. The patient-specific generated 3D cardiac models considerably improved the quality of treatment that patients received.

12.3.2   Cardiac Image Acquisition Techniques

Cardiac 3D printing is a challenging task, and the accuracy of the generated model relies heavily on the quality of the captured cardiac images. It has to be considered that the cardiac structure is complicated with different anatomical features such as chambers, atrial, vessels, atrioventricular valves, papillary muscles, trabeculae carneae, and parts of the coronary circulation (Uccheddu et al. 2018). Accurate 3D manufacturing of these organs requires high-quality image acquisition which makes precise cardiac segmentation possible.

The main image acquisition techniques in cardiovascular surgery and intervention are Computed Tomography (CT), Magnetic Resonance Imaging (MRI), and echocardiography (Schmauss et al. 2015). Each of these imaging techniques is suitable for 3D printing of different parts of the heart. While for 3D printing of the heart chambers, ventricular septal defects, and large vessels CT scans and MRIs are more suitable, an echocardiogram is the preferred imaging technique for 3D printing of interatrial septum, heart valves, and sub-valvular apparatus (Valverde 2017).

Furthermore, different cardiac image acquisition techniques have some advantages and disadvantages. As the quality of the derived images in MRIs and CTs could be sub-optimal, the homogeneous opacification of cardiac is an essential step to provide an accurate ROI before 3D printing (Yoo et al. 2021). MRIs are preferable to CT scans, as the contrast enhancement process could be performed more homogeneously on MRIs; however, the image acquisition time is longer in MRI (Yoo et al. 2021). Although 3D echocardiograms can provide a 3D vision of cardiac structure and are useful in the reconstruction of the valves' morphology, a considerable amount of artifact occurs which constrain their application in cardiac 3D printing (Lang et al. 2018).

12.4   CURRENT MEDICAL IMAGE SEGMENTATION TECHNIQUES

Image segmentation is defined as the technique of dividing an image into multiple regions, each of which is made up of a group of pixels (or voxels) that correspond to a specific meaningful structure (Robinson et al. 2019). It can also be defined as the process of drawing boundaries between separate semantic entities in an image. The purpose of image segmentation is to categorize pixels of similar characteristics into groups that make image analysis and understanding easier.

Image segmentation has many applications in different fields. In the medical domain, it is used for tumor localization (Wu et al. 2014), cancerous regions detection (Singh and Gupta 2015), and vessel thickness estimation (Yan, Yang, and Cheng 2018). Image segmentation is also used in surveillance applications such as pedestrian detection (Leibe, Seemann, and Schiele 2005) and traffic surveillance (Friedman and Russell 2013). Other application areas of image segmentation are face detection in security applications (Yadav and Nain 2015), and scene understanding (Li, Socher, and Fei-Fei 2009).

There are three main categories of image segmentation: semantic segmentation, instance segmentation, and panoptic segmentation. Semantic segmentation in which multiple objects of a single category are labeled as one entity is mainly used in medical image segmentation. These techniques are reviewed in the rest of this section. More information about instance and panoptic segmentation techniques can be found in (Hafiz and Bhat 2020; Kirillov et al. 2019).

12.4.1   SEMANTIC SEGMENTATION TECHNIQUES

Image segmentation techniques can be classified into two main categories: classical computer vision approaches and deep learning-based methods. Classical image segmentation techniques include a wide range of methods such as thresholding (Tobias and Seara 2002), edge detection (Senthilkumaran and Rajesh 2009), histogram-based approaches (Shapiro and Stockman 2001), feature space clustering methods (Comaniciu and Meer 1997), region-based approaches (Freixenet et al. 2002), and graph partitioning methods (Peng, Zhang, and Zhang 2013). These methods incorporate digital image processing techniques along with mathematics to extract features and segment the image. These techniques are usually fast; however, their performance is reliant on the knowledge and expertise of domain experts to extract features from the images.

The dependence on human experts has been decreased due to the advancement of deep learning algorithms. Deep learning is a type of machine learning approach that mimics the learning procedure of the human brain's biological neural system. Convolutional neural networks (CNNs) have shown remarkable performance in image analysis and computer vision tasks among deep neural networks. The convolutional layer is the primary building block in CNNs, and it is responsible for extracting features from unstructured data such as images, signals, and videos. By utilizing filters, this layer automatically detects the existing patterns in data, allowing for data analysis without requiring manual feature engineering (Wu 2017; Samadiani et al. 2021). CNN-based data analysis networks have been enhanced and improved since the development of CNNs to enhance their performance. These models are widely used in various image and signal analysis tasks, including classification (Wang, Yang, et al. 2016; Fayyazifar 2021; Fayyazifar et al. 2020), segmentation (Dolz et al. 2018; Minaee et al. 2021), and image reconstruction (Gupta et al. 2018).

Deep learning approaches were first developed for image classification tasks; however, their superior performance encouraged researchers to utilize them for image segmentation tasks. In this section, deep learning-based methods for general image segmentation, medical image segmentation, and specifically, cardiac segmentation are summarized. These segmentation techniques can be grouped into Fully Convolutional Networks, Regional Convolutional Neural Networks, Recurrent Neural Networks, Encoder-decoder Models, enhancement mechanisms, Transformers, and Ensemble models. Table 12.1 summarizes all studies which utilized these segmentation methods in cardiology and other domains.

12.4.2 Fully Convolutional Networks (FCNs)-Based Segmentation Techniques

Well-known FCN classification models were first unitized for image segmentation (Long, Shelhamer, and Darrell 2015). They modified existing AlexNet (Krizhevsky, Sutskever, and Hinton 2017), VGG16 (Simonyan and Zisserman 2014), and GoogLeNet (Szegedy et al. 2015), and fine-tuned them for pixel-wise image segmentation. Since then, FCN models have been employed in different medical image segmentation tasks such as retinal image segmentation (Feng et al. 2017), bone surface segmentation (Villa et al. 2018), and skin lesion segmentation (Kaymak, Kaymak, and Ucar 2020). In the cardiovascular domain, FCNs have been used for coronary arteries segmentation (Shen et al. 2019), ventricular segmentation (Lieman-Sifry et al. 2017), and myocardium segmentation (Fahmy et al. 2019). 3D FCN has also been used for ventricular segmentation from MRI (Yang et al. 2017).

TABLE 12.1

AI-Assisted Image Segmentation Techniques and Their Application

Method	Approach/Reference	Segmented Organ/Object		Data Type/Dataset
		Cardiology	Others	
Fully convolutional networks (FCNs)	Long, Shelhamer, and Darrell (2015)	✗	Scene's objects	RGB (PASCAL VOC)
	Feng et al. (2017)	✗	Retinal	RGB (DRIONS-DB)
	Villa et al. (2018)	✗	Bone surface	2D ultrasound
	Kaymak, Kaymak, and Ucar (2020)	✗	Skin lesion	RGB (ISIC)
	Shen et al. (2019)	Coronary arteries	✗	Coronary computed tomography angiography
	Lieman-Sifry et al. (2017) and Yang et al. (2017)	Ventricle	✗	CMRI[a]
	Fahmy et al. (2019)	Myocardium	✗	Coronary computed tomography angiography
R-CNN-based models	R-CNN: Girshick et al. (2014)		Scene's objects	CMRI
	Fast R-CNN: Girshick (2015)		Scene's objects	CMRI
	Faster R-CNN: Ren et al. (2017)		Scene's objects	RGB (PASCAL VOC)
	Mask R-CNN: He et al. (2017)		Scene's objects	RGB (COCO dataset)
	MaskLab: Chen, Hermans, et al. (2018)		Scene's objects	RGB (COCO dataset)
	Sheng et al. (2020)		Lymphoma	Blood smear slides
	Shu et al. (2020)		Multiorgan (heart, left and right lungs)	CT scan[b]
	Zhang et al. (2020)	Left ventricle	Breast cancer	MRI
	Qadeer, Shah, and Sharif (2021) and Hsu (2019)	Cardiac localization	✗	CMRI, Echocardiography
	NF-RCNN: Kermani et al. (2020)		✗	CMRI

(*Continued*)

TABLE 12.1 (Continued)
AI-Assisted Image Segmentation Techniques and Their Application

Method	Approach/Reference	Segmented Organ/Object		Data Type/Dataset
		Cardiology	Others	
RNN-based models	Li et al. (2020)	✗	Tooth root	Cone beam CT scan
	Zheng and Yi (2012)	✗	Brain	MRI
	Xie et al. (2016)	✗	Muscle perimysium	Microscopy images
	Bai et al. (2018)	Aorta	✗	MRI
	Zhang et al. (2018)	Ventricle, myocardium	✗	CMRI
	Xu et al. (2018)	Myocardial scars	✗	CMRI
Encoder-decoder based models	SegNet: Badrinarayanan, Kendall, and Cipolla (2017)	✗	Road scene, indoor scene's objects	RGB (CamVid, LabelMe)
	U-Net: Ronneberger, Fischer, and Brox (2015)	✗	Cell segmentation	Gray-scale images (biological microscopy images)
	D-UNet: Zhou, Huang, et al. (2019)	✗	Chronic stroke lesion	MRI
	Isensee et al. (2017)	Ventricle	✗	CMRI
	MFP-Unet: Moradi et al. (2019) and Tao et al. (2019)			Echocardiography
	Xia et al. (2018)	Atria	✗	Gadolinium-enhanced Cardiac MRI
Transformer models	Merkow et al. (2016)	Coronary artery	✗	MRI, CT
	Visual transformers (ViT): Dosovitskiy et al. (2020)	✗	Scene understanding	RGB (ImageNet)
	MedT: Jeya Maria Jose Valanarasu (2021)	✗	Brain, gland, multi-organ nucleus	Ultrasound images / Microscopic images

(Continued)

TABLE 12.1 (*Continued*)
AI-Assisted Image Segmentation Techniques and Their Application

Method		Approach/Reference	Segmented Organ/Object		Data Type/Dataset
			Cardiology	Others	
Enhancement mechanism	Residual mechanism	Shehab et al. (2021)	✗	Brain tumor	MRI
		Alom et al. (2018)	✗	Nuclei	Histopathological images
		RIC-Unet: Zeng et al. (2019)	✗	Nuclei	Histopathological images
		FR-NET: Chen, Fang, and Liu (2019)	Left ventricle	✗	CMRIs
		Zyuzin et al. (2020)	Left ventricle, left atrium	✗	2D Echocardiography
		Liu, Feng, and Yang (2019)	Right ventricle	✗	CMRI
	Attention mechanism	Mnih, Heess, and Graves (2014)	✗	Digits	MNIST digits dataset
		Attention U-Net: Oktay et al. (2018)	✗	Pancreas	Abdominal CT scan
		3D attention U-Net: Islam et al. (2019)	✗	Brain tumor	MRI
		MA-Unet: Cai and Wang (2022)	✗	Lung/nodule localization and oesophageal cancer	CT scan
		RIANet: Tong et al. (2019)	Myocardium, left and right ventricles	✗	MRI
		Chen, Yang, et al. (2018)	Left atrium, atrial scars	✗	CMRI

(*Continued*)

TABLE 12.1 (*Continued*)
AI-Assisted Image Segmentation Techniques and Their Application

Method	Approach/Reference	Segmented Organ/Object		Data Type/Dataset
		Cardiology	Others	
Dilation mechanism	DeepLabv2: Chen, Papandreou, Kokkinos, et al. (2017)	✗	Scene's objects	RGB (PASCAL VOC)
	DeepLabv3: Chen, Papandreou, Schroff et al. (2017)	✗	Scene's objects	RGB (PASCAL VOC)
	DeepLabv3+: Chen, Zhu, et al. (2018)	✗	Scene's objects	RGB (PASCAL VOC)
	Chen, Xuan, et al. (2021)	✗	Lung	CT scan
	Wang et al. (2019)	✗	Prostates	MRI
	Vesal, Ravikumar, and Maier (2018)	Left atrium	✗	CMRI
Enhancement mechanism	Borodin and Senyukova (2018)	Right ventricle	✗	CMRI
Multi-scale and pyramid	UNet++: Zhou, Siddiquee, et al. (2019)	✗	Cell, nuclei, brain tumor, liver, lung nodule	Microscopy, MRI, CT
	APCNet: He et al. (2019)	✗	Scene's objects	RGB (PASCAL VOC)
	Fan et al. (2020)	✗	Liver	CT scan
	Zreik et al. (2018)	Left ventricle	✗	Coronary computed tomography angiography
	MDFA-Net: Li et al. (2021)	Left and right ventricles	✗	CMRI
	PLANet: Liu et al. (2021)	Left ventricler, myocardium	✗	2D Echocardiography

(*Continued*)

TABLE 12.1 (*Continued*)
AI-Assisted Image Segmentation Techniques and Their Application

| Method | Approach/Reference | Segmented Organ/Object | | Data Type/Dataset |
		Cardiology	Others	
Ensemble models	TransBTS: Wang, Chen, et al. (2021)	✗	Brain tumor	3D MRIs
	Unet CNN-transformer TransUnet: Chen, Lu, et al. (2021)	✗	Abdominal	CT scan
	transformers with Unet-shaped TF-Unet: Fu et al. (2022)	Left and right ventricles ✗		CMRI
	BERT CNN Reynaud et al. (2021)	Left ventricle	✗	Echocardiography
	Unet transformer TransNUNet: Yang and Tian (2022)	Whole heart	✗	CT scan
	U-Net and Faster RCNN Nurmaini et al. (2022)	Atrial, Ventricular, Atrioventricular	✗	Echocardiography

a Cardiac magnetic resonance imaging.
b Computerized tomography scan.

12.4.3 REGIONAL CONVOLUTIONAL NEURAL NETWORKS-BASED SEGMENTATION TECHNIQUES

The concept of Regional Convolutional Neural networks (R-CNNs) was proposed in (Girshick et al. 2014). In this method, a selective search is utilized for the generation of category-independent regions. These region proposals are passed through a CNN network which extracts fixed-length feature vectors from each region, and then a set of class-specific linear SVMs are used for final segmentation. To enhance the performance of R-CNN, several improved versions of R-CNN have been developed. Fast R-CNN (Girshick 2015) and Faster R-CNN (Ren et al. 2017) reduced the computation time of R-CNN, by employing the VGG16 network as a feature extractor and introducing a new Region Proposal Network which shares full-image convolutional features with the detection network, respectively. Mask R-CNN (He et al. 2017) improved the performance of Faster R-CNN by predicting an object mask in addition to the available bounding box recognition. MaskLab (Chen, Hermans, et al. 2018), the other refinement of Faster R-CNN, provides semantic segmentation logits for pixel-wise classification and logits for predicting pixels' direction toward their instance center. These predicted boxes enable more accurate object localization. R-CNN-based networks have been utilized for different medical image segmentation tasks. Faster R-CNN has proven effective in lymphoma segmentation using blood smear slides (Sheng et al. 2020). Shu et al. (2020) proposed a refined Mask R-CNN for multiorgan segmentation using CT data. Mask R-CNN has also been used for breast cancer localization in MRI (Zhang et al. 2020). In cardiology, Faster R-CNN has been used for automatic left ventricular segmentation in MRI data (Qadeer, Shah, and Sharif 2021) and cardiac ultrasound images (Hsu 2019). Kermani et al. (2020) proposed a Network-in-network Faster-RCNN (NF-RCNN) for cardiac segmentation and right ventricle wall motion abnormality detection in MRI.

12.4.4 RECURRENT NEURAL NETWORKS-BASED SEGMENTATION TECHNIQUES

Recurrent Neural Networks (RNNs) (Sherstinsky 2020) are a type of neural network with the ability to remember the previous state of data and thereby are suitable for tasks with sequential data. RNNs are used in semantic segmentation as they can model the short-term and long-term dependencies among pixels which may lead to more accurate segmentation. Long-Short-Term Memory (LSTM) models (Graves 2012), a type of RNN which uses both short-term and long-term memory to model dependencies in input data, proved effective in semantic segmentation.

RNN-based models are incorporated into medical image segmentation for a variety of clinical conditions including tooth root segmentation using CT images (Li et al. 2020), brain image segmentation on MRI (Zheng and Yi 2012), and muscle perimysium segmentation (Xie et al. 2016). In the cardiac domain, RNN-based models are used for aortic segmentation (Bai et al. 2018), ventricular myocardial segmentation (Zhang et al. 2018), and localization of myocardial scars in Cardiac MRI (Xu et al. 2018).

12.4.5   ENCODER-DECODER-BASED SEGMENTATION TECHNIQUES

The idea behind encoder-decoder networks is to encode the input image into feature vectors by convolutional layers and decode the feature vectors into a semantic
segmentation mask. SegNet (Badrinarayanan, Kendall, and Cipolla 2017) is one of
the first encoder-decoder networks that was proposed for image segmentation. This
network uses 13 layers of VGG16 as the encoder network and a multilayer deconvo-
lutional network that generates class probabilities at a pixel level. Encoder-decoder
networks proved effective in medical image segmentation. Ronneberger, Fischer, and
Brox (2015) proposed a U-shaped architecture for the segmentation of biological
microscopy images. Through skip connections, this architecture combines high-level
feature vectors (from each layer in the decoder) and low-level detailed feature vectors (from the corresponding encoder layer), thereby providing accurate semantic
segmentation performance.

The proposed 2D U-Net models discard the existing 3D information of medical images; on the other hand, 3D U-Net models are computationally expensive. To
address this problem, Dimension-fusion-UNet (D-UNet) (Zhou, Huang, et al. 2019)
combines 2D and 3D convolution in the encoding stage of U-Net. In cardiology,
U-Net and its variants have been used on different imaging techniques including
MRI, CT, and echocardiography, and for segmentation of different cardiac segments
such as ventricles (Isensee et al. 2017; Moradi et al. 2019; Tao et al. 2019), and atria
(Xia et al. 2018).

12.4.6   ENHANCEMENT MECHANISMS FOR IMAGE SEGMENTATION TECHNIQUES

12.4.6.1   Residual Enhancement Models

Deep residual networks (ResNet) (He et al. 2016) were proposed to address the problem of vanishing gradients in deep networks. He et al. (2016) showed that despite
the common assumption that increasing the number of convolution layers in a CNN
model improves the model performance when the model becomes very deep, its performance degrades. To solve this, they proposed an identity mapping/skip connection,
which facilitates the flow of information in training. ResNet-based models were used
in many medical image segmentation applications such as brain tumor segmentation
from MRIs (Shehab et al. 2021), and nuclei segmentation from high-resolution his-
topathological images (Alom et al. 2018). In RIC-Unet (Residual-Inception-Channel
attention-Unet) (Zeng et al. 2019), proposed for nuclei segmentation, the convolution
layers of U-Net in the encoder section are replaced by RI (residual inception) blocks,
and the decoder section incorporates DC block (Deconvolution-Channel-block) that
provides nuclei mask.

In cardiovascular conditions, ResNet-based models were used for different cardiac structure segmentation. FR-NET (Chen, Fang, and Liu 2019) utilized ResNet
for left ventricular segmentation from cardiac MRIs. Zyuzin et al. (2020) integrated
a residual block into U-Net for the left ventricle and left atrial segmentations in 2D
echocardiograms. Residual-based U-Net has also been used for right ventricular segmentation in Cardiac MRI (Liu, Feng, and Yang 2019).

12.4.6.2 Attention-Based Enhancement Models

Attention-based networks use a mechanism to assist a model to focus on regions of input that contain more discriminative features and discard sections with less important features. The attention mechanism was first introduced by Mnih, Heess, and Graves (2014) on the MNIST digits dataset. Attention U-Net (Oktay et al. 2018) integrates an attention gate into U-Net and produces an attention coefficient which assists the U-Net model to identify regions with salient features and deactivate the responses to irrelevant segmentation regions. This model accurately segmented the pancreas in abdominal CT images. 3D attention U-Net has been used for brain tumor segmentation and survival prediction (Islam et al. 2019). MA-Unet (Cai and Wang 2022) improved the performance of base U-Net by incorporating two multi-scale attention modules; channel attention and spatial attention, to model the dependencies between different channels and encode the high-level context information into local features, respectively. Their model was successful in semantic segmentation of lung CT scans for nodule localization and oesophageal cancer segmentation.

The attention mechanism has applications in different cardiac segmentation tasks. RIANet (Tong et al. 2019) was developed for myocardial, and left and right ventricular segmentations from cardiac MRI data. RIANet integrated a recurrent feedback structure which uses forward and backward connections between different layers of the same resolution. In addition, an interleaved attention (IA) block was incorporated to fuse the spatial information of high-level features and semantic information of low-level features in an interleaved manner. Chen, Yang, et al. (2018) proposed a recursive attention model for left atrium and atrial scars segmentation, using cardiac MRI data.

12.4.6.3 Dilated Convolution-Based Enhancement Models

Dilated convolutional models were developed to reduce the information loss caused by pooling operation in standard convolutional models. Dilated convolution adds a new parameter, dilated rate, to convolutional layers which defines the expansion rate in the receptive field of convolution layers. Dilation reduces the size of a network while maintaining the spatial dimension of feature maps. DeepLabv2 (Chen, Papandreou, Kokkinos, et al. 2017) proposed to employ dilation in ResNet 101 and showed performance improvement in semantic segmentation. In the later version, DeepLabv3 (Chen, Papandreou, Schroff, et al. 2017), different modules that incorporate parallel dilated convolutions with different dilatation rates, were designed. The latest work of the DeepLab family, DeepLabv3+ (Chen, Zhu, et al. 2018), incorporates dilation in an encoder-decoder architecture where the decoder was the model developed in Deeplabv3. Dilated convolution in conjunction with U-Net was used in (Chen, Xuan, et al. 2021) for the segmentation of lung CT scans. Dilatation in fully CNNs was used by Wang et al. (2019) for the segmentation of prostates in MRIs. In cardiology, dilated U-Net was used for left atrial (Vesal, Ravikumar, and Maier 2018) and right ventricular (Borodin and Senyukova 2018) segmentation in cardiac MRIs.

12.4.6.4   Multi-Scale and Pyramid Enhancement Models

In semantic segmentation, regions of interest may have varying scales. Multi-scale networks are proposed to provide accurate segmentation for varying size objects. These
models combine low-level and high-level features, and thereby, lead to more accurate
segmentation. UNet++ (Zhou, Siddiquee, et al. 2019) modified U-Net by combining
different U-Net-based models with varying depth, as well as adding nested and dense
skip connections to aggregate features from different semantic levels in the decoder.
Adaptive Pyramid Context Network (APCNet) (He et al. 2019) used Global-guided
Local Affinity (GLA) to adaptively construct multi-scale context features. This GLA
is used as a guide to estimate the local affinity coefficients of different regions in
the image and calculate the context vector. Fan et al. (2020) proposed a multiscale
attention network for liver segmentation from CT scans. Multiscale networks have
applications in cardiac image segmentation. Zreik et al. (2018) used multiscale CNN
for left ventricular myocardial segmentation for the diagnosis of significant coronary
artery stenosis. MDFA-Net (Li et al. 2021) aggregates features from multiscale dual
path for cardiac segmentation from cardiac MRIs. Pyramid local attention network
(PLANet) (Liu et al. 2021) was proposed to address the challenging task of left ventricular and myocardial segmentation in 2D echocardiograms. PLANet refines feature vectors by extracting information from compact and neighboring contexts.

12.4.7   Transformer Models for Image Segmentation

Transformers were initially developed by Google Brain for a sequence-to-sequence
modeling task (Vaswani et al. 2017). Transformers addressed the limitation of residual models in processing long inputs. Specifically, residual models such as LSTMs
tend to associate less importance with past timeslots compared to recent slots. To
resolve this limitation, transformers employ a self-attention mechanism to associate an importance weight to different sections of input, regardless of their presence time. To achieve this, multi-head attention was proposed, which collectively
considers information from all subspace representations and at different positions
(Vaswani et al. 2017). Visual transformers (ViT) (Dosovitskiy et al. 2020) were proposed to extend the application of transformers to image analysis tasks. In ViT, the
input image is portioned into patches, and a sequence of flattened 2D patches along
with positional embeddings are served as the input to the model. ViT indicated that
no reliance on CNN is required to achieve a high performance in image classification and a pure visual transformer can outperform CNN models. Transformers were
used in different image classification and segmentation applications. MedT (medical
transformer) (Jeya Maria Jose Valanarasu 2021) uses a gated axial attention which
adds a new control mechanism to the existing self-attention module. Specifically,
self-attention is decomposed into two attention modules: a self-attention performing
on the height axis and another self-attention performing on the width axis of feature
maps. MedT was evaluated on three medical image segmentation tasks, brain anatomy segmentation from ultrasound images, gland segmentation, and multi-organ
nucleus segmentation from microscopic images. Visual transforms are usually
combined with other segmentation models; thus, we review them in the Ensemble
model's section.

12.4.8 ENSEMBLE MODELS FOR IMAGE SEGMENTATION

Ensemble models combine two or more of the base models, described in previous sections, to improve the performance of segmentation tasks. Ensemble models have been used in different medical image segmentation applications. TransBTS (Wang, Chen, et al. 2021) was developed for tumor brain segmentation from 3D MRIs, through the combination of transformers and encoder-decoder models. TransUnet (Chen, Lu, et al. 2021) is proposed based on the U-Net model; however, in the encoder section of U-Net, a hybrid model of CNN-transformer is used to encode input images. The CNN model extracts features from input images and forms feature maps, then the transformer is applied to feature maps to model dependencies in them.

In cardiac segmentation, the impact of combining transformers with U-Net-shaped architecture has been studied for left and right ventricular segmentations from cardiac MRIs (Fu et al. 2022). In Bidirectional Encoder Representations from Transformers (BERT)-based model which provides spatio-temporal reasoning was ensembled with a CNN-based encoder for ejection fraction estimation from echocardiography images (Reynaud et al. 2021). TransNUNet (Yang and Tian 2022), an improved version of TransUnet (Chen, Lu, et al. 2021), was developed for whole heart segmentation from CT scans. This model introduced a new attention mechanism on TransUnet. Nurmaini et al. (Nurmaini et al. 2022) combined a U-Net and Faster RCNN model for atrial, ventricular, and atrioventricular segmentations and defect detection in echocardiograms.

12.5 ROLE OF AI IN IMPROVING CARDIAC 3D PRINTING

12.5.1 IMPROVED 3D PRINTING DESIGN APPROACHES

The design phase plays a critical role in printing a defect-free 3D model. As medical devices are constructed to be patient-specific and to be embedded in the body, they require extra caution to fabricate flawlessly. As stated before, in the design stage, the acquired cardiac images should be segmented, using a segmentation technique, and then this segmented cardiac structure needs to be converted into suitable 3D mesh prior to extracting the STL file format. There are several challenges in every step of the design phase; however, it has been found that AI approaches can resolve these obstacles.

Firstly, whole heart segmentation and thereby precise mesh construction are challenging tasks due to the complexity of cardiac structures (Uccheddu et al. 2018). As mentioned in Section 12.1, accurate segmentation of cardiac structure (in cardiac images) is the first and most labor-intensive task in manufacturing effective 3D cardiac models. Different AI techniques have been proposed (see Section 5.2) to automate the segmentation process of cardiac images, and thereby, reduce the dependency of 3D models on cardiology-expert image interpretation and segmentation. While most existing AI-based cardiac segmentation models only focused on the segmentation of some parts of the cardiac structure (Table 12.1), very few studies have explored the problem of automatic whole heart segmentation (Payer et al. 2017; Yang et al. 2018). Thus, novel AI segmentation techniques are required to further improve the process of whole heart segmentation.

Secondly, another limiting factor in accurate cardiac 3D printing relates to the inefficiency of imaging techniques. As stated in Section 12.4, different cardiac image acquisition techniques are suitable for 3D printing of different parts of the cardiac structure. For instance, heart valves are better captured in echocardiograms, while MRI is more often used for large vessel imaging. Using multimodal images could greatly assist in obtaining the perfect segment by providing complementary information on related anatomic details (Samadiani et al. 2022; Harb et al. 2019). To this end, AI can be used to fuse different image modalities and provide a complete view of the whole heart which accordingly will lead to 3D printing of the cardiovascular system. A few initial research has been conducted in this area to fuse echocardiograms and CT scans (Gosnell et al. 2016), but further research could be performed in employing AI techniques for combining other imaging techniques and developing automatic cardiac segmentation algorithms.

Another constraining factor in the design phase of 3D cardiac printing relates to the accuracy of AI-based segmentation techniques. In Section 12.5.2, we reviewed different segmentation methods including the state-of-the-art transformers and their combination with well-performing UNet-shaped models. Different AI segmentation methods could be integrated to improve the accuracy of these segmentation methods. For instance, the CNN block in the well-known TransUnet (Chen, Lu, et al. 2021) which is responsible for generating feature maps from the input image, prior to feeding them as image patches into the transformer block, can be replaced by the residual-inception blocks that were proposed in RIC-Unet (Zeng et al. 2019). As each of these two models is associated with high segmentation accuracy in other medical image segmentation tasks, one could expect a performance boost for cardiac image segmentation by the combination of these two well-performing AI models. An ensemble of other basic segmentation models (developed for other medical applications) could be investigated and evaluated for cardiac segmentation. One possible strategy would be to employ Neural Architecture Search (NAS) algorithms to perform a search on predefined operations and discover the optimum model. These NAS search algorithms have been previously employed for discovering the optimum neural model for biomedical signal classification and have shown promising results (Fayyazifar et al. 2020, 2023; Fayyazifar 2021). One could advance these NAS-based search algorithms for segmentation task and discover the optimum AI-assisted model for whole cardiac segmentation.

Furthermore, the available CAD software cannot accurately convert a series of 2D images into 3D models since it should identify a fixed relationship between every point of the current image and the corresponding element on other images (Otton et al. 2017). Failure in observing a relationship may need an estimation/interpolation of hidden points, leading to artifact creation. Thus, it necessitates a post-processing step to refine the model by correcting errors and removing redundant artifacts, such as streak artifacts, blooming, and beam hardening, as they also appear on the 3D model (Uccheddu et al. 2018). Moreover, other properties, such as myocardial structure and patient-specific characteristics (e.g., pathologies), are required to be added manually after constructing a 3D mesh. Lastly, breaking down and converting the 3D constructed mesh into different layers and variable-size triangles may add additional errors to the STL files. This is due to the incompatibility between the size and shape of regions on the cardiac mesh and the converted triangles. As a result, not handling these errors in the STL file may result in a porous 3D-printed heart.

The field of design for AM (DfAM) is specifically defined to create, optimize, and refine part features to benefit from all AM advantages (Jiang et al. 2022). Recently, AI algorithms have shown effectiveness in improving DfAM and addressing the existing challenges by pre-assessing of 3D models and providing solutions for designing optimized parts. As a result, they could minimize the cost by reducing the post-processing requirements to achieve the desired properties of the product. Developed design solutions could be employed to handle these challenges in cardiology. For instance, Decker et al. (2021) proposed an RF-based framework to provide a model for designing 3D freeform objects with limited human interaction in creating a model. They could achieve good results whilst training the model with a few STL meshes. Adapting such a system could generate an optimized and error-free STL file of a heart without expert/medical staff supervision. Yao et al. developed a hybrid approach using SVM and hierarchical clustering to decipher a group of AM design features for particular designs (Yao, Moon, and Bi 2017). This system could reduce the design errors in cardiac prototypes, by specifying the desired object properties, customized for every patient. Some optimization design methods that consider customization on both macro and mesoscale, like the technique proposed by Tang et al. (2021), have the potential to facilitate the heart 3D model construction by non-experts in the field. This method allows customized 3D model printing, through freedom in adjusting different object characteristics such as size and weight. This framework uses a statistical-based machine learning method called the Gaussian process to generate a fully customized 3D part. In the cardiology domain, researchers have utilized AI to develop software specifically designed for cardiac 3D printing (Razeghi et al. 2020) to handle multimodal imaging data and create an accurate 3D model. It is also a platform to apply machine learning and image processing techniques on the cardiac data to register the heart, segment the ROI, and track the cardiac motion.

AI techniques, such as object recognition methods have significant potential to be used to recognize endocardial details accurately and estimate their sizes on the segmented structure. This could be used as an automatic pre-processing step, facilitating 3D printing of cardiac which contains different sub-structures. For instance, if coronary arteries are localized, a more precise 3D cardiac mesh would be created, and thereby, more accurate CAD designs formed, leading to an anatomically accurate printed cardiac model. Moreover, embedding such an AI module would help avoid the manual post-processing of the model for determining the heart specifications. Therefore, cardiologists could benefit from using computer vision and object recognition methods to save cost and the medical staff's time by creating a 3D-printed heart with desirable features.

12.5.2 Robust In-Situ Monitoring of 3D Printing Process

Advanced manufactured parts are used in mission-critical components by various industries, such as aerospace, biomedical, defence, and space. Although these industries consent to pay a premium price for AM parts, which are typically lighter, much more optimized in terms of geometry, and customized, they also demand exacting

standards in quality. Different anomalies, including spreading, surface and geome-
try-related defects, sub-surface and melt-pool-related flaws, may appear on the final
3D printed product which might affect the part's quality and its properties (Mostafaei
et al. 2022). That is because, with dozens of process parameters that can be changed
in an AM machine, a 3D printer creates numerous solidification pathways for any
given part (Chen and Zhao 2016). This creates multiple causes and effects scenarios that often require extensive trial and error to reach an optimal processing window. As a result, improper selection of process parameters could lead to generating
some defects on the printed part. Discovering the optimum combinations of process
parameters is crucial to fabricate a defect-free AM part, but it is a time-consuming
and labor-intensive task. Moreover, due to the layer-by-layer procedure of 3D printing, the defects may derive from the layer-wise melting process, meaning they can
extend from one layer to the subsequent layers. It consequently causes the entire
printing to fail. Thus, in-situ monitoring of the printing process during the creation
of different layers could assist in analyzing the process in every layer and preventing
the propagation of the defects to the final build by adjusting the process parameters
in real-time (Wang et al. 2020).

The same challenges exist in the manufacturing of medical components. For
instance, fabricating metal-based implants and porous structures of bone scaffolds may fail due to various anomalies, including inappropriate residual stress,
microstructure, and surface roughness (Du Plessis, Yadroitsava, and Yadroitsev
2020). These errors arise due to the stochastic metal-based AM process and
experimental uncertainties, transient phenomena of heat transfer (Azarniya et al.
2019), and geometrical effects with inadequately defined powder thermal properties (Heeling and Wegener 2018). As in medical applications, high printing accuracy is required, considerable efforts must be dedicated to mitigating the impact
of unavoidable geometry flaws. Indeed, the final product must be printed with
precise dimensions to obtain an accurate geometry described by the CAD data.
Dallago et al. (2019) have studied the geometrical defects and their impact on the
properties of porous structures.

To the best of our knowledge, in the cardiac 3D printing domain, no studies have
focused on analyzing different anomalies arising during the printing process of cardiac components. Sun et al. (Sun et al. 2019) claimed that there is no comprehensive
study on measuring similarities between 3D-printed hearts and the STL file. Such a
comparison leads to discovering geometry flaws and dimensional errors. However, an
acceptable accuracy of around 1mm has been achieved by comparing the 3D-printed
heart models with the original images (Wang et al. 2020).

AI techniques, including machine learning methods, improve the in-situ
monitoring process to detect anomalies and discover the optimum set of process
parameters. As mentioned earlier, the machine learning models have great potential for analyzing the large volume of data in multiple modalities, such as images,
videos, and signals that can be collected from the monitoring sensors embedded in the 3D printer machines. The data mining process can be accomplished
in an open-loop or closed-loop (feedback control) workflow (Zhu et al. 2021).
Open-loop refers to the AI-based framework, which studies the captured data
in an offline process either before printing or after completion. For example, a

process parameter optimization algorithm (Lapointe et al. 2022) that is executed at the starting point of printing a custom product to initialize the process by providing the optimal parameters is an open-loop approach. A predictive anomaly detection method is another example of an open-loop algorithm that evaluates the quality of the final 3D-printed component (Wang and Cheung 2022; Cannizzaro et al. 2021). Geometric compensation approaches are also open-loop techniques as they evaluate the geometry of the manufactured part by comparing it against the CAD profile. For example, the ANN-based geometry compensation method was developed by Chowdhury and Anand to study the thermal deformations in the final product (Chowdhury and Anand 2016).

In contrast, closed-loop printing investigates the data in real-time to predict the probabilistic anomalies and adjust the process parameters accordingly as the printing progresses layer by layer (Wang and Cheung 2022). For instance, a control framework for liquid metal jet printing has been developed using a machine learning model and stroboscopic images to adjust the input voltage (Wang et al. 2018). Deep reinforcement learning is an impressive solution to implementing closed-loop algorithms in different applications (Ji et al. 2022; Ogoke and Farimani 2021). The developed AI-based open-loop and closed-loop algorithms for in-situ monitoring of AM machines have the potential to be extended to medical and cardiac applications, to improve the printing process of medical and biomedical devices. A multi-objective optimization algorithm for a binder jetting machine that was proposed to decipher the relationship between process parameters and part density for Cobalt Chrome Molybdenum (Co-Cr-Mo) material (Onler et al. 2022), could also be extended to optimize the printing parameters required for building dental implants by Co-Cr-Mo alloys. A deep CNN approach that examined the performance of the printed porous structures could be employed to measure the performance of bone scaffolds (Johnson et al. 2022). Some researchers have already applied ML methods to 3D printing in medicine (Ji et al. 2022; Guo and Zhou 2021). An ML-based framework, including k-NN, SVM, decision trees, and Random Forest (RF), was leveraged to predict the strength of 3D-printed orthopedic cortical screws on a fused deposition process (Agarwal, Singh, and Gupta 2022).

Robotics systems as a subset of AI are increasingly being developed for two different 3D printing applications: a hybrid system of the robotic surgeon to safely apply 3D printing to human organs and tissues (Zhu et al. 2021) and constructing patient-specific 3D printed robotics (Desai et al. 2019). These 3D printing robotics-driven approaches have been widely investigated for cardiac applications. A robotic dispensing 3D printer was employed to construct an orientation-controlled 3D cardiac tissue with a vascular network (Tsukamoto, Akagi, and Akashi 2020). Wang, Min, and Xiong (2015) presented a robust system using a robotic arm and an extruder tool to facilitate printing along trajectories on any curved surface, including cardiac therapeutic components. A 3D-printed slave master robot has been fabricated to reduce the recovery time and improve procedural facilitation by processing MRI images of endovascular interventions (Abdelaziz et al. 2019). A comprehensive review of surgical robotics and robotic perception of in-situ tissue engineering can be found in (Zhu et al. 2021; Roy, Saxena, and Pandey 2018).

12.6   UNRESOLVED CHALLENGES IN CARDIAC 3D PRINTING

To date, AI has assisted cardiac 3D printing by resolving the existing challenges in design and printing process of 3D printers. As stated in Section 12.5, AI has been used for developing an accurate image segmentation and real-time control of print processes. However, there are still unresolved complications in cardiac 3D printing that AI has not addressed yet. In the following subsections, we list three obstacles which should be tackled to improve the process of cardiac 3D printing.

12.6.1   COMPUTATIONAL MODELING AND TISSUE DIFFERENCE

As stated in Section 12.4, the cardiovascular system is a complicated structure with different chambers, vessels, and valves. 3D-printed cardiac models should be able to replicate both the appearance of the heart and vascular structures as well as the mechanical behaviors of the heart such as tissue deformation's law, blood flow dynamics, and their interaction. Although considerable success has been achieved in static modelling of the heart, the dynamic behavior of the cardiovascular system has not been well developed. This is mainly due to the inability to find material that can intimate the nonlinear behavior of biological tissue. The material used in computational modeling is mostly constructed by performing in vitro tests on animal tissue or human cadavers (Martin and Sun 2012). However, it should be noted that the mechanical properties of living human tissue are different from those of animal tissue and cadavers (Wang, Qian, et al. 2021).

12.6.2   DEFICIENCIES IN CARDIAC IMAGING TECHNIQUES AND AI-ASSISTED IMAGE SEGMENTATION METHODS

In cardiovascular healthcare, various imaging technologies are utilized; however, the quality of these images is sub-optimal. The complexity in the geometry of the coronary arteries constrains the quality of cardiac images (Stepniak et al. 2020). Also, some artifacts associated with cardiac motion may arise on the output image. This makes the process of converting a series of 2D images to a 3D model more likely to fail, as establishing a relationship between different 2D images is more challenging in the presence of artifacts. These issues constrain the performance of AI-assisted cardiac segmentation algorithms, and thereby lead to some errors in the final cardiac 3D-printed model.

The other deficiency of AI-assisted cardiac segmentation methods relates to lack of annotated images (i.e. segmented cardiac images). Indeed, these methods require a huge amount of training data, segmented by human experts, to achieve high accuracy. As this is a labor-intensive task which requires a high level of medical training, efforts toward developing an accurate AI-based cardiac segmentation algorithm with minimum reliance on data annotation should be continued.

12.6.3   PRIVACY AND CONFIDENTIALITY

Concerns for privacy and confidentiality of medical data are other considerations for applying AI for the manufacturing of cardiac 3D printed models. To accomplish the task of cardiac 3D printing, patients' confidential healthcare data should be shared among AI experts, designers, and 3D printing operators. A cloud-sharing approach is commonly used to provide data access for all personnel involved in the project. This raises the privacy issue, as there is a risk of sensitive information leakage over the cloud or in servers used for data storage. The risks associated with data sharing and medical-legal liability must be addressed prior to implementing large-scale AI applications in medical imaging and intervention (Wang, Qian, et al. 2021).

12.7 FUTURE RESEARCH DIRECTION

AI can be utilized in different stages of cardiac 3D printing procedures such as image segmentation, AM design, and in-situ monitoring. Although AI has been used to address these challenges, several obstacles still exist. The future directions to tackle these challenges are summarized as follows.

12.7.1 GENERATIVE DESIGN OF CARDIAC 3D PRINTED HEART

Several existing challenges pose limitations on widely using cardiac 3D printed models. Currently, the materials used for fabricating 3D-printed cardiac models are expensive. This is due to the required elasticity features in printing material which provides similar characteristics to human tissue. If the expensive materials could be replaced with cheap and still high-quality ones, the AM process would be more cost-effective. In the literature, there are no sufficient investigations on the suitability of the cheap materials that can be used for creating 3D heart models. Indeed, the mechanical properties of these materials have not been studied yet to examine their feasibility.

The generative design which provides new pathways to freely design complex geometries with desirable characteristics can address the lack of investigation of new cheap materials. The AI-based infrastructure of the generative design approaches iteratively explores the design process to examine the proper materials for every geometry. It could result in the creation of various design-optimized solutions with the given desirable constraints (McCormack, Dorin, and Innocent 2004). The initial investigation for using generative design in a few applications such as the design of mechanical spring (Tutum et al. 2018) and door handles (De Crescenzio, Fantini, and Asllani 2022) has been done. In these works, up-to-date AI methods were employed to enhance the existing generative design's performance. This showed the potential of using AI for improving 3D printing procedures. For instance, Forte was introduced as a sketch-based and real-time user interface to facilitate the design stage of AM processes by issuing an optimized topology of the object (Chen, Tao, et al. 2018). Only a few studies have explored the potential of generative design in 3D heart modelling. As an example, Ahmed et al. (Rahman 2021) have employed a generative design approach to examine new materials and corresponding mechanical heart valves for improving blood flow. In future works, the possibility of using generative design methods created by newly released AI techniques in cardiac 3D printing should be thoroughly investigated.

Furthermore, determining the desirable properties of a cardiac 3D model like its weight and strength is a challenging task that requires medical expert personnel to monitor the 3D design phase. As a future direction, the generative design could be used to advance 3D cardiac printing by enabling the fabrication of 3D hearts with suitable features. Currently, some frameworks utilizing generative design for medical applications are being developed to build a product with preferable properties. For example, using a generative design process coupled with the genetic algorithm could help define multifunctional personalization for creating resistant composites without incorporating bioactives (He et al. 2021). The bacterial biofilm coverage of the newly manufactured composites was reduced by ~75% compared to well-known silicone rubbers, performing under given load constraints. In another work, proposing a generative design based on the Voronoi diagram could improve the creation of biomimetic customized cranial prostheses (Sharma et al. 2021). This design process enabled the manufacturer to build a lightweight prosthesis in which the quasi-self-supporting fabrication was feasible. Future works could examine the potential of these systems for creating 3D cardiac structures with appropriate features like the fabrication of a stronger and lighter 3D heart.

12.7.2 DYNAMIC CARDIAC 3D PRINTING

As stated in Section 12.7.2, current 3D printing technologies generate static heart models rather than a dynamic organ, thus, only morphological cardiac features can be assessed rather than cardiac hemodynamics. As printing technologies develop, 3D models of the beating heart should be produced, which allow the detection of anatomical and physiological changes during the cardiac cycle (Wang, Qian, et al. 2021). These models are also beneficial in determining the function evaluation standards for specific parameters in coronary artery disease (Hammel et al. 2022). Moreover, cutting-edge technologies, such as surgeon robotics, virtual and augmented reality (VR/AR), and other modern approaches can unlock new pathways to improve the 3D printing of dynamic cardiac models (Goo, Park, and Yoo 2020; Rad et al. 2022). Future research in cardiac 3D printing should focus on developing dynamic models. These models will improve many aspects of cardiovascular health by delivering excellent visualizations for understanding the endocardial details of the heart, facilitating the cardiac design for 3D printing, and also providing an excellent instrument for the education of cardiac interventionalists.

12.7.3 ACTIVE AND SELF-SUPERVISED AI ALGORITHMS FOR CARDIAC 3D PRINTING

Although deep learning has demonstrated flawless performance in data segmentation, most methods require labeled training data to achieve the highest accuracy. As labeling the clinical data is significantly labor-intensive and requires a physician's expertise, the development of AI techniques with minimum reliance on expert annotation would improve the cardiac 3D printing procedure. As a future work in cardiac image segmentation, the potential of active learning methods which proactively learn the training data labels and seek the subset of samples to be annotated next from a collection of unlabeled data should be evaluated (Budd, Robinson, and Kainz 2021). These learning algorithms could outperform the existing AI techniques while reducing the expensive labeling burden of medical images, specifically cardiac images. The existing active learning-based semantic segmentation techniques, used in scene understanding applications, could be utilized for cardiac image segmentation (Siddiqui, Valentin, and

Nießner 2020). Furthermore, a combination of active segmenting and reinforcement learning, evaluated in semantic urban scene understanding (Casanova et al. 2020), can be employed in a closed-loop algorithm for the 3D printing of cardiac models.

12.8 CONCLUSIONS

This chapter introduced 3D printers, AM processes, and their applications in medical imaging, specifically cardiovascular healthcare. We summarized the required steps for cardiac 3D printing as: cardiac image acquisition using imaging technologies, segmentation of cardiac structures in the acquired images, CAD conversion, STL file extraction, manufacturing, and quality control. As cardiac segmentation is a fundamental step in providing an accurate 3D model, we reviewed the novel AI segmentation techniques and analyzed their potential to be applied in cardiac image segmentation. Moreover, a review of existing challenges in the 3D printing of cardiac structures was provided, and the role of AI in mitigating these issues was summarized. AI can be utilized to improve the design stage in cardiac 3D printing by developing more accurate segmentation algorithms which can segment the whole heart structure. AI can also suggest novel methods for combining different cardiac imaging techniques to facilitate the cardiac segmentation. DfAM can be improved through AI to provide optimized objects. Furthermore, AI can enhance the in situ-monitoring process by detecting anomalies in the early stage of cardiac fabrication and discovering an optimum set of process parameters which leads to fewer geometry flaws. Despite the success of AI in improving cardiac 3D printing, some challenges remain, such as limitations in fabricating dynamic cardiac structures and the reliance of AI-assisted segmentation methods on a large proportion of annotated training data. To resolve these in the future, we provided some research directions. To this end, we propose that generative design techniques can be utilized for cardiac 3D printing. As these approaches have demonstrated significant improvement in the design phase of 3D printing of other objects, they could help optimize the 3D design of the complex heart structure. In addition, we suggest potential pathways for improving dynamic cardiac 3D printing, such as employing cutting-edge techniques like virtual and augmented reality (VR/AR) in the fabrication of cardiac structures. Ultimately, we advise utilizing active and self-supervised learning AI techniques for reducing the reliance on human experts in the data preparation stage (for heart image segmentation) and improving the closed-loop real-time control of cardiac 3D printing.

REFERENCES

Abdelaziz, Mohamed E. M. K., Dennis Kundrat, Marco Pupillo, Giulio Dagnino, Trevor M. Y. Kwok, Wenqiang Chi, Vincent Groenhuis, Françoise J. Siepel, Celia Riga, Stefano Stramigioli, and Guang-Zhong Yang. 2019. "Toward a versatile robotic platform for fluoroscopy and MRI-guided endovascular interventions: A pre-clinical study." *2019 IEEE/RSJ International Conference on Intelligent Robots and Systems (IROS)*, Macau, 3–8 November 2019.

Abdulhameed, Osama, Abdulrahman Al-Ahmari, Wadea Ameen, and Syed Hammad Mian. 2019. "Additive manufacturing: Challenges, trends, and applications." *Advances in Mechanical Engineering* 11(2):1687814018822880. doi: 10.1177/1687814018822880.

Agarwal, Raj, Jaskaran Singh, and Vishal Gupta. 2022. "Predicting the compressive strength of additively manufactured PLA-based orthopedic bone screws: A machine learning framework." *Polymer Composites* 43(8):5663–5674. doi: 10.1002/pc.26881.

Alom, Md Zahangir, Chris Yakopcic, Tarek M. Taha, and Vijayan K. Asari. 2018. "Nuclei segmentation with recurrent residual convolutional neural networks based U-Net (R2U-Net)." *NAECON 2018-IEEE National Aerospace and Electronics Conference,* Dayton, OH.

Artec3D. 2022. "Using 3D scanning and printing to help children with ear deformities." https://www.artec3d.com/cases/prosthetic-3d-printed-ear-implants.

Astm, I. 2010. "ASTM F2792-10: Standard terminology for additive manufacturing technologies." ASTM International.

Azarniya, Abolfazl, Xabier Garmendia Colera, Mohammad J. Mirzaali, Saeed Sovizi, Flavio Bartolomeu, Wessel W. Wits, Chor Yen Yap, Joseph Ahn, Georgina Miranda, and Filipe Samuel Silva. 2019. "Additive manufacturing of Ti-6Al-4V parts through laser metal deposition (LMD): Process, microstructure, and mechanical properties." *Journal of Alloys and Compounds* 804:163–191.

Badrinarayanan, Vijay, Alex Kendall, and Roberto Cipolla. 2017. "Segnet: A deep convolutional encoder-decoder architecture for image segmentation." *IEEE Transactions on Pattern Analysis and Machine Intelligence* 39(12):2481–2495.

Bai, Wenjia, Hideaki Suzuki, Chen Qin, Giacomo Tarroni, Ozan Oktay, Paul M. Matthews, and Daniel Rueckert. 2018. "Recurrent neural networks for aortic image sequence segmentation with sparse annotations." *International Conference on Medical Image Computing and Computer-Assisted Intervention*, Granada.

Balamurugan, Pandian, and Natarajan Selvakumar. 2021. "Development of patient specific dental implant using 3D printing." *Journal of Ambient Intelligence and Humanized Computing* 12(3):3549–3558.

Borodin, Gregory, and Olga Senyukova. 2018. "Right ventricle segmentation in cardiac MR images using U-Net with partly dilated convolution." *International Conference on Artificial Neural Networks*, Rhodes.

Budd, Samuel, Emma C. Robinson, and Bernhard Kainz. 2021. "A survey on active learning and human-in-the-loop deep learning for medical image analysis." *Medical Image Analysis* 71:102062. doi: 10.1016/j.media.2021.102062.

Cai, Yutong, and Yong Wang. 2022. "Ma-unet: An improved version of unet based on multi-scale and attention mechanism for medical image segmentation." *Third International Conference on Electronics and Communication; Network and Computer Technology (ECNCT 2021)*, Xiamen.

Cannizzaro, Davide, Antonio Giuseppe Varrella, Stefano Paradiso, Roberta Sampieri, Enrico Macii, Edoardo Patti, and Santa Di Cataldo. 2021. "Image analytics and machine learning for in-situ defects detection in additive manufacturing." *2021 Design, Automation & Test in Europe Conference & Exhibition (DATE)*, Grenoble.

Casanova, Arantxa, Pedro O. Pinheiro, Negar Rostamzadeh, and Christopher J. Pal. 2020. "Reinforced active learning for image segmentation." arXiv preprint arXiv:2002.06583.

Chen, Han, and Yaoyao Fiona Zhao. 2016. "Process parameters optimization for improving surface quality and manufacturing accuracy of binder jetting additive manufacturing process." *Rapid Prototyping Journal* 22(3):527–538.

Chen, Jieneng, Yongyi Lu, Qihang Yu, Xiangde Luo, Ehsan Adeli, Yan Wang, Le Lu, Alan L. Yuille, and Yuyin Zhou. 2021. "Transunet: Transformers make strong encoders for medical image segmentation." arXiv preprint arXiv:2102.04306.

Chen, Jun, Guang Yang, Zhifan Gao, Hao Ni, Elsa Angelini, Raad Mohiaddin, Tom Wong, Yanping Zhang, Xiuquan Du, and Heye Zhang. 2018. "Multiview two-task recursive attention model for left atrium and atrial scars segmentation." *International Conference on Medical Image Computing and Computer-Assisted Intervention*, Grenoble.

Chen, Kuan-bing, Ying Xuan, Ai-jun Lin, and Shao-hua Guo. 2021. "Lung computed tomography image segmentation based on U-Net network fused with dilated convolution." *Computer Methods and Programs in Biomedicine* 207:106170.

Chen, Liang-Chieh, Alexander Hermans, George Papandreou, Florian Schroff, Peng Wang, and Hartwig Adam. 2018. "Masklab: Instance segmentation by refining object detection with semantic and direction features." *Proceedings of the IEEE Conference on Computer Vision and Pattern Recognition*, Salt Lake City, UT.

Chen, Liang-Chieh, George Papandreou, Florian Schroff, and Hartwig Adam. 2017. "Rethinking atrous convolution for semantic image segmentation." arXiv preprint arXiv:1706.05587.

Chen, Liang-Chieh, George Papandreou, Iasonas Kokkinos, Kevin Murphy, and Alan L. Yuille. 2017. "Deeplab: Semantic image segmentation with deep convolutional nets, atrous convolution, and fully connected CRFs." *IEEE Transactions on Pattern Analysis and Machine Intelligence* 40(4):834–848.

Chen, Liang-Chieh, Yukun Zhu, George Papandreou, Florian Schroff, and Hartwig Adam. 2018. "Encoder-decoder with atrous separable convolution for semantic image segmentation." *Proceedings of the European Conference on Computer Vision (ECCV)*, Munich.

Chen, Mingqiang, Lin Fang, and Huafeng Liu. 2019. "FR-NET: Focal loss constrained deep residual networks for segmentation of cardiac MRI." *2019 IEEE 16th International Symposium on Biomedical Imaging (ISBI 2019)*, Venice.

Chen, Xiang'Anthony', Ye Tao, Guanyun Wang, Runchang Kang, Tovi Grossman, Stelian Coros, and Scott E. Hudson. 2018. "Forte: User-driven generative design." *Proceedings of the 2018 CHI Conference on Human Factors in Computing Systems*, Montreal.

Chowdhury, Sushmit, and Sam Anand. 2016. "Artificial neural network based geometric compensation for thermal deformation in additive manufacturing processes." *International Manufacturing Science and Engineering Conference*, Virginia.

Comaniciu, Dorin, and Peter Meer. 1997. "Robust analysis of feature spaces: Color image segmentation." Proceedings of IEEE Computer Society Conference on Computer Vision and Pattern Recognition.

Dallago, Michele, Bart Winiarski, Filippo Zanini, Simone Carmignato, and Matteo Benedetti. 2019. "On the effect of geometrical imperfections and defects on the fatigue strength of cellular lattice structures additively manufactured via selective laser melting." *International Journal of Fatigue* 124:348–360. doi: 10.1016/j.ijfatigue.2019.03.019.

Dankowski, Rafał, Artur Baszko, Michael Sutherland, Ludwik Firek, Piotr Kałmucki, Katarzyna Wróblewska, Andrzej Szyszka, Adam Groothuis, and Tomasz Siminiak. 2014. "3D heart model printing for preparation of percutaneous structural interventions: Description of the technology and case report." *Kardiologia Polska (Polish Heart Journal)* 72(6):546–551.

De Crescenzio, Francesca, Massimiliano Fantini, and Edison Asllani. 2022. "Generative design of 3D printed hands-free door handles for reduction of contagion risk in public buildings." *International Journal on Interactive Design and Manufacturing (IJIDeM)* 16(1):253–261.

DebRoy, Tarasankar, Huiliang L. Wei, James S. Zuback, Tuhin Mukherjee, John W. Elmer, John O. Milewski, Allison Michelle Beese, A de Wilson-Heid, Amitava De, and W. Zhang. 2018. "Additive manufacturing of metallic components-process, structure and properties." *Progress in Materials Science* 92:112–224.

Decker, Nathan, Mingdong Lyu, Yuanxiang Wang, and Qiang Huang. 2021. "Geometric accuracy prediction and improvement for additive manufacturing using triangular mesh shape data." *Journal of Manufacturing Science and Engineering* 143(6):1–37.

Desai, Jaydev P., Jun Sheng, Shing Shin Cheng, Xuefeng Wang, Nancy J. Deaton, and Nahian Rahman. 2019. "Toward patient-specific 3D-printed robotic systems for surgical interventions." *IEEE Transactions on Medical Robotics and Bionics* 1(2):77–87. doi: 10.1109/TMRB.2019.2912444.

Di Rosa, Luigi. 2022. "3D printing for aesthetic and reconstructive breast surgery." In *3D Printing in Plastic Reconstructive and Aesthetic Surgery*, pp. 91–100. Springer: Berlin, Heidelberg.

Dolz, Jose, Karthik Gopinath, Jing Yuan, Herve Lombaert, Christian Desrosiers, and Ismail Ben Ayed. 2018. "HyperDense-Net: A hyper-densely connected CNN for multi-modal image segmentation." *IEEE Transactions on Medical Imaging* 38(5):1116–1126.

Dosovitskiy, Alexey, Lucas Beyer, Alexander Kolesnikov, Dirk Weissenborn, Xiaohua Zhai, Thomas Unterthiner, Mostafa Dehghani, Matthias Minderer, Georg Heigold, and Sylvain Gelly. 2020. "An image is worth 16×16 words: Transformers for image recognition at scale." arXiv preprint arXiv:2010.11929.

Du Plessis, Anton, Ina Yadroitsava, and Igor Yadroitsev. 2020. "Effects of defects on mechanical properties in metal additive manufacturing: A review focusing on X-ray tomography insights." *Materials & Design* 187:108385.

Fahmy, Ahmed S., Hossam El-Rewaidy, Maryam Nezafat, Shiro Nakamori, and Reza Nezafat. 2019. "Automated analysis of cardiovascular magnetic resonance myocardial native T1 mapping images using fully convolutional neural networks." *Journal of Cardiovascular Magnetic Resonance* 21(1):1–12.

Fan, Tongle, Guanglei Wang, Yan Li, and Hongrui Wang. 2020. "Ma-net: A multi-scale attention network for liver and tumor segmentation." *IEEE Access* 8:179656–179665.

Farooqi, Kanwal M., Omar Saeed, Ali Zaidi, Javier Sanz, James C. Nielsen, Daphne T. Hsu, and Ulrich P. Jorde. 2016. "3D printing to guide ventricular assist device placement in adults with congenital heart disease and heart failure." *JACC: Heart Failure* 4(4):301–311.

Fayyazifar, Najmeh, Girish Dwivedi, David Suter, Selam Ahderom, Andrew Maiorana, Owen Clarkin, Saad Balamane, Nishita Saha, Benjamin King, Martin S. Green, Mehrdad Golian, and Benjamin J. W. Chow. 2023. "A novel convolutional neural network structure for differential diagnosis of wide QRS complex tachycardia." *Biomedical Signal Processing and Control* 81:104506. doi: 10.1016/j.bspc.2022.104506.

Fayyazifar, Najmeh. 2021. "An accurate CNN architecture for atrial fibrillation detection using neural architecture search." *2020 28th European Signal Processing Conference (EUSIPCO)*, Amsterdam, Netherlands.

Fayyazifar, Najmeh, Selam Ahderom, David Suter, Andrew Maiorana, and Girish Dwivedi. 2020. "Impact of neural architecture design on cardiac abnormality classification using 12-lead ECG signals." *2020 Computing in Cardiology*, Rimini, Italy.

Feng, Zhongwei, Jie Yang, Lixiu Yao, Yu Qiao, Qi Yu, and Xun Xu. 2017. "Deep retinal image segmentation: A FCN-based architecture with short and long skip connections for retinal image segmentation." *International Conference on Neural Information Processing*, Long Beach, CA.

Fina, Fabrizio, Simon Gaisford, and Abdul W Basit. 2018. "Powder bed fusion: The working process, current applications and opportunities." *3D printing of pharmaceuticals*:81–105.

Freixenet, Jordi, Xavier Munoz, David Raba, Joan Martí, and Xavier Cufí. 2002. "Yet another survey on image segmentation: Region and boundary information integration." *European Conference on Computer Vision*, Copenhagen, Denmark.

Friedman, Nir, and Stuart Russell. 2013. "Image segmentation in video sequences: A probabilistic approach." arXiv preprint arXiv:1302.1539.

Fu, Zhenyin, Jin Zhang, Ruyi Luo, Yutong Sun, Dongdong Deng, and Ling Xia. 2022. "TF-Unet: An automatic cardiac MRI image segmentation method." *Mathematical Biosciences and Engineering* 19(5):5207–5222.

Giannopoulos, Andreas A., Leonid Chepelev, Adnan Sheikh, Aili Wang, Wilfred Dang, Ekin Akyuz, Chris Hong, Nicole Wake, Todd Pietila, Philip B. Dydynski, Dimitrios Mitsouras, and Frank J. Rybicki. 2015. "3D printed ventricular septal defect patch: A primer for the 2015 Radiological Society of North America (RSNA) hands-on course in 3D printing." *3D Printing in Medicine* 1(1):3. doi: 10.1186/s41205-015-0002-4.

Gibson, Ian, David Rosen, and Brent Stucker. 2015. "Vat photopolymerization processes." In *Additive Manufacturing Technologies*, pp. 63–106. Springer: Berlin, Heidelberg.

Girshick, Ross. 2015. "Fast R-CNN." *Proceedings of the IEEE International Conference on Computer Vision*, Santiago.

Girshick, Ross, Jeff Donahue, Trevor Darrell, and Jitendra Malik. 2014. "Rich feature hierarchies for accurate object detection and semantic segmentation." *Proceedings of the IEEE Conference on Computer Vision and Pattern Recognition*.

Goo, Hyun Woo, Sang Joon Park, and Shi-Joon Yoo. 2020. "Advanced medical use of three-dimensional imaging in congenital heart disease: Augmented reality, mixed reality, virtual reality, and three-dimensional printing." *Korean Journal of Radiology* 21(2):133–145.

Gosnell, Jordan, Todd Pietila, Bennett P. Samuel, Harikrishnan K. N. Kurup, Marcus P. Haw, and Joseph J. Vettukattil. 2016. "Integration of computed tomography and three-dimensional echocardiography for hybrid three-dimensional printing in congenital heart disease." *Journal of Digital Imaging* 29(6):665–669.

Graves, Alex. 2012. "Long short-term memory." In *Supervised Sequence Labelling with Recurrent Neural Networks*, pp. 37–45. Springer: Berlin, Heidelberg.

Guo, Zipeng, and Chi Zhou. 2021. "Recent advances in ink-based additive manufacturing for porous structures." *Additive Manufacturing* 48:102405. doi: 10.1016/j.addma.2021.102405.

Gupta, Harshit, Kyong Hwan Jin, Ha Q. Nguyen, Michael T. McCann, and Michael Unser. 2018. "CNN-based projected gradient descent for consistent CT image reconstruction." *IEEE Transactions on Medical Imaging* 37(6):1440–1453.

Gushchina, Marina O., Yulia O. Kuzminova, Egor A. Kudryavtsev, Konstantin D. Babkin, Valentina D. Andreeva, Stanislav A. Evlashin, and Evgeniy V. Zemlyakov. 2022. "Effect of scanning strategy on mechanical properties of Ti-6Al-4V alloy manufactured by laser direct energy deposition." *Journal of Materials Engineering and Performance* 31(4):2783–2791.

Hafiz, Abdul Mueed, and Ghulam Mohiuddin Bhat. 2020. "A survey on instance segmentation: State of the art." *International Journal of Multimedia Information Retrieval* 9(3):171–189.

Haleem, Abid, Mohd Javaid, Rizwan Hasan Khan, and Rajiv Suman. 2020. "3D printing applications in bone tissue engineering." *Journal of Clinical Orthopaedics and Trauma* 11:S118–S124.

Hammel, Johannes, Lorenz Birnbacher, Marcus R. Makowski, Franz Pfeiffer, and Daniela Pfeiffer. 2022. "Absolute iodine concentration for dynamic perfusion imaging of the myocardium: Improved detection of poststenotic ischaemic in a 3D-printed dynamic heart phantom." *European Radiology Experimental* 6(1):51. doi: 10.1186/s41747-022-00304-x.

Harb, Serge C., Leonardo L. Rodriguez, Marija Vukicevic, Samir R. Kapadia, and Stephen H. Little. 2019. "Three-dimensional printing applications in percutaneous structural heart interventions." *Circulation: Cardiovascular Imaging* 12(10):e009014.

He, Junjun, Zhongying Deng, Lei Zhou, Yali Wang, and Yu Qiao. 2019. "Adaptive pyramid context network for semantic segmentation." *Proceedings of the IEEE/CVF Conference on Computer Vision and Pattern Recognition*, Long Beach, CA.

He, Kaiming, Georgia Gkioxari, Piotr Dollár, and Ross Girshick. 2017. "Mask R-CNN." *Proceedings of the IEEE International Conference on Computer Vision*.

He, Kaiming, Xiangyu Zhang, Shaoqing Ren, and Jian Sun. 2016. "Deep residual learning for image recognition." *Proceedings of the IEEE Conference on Computer Vision and Pattern Recognition*, Las Vegas, NV.

He, Yinfeng, Meisam Abdi, Gustavo F. Trindade, Belén Begines, Jean-Frédéric Dubern, Elisabetta Prina, Andrew L. Hook, Gabriel Y. H. Choong, Javier Ledesma, Christopher J. Tuck, Felicity R. A. J. Rose, Richard J. M. Hague, Clive J. Roberts, Davide S. A. De

Focatiis, Ian A. Ashcroft, Paul Williams, Derek J. Irvine, Morgan R. Alexander, and Ricky D. Wildman. 2021. "Exploiting generative design for 3D printing of bacterial biofilm resistant composite devices." *Advanced Science* 8(15):2100249. doi: 10.1002/advs.202100249.

Heeling, Thorsten, and Konrad Wegener. 2018. "The effect of multi-beam strategies on selective laser melting of stainless steel 316L." *Additive Manufacturing* 22:334–342.

Hsu, Wei-Yen. 2019. "Automatic left ventricle recognition, segmentation and tracking in cardiac ultrasound image sequences." *IEEE Access* 7:140524–140533.

Hui, Rex W. H. 2017. "Three-dimensional printing for patient counseling." *Journal of Surgical Oncology* 116(7):961–961.

Hussein, Nabil, Pascal Voyer-Nguyen, Sharon Portnoy, Brandon Peel, Eric Schrauben, Christopher Macgowan, and Shi-Joon Yoo. 2020. "Simulation of semilunar valve function: computer-aided design, 3D printing and flow assessment with MR." *3D Printing in Medicine* 6(1):1–9.

Isensee, Fabian, Paul F. Jaeger, Peter M. Full, Ivo Wolf, Sandy Engelhardt, and Klaus H. Maier-Hein. 2017. "Automatic cardiac disease assessment on cine-MRI via time-series segmentation and domain specific features." *International Workshop on Statistical Atlases and Computational Models of the Heart*, Singapore.

Islam, Mobarakol, V. S. Vibashan, Valanarasu Jeya Maria Jose, Navodini Wijethilake, Uppal Utkarsh, and Hongliang Ren. 2019. "Brain tumor segmentation and survival prediction using 3D attention UNet." *International MICCAI Brainlesion Workshop*, Shenzhen.

Jeya Maria Jose Valanarasu, Poojan Oza, Ilker Hacihaliloglu, Vishal M. Patel. 2021. "Medical transformer: Gated axial-attention for medical image segmentation." *International Conference on Medical Image Computing and Computer-MICCAI 2021*.

Ji, Qinglei, Mo Chen, Xi Vincent Wang, Lihui Wang, and Lei Feng. 2022. "Optimal shape morphing control of 4D printed shape memory polymer based on reinforcement learning." *Robotics and Computer-Integrated Manufacturing* 73:102209.

Jiang, Jingchao, Yi Xiong, Zhiyuan Zhang, and David W. Rosen. 2022. "Machine learning integrated design for additive manufacturing." *Journal of Intelligent Manufacturing* 33(4):1073–1086. doi: 10.1007/s10845-020-01715-6.

Johnson, Kyle L., Demitri Maestas, John M. Emery, Mircea D. Grigoriu, Matthew D. Smith, and Carianne Martinez. 2022. "Failure classification of porous additively manufactured parts using deep learning." *Computational Materials Science* 204:111098. doi: 10.1016/j.commatsci.2021.111098.

Kaymak, Ruya, Cagri Kaymak, and Aysegul Ucar. 2020. "Skin lesion segmentation using fully convolutional networks: A comparative experimental study." *Expert Systems with Applications* 161:113742.

Kermani, Saeed, Mostafa Ghelich Oghli, Ali Mohammadzadeh, and Raheleh Kafieh. 2020. "NF-RCNN: Heart localization and right ventricle wall motion abnormality detection in cardiac MRI." *Physica Medica* 70:65–74.

Kirillov, Alexander, Kaiming He, Ross Girshick, Carsten Rother, and Piotr Dollár. 2019. "Panoptic segmentation." *Proceedings of the IEEE/CVF Conference on Computer Vision and Pattern Recognition*, Long Beach, CA.

Kozakiewicz, Marta, Karol P. Nartowski, Aleksandra Dominik, Katarzyna Malec, Anna M. Gołkowska, Adrianna Złocińska, Małgorzata Rusińska, Patrycja Szymczyk-Ziółkowska, Grzegorz Ziółkowski, and Agata Górniak. 2021. "Binder jetting 3D printing of challenging medicines: From low dose tablets to hydrophobic molecules." *European Journal of Pharmaceutics and Biopharmaceutics* 170:144–159.

Krizhevsky, Alex, Ilya Sutskever, and Geoffrey E. Hinton. 2017. "Imagenet classification with deep convolutional neural networks." *Communications of the ACM* 60(6):84–90.

Kuk, Mariya, Dimitris Mitsouras, Karin E. Dill, Frank J. Rybicki, and Girish Dwivedi. 2017. "3D printing from cardiac computed tomography for procedural planning." *Current Cardiovascular Imaging Reports* 10(7):1–9.

Lang, Roberto M., Karima Addetia, Akhil Narang, and Victor Mor-Avi. 2018. "3-Dimensional echocardiography: Latest developments and future directions." *JACC: Cardiovascular Imaging* 11(12):1854–1878.

Lapointe, Simon, Gabe Guss, Thomas Michael Reese, Maria Strantza, Manyalibo J. Matthews, and Clara L. Druzgalski. 2022. "Photodiode-based machine learning for optimization of laser powder bed fusion parameters in complex geometries." *Additive Manufacturing* 53:102687.

Lee, Jung-Seob, Byoung Soo Kim, Donghwan Seo, Jeong Hun Park, and Dong-Woo Cho. 2017. "Three-dimensional cell printing of large-volume tissues: Application to ear regeneration." *Tissue Engineering Part C: Methods* 23(3):136–145.

Leibe, Bastian, Edgar Seemann, and Bernt Schiele. 2005. "Pedestrian detection in crowded scenes." *2005 IEEE Computer Society Conference on Computer Vision and Pattern Recognition (CVPR'05)*, San Diego, CA.

Li, Feiyan, Weisheng Li, Sheng Qin, and Linhong Wang. 2021. "MDFA-Net: Multiscale dual-path feature aggregation network for cardiac segmentation on multi-sequence cardiac MR." *Knowledge-Based Systems* 215:106776.

Li, Qingqing, Ke Chen, Lin Han, Yan Zhuang, Jingtao Li, and Jiangli Lin. 2020. "Automatic tooth roots segmentation of cone beam computed tomography image sequences using U-net and RNN." *Journal of X-Ray Science and Technology* 28(5):905–922.

Lieman-Sifry, Jesse, Matthieu Le, Felix Lau, Sean Sall, and Daniel Golden. 2017. "FastVentricle: Cardiac segmentation with ENet." *International Conference on Functional Imaging and Modeling of the Heart*, Toronto.

Little, Stephen H., Marija Vukicevic, Eleonora Avenatti, Mahesh Ramchandani, and Colin M. Barker. 2016. "3D printed modeling for patient-specific mitral valve intervention: Repair with a clip and a plug." *JACC: Cardiovascular Interventions* 9(9):973–975.

Liu, Fei, Kun Wang, Dan Liu, Xin Yang, and Jie Tian. 2021. "Deep pyramid local attention neural network for cardiac structure segmentation in two-dimensional echocardiography." *Medical Image Analysis* 67:101873.

Liu, Zexiong, Yuhong Feng, and Xuan Yang. 2019. "Right ventricle segmentation of cine MRI using residual U-net convolutinal networks." *2019 20th International Conference on Parallel and Distributed Computing, Applications and Technologies (PDCAT)*, Gold Coast.

Long, Jonathan, Evan Shelhamer, and Trevor Darrell. 2015. "Fully convolutional networks for semantic segmentation." *Proceedings of the IEEE Conference on Computer Vision and Pattern Recognition*, Boston, MA.

Mahmud, Zaheri, Mahbub Hassan, Anwarul Hasan, and Vincent G. Gomes. 2021. "3D printed nanocomposites for tailored cardiovascular tissue constructs: A minireview." *Materialia* 19:101184.

Martin, Caitlin, and Wei Sun. 2012. "Biomechanical characterization of aortic valve tissue in humans and common animal models." *Journal of Biomedical Materials Research Part A* 100(6):1591–1599.

McCormack, Jon, Alan Dorin, and Troy Innocent. 2004. "Generative design: A paradigm for design research." In Redmond, J., D. Durling, and A. de Bono (eds.), *Futureground - DRS International Conference 2004*, 17–21 November, Melbourne.

Merkow, Jameson, Alison Marsden, David Kriegman, and Zhuowen Tu. 2016. "Dense volume-to-volume vascular boundary detection." *International Conference on Medical Image Computing and Computer-Assisted Intervention*, Athens.

Minaee, Shervin, Yuri Y. Boykov, Fatih Porikli, Antonio J. Plaza, Nasser Kehtarnavaz, and Demetri Terzopoulos. 2021. "Image segmentation using deep learning: A survey." *IEEE Transactions on Pattern Analysis and Machine Intelligence*.

Mnih, Volodymyr, Nicolas Heess, and Alex Graves. 2014. "Recurrent models of visual attention." *Advances in Neural Information Processing Systems* 27:2204–2212.

Moradi, Shakiba, Mostafa Ghelich Oghli, Azin Alizadehasl, Isaac Shiri, Niki Oveisi, Mehrdad Oveisi, Majid Maleki, and Jan Dhooge. 2019. "MFP-Unet: A novel deep learning based approach for left ventricle segmentation in echocardiography." *Physica Medica* 67:58–69.

Mostafaei, Amir, Cang Zhao, Yining He, Seyed Reza Ghiaasiaan, Bo Shi, Shuai Shao, Nima Shamsaei, Ziheng Wu, Nadia Kouraytem, Tao Sun, Joseph Pauza, Jerard V. Gordon, Bryan Webler, Niranjan D. Parab, Mohammadreza Asherloo, Qilin Guo, Lianyi Chen, and Anthony D. Rollett. 2022. "Defects and anomalies in powder bed fusion metal additive manufacturing." *Current Opinion in Solid State and Materials Science* 26(2):100974. doi: 10.1016/j.cossms.2021.100974.

Nasiri, Sara, and Mohammad Reza Khosravani. 2021. "Machine learning in predicting mechanical behavior of additively manufactured parts." *Journal of Materials Research and Technology* 14:1137–1153.

Nurmaini, Siti, Bayu Adhi Tama, Muhammad Naufal Rachmatullah, Annisa Darmawahyuni, Ade Iriani Sapitri, Firdaus Firdaus, and Bambang Tutuko. 2022. "An improved semantic segmentation with region proposal network for cardiac defect interpretation." *Neural Computing and Applications* 34(5):1–14.

Ogoke, Francis, and Amir Barati Farimani. 2021. "Thermal control of laser powder bed fusion using deep reinforcement learning." *Additive Manufacturing* 46:102033.

Oktay, Ozan, Jo Schlemper, Loic Le Folgoc, Matthew Lee, Mattias Heinrich, Kazunari Misawa, Kensaku Mori, Steven McDonagh, Nils Y. Hammerla, and Bernhard Kainz. 2018. "Attention U-net: Learning where to look for the pancreas." arXiv preprint arXiv:1804.03999.

Onler, Recep, Ahmet Selim Koca, Baris Kirim, and Emrecan Soylemez. 2022. "Multi-objective optimization of binder jet additive manufacturing of Co-Cr-Mo using machine learning." *The International Journal of Advanced Manufacturing Technology* 119(1):1091–1108. doi: 10.1007/s00170-021-08183-z.

Otton, James M., Nicolette S. Birbara, Tarique Hussain, Gerald Greil, Thomas A. Foley, and Nalini Pather. 2017. "3D printing from cardiovascular CT: A practical guide and review." *Cardiovascular Diagnosis and Therapy* 7(5):507.

Pagac, Marek, Jiri Hajnys, Quoc-Phu Ma, Lukas Jancar, Jan Jansa, Petr Stefek, and Jakub Mesicek. 2021. "A review of vat photopolymerization technology: Materials, applications, challenges, and future trends of 3D printing." *Polymers* 13(4):598.

Pantermehl, Sven, Steffen Emmert, Aenne Foth, Niels Grabow, Said Alkildani, Rainer Bader, Mike Barbeck, and Ole Jung. 2021. "3D printing for soft tissue regeneration and applications in medicine." *Biomedicines* 9(4):336.

Payer, Christian, Darko Štern, Horst Bischof, and Martin Urschler. 2017. "Multi-label whole heart segmentation using CNNs and anatomical label configurations." *International Workshop on Statistical Atlases and Computational Models of the Heart*, Quebec City.

Peng, Bo, Lei Zhang, and David Zhang. 2013. "A survey of graph theoretical approaches to image segmentation." *Pattern Recognition* 46(3):1020–1038.

Phillips, Alistair B. M., Phillip Nevin, Avni Shah, Vincent Olshove, Ruchira Garg, and Evan M. Zahn. 2016. "Development of a novel hybrid strategy for transcatheter pulmonary valve placement in patients following transannular patch repair of tetralogy of fallot." *Catheterization and Cardiovascular Interventions* 87(3):403–410.

Piłczyńska, Katarzyna. 2022. "Material jetting." In Joanna Izdebska-Podsiadly (ed.), *Polymers for 3D Printing*, pp. 91–103. Elsevier: Amsterdam, Netherlands.

Qadeer, Nauman, Jamal Hussain Shah, and Muhammad Sharif. 2021. "Automated localization and segmentation of left ventricle in cardiac MRI using faster R-CNN." 2021 International Conference on Frontiers of Information Technology (FIT), Islamabad.

Rad, Arian Arjomandi, Robert Vardanyan, Aleksandra Lopuszko, Christina Alt, Ingo Stoffels, Bastian Schmack, Arjang Ruhparwar, Konstantin Zhigalov, Alina Zubarevich, and Alexander Weymann. 2022. "Virtual and augmented reality in cardiac surgery." *Brazilian Journal of Cardiovascular Surgery* 37(01):123–127.

Rahman, Sajjad Ahmed, Maisha Tarannum, Md. Shahriar Aquib. 2021. "Simulation on mechanical heart valve: Generative design and blood flow analysis of a mechanical heart valve." *Fifth International Conference on Industrial & Mechanical Engineering and Operations Management (IMEOM 2022),* 26–28 December, Dhaka, Bangladesh.

Rahman, Ziyaur, Naseem A. Charoo, Mathew Kuttolamadom, Amir Asadi, and Mansoor A. Khan. 2020. "Printing of personalized medication using binder jetting 3D printer." In Joel Faintuch and Salomao Faintuch (eds.), *Precision Medicine for Investigators, Practitioners and Providers*, pp. 473–481. Academic Press: Cambridge, MA.

Razeghi, Orod, José Alonso Solís-Lemus, Angela W. C. Lee, Rashed Karim, Cesare Corrado, Caroline H. Roney, Adelaide de Vecchi, and Steven A. Niederer. 2020. "CemrgApp: An interactive medical imaging application with image processing, computer vision, and machine learning toolkits for cardiovascular research." *SoftwareX* 12:100570. doi: 10.1016/j.softx.2020.100570.

Ren, Shaoqing, Kaiming He, Ross Girshick, and Jian Sun. 2017. "Faster R-CNN: Towards real-time object detection with region proposal networks." *IEEE Transactions on Pattern Analysis and Machine Intelligence* 39(6):1137–1149.

Reynaud, Hadrien, Athanasios Vlontzos, Benjamin Hou, Arian Beqiri, Paul Leeson, and Bernhard Kainz. 2021. "Ultrasound video transformers for cardiac ejection fraction estimation." *International Conference on Medical Image Computing and Computer-Assisted Intervention*, Strasbourg.

Robinson, Robert, Vanya V. Valindria, Wenjia Bai, Ozan Oktay, Bernhard Kainz, Hideaki Suzuki, Mihir M. Sanghvi, Nay Aung, José Miguel Paiva, and Filip Zemrak. 2019. "Automated quality control in image segmentation: Application to the UK Biobank cardiovascular magnetic resonance imaging study." *Journal of Cardiovascular Magnetic Resonance* 21(1):1–14.

Ronneberger, Olaf, Philipp Fischer, and Thomas Brox. 2015. "U-net: Convolutional networks for biomedical image segmentation." International Conference on Medical Image Computing and Computer-Assisted Intervention, Munich.

Roy, Abhishek, Varun Saxena, and Lalit M. Pandey. 2018. "3D printing for cardiovascular tissue engineering: A review." *Materials Technology* 33(6):433–442. doi: 10.1080/10667857.2018.1456616.

Salmi, Mika. 2021. "Additive manufacturing processes in medical applications." *Materials* 14(1):191.

Samadiani, Najmeh, Guangyan Huang, Yu Hu, and Xiaowei Li. 2021. "Happy emotion recognition from unconstrained videos using 3D hybrid deep features." *IEEE Access* 9:35524–35538. doi: 10.1109/ACCESS.2021.3061744.

Samadiani, Najmeh, Guangyan Huang, Wei Luo, Chi-Hung Chi, Yanfeng Shu, Rui Wang, and Tuba Kocaturk. 2022. "A multiple feature fusion framework for video emotion recognition in the wild." *Concurrency and Computation: Practice and Experience* 34(8):e5764.

Sandhu, Kamalpreet, Sunpreet Singh, Chander Prakash, Neeta Raj Sharma, and Karuppasamy Subburaj. 2022. *Emerging Applications of 3D Printing During CoVID 19 Pandemic.* Springer: Berlin, Heidelberg.

Sardari Nia, Peyman, Samuel Heuts, Jean Daemen, Peter Luyten, Jindrich Vainer, Jan Hoorntje, Emile Cheriex, and Jos Maessen. 2017. "Preoperative planning with three-dimensional reconstruction of patient's anatomy, rapid prototyping and simulation for endoscopic mitral valve repair." *Interactive Cardiovascular and Thoracic Surgery* 24(2):163–168.

Schmauss, Daniel, Christoph Schmitz, Amir Koshrow Bigdeli, Stefan Weber, Nicholas Gerber, Andres Beiras-Fernandez, Florian Schwarz, Christoph Becker, Christian Kupatt, and Ralf Sodian. 2012. "Three-dimensional printing of models for preoperative planning and simulation of transcatheter valve replacement." *The Annals of Thoracic Surgery* 93(2):e31–e33.

Schmauss, Daniel, Nicolas Gerber, and Ralf Sodian. 2013. "Three-dimensional printing of models for surgical planning in patients with primary cardiac tumors." *The Journal of Thoracic and Cardiovascular Surgery* 145(5):1407–1408.

Schmauss, Daniel, Sandra Haeberle, Christian Hagl, and Ralf Sodian. 2015. "Three-dimensional printing in cardiac surgery and interventional cardiology: A single-centre experience." *European Journal of Cardio-Thoracic Surgery* 47(6):1044–1052.

Senthilkumaran, Natarajan, and Reghunadhan Rajesh. 2009. "Image segmentation: A survey of soft computing approaches." *2009 International Conference on Advances in Recent Technologies in Communication and Computing*, Kottayam.

Shapiro, Linda G., and George C. Stockman. 2001. *Computer Vision*, vol. 3. Prentice Hall: Upper Saddle River, NJ.

Sharma, Neha, Daniel Ostas, Horatiu Rotar, Philipp Brantner, and Florian Markus Thieringer. 2021. "Design and additive manufacturing of a biomimetic customized cranial implant based on voronoi diagram." *Frontiers in Physiology* 12:647923.

Shehab, Lamia H., Omar M. Fahmy, Safa M. Gasser, and Mohamed S. El-Mahallawy. 2021. "An efficient brain tumor image segmentation based on deep residual networks (ResNets)." *Journal of King Saud University-Engineering Sciences* 33(6):404–412.

Shen, Ye, Zhijun Fang, Yongbin Gao, Naixue Xiong, Cengsi Zhong, and Xianhua Tang. 2019. "Coronary arteries segmentation based on 3D FCN with attention gate and level set function." *IEEE Access* 7:42826–42835.

Sheng, Biaosheng, Mei Zhou, Menghan Hu, Qingli Li, Li Sun, and Ying Wen. 2020. "A blood cell dataset for lymphoma classification using faster R-CNN." *Biotechnology & Biotechnological Equipment* 34(1):413–420.

Sherstinsky, Alex. 2020. "Fundamentals of recurrent neural network (RNN) and long short-term memory (LSTM) network." *Physica D: Nonlinear Phenomena* 404:132306.

Shu, Jian-Hua, Fu-Dong Nian, Ming-Hui Yu, and Xu Li. 2020. "An improved mask R-CNN model for multiorgan segmentation." *Mathematical Problems in Engineering* 2020: 351725.

Siddiqui, Yawar, Julien Valentin, and Matthias Nießner. 2020. "Viewal: Active learning with viewpoint entropy for semantic segmentation." *Proceedings of the IEEE/CVF Conference on Computer Vision and Pattern Recognition, Seattle*, WA.

Simonyan, Karen, and Andrew Zisserman. 2014. "Very deep convolutional networks for large-scale image recognition." arXiv preprint arXiv:1409.1556.

Singh, Anuj Kumar, and Bhupendra Gupta. 2015. "A novel approach for breast cancer detection and segmentation in a mammogram." *Procedia Computer Science* 54:676–682.

Stepniak, Karolina, Ali Ursani, Narinder Paul, and Hani Naguib. 2020. "Novel 3D printing technology for CT phantom coronary arteries with high geometrical accuracy for biomedical imaging applications." *Bioprinting* 18:e00074. doi: 10.1016/j.bprint.2020. e00074.

Sun, Zhonghua, Ivan Lau, Yin How Wong, and Chai Hong Yeong. 2019. "Personalized three-dimensional printed models in congenital heart disease." *Journal of Clinical Medicine* 8(4):522.

Szegedy, Christian, Wei Liu, Yangqing Jia, Pierre Sermanet, Scott Reed, Dragomir Anguelov, Dumitru Erhan, Vincent Vanhoucke, and Andrew Rabinovich. 2015. "Going deeper with convolutions." Proceedings of the *IEEE Conference on Computer Vision and Pattern Recognition*, Boston, MA.

Tang, Yunlong, Guoying Dong, Yi Xiong, and Qiusen Wang. 2021. "Data-driven design of customized porous lattice sole fabricated by additive manufacturing." *Procedia Manufacturing* 53:318–326. doi: 10.1016/j.promfg.2021.06.035.

Tao, Qian, Wenjun Yan, Yuanyuan Wang, Elisabeth H. M. Paiman, Denis P. Shamonin, Pankaj Garg, Sven Plein, Lu Huang, Liming Xia, and Marek Sramko. 2019. "Deep learning-based method for fully automatic quantification of left ventricle function from cine MR images: A multivendor, multicenter study." *Radiology* 290(1):81–88.

Tobias, Orlando José, and Rui Seara. 2002. "Image segmentation by histogram thresholding using fuzzy sets." *IEEE Transactions on Image Processing* 11(12):1457–1465.

Tong, Qianqian, Caizi Li, Weixin Si, Xiangyun Liao, Yaliang Tong, Zhiyong Yuan, and Pheng Ann Heng. 2019. "RIANet: Recurrent interleaved attention network for cardiac MRI segmentation." *Computers in Biology and Medicine* 109:290–302.

Tsukamoto, Yoshinari, Takami Akagi, and Mitsuru Akashi. 2020. "Vascularized cardiac tissue construction with orientation by layer-by-layer method and 3D printer." *Scientific Reports* 10(1):5484. doi: 10.1038/s41598-020-59371-y.

Tutum, Cem C., Supawit Chockchowwat, Etienne Vouga, and Risto Miikkulainen. 2018. "Functional generative design: An evolutionary approach to 3D-printing." *Proceedings of the Genetic and Evolutionary Computation Conference*, Kyoto.

Uccheddu, Francesca, Monica Carfagni, Lapo Governi, Rocco Furferi, Yary Volpe, and Erica Nocerino. 2018. "3D printing of cardiac structures from medical images: An overview of methods and interactive tools." *International Journal on Interactive Design and Manufacturing (IJIDeM)* 12(2):597–609.

Valverde, Israel. 2017. "Three-dimensional printed cardiac models: Applications in the field of medical education, cardiovascular surgery, and structural heart interventions." *Revista Española de Cardiología (English Edition)* 70(4):282–291.

Vaswani, Ashish, Noam Shazeer, Niki Parmar, Jakob Uszkoreit, Llion Jones, Aidan N. Gomez, Łukasz Kaiser, and Illia Polosukhin. 2017. "Attention is all you need." *Advances in Neural Information Processing Systems* 30:5998-6008.

Vesal, Sulaiman, Nishant Ravikumar, and Andreas Maier. 2018. "Dilated convolutions in neural networks for left atrial segmentation in 3D gadolinium enhanced-MRI." *International Workshop on Statistical Atlases and Computational Models of the Heart*, Granada.

Villa, Mateo, Guillaume Dardenne, Maged Nasan, Hoel Letissier, Chafiaa Hamitouche, and Eric Stindel. 2018. "FCN-based approach for the automatic segmentation of bone surfaces in ultrasound images." *International Journal of Computer Assisted Radiology and Surgery* 13(11):1707–1716.

Wang, Bo, Yang Lei, Sibo Tian, Tonghe Wang, Yingzi Liu, Pretesh Patel, Ashesh B. Jani, Hui Mao, Walter J. Curran, and Tian Liu. 2019. "Deeply supervised 3D fully convolutional networks with group dilated convolution for automatic MRI prostate segmentation." *Medical Physics* 46(4):1707–1718.

Wang, Chengcheng, Xipeng P. Tan, Shu Beng Tor, and Chu Sing Daniel Lim. 2020. "Machine learning in additive manufacturing: State-of-the-art and perspectives." *Additive Manufacturing* 36:101538.

Wang, Dee Dee, Marvin Eng, Adam Greenbaum, Eric Myers, Michael Forbes, Milan Pantelic, Thomas Song, Christina Nelson, George Divine, and Andrew Taylor. 2016. "Predicting LVOT obstruction after TMVR." *JACC: Cardiovascular Imaging* 9(11):1349–1352.

Wang, Dee Dee, Zhen Qian, Marija Vukicevic, Sandy Engelhardt, Arash Kheradvar, Chuck Zhang, Stephen H. Little, Johan Verjans, Dorin Comaniciu, and William W. O'Neill. 2021. "3D printing, computational modeling, and artificial intelligence for structural heart disease." *Cardiovascular Imaging* 14(1):41–60.

Wang, Jiang, Yi Yang, Junhua Mao, Zhiheng Huang, Chang Huang, and Wei Xu. 2016. "CNN-RNN: A unified framework for multi-label image classification." *Proceedings of the IEEE Conference on Computer Vision and Pattern Recognition*, Las Vegas, NV.

Wang, Ruoxin, and Chi Fai Cheung. 2022. "CenterNet-based defect detection for additive manufacturing." *Expert Systems with Applications* 188:116000.

Wang, Tianjiao, Tsz-Ho Kwok, Chi Zhou, and Scott Vader. 2018. "In-situ droplet inspection and closed-loop control system using machine learning for liquid metal jet printing." *Journal of Manufacturing Systems* 47:83–92. doi: 10.1016/j.jmsy.2018.04.003.

Wang, Wenxuan, Chen Chen, Meng Ding, Hong Yu, Sen Zha, and Jiangyun Li. 2021. "Transbts: Multimodal brain tumor segmentation using transformer." *International Conference on Medical Image Computing and Computer-Assisted Intervention*, Strasbourg.

Wang, Zeyu, James K. Min, and Guanglei Xiong. 2015. "Robotics-driven printing of curved 3D structures for manufacturing cardiac therapeutic devices." *2015 IEEE International Conference on Robotics and Biomimetics (ROBIO)*, Zhuhai, 6–9 December 2015.

Wu, Jianxin. 2017. "Introduction to convolutional neural networks." *National Key Lab for Novel Software Technology*, Nanjing University, China 5(23):495.

Wu, Wei, Albert Y. C. Chen, Liang Zhao, and Jason J. Corso. 2014. "Brain tumor detection and segmentation in a CRF (conditional random fields) framework with pixel-pairwise affinity and superpixel-level features." *International Journal of Computer Assisted Radiology and Surgery* 9(2):241–253.

Wu, Yongjian, Vladimiro L. Vida, Minwen Zheng, and Jian Yang. 2021. "Progress and prospects of cardiovascular 3D printing." In Jian Yang, Alex Pui-Wai Lee, Vladimiro L. Vida (eds.), *Cardiovascular 3D Printing*, pp. 179–185. Springer Nature: Singapore

Xia, Qing, Yuxin Yao, Zhiqiang Hu, and Aimin Hao. 2018. "Automatic 3D atrial segmentation from GE-MRIs using volumetric fully convolutional networks." *International Workshop on Statistical Atlases and Computational Models of the Heart*, Granada.

Xie, Yuanpu, Zizhao Zhang, Manish Sapkota, and Lin Yang. 2016. "Spatial clockwork recurrent neural network for muscle perimysium segmentation." *International Conference on Medical Image Computing and Computer-Assisted Intervention*, Athens.

Xu, Chenchu, Lei Xu, Zhifan Gao, Shen Zhao, Heye Zhang, Yanping Zhang, Xiuquan Du, Shu Zhao, Dhanjoo Ghista, and Huafeng Liu. 2018. "Direct delineation of myocardial infarction without contrast agents using a joint motion feature learning architecture." *Medical Image Analysis* 50:82–94.

Yadav, Shalini, and Neeta Nain. 2015. "Fast face detection based on skin segmentation and facial features." *2015 11th International Conference on Signal-Image Technology & Internet-Based Systems (SITIS)*, Bangkok.

Yamada, Toshiyuki, Motohiko Osaka, Tomoya Uchimuro, Ryogen Yoon, Toshiaki Morikawa, Maki Sugimoto, Hisao Suda, and Hideyuki Shimizu. 2017. "Three-dimensional printing of life-like models for simulation and training of minimally invasive cardiac surgery." *Innovations* 12(6):459–465.

Yan, Zengqiang, Xin Yang, and Kwang-Ting Cheng. 2018. "Joint segment-level and pixel-wise losses for deep learning based retinal vessel segmentation." *IEEE Transactions on Biomedical Engineering* 65(9):1912–1923.

Yang, Chaoyu, Yunru Yu, Xiaocheng Wang, Qiao Wang, and Luoran Shang. 2021. "Cellular fluidic-based vascular networks for tissue engineering." *Engineered Regeneration* 2:171–174.

Yang, Guang, Xiahai Zhuang, Habib Khan, Shouvik Haldar, Eva Nyktari, Lei Li, Ricardo Wage, Xujiong Ye, Greg Slabaugh, and Raad Mohiaddin. 2018. "Fully automatic segmentation and objective assessment of atrial scars for long-standing persistent atrial fibrillation patients using late gadolinium-enhanced MRI." *Medical Physics* 45(4):1562–1576.

Yang, Xiaoniu, and Xiaolin Tian. 2022. "TransNUNet: Using attention mechanism for whole heart segmentation." *2022 IEEE 2nd International Conference on Power, Electronics and Computer Applications (ICPECA)*, Shenyang.

Yang, Xin, Cheng Bian, Lequan Yu, Dong Ni, and Pheng-Ann Heng. 2017. "Class-balanced deep neural network for automatic ventricular structure segmentation." *International Workshop on Statistical Atlases and Computational Models of the Heart*, Quebec City.

Yao, Xiling, Seung Ki Moon, and Guijun Bi. 2017. "A hybrid machine learning approach for additive manufacturing design feature recommendation." *Rapid Prototyping Journal* 23(6):983–997.

Yoo, Shi Joon, Nabil Hussein, Brandon Peel, John Coles, Glen S. van Arsdell, Osami Honjo, Christoph Haller, Christopher Z. Lam, Mike Seed, and David Barron. 2021. "3D modeling and printing in congenital heart surgery: Entering the stage of maturation." *Frontiers in Pediatrics* 9:621672.

Zeng, Zitao, Weihao Xie, Yunzhe Zhang, and Yao Lu. 2019. "RIC-Unet: An improved neural network based on Unet for nuclei segmentation in histology images." *IEEE Access* 7:21420–21428.

Zhang, Dongqing, Ilknur Icke, Belma Dogdas, Sarayu Parimal, Smita Sampath, Joseph Forbes, Ansuman Bagchi, Chih-Liang Chin, and Antong Chen. 2018. "A multi-level convolutional LSTM model for the segmentation of left ventricle myocardium in infarcted porcine cine MR images." *2018 IEEE 15th International Symposium on Biomedical Imaging (ISBI 2018)*, Washington, DC.

Zhang, Yang, Siwa Chan, Vivian Youngjean Park, Kai-Ting Chang, Siddharth Mehta, Min Jung Kim, Freddie J. Combs, Peter Chang, Daniel Chow, and Ritesh Parajuli. 2020. "Automatic detection and segmentation of breast cancer on MRI using mask R-CNN trained on non-fat-sat images and tested on fat-sat images." *Academic Radiology* 29(Suppl. 1):S135–S144.

Zheng, Bochuan, and Zhang Yi. 2012. "A new method based on the CLM of the LV RNN for brain MR image segmentation." *Digital Signal Processing* 22(3):497–505.

Zheng, Yong-Chao, Yuan-Yuan Zhao, Mei-Ling Zheng, Shi-Lu Chen, Jie Liu, Feng Jin, Xian-Zi Dong, Zhen-Sheng Zhao, and Xuan-Ming Duan. 2019. "Cucurbit [7] uril-carbazole two-photon photoinitiators for the fabrication of biocompatible three-dimensional hydrogel scaffolds by laser direct writing in aqueous solutions." *ACS Applied Materials & Interfaces* 11(2):1782–1789.

Zhou, Yongjin, Weijian Huang, Pei Dong, Yong Xia, and Shanshan Wang. 2019. "D-UNet: A dimension-fusion U shape network for chronic stroke lesion segmentation." *IEEE/ACM Transactions on Computational Biology and Bioinformatics* 18(3):940–950.

Zhou, Zongwei, Md Mahfuzur Rahman Siddiquee, Nima Tajbakhsh, and Jianming Liang. 2019. "Unet++: Redesigning skip connections to exploit multiscale features in image segmentation." *IEEE Transactions on Medical Imaging* 39(6):1856–1867.

Zhu, Zhijie, Daniel Wai Hou Ng, Hyun Soo Park, and Michael C. McAlpine. 2021. "3D-printed multifunctional materials enabled by artificial-intelligence-assisted fabrication technologies." *Nature Reviews Materials* 6(1):27–47.

Zhuo, Peng, Shuguang Li, Ian A. Ashcroft, and Arthur I. Jones. 2021. "Material extrusion additive manufacturing of continuous fibre reinforced polymer matrix composites: A review and outlook." *Composites Part B: Engineering* 224:109143.

Zreik, Majd, Nikolas Lessmann, Robbert W. van Hamersvelt, Jelmer M. Wolterink, Michiel Voskuil, Max A. Viergever, Tim Leiner, and Ivana Išgum. 2018. "Deep learning analysis of the myocardium in coronary CT angiography for identification of patients with functionally significant coronary artery stenosis." *Medical Image Analysis* 44:72–85.

Zyuzin, Vasily, Andrey Mukhtarov, Denis Neustroev, and Tatiana Chumarnaya. 2020. "Segmentation of 2D echocardiography images using residual blocks in U-net architectures." *2020 Ural Symposium on Biomedical Engineering, Radioelectronics and Information Technology (USBEREIT)*, Yekaterinburg.

13 COVID-19 and Pneumonia Detection System Using Deep Learning with Chest X-Ray Images

Mohammad Afiq Hassan,
Rahimi Zahari, and Dina Shona Laila
University Teknologi Brunei

13.1 INTRODUCTION

The world has suffered from the Corona Virus Disease 2019 (COVID-19) pandemic for 3 years between 2020 and 2022. While the knowledge about the virus variants, symptoms, tests and treatment have improved, and vaccines have been developed, and during the writing of this article the pandemic has been declared as finished, the disease still exists and is still affecting thousands of people worldwide. New variants are also still evolving. The daily rate of infections is still quite high although most countries no more do formal counting of the cases. Some countries are still dealing with the impacts of the pandemic and are still struggling to recover; thus, we have to stay vigilant towards this disease or similar possible pandemic in the future.

The symptoms of severe acute respiratory syndrome coronavirus 2 (SARS-CoV-2) infection could range from nothing to severe. Many symptoms can be experienced by the patient but some of them do not have any symptoms at all. The most common symptoms include fever, body ache, dry cough, fatigue, chills, headache, sore throat, loss of appetite, and loss of smell. The more severe symptoms include trouble in breathing, chest pain, high fever, and severe cough. COVID-19 can affect the patient's lungs, which can cause severe symptoms. When SARS-CoV-2 enters the patient's body through the nose, mouth, and eyes, it multiplies and replaces the healthy cell with new viruses. The viruses then travel to the lungs and could cause damage to the lungs. Several test methods have been used to detect coronavirus in a person's body. The most commonly used method is called Reverse Transcription–Polymerase Chain Reaction (RT-PCR) or also known as the *swab test*, which detects the SARS-CoV-2 RNA in the upper and lower respiratory specimens. Figure 13.1 shows the whole process for the swab test. A person who is infected by the virus is called COVID-19 positive, otherwise is a COVID-19 negative. For patients with

DOI: 10.1201/9781003346678-13

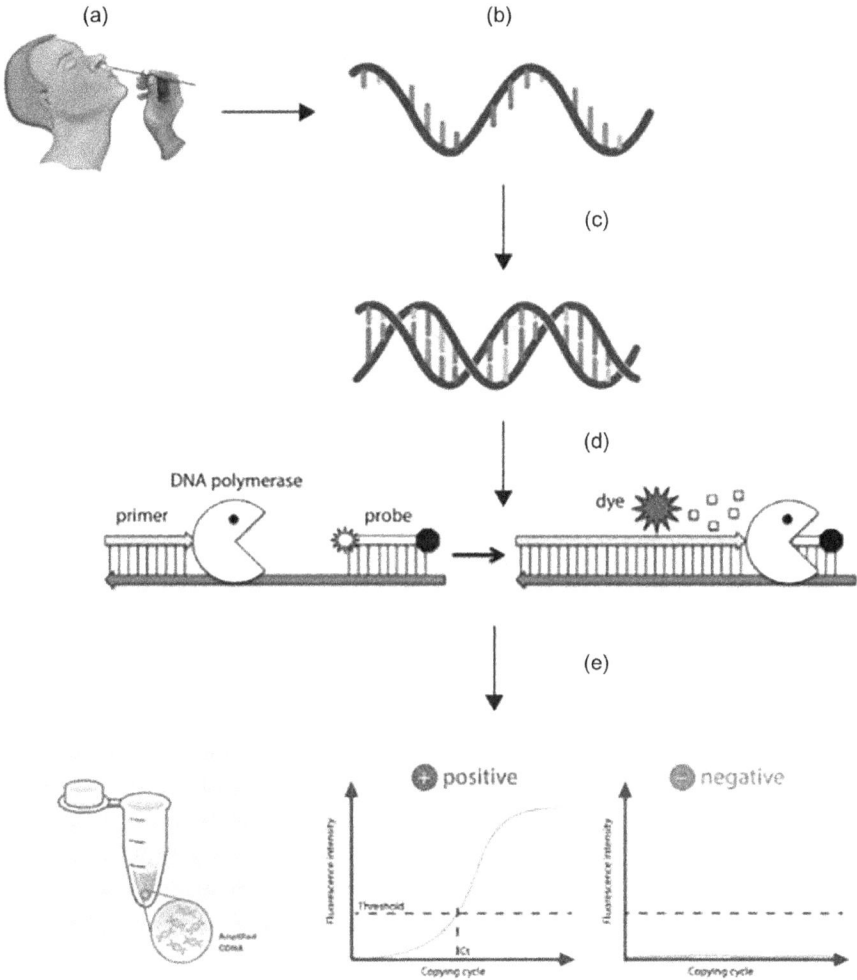

FIGURE 13.1 RT-PCR processes. (a) Sample collection, (b) RNA extraction, (c) reverse transcription, (d) RT-PCR amplification, and (e) result.

more severe symptoms, particularly those that show difficulty in breathing, or for those with underlying conditions, the medical imaging needs to be carried out.

A severe infection could cause respiratory failure and damage to the internal organs, especially the lungs. Thus, diagnosis needs to be done using radiological examination (X-ray, Computer Tomography (CT) scan or Magnetic Resonance Imaging (MRI)) [2]. The challenge is that COVID-19 markers are difficult to distinguish from some other diseases such as pneumonia, and not every health center has a radiologist to make the diagnosis [16]. This is particularly true in many least developed countries and developing countries.

The aim of this project is to develop a COVID-19 and pneumonia detection system, applying digital image processing with deep learning, that can be used to assist radiologists in performing diagnosis for COVID-19 and pneumonia, based on the patient's chest X-ray. This research proposes the use of the chest X-ray (CXR) images, as X-ray is the most commonly available and affordable medical images [7]. The focus is to accurately distinguish between COVID-19, pneumonia and healthy CXR. The Visual Geometry Group version 16 (VGG-16) convolutional neural network (CNN) algorithm and its variants are exploited [18]. The accuracy graph, confusion matrix and performance metrics are produced for evaluation and analysis in searching for better approaches for the final model.

Moreover, this research offers one step further solution, proposing the Graphical User Interface (GUI) design and the embedded system implementation, allowing to build an affordable but reliable portable detection system to assist radiologist to diagnose COVID-19 and pneumonia through patients CXR more accurately in shorter time and to help other medical practitioners to make emergency diagnosis of the diseases, in the absent of a radiologist or to help existing radiologist to verify their diagnosis.

The rest of this chapter is structured as follows. In Section 13.2, a review on radiology examination using CXR is presented. In Section 13.3, a brief summary on CNN, which is the basis of the algorithm applied in the this study, is provided. In Section 13.4, methodology for creating the model of this study is stated here such as specific type of algorithms for the programming and methods on analyzing the model. Many approaches had been done for investigating the most accurate and reliable model. Therefore, different results were shown in this section in terms of the accuracy graph and confusion matrix. In Section 13.5, analysis had been done for each of the results and the final model was chosen from all the approaches. Some problems and challenges faced on implementing the GUI and hardware are also stated in Section 13.6. In Section 13.7, the conclusion for the study is written here based on the result obtained as well as the limitations and recommendations.

13.2   RADIOLOGY EXAMINATION OF COVID-19 USING CXR

While either X-ray, CT-scan, and MRI can be used in the diagnosis, X-ray method is regarded as safer and more available in most hospitals or medical centres. For example, a patient with positive COVID-19 can contaminate the equipment for CT-scan, whereas an X-ray image can be taken while the patient is standing or lying down on a specifically assigned bed [7]. This will reduce the problem of contaminating the equipment used for the procedure. In addition, CT-scan machine is also not available in most general health centers due to its high costs, and the exposure is quite dangerous, whereas CXR is less expensive, possesses lower risk, and is less time-consuming method [7]. Therefore, CXR is the most suitable to use, particularly for the COVID-19 detection.

Furthermore, there have been studies that show that there are patterns to identify COVID-19 in CXR. A study from Ref. [6] concluded that the patterns usually show mixed ground glass opacity in bilateral peripheral middle and lower lung zones. Figure 13.2 shows CXR of COVID-19, and the pattern shows the ground glass opacity in the mid and lower zone of the lungs (labelled by white arrows) taken from a 50-year-old patient [4]. However, the ground glass opacity also appears in patients

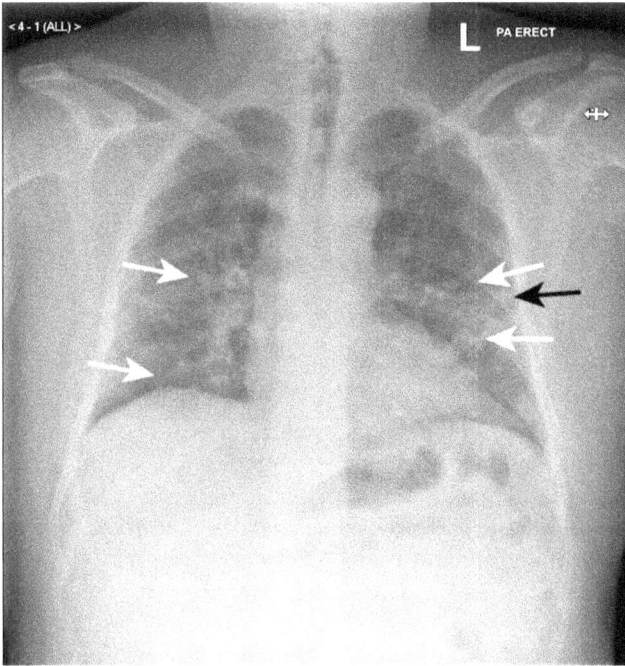

FIGURE 13.2 Pattern for COVID-19 in CXR.

who suffer lung cancer, hemorrhage, inflammatory lung disease, vasculitides, pulmonary aspiration, pulmonary oedema, and common pneumonia [4].

Nevertheless, it was proved by the American College of Radiology (ACR) in the US that medical imaging cannot accurately determine the pattern if the image of the lung contained coronavirus or other diseases [12]. The radiologist can detect the difference in the patterns from normal CXR, but the signs might be from other diseases. Then, in Ref. [12], it is also stated that positive patients can be detected wrongly as normal, whereas in Ref. [3] it is also stated that 60% was diagnosed as normal CXR in COVID-19, which shows that a more accurate detection method is needed.

13.3 CNN AND VGG-16 ARCHITECTURE

CNN is a type of deep learning algorithm that usually is used in digital image processing and vision, and it contains many layers of perceptron. Perceptron is a block of single neural network that contains the input, weights, bias, net sum, and output parts. The CNN overall contains three layers such as input, output, and multiple hidden layers (between input and output layers). For the hidden layers, there are layers of multi-convolution, pooling, fully connected layers which can be shown in Figure 13.3 [11].

CNN architecture contains an input and an output layer with many hidden layers in between. The hidden layers contain multiple convolution layers, as well as some additional layers such as pooling, fully connected and activation function layers. Several

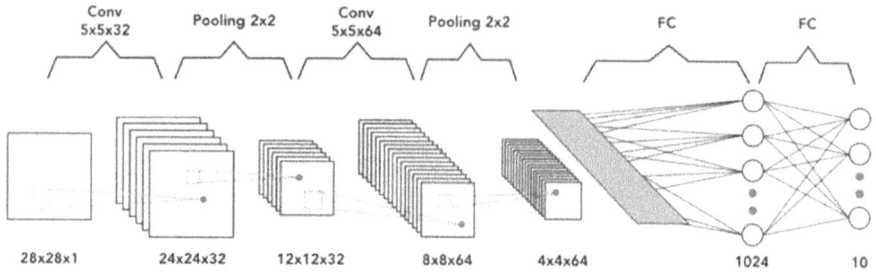

FIGURE 13.3   Basic CNN layers.

CNN architectures have been introduced such as the AlexNet, VGG-16, ResNet and GoogleNet. All the models use the same combination of layers such as convolution, pooling, and fully connected layers but the numbers of the layers are different.

Various deep learning algorithms/architectures have been developed by the researchers for specific applications. Some popular architectures such as CNNs, deep belief networks (DBN), recurrent neural networks (RNN) and deep neural networks (DNN) are implemented in predictive models. Moreover, the amount of the available data is rising, which can increase the accuracy of the deep learning models, as well as faster on producing the outputs. For this study, CNN algorithm, particularly the VGG-16 architecture, will be implemented for the COVID-19 and pneumonia detection model with the datasets of CXR.

13.3.1   VGG-16 Layers

VGG-16 is a CNN model created by Karen Simonyan and Andrew Zisserman from the University of Oxford [18]. The architecture contains 16 layers, comprises convolution, max pooling and fully connected layers, as shown in Figure 13.4. Moreover, the model was ranked the top 5 in the 2014 ImageNet challenge with a test accuracy about 92.7%.

Figure 13.5 shows the processes inside the VGG-16 layers when an input image is feed into the layers. First, an image with a size of $224 \times 224 \times 3$ RGB pixels is used as the input. This image passes to the second layer, which is the first convolution layer, with 3×3 kernel size and 64 filters. As a result, the input image dimensions change to $224 \times 224 \times 64$. Then the input image goes to the first Max pooling layer in the VGG-16 architecture with filter size of 3×3 and a stride of two. As a result, the input image dimension changes to $112 \times 112 \times 64$. According to Figure 13.5 [18], the next layers comprise multiple convolution and max pooling layers for the input image to undergo. At the end, the final image dimensions become $7 \times 7 \times 512$.

Next, the output image ($7 \times 7 \times 512$) is being flatten in a dense and fully connected layers with 25,088 units/feature maps (the multiplication of the image dimensions is equal to 25,088). Then, two more fully connected layers are passed with 4,096 units and then the final fully connected layers with 1,000 ways of softmax [18]. However, this study used only two ways softmax because two outputs, i.e. either COVID-19 positive or COVID-19 negative, are expected.

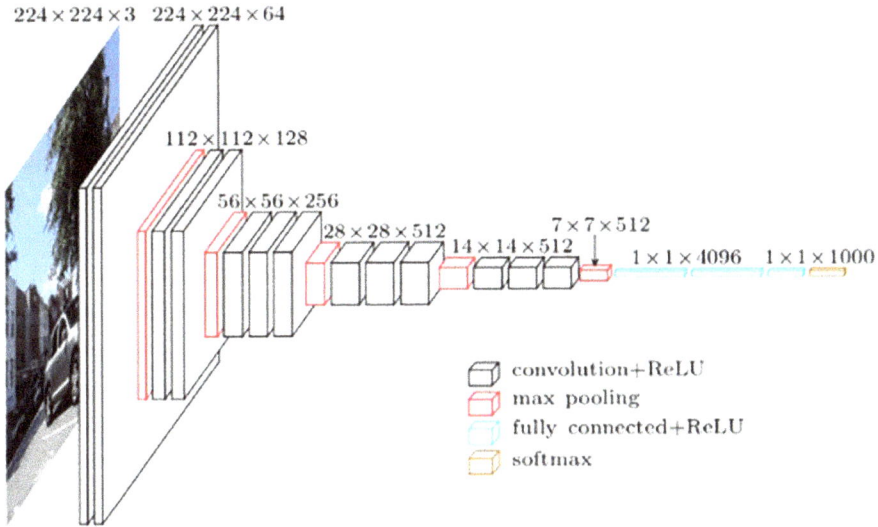

FIGURE 13.4 VGG-16 layers.

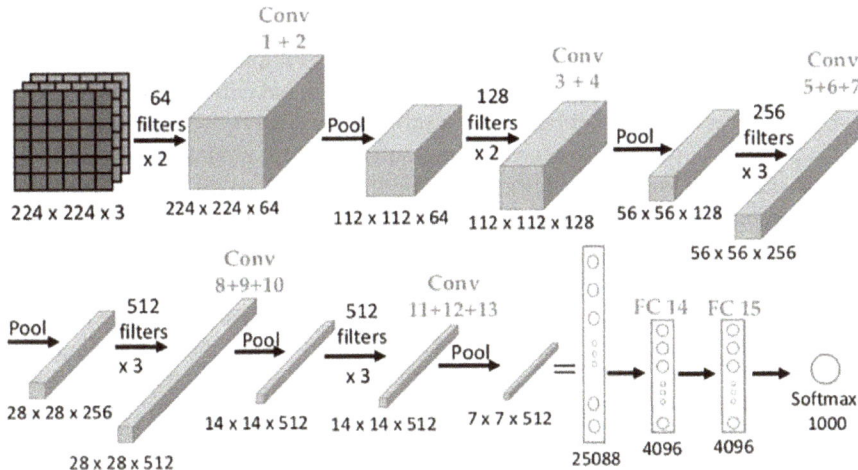

FIGURE 13.5 Process in VGG-16 layers.

13.3.2 Optimizers

After all the layers are passed, the datasets need to be compiled using Keras library with specific optimizer, loss, and metric functions. Optimizers in deep learning are algorithms or techniques that can reduce the losses in the model by changing the neural network's weights and learning rate. There are many types of optimizers that can be implemented in the model such as Gradient Descent, Stochastic Gradient Descent, Mini-Batch Gradient Descent, Adagrad, AdaDelta, Adam, and so on [15].

The equation below shows on how the optimizers work with the weight during the training process. Every iteration or epoch, the weight is reducing in order to reach the true value. The old weight is deducted with the product of learning rate and gradient. The learning rate is like a step for the value to reach the true value, and it is preferable to set it low. On the other hand, the gradient is actually the derivative of the loss against the old weight. If there is more than one iteration or epoch, the gradient and the weight are going to be updated until the new weight reached the optimized value.

13.3.3 Loss Function and Activation Function for Output Layer

For the activation function, Softmax was chosen for the final model, and it will be used in the output layers of the classification. This is because Softmax is used for multi-classification problem (two or more outputs) and the sum of the results' probabilities equal to 1. During prediction, if the probability of the input CXR image higher at the first-class type (COVID), the image is classified as COVID CXR. The probability for the input image will be lower for the second-class type (normal). For Softmax, the sum of both probabilities will be one, and this is the reason on the use of Softmax in the output layers of the final model.

Compared to other activation function like sigmoid, it is used for binary classification problem (one or two outputs), and the sum of the result's probabilities is not equal to 1. During prediction, the probability might be high for both classes. Furthermore, sigmoid also used for multi-labels from multiple classes, which means there are two outputs are true values. In summary, sigmoid can be implemented for non-mutually exclusive model because the probabilities can be high for both output classes. Softmax can be implemented for mutually exclusive model because only one class will have high probability. For loss function, it depends on the type of activation function on output layers. For sigmoid, it suits with Binary cross entropy loss function because it can be used for getting two outputs only (0 or 1) or it can be used for two or more labels (not mutually exclusives). Softmax is suitable to use with categorical cross entropy because it is used for predicting a single true value from multiple classes (mutually exclusive). Form these, the final model implemented Softmax for the output layer of the VGG16 architecture. Then, the type of loss function used is categorical loss function in order to normalize only one true value from the output layer.

13.3.4 Compiling the Training Model

Moreover, the callback function is also used for the model where it controls and monitors when training the model. There are two types of callbacks which are *Modelcheckpoint* and *Early Stopping*. *Modelcheckpoint* is used for saving the model for every epoch [19]. If the accuracy is high, the model will be saved, and if the accuracy is low, the model will not save it until the accuracy is higher than the existing accuracy. Then, *Earlystopping* is used to stop during compiling the model when it is not improving.

13.3.5   Experimental Evaluation Using Confusion Matrix

After the trained model has successfully been completed and built, evaluation can be made to determine the performance of the trained model. To evaluate it, a confusion matrix will be implemented which can be shown in Figure 13.6.

Confusion matrix is used to evaluate the performance of the classification model. It contains four parameters such as True Positive (TP), False Positive (FP), False Negative (FN), and True Negative (TN) [1].

According to Figure 13.6, TP is where the predicted values match correctly with the actual values. The actual and predicted values are positive. TN is also similar to TP but the actual and predicted values are negative. On the other hand, FP and FN are types of error where the actual values are falsely predicted. For FP, the actual values are negative, but the model predicted positive and vice versa for FN. Figure 13.6 shows the graph for 2×2 matrix which means that there are only two classifications. If there are three classifications such as dog, cat or bird, the confusion matrix becomes 3×3 matrix.

13.3.6   VGG-16 in Other Applications

A study by Ref. [10] where VGG-16 layers with transfer learning were used to build a brain image classification. The datasets were images of the brain obtained from Harvard medical school. The analysis was done using performance metrics such as sensitivity, specificity, and accuracy, similar to the previous studies. The researcher compared with other previous models that used the same datasets. It was concluded that the VGG-16 layers had higher accuracy based on the metrics.

Lastly, a system for diagnosing papillary thyroid carcinoma was invented using the VGG-16 and InceptionV3 layers [8]. The datasets were about 279 cytological images of thyroid nodules which were divided into two sets such as training and

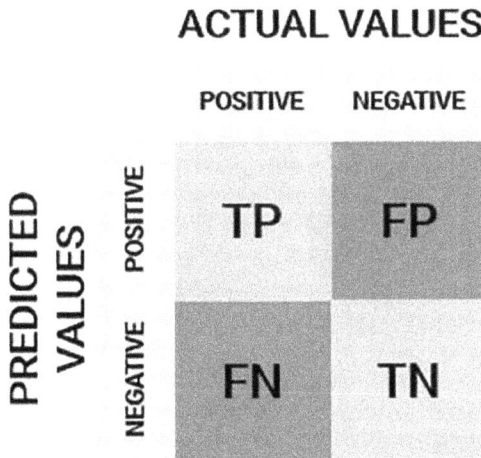

FIGURE 13.6   Confusion matrix for 2×2 matrix (two outputs).

testing sets. The overall accuracy was obtained for both VGG-16 and InceptionV3 about 95% and 87.5%, respectively which can be shown in Table 13.1. As a result, the studies above showed that VGG-16 can produce higher accuracy rates for image classification system.

13.4 COVID-19 AND PNEUMONIA DETECTION USING VGG-16

For the above consideration, VGG-16 is chosen to be used in this study. In this section, we present the use of VGG-16 architecture for COVID-19 and pneumonia detection. The VGG-16 consists of 12 convolution and max pooling layers, as well as 3 fully connected layers and the output layer. From the previous studies, methods for data collection and analysis of the model were discovered. For the data collection, the study will be using datasets of CXR images that can be found from the open source such as Kaggle [13,14] and GitHub [5].

Firstly, a VGG-16 algorithm was implemented for the image classification model. The layers were used from scratch and trained from the first layer, without any parameters or weights from the previous task saved in the model. For CXR, a total of 3,829 images of CXR from COVID, normal and pneumonia-confirmed cases, respectively, were collected from online repositories [5,13,14] for training and testing processes.

The general workflow is illustrated in Figure 13.7. The grouping of the datasets is shown in Table 13.2 for training and testing the model. In addition, the images for training were divided again to 70% for training and 30% for validation.

Note that it is important to consider the three cases, rather than only COVID-19 and normal cases, as pneumonia is also quite common in least developed and

TABLE 13.1
VGG-16   Versus Inception V3

Model	VGG-16 (%)	Inception V3 (%)
Accuracy	97.66	92.75
Sensitivity	100	98.55
Specificity	94.91	86.44
Positive prediction value	95.83	89.47
Negative prediction value	100	98.08

TABLE 13.2
Data Sets for the Model

CXR	Training	Testing
COVID	1,029	114
Normal	1,207	134
Pneumonia	1,211	134

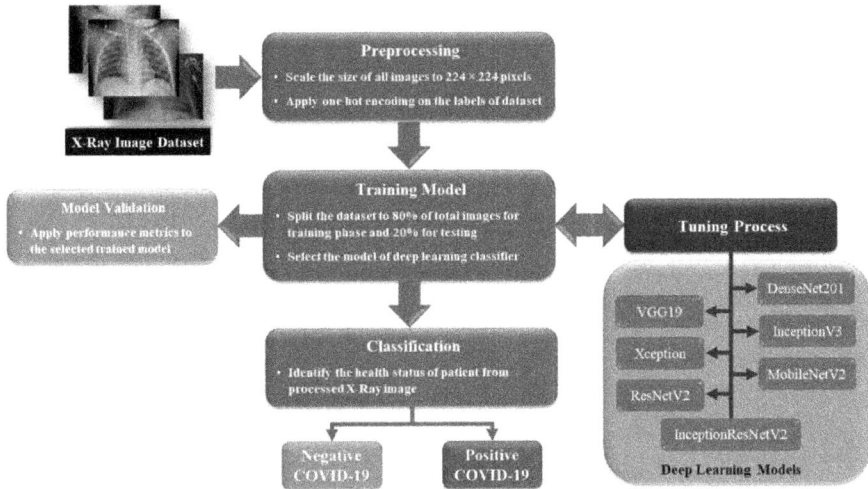

FIGURE 13.7 Workflow for the system.

developing countries. Correctly diagnosing patients is crucial to deciding the right action and further treatment to be given to the patients.

After the dataset has been formed, preprocessing of data needs to be carried out before feeding them into the deep learning algorithm. The first stage of the image pre-processing is to resize all the images to the same size. Following the literature, all the images will be resized to 224 × 224 pixels. The size or dimension can be increased further to get better accuracy, but it will also increase the training computational cost and time.

For the final model, the VGG-16 was built from scratch and the chosen parameters are given in Table 13.3. The number of epochs was set to tenth iterations and steps per epoch were decided based on the formula from Ref. [9]. The batch size is set to 32 which means every epoch/iteration will train 32 images. For this system, from a CXR image input, the expected output is the correct classification of either COVID, normal or pneumonia, which means there is only one, mutually exclusive, answer.

For training the model, the phyton programming will be used due to the libraries are suitable for creating a complex deep learning model such as Keras and TensorFlow library. The datasets are going to be loaded using the library and fed

TABLE 13.3
Hyper Parameters for the Model

Optimizer	Adam
Learning rate	0.0001
Loss function	Categorical crossentrophy
Output activation	Softmax

to the deep learning model. Then, the software for writing the programming is in Jupyter Notebook. It was chosen for this study because it is an open-source web application that is compatible with more than 40 programming languages and it can be installed directly through Python [17]. A computing machine is going to be used for the implementation of the CNN with a processor configuration, Intel Core i5–8300H CPU @ 2.30 GHz and 16 GB RAM.

Three classes were implemented to the VGG-16 model, i.e., COVID, normal and pneumonia CXR images. The training shows correct results where the accuracy was increasing and the losses were decreasing, as shown in Figure 13.8a. Then, the confusion matrix in Figure 13.8b shows that from the 114, 134 and 134 testing set of images, 103, 131 and 128 images were predicted accurately as COVID, normal and pneumonia, respectively.

To improve the reliability of the system, preprocessing is applied to the CXR images, by converting the images to BGR from the normal RGB state, as well as the color channel was zero-centered, as seen in Figure 13.9.

The confusion matrix in Figure 13.10b shows that correct predictions are slightly higher compared to those without the preprocess function. About 111, 130 and 128 images were correctly predicted as COVID, normal and pneumonia, respectively.

Further improvement is done using a pretrained model, which is a model invented by earlier researchers to solve a similar classification problem. The model can be

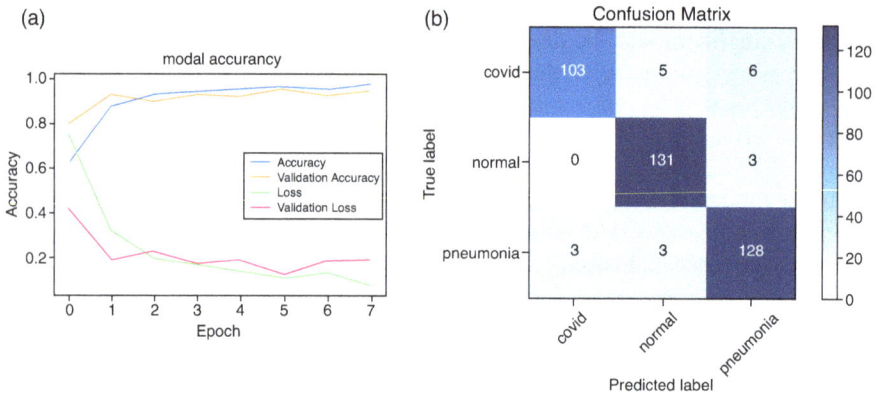

FIGURE 13.8 Three classes VGG-16 from scratch (a) training accuracy; (b) confusion matrix.

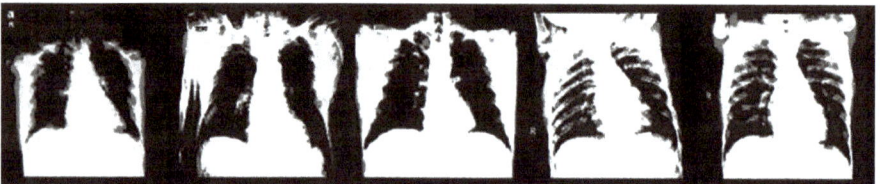

FIGURE 13.9 CXR images with the preprocess function.

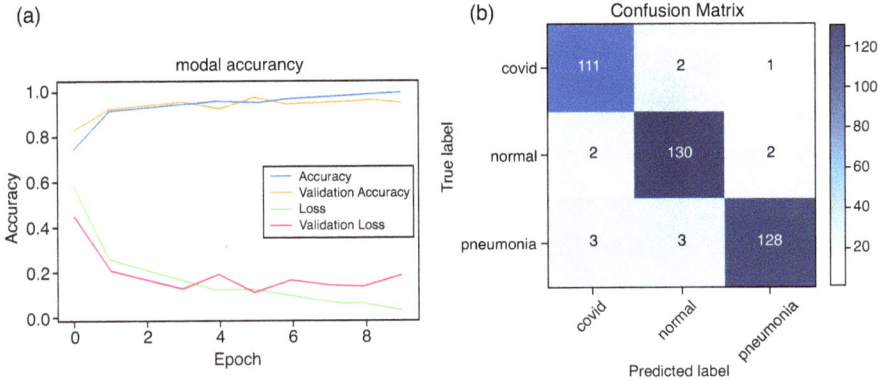

FIGURE 13.10 Three classes VGG16 with preprocess function (a) training accuracy and (b) confusion.

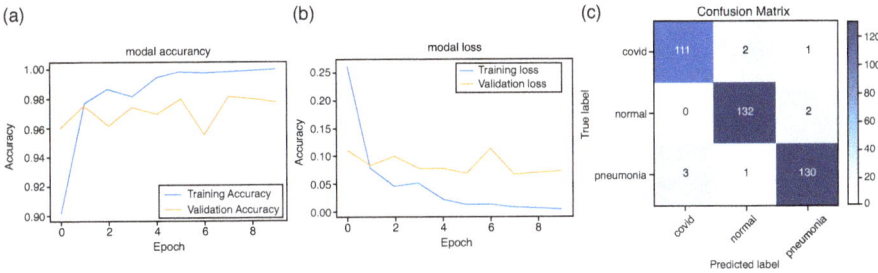

FIGURE 13.11 Three classes VGG16 pretrained model (a) accuracy graph, (b) loss graph and (c) confusion matrix.

implemented for different problems, known as transfer learning, where the pretrained model is reused for other problems. It is shown in Figure 13.11 that correct predictions are slightly higher from previous approaches, with 111, 132 and 130 correct image predictions as COVID, normal and pneumonia, respectively.

13.5 RESULTS AND DISCUSSION

Using the confusion matrix, the performance metric was calculated for the VGG-16 model. The parameters, such as precision, recall, F_1-score and accuracy were obtained higher than 95% overall. For the VGG-16 from scratch, even though the result was not too bad, some images with distortion were wrongly predicted. This happens when the input image was tilted to the right and another input image was added with noise. The model predicted those cases as normal CXR, whereas in fact they are COVID CXR, as shown in Figure 13.12. After preprocessing was applied, the same set of input images are now predicted correctly as shown in Figure 13.13.

With the pre-trained approach, fine-tuning was carried out to classify the images for COVID, normal and pneumonia. After compiling the model, the result shows

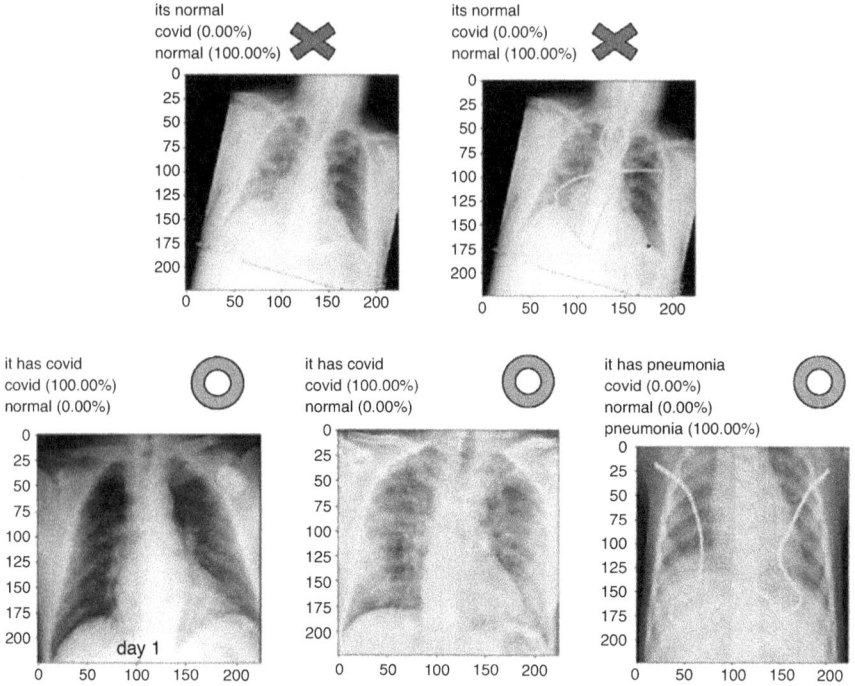

FIGURE 13.12   Prediction using VGG-16 from scratch.

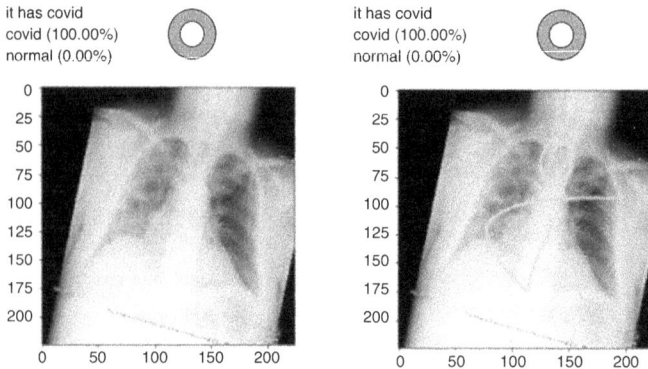

FIGURE 13.13   Prediction using VGG-16 from scratch with preprocess function.

that the training accuracy was improved, and the loss was reduced as the Epoch increased. Using the confusion matrix, the performance metrics were calculated, showing that this approach produces better results, as demonstrated in Figure 13.14.

As the pre-trained model has been already trained with different types of images from the 'ImageNet', the layers are already intelligent enough to learn the new

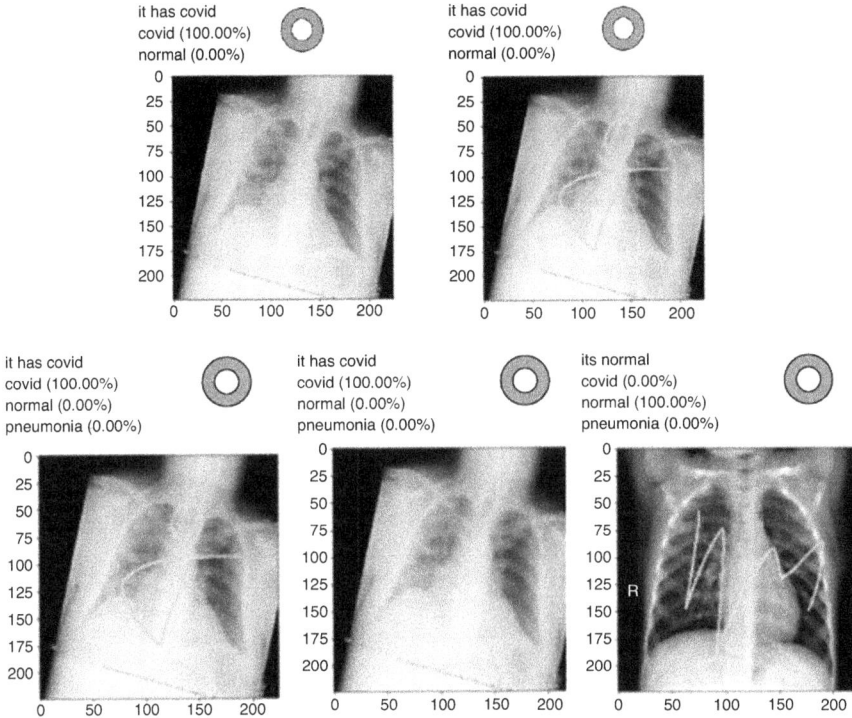

FIGURE 13.14   Prediction using VGG-16 pretrained.

datasets which leads to higher accuracy. Another advantage is the shorter train-ing time using the model as training is needed only for partial parts of the layers. Furthermore, the prediction is also able to get the correct label for examples shown in Figure 13.13, which is a significant improvement compared to the VGG-16 model from scratch, both with and without the preprocess function of the images.

Table 13.4 shows all the performance metrics of the approaches based on the values obtained from the confusion matrix. It shows that the pre-trained model has better percentages compared to the others. This approach also requires much shorter training time.

TABLE 13.4
Datasets for the Various VGG-16   Models

VGG-16	Accuracy (%)	Precision (%)	Recall (%)	F_1-Square (%)	Specificity (%)	Training Time (Seconds)
Scratch	95	97	90	94	96	624
Preprocess	97	96	97	97	99	640
Pretrained	98	97	97	97	99	390

13.6   GUI AND HARDWARE IMPLEMENTATION

The study also implemented the use of a graphical user interface (GUI) for the others to use the program without compiling the real code. To make this, a standard interface from Python called Tkinter. The GUI is going to have buttons to choose the image and to classify the image using the model created. On the other hand, an affordable small computer is also implemented called Jetson Nano 2 GB developer kit for running the GUI. This can make the project portable and easier to use anywhere due to its size is small, provided the CXR image of the patient is available. Figure 13.15 shows the board with a memory size of 2 GB 64-bit LPDDR4 25.6 GB/s, including a GPU of 128-core NVIDIA Maxwell™ GPU and CPU of Quad-core ARM A57 @ 1.43 GHz CPU. The operating system is officially using the Linux4Tegra, based on Ubuntu 18.04.

The GUI was successfully created using the Tkinter library with the interface shown in Figure 13.16. It contains two buttons for uploading an image and for classifying the selected image. It can be run if the device has Python software with specific libraries for compiling the model.

For the hardware implementation, the Jetson Nano 2GB was used to implement the GUI. However, the processor can only run the simple algorithm due to the limited memory. The model needs to be optimized in terms of the dimensions and layers of the architecture without reducing the overall accuracy of the trained model. Another type of Jetson with higher memory capacity can be chosen for improved implementation.

FIGURE 13.15   Jetson Nano 2GB development kit.

Image Classification

covid: 0.00% normal: 0.00% pneumonia: 100.00%

Detected Pneumonia

Description:
This is a GUI for CXR image
classification between
Covid, normal and pneumonia.

Steps:
1. Upload the image
2. Clasify the image

Model = VGG16
Optimizer = Adam(0.0001)
Output activation function = softmax
Lost Function = Categorical

Done by: Mohammad Afiq

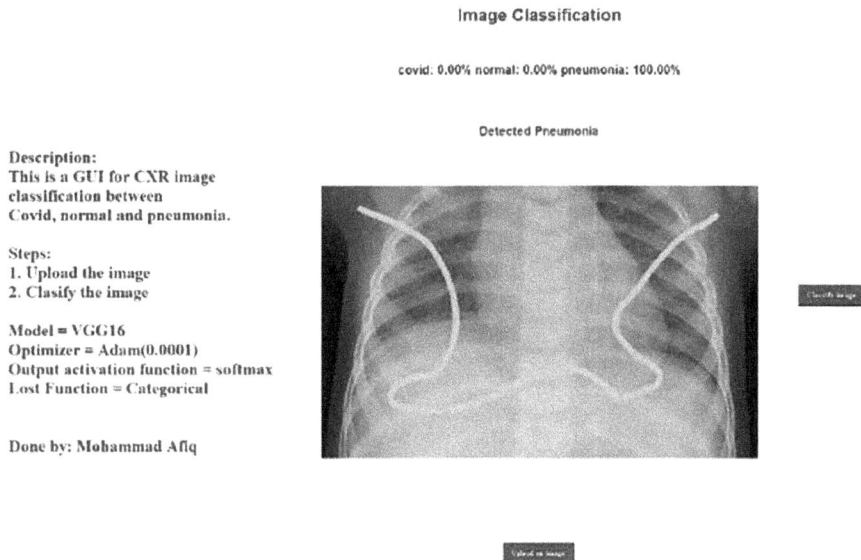

FIGURE 13.16 The GUI interface.

13.7 CONCLUSION

The real implementation of VGG-16 for COVID-19 and pneumonia detection has been done using the CXR datasets and approaches from previous researchers in building an accurate detection system. The two diseases are considered as not only COVID-19, but pneumonia is also quite prevalent.

It has been demonstrated that the VGG-16 layers with a pretrained model were the best method for the final model based on the analysis of this study. Using the selected method, the study successfully produced about 98% accuracy with 99% specificity, and training time of 390 seconds. However, more study needs to be done regarding the robustness of the model, as it uses Imagenet which was not pre-trained for medical images like CXR. Furthermore, a simple GUI was successfully created to make the system user-friendly for the radiologist or doctors. The next stage of this research is to implement the algorithm in a processor with larger memory capacity. This will allow to build an affordable but reliable portable detection system to assist radiologist to diagnose COVID-19 and pneumonia through patients' CXR more accurately in a shorter time, and to help other medical practitioners to make emergency diagnosis of the diseases, in the absence of radiologist, which is expected to help compensating the situation in least developed and develop countries.

REFERENCES

1. B. Aniruddha. Everything you should know about confusion matrix for machine learning. https://www.analyticsvidhya.com/blog/2020/04/confusion-matrix-machine-learning/, 2018.
2. E. M. Chamorroa, A. D. Tasconb, L. I. Sanza, S. O. Velezb, and S. B. Nacenta. Radiologic diagnosis of patients with COVID-19. *Radiologia (English Edition)*, 63:56–73, 2021.
3. T. B. Chandra, K. Verma, B. Singh, D. Jain, and S. Netam. Coronavirus disease (COVID-19) detection in chest X-ray images using majority voting based classifier ensemble. *Expert Systems with Applications*, 165:113909, 2020.
4. J. Cleverley, J. Piper, and M. Jones. The role of chest radiography in confirming COVID-19 pneumonia. *BMJ*, 370:m2426, 2020.
5. J. P. Cohen, P. Morrison, and L. Dao. COVID-19 image data collection, 2020. https://arxiv.org/abs/2003.11597.
6. M. Durrani, I. Haq, and A. Yousaf. Chest X-rays findings in COVID 19 patients at a university teaching hospital: A descriptive study. *Pakistan Journal of Medical Sciences*, 36:S22–S26, 2020.
7. M. Gaeta, G. Cicero, M. A. Marino, T. D'Angelo, E. Mormina, S. Mazziotti, A. Blandino, G. Siracusano, A. La Corte, M. Chiappini, and G. Finocchio. Effectiveness of baseline and post-processed chest X-ray in nonearly COVID-19 patients. medRxiv, 2020.
8. Q. Guan, Y. Wang, B. Ping, D. Li, J. Du, Q. Yu, H. Lu, X. Wan, and Jun X. Deep convolutional neural network VGG-16 model for differential diagnosing of papillary thyroid carcinomas in cytological images: A pilot study. *Journal of Cancer*, 10:4876–4882, 2019.
9. B. Jason. Difference between a batch and an epoch in a neural network, 2018. https://machinelearningmastery.com/difference-between-a-batch-and-an-epoch/.
10. T. Kaur and T. K. Gandhi. Automated brain image classification based on VGG-16 and transfer learning. In *2019 International Conference on Information Technology (ICIT)*, Bhubaneswar, India, pp. 94–98, 2019.
11. M. R. Khan. Convolutional neural network: In a nut shell, 2018. https://engmrk.com/convolutional-neural-network-3/.
12. A. Krishnaraj and A. Matsumoto. COVID-19 and imaging: Why CT scans and X-Rays have a limited role in diagnosing coronavirus, 2020. https://blog.radiology.virginia.edu/coid-19-and-imaging/.
13. P. Mooney. Chest X-ray images (pneumonia), 2018. https://www.kaggle.com/paultimothymooney/chest-xray-pneumonia.
14. P. Patel. Chest X-ray (COVID-19 & pneumonia), 2018. https://www.kaggle.com/datasets/prashant268/chest-xray-covid19-pneumonia.
15. S. Ruder. An overview of gradient descent optimization algorithms, 2016. https://arxiv.org/abs/1609.04747.
16. P. Sadhukhan, M. T. Ugurlu, and M. O. Hoque. Effect of COVID-19 on lungs: Focusing on prospective malignant phenotypes. *Cancers*, 12:3822, 2020.
17. D. Sejuti. Why Jupyter notebooks are so popular among data scientists, 2021. https://analyticsindiamag.com/why-jupyter-notebooks-are-so-popular-among-data-scientists/.
18. K. Simonyan and A. Zisserman. Very deep convolutional networks for large scale image recognition, 2014. https://arxiv.org/abs/1409.1556.
19. S. Sinha, T. N. Singh, V. Singh, and A. K. Verma. Epoch determination for neural network by self-organized map (SOM). *Computational Geosciences*, 14:199–206, 2010.

14 Scattering Operators and High-Order Statistics along with Elastography to Identify and Characterize Salivary Gland Abnormalities

Thibaud Berthomier
ENSTA-Bretagne
CHU de la Cavale Blanche

Ali Mansour
ENSTA-Bretagne

*Luc Bressollette, Clément Hoffman,
and Sandrine Jousse-Joulin*
CHU de la Cavale Blanche

14.1 INTRODUCTION

The rheumatology department of the Brest Hospital collects ultrasound images of the salivary glands and they are interested in the detection of the Gougerot-Sjögren syndrome.[1] Their database contains mainly two types of images obtained by:

- **Ultrasound imaging in mode B (brightness)**: *i.e.* a two-dimensional image is formed by juxtaposition of a large number of lines each corresponding to an ultrasound in mode A (Amplitude). The gray level, called brightness, represents the amplitude of the echoes.
- **Elastography (stiffness)**: The major methods of elastography are the quasi-static methods (with external static stress) and Dynamic or Pulse method (measuring the propagation speed of shear waves.)

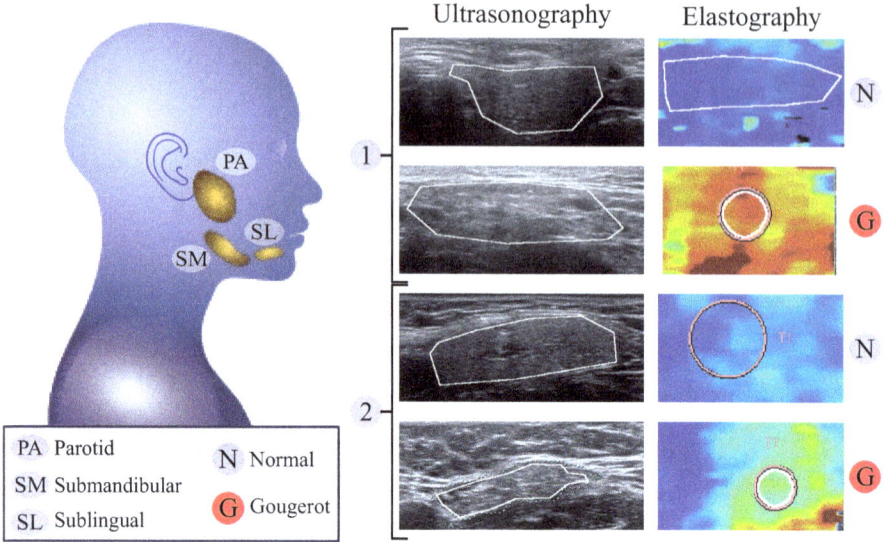

FIGURE 14.1 Ultrasonography and Elastography of normal salivary glands or those with the Gougerot's syndrome.

The salivary glands produce saliva which is an essential element for making speech, chewing, tasting, swallowing and having good oral hygiene [7]. We can distinguish three main salivary glands as shown in Figure 14.1:

1. The parotid glands, located below the ears, are the most voluminous.
2. Submandibular (or submandibular) glands can be found under the jaw.
3. Sublingual glands are under the tongue.

To produce more than a liter of saliva per day, the three mentioned glands are helped by accessory glands (microscopic) disseminated in the oral mucosa. The salivary glands can be the site of various ailments, of an infectious nature, lithiasis (presence of stony concretions, calculi, in the ducts of the salivary glands), tumor and/or immunological. These conditions can lead to increase, sometimes painful, the volume of the glands or generate a functional disorder, either by insufficiency or by hyperfunction.

In our project, we are interested in the detection of the Gougerot-Sjögren syndrome [2,17] which is an autoimmune disease: cells of the immune system, called lymphocytes, and antibodies secreted by these same cells (lymphocytes), attack the glands and cause them to malfunction. This disease mainly results in dry mouth and/or eyes which means that the glands are no longer able to produce sufficient tear and saliva [17]. Patients included in our study are generally suffering from the Sicca syndrome[2] and have been diagnosed for a suspected Gougerot-Sjögren syndrome. Without this autoimmune disease, the examined glands are considered normal and the sicca syndrome may be the result of other factors. As can be seen in Figure 14.1, the ultrasound of a normal gland shows a rather homogeneous tissue; whereas a gland affected by the Gougerot-Sjögren syndrome presents more heterogeneous and more echogenic textures [4,26,50]. Indeed, diseased glands, especially the parotids, seem to be harder than normal glands as shown in Figure 14.1.

In medicine, two statistical criteria are generally used to evaluate the accuracy of the diagnosis:

- The diagnosis sensitivity is the ability to detect infected patients by the Gougerot-Sjörden disease [49]:

$$\text{Sensitivity} = \frac{\text{TP}}{\text{TP} + \text{FN}} \tag{14.1}$$

where TP stands for the number (or percentage) of True Positives and FN represents the False Negatives.

- The diagnosis specificity corresponds to the detection rate of patients not affected by the disease. It is estimated from the number of True Negatives (TN) and False Positives (FP) by:

$$\text{Specificity} = \frac{\text{TN}}{\text{TN} + \text{FP}} \tag{14.2}$$

The authors of Ref. [26] carried out comparative evaluation studies on the detection (by medical experts) of Gougerot-Sjödren syndrome by ultrasound. The 11 studies mentioned in this publication (between 1996 and 2012) show an average sensitivity of 77.9% (ranging from 46.6% to 90%) and an average specificity of 86.7% (ranging from 73% to 100%). In another publication [14], the authors are interested in more recent studies (from 2010 to 2015) which present an average sensitivity of 71.3% (from 63% to 91%) and an average specificity of 91.9% (83%–98%). In these studies, the parotid and submandibular glands are mixed.

Recent studies have examined the impact of elastography on the diagnosis and they showed that the elastometry (mean values of elastography in a region of interest) of the pathological parotid glands has higher values (indicating a tougher gland) than in the case of normal glands [27,50]. The study presented in Ref. [27] showed that the elastometry of parotid glands was used to distinguish between healthy and infected patients with a sensitivity of 81% and a specificity of 67%. On the submandibular glands and using only elastometry [27,50], experts were not able to separate the two groups of patients. More recently, deep learning was successfully applied to identify gland tumours [45] and to segment parotid gland [41,44] but using computed tomography (CT) imagery, which offers much better image quality and resolution than ultrasonography and elastography.

In this project, our main objectives are:

1. Automatic detection of the Gougerot-Sjögren syndrome;
2. To enhance the importance of elastography in the diagnosis of such disease;

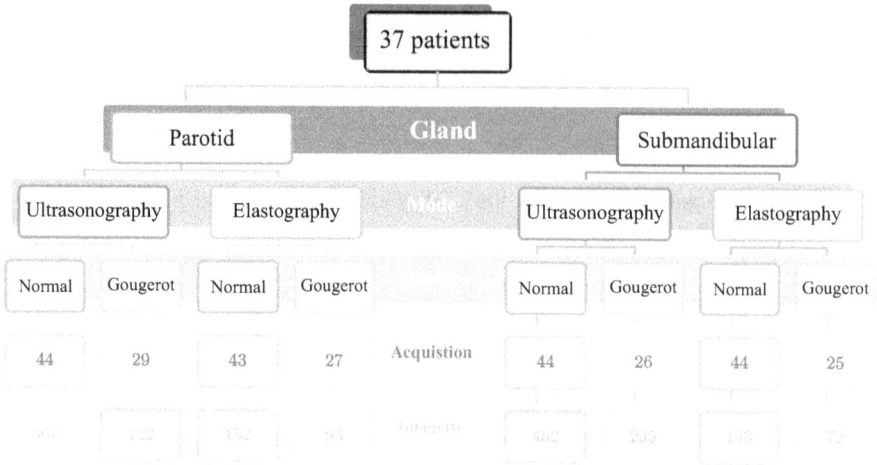

FIGURE 14.2   The 37 patients participated in our salivary gland database: number of acquisitions and small images extracted from these acquisitions.

Classification rate, sensitivity and specificity will be the three main criteria to evaluate our approach to ultrasound and elastography clinical data. Figure 14.2 presents our database which is made up of ultrasound images of parotid and submandibular glands collected from 37 patients.

14.2   SCATTERING OPERATORS (SO)

Our approach consists of applying clustering algorithms on reduced features extracted from our ultra-acoustic images using two major approaches: Wavelets (*i.e.* Scattering Operators) and High Order Statistics. To help the readers easily follow the proposed method, we will briefly introduce, hereinafter, the concept and main ideas of the major used approaches.

An acoustic image (ultrasound or elastography) often has different textures depending on the observed tissues (organs, vessels, tendons, *etc.*) and their state of health (healthy, benign, malignant, fibrosis, *etc.*). These images often suffer from small local deformations due to the characteristics of the tissue and the acquisition process (several approximations are made during the image reconstruction). The general idea of the Scattering Operators (SO) is to reduce the variability between images of the same class (same type and state of the observed tissue) while maintaining variability among different classes. To reach that goal, SO transforms the representation space of the images to become invariant to small deformations and translations.

14.2.1   MAIN CONCEPT

The authors of Ref. [8] introduce the concept of SO as a powerful method to improve automatic classification techniques for signals (and images). For example, let us consider a cloud of points representing mixed images of two different classes; then, the

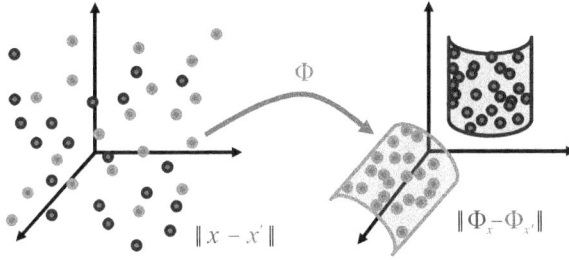

FIGURE 14.3 The transformation by a function Φ of the representation spaces of two class images.

objective of SO is to find a function Φ that will change the point representations to distinguish the two hidden classes (see Figure 14.3). The function Φ should be selected in such a way as to immune the extracted features from any previous image transformations, such as translation, rotation and scaling. Using the Lipschitz continuity, Mallat suggests that Φ should satisfy the following constraint:

$$\exists C > 0 \left| \; \|\Phi_{xr} - \Phi_x\| \le C\|x\|\sup_u |\Delta\tau(u)| \right. \tag{14.3}$$

$\forall x$ & τ, where τ represents the image deformation, $x\tau$ is the obtained image by the τ transformation, $\Delta\tau$ is the gradient tensor and its norm is given by $|\Delta\tau|$, and $|\Delta\tau(u)|$ stands for the global deformation of the image.

14.2.2 MAJOR OPERATORS

SO convolves each image with scaled and oriented wavelets. Each wavelet is obtained by rotating and scaling the mother wavelet ψ, see Figure 14.4:

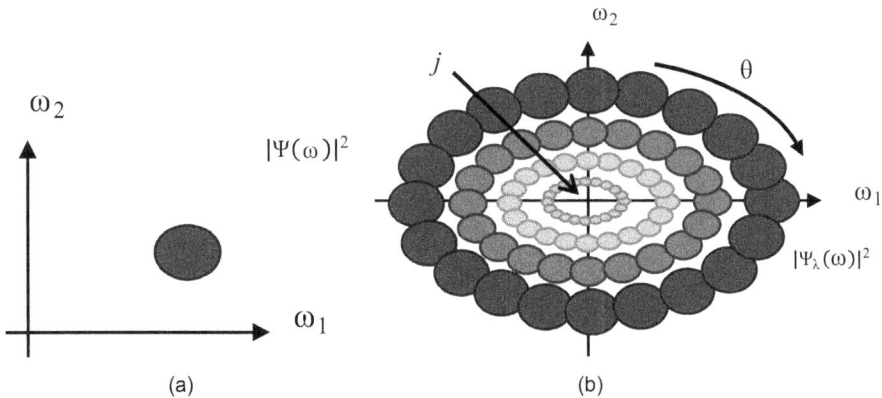

FIGURE 14.4 Frequency representations of the mother wavelet (a) and the daughter wavelets (b) obtained by scaling and rotation of their mother.

$$\psi_\lambda(u) = 2^{-2j}\psi\left(2^{-j}r_\theta^{-1}u\right) \tag{14.4}$$

where $\lambda = 2^jr_\theta$, $u = (u_1, \ u_2)^T$ is the position vector, k the scale and $r_\theta = \begin{pmatrix} \cos(\theta) & -\sin(\theta) \\ \sin(\theta) & \cos(\theta) \end{pmatrix}$ is a rotation matrix related to the angle θ. In our approach, we used the Morlet (or Gabor) wavelet as the mother wavelet for our algorithm [36].

For real images, one can only consider positive rotation angles $\theta = \dfrac{l\pi}{L}$, with $0 \le l \le L$ and L is the number of rotations. To cover the whole frequency space, one should also introduce a low-pass filter, called the father wavelet defined as follows:

$$\Phi_J(u) = 2^{-2J}\Phi\left(2^{-J}u\right) \tag{14.5}$$

where J stands for the maximum scale level of the low-pass filters. Therefore, the wavelet transformation on a discrete image x consists of filtering that image by the following filter bank:

$$\left\{x*\Phi_J, \left(x*\psi_\lambda\right)\right\} \tag{14.6}$$

where $*$ is the convolution product, $\lambda = 2^jr_\theta$, $0 \le j \le J$, $\theta = \dfrac{l\pi}{L}$, and $0 \le l \le L$. The SO coefficients are obtained as follows:

$$S[\lambda]x = |x*\psi_\lambda|*\Phi_J \tag{14.7}$$

The transformation of order m of x is given by:

$$S[p]x = \left\| |x*\psi_{\lambda_1}|*\cdots\psi_{\lambda_m}\right|*\Phi_J\right| \tag{14.8}$$

Here $p = (\lambda_1, \lambda_2,..., \lambda_m)$. To reduce the number of obtained coefficients, a pooling procedure should be applied, then the final SO coefficients are given as follows:

$$S_x = \left(\bar{S}[\varnothing]x, \bar{S}[\lambda_1]x, \bar{S}[\lambda_1,\lambda_2]x, \text{etc}\right) \tag{14.9}$$

where $\bar{S}[p]x$ is the average of the all coefficients obtained with the set p, see Figure 14.5.

According to Ref. [8], the energy represented by the mth decomposition is decreasing inversely to m, they prove as well that the fourth order is equivalent to 99% of the total energy of the image.

The coefficient of SO can be represented in the normalized frequency domain, $(\omega_1,\omega_2) = \dfrac{2\pi}{F_s}(f_1,f_2)$; where F_s is the sampling frequency, see Figure 14.6.

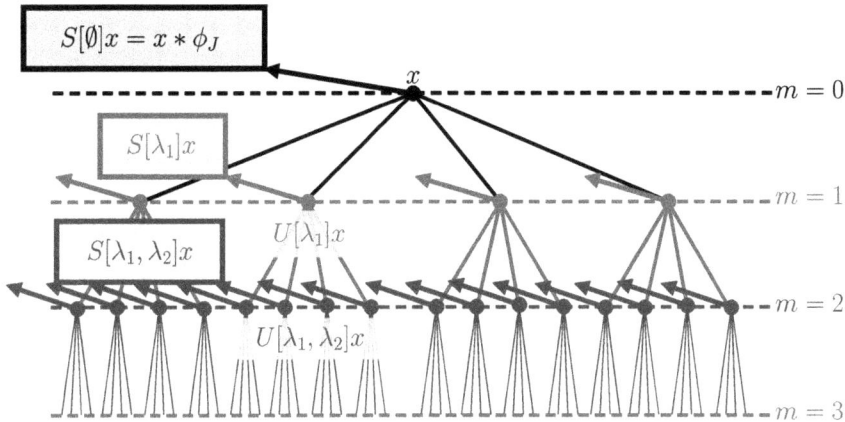

FIGURE 14.5 A third level decomposition of SO using two scale levels and two orientations.

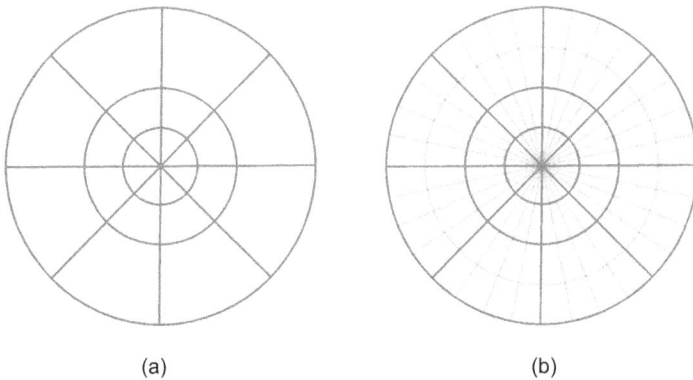

(a)   (b)

FIGURE 14.6 Frequency representation of first order (a) and second order (b) SO decomposition.

14.2.3 ROTATION AND SCALING INVARIANCE

To make the SO invariant to image rotations, the authors of Ref. [53] modify the first-order coefficient of SO as follows:

$$S[j_1] = U[j_1, \theta_1] x \circledast \Phi_J \tag{14.10}$$

where $\circledast$ stands for the convolution product done over a specific orientation. In practice, the authors propose to average the coefficients obtained with different orientations at the same scale level. For the high order, the authors suggest using tridimensional wavelets, for further information please see Ref. [53].

Finally, to obtain a scale invariance, the authors of Ref. [53] propose to average the coefficients obtained by considering different scale levels.

14.3   HIGH ORDER STATISTICS

Stochastic signal processing (one-dimensional or two-dimensional like images) is based on probabilistic models and theory. Most conventional signal processing techniques are based on a simplified description of signals using moments of order one or two (mean, variance, standard deviation, or correlation). However, new applications with their complex data encourage scientists and engineers to develop new models and approaches using statistics of order greater than two. These statistics, developed primarily for astronomy, seismic prospecting and communications, provide extra information not possible with classic signal processing tools and that can be used to resolve certain problems not accessible before, such as blind source separation [13,29,46]. These new approaches and methods have been introduced in many fields of Ref. [37] such as geophysics, speech [23, 40], sonar, radar [38] and more recently image processing [34,52] or biomedical image processing [5,6].

14.3.1   SPATIAL MOMENTS AND CUMULANTS

For an image I, the qth order spatial moment is defined as follows:

$$\mathcal{M}_q[I](i,j) = E\left[I(m,n)I(m+i_1,n+j_1)\cdots I\left(m+i_{q-1},j_{q-1}\right)\right] \qquad (14.11)$$

where E stands for the expectation, $i = (i_1, i_2, \ldots, i_{q-1})$ and $j = (j_1, j_2, \ldots, j_{q-1})$ are the delay vectors, m and n are the pixel indexes of I. In our study, we modified the images to be zero mean and we used the high-order statistic estimators proposed in Refs. [39, 40]. Based on the second-order moment, one can define the autocorrelation as follows:

$$\hat{C}_2[I](i_1,j_1) = \hat{\mathcal{M}}_2[I](i_1,j_1) \qquad (14.12)$$

In a similar way, the bi-correlation (resp. the triple-correlation) can be defined using the third (resp. the fourth) order moment: $\hat{C}_3[I]((i_1,i_2),(j_1,j_2)) = \hat{\mathcal{M}}_3[I]((i_1,i_2),(j_1,j_2))$ and $\hat{C}_4[I](i_1,i_2) = \hat{\mathcal{M}}_4[I]((i_1,i_2),(j_1,j_2))$. We should notice that the bi-correlation and the triple-correlation are tensors and not matrices and therefore they cannot be presented in 3D images. In our project, we considered projected versions of these values, see Figure 14.7.

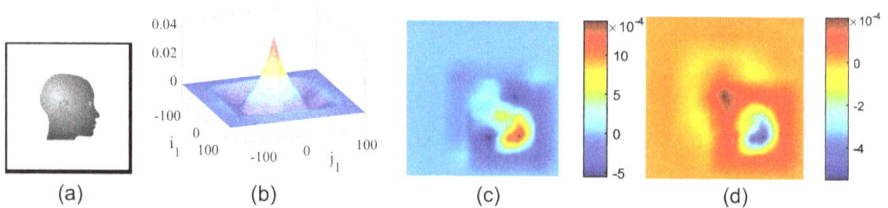

(a)   (b)   (c)   (d)

FIGURE 14.7   A $M \times N$ image I and its projected versions of high-order correlations. (a) Original image, (b) autocorrelation, (c) $\hat{C}_3[I]((M/2,i),(N/2,j))$, and (d) $\hat{C}_4[I]((0,M/2,i),(0,N/2,j))$

14.3.2 HIGH ORDER STATISTICS FEATURES

The size of the tensor is strictly related to the order of the order of the multiple-correlation (*i.e.* a correlation of a general order higher than two). Above the third order, the required memory space can in particular become important and/or it can require huge computation times: for an image of 64×64 pixels, its bicorrelation will include more than 260 million values. Therefore, it becomes a complicated task, if not impossible, to calculate all the values. To tackle this problem, we can downsample the images or reduce their size. However, even by reducing the image to 16×16 pixels, the bicorrelation still contains almost a million points. Another solution has been proposed in the literature, Cardoso notably proposes a technique to reduce the fourth-order tensor of statistical cumulants by a matrix generated from its eigen matrices (Joint Approximation Diagonalization of Eigenmatrices [9,10]). The authors of [18] also present an algorithm for simultaneously diagonalizing tensor slices of order three (Simultaneous Thrid-Order Tensor Diagonalization). Independently, Comon proposes to diagonalize the tensor of order three by maximizing the tensor trace [12,13]. These approaches perform a tensor contraction (*e.g.* a change from a tensor to a matrix). These approaches can be suitable for easily using multicorrelations in our project.

To reduce the complexity of our algorithm, we decided to opt for a simpler approach which consists in estimating the multiple-correlations only according to the last two indexes (i_{q-1}, j_{q-1}) of the tensor of order q. By doing so, an input image of size $M \times N$ will generate images of size $(2M-1) \times (2N-1)$: $\hat{C}_2[I](i,j), \hat{C}_3[I](0,0,i,j), \hat{C}_4[I](0,0,i,j)$ *etc.* For each image, we would therefore almost multiply by four the number of coefficients. To avoid any memory and processing problem, we should further reduce the total number of coefficients. In the space formed by the last two indexes, we selected some privileged axes (based on a heuristical study conducted on our database). The choice of these axes fell on those which seem to be the most basic and intuitive: three horizontal axes, three vertical axes and two diagonals. Ten privileged axes have been finally adopted, see Figure 14.8. To explain our strategy, let us consider the following

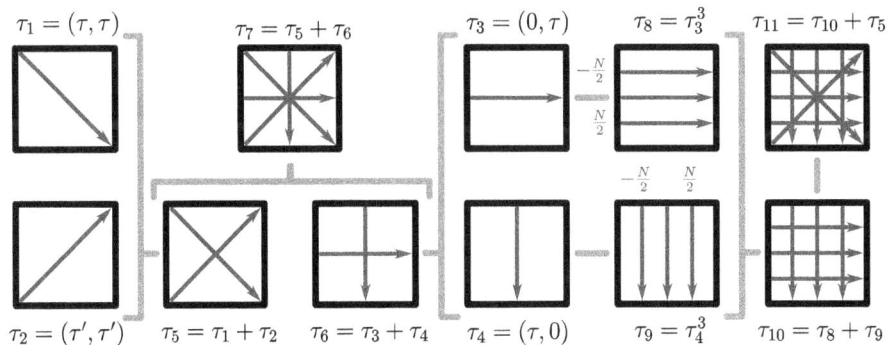

FIGURE 14.8 Privileged axes used to estimate the tensor of high-order statistics. τ stands for the index delay and τ_j means that the multi-correlation has been applied on a 90° rotated version of the original image.

example where we evaluate the bi correlation according to the third and the sixth axes, *i.e.* τ_3 and τ_6:

$$\hat{C}_3[I](\tau_3) = \hat{C}_3[I](0,0,0,\tau) = \frac{1}{N^2} \sum_{m=0}^{N-1} \sum_{n=0}^{N-1} I(m,n)I(m,n)I(m,n+\tau) \quad (14.13)$$

$$\hat{C}_3[I](\tau_6) = \left(\hat{C}_3[I](\tau_3), \hat{C}_3[I](\tau_4)\right)^T = \left(\hat{C}_3[I](0,0,0,\tau), \hat{C}_3[I](0,0,\tau,0)\right)^T \quad (14.14)$$

where $|\tau| \leq N-1$. Using these privileged axes, we can obtain up to 11 features to describe an ultra-sound image.

14.4 SCATTERING OPERATORS AND HOS FEATURES

In previous sections, we presented two approaches to extract features from biomedical images (*i.e.* Scattering Operators, SO, and High Order Statistics, HOS). It is clear that the two approaches don't seek similar features. Indeed, while SO extract features based on the details of the images, HOS looks for features based on the statistical properties of the images. To explore and identify better different objects represented in our ultrasound images, we opted to concatenate the two kinds of the obtained features. Due to their different nature, several pre-processing steps are required to properly combine these features.

14.4.1 A CONCATENATION OF SCATTERING OPERATORS AND HOS FEATURES

The SO can be considered as a multi-layer network of convolutions (see Figure 14.5) whose architecture depends on a certain number of parameters: scales, orientations, and decomposition orders. This last parameter corresponds to the resolution depth of the images and is therefore linked to the number of layers of the network. The first order, which represents low-frequency information, consists of a single coefficient representing the coefficient of approximation. The second layer of the network is nothing but the result of transforms in classical wavelets followed by a modulus and a Gaussian low-pass filter (and an aggregation step if necessary). The number of second-order coefficients is the product of the number of scales and the number of guidance. The peculiarity of SO compared to a classical wavelet transform is based on orders of decomposition greater than two: a new wavelet transform (associated with the modulus and the low pass) is applied to each of the moduli of the wavelet transforms of the previous layer of the network. This process, which can be repeated as needed, achieves a higher level of details in the examined image. In practice (and during simulations), we are limited to order two, which our simulations show that it is often sufficient to obtain good performance. Figure 14.9 illustrates the composition of the features obtained with SO. In addition, it should be noted that the energy of the coefficients decreases exponentially with the order [8] (hence the decreasing intensity of the blue color in Figure 14.9).

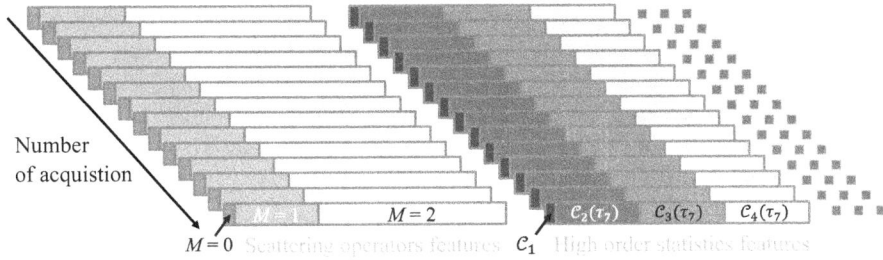

FIGURE 14.9 A combination of High order statistics (HOS) and Scattering Operators (SO) features. M stands for the order of the scattering operators and $C_i(\tau)$ is the ith order cumulant.

Concerning the features based on the HOS of the images, we therefore suggest the estimation of the multiple- correlations of the small size images in several orders and different combinations of privileged axes. The obtained features with the HOS are also shown in Figure 14.9. During the simulations, we will try in particular to determine the most discriminant axes or orders. By their definitions, we can see that these two descriptor types (SO and HOS) are of a different nature and that, in order to combine them, they must certainly be transformed before proceeding to the classification step.

14.4.2 FEATURE STANDARDIZATION

In data analysis, we may often have to associate various heterogeneous information. To standardize our data and homogenize different collected numerical values of our features, we used two well-known approaches:

1. **Zero-mean and unit variance**: in this step, we considered various collected feature values as the outcomes of a random variable x, and we normalize that variable as follows:

$$x = \frac{x - \bar{x}}{\sigma_x} \tag{14.15}$$

   where $\bar{x}$ stands for the average of x and σ_x is the standard deviation.

2. **Min-Max**: using the extremums (the minimum $\min_x$ and maximum $\max_x$ values) of x, one can push x to be in [0, 1]:

$$x = \frac{x - \min_x}{\max_x - \min_x} \tag{14.16}$$

14.5 IMAGE PROCESSING AND PARAMETER TUNING

To test and optimize our algorithms and classifiers, we use a "KTH-TIPS" database (Kungliga Tekniska Högskolan; *i.e.* the Royal Institute of Technology in Stockholm) which is a database of Textured images under varying Illumination, Pose and Scale (TIPS) [21]. We should highlight that the KTH-TIPS, which contains ten different kinds of objects (or clusters), is inspired from a wider database "CUReT" (Columbia-Utrecht Reflectance and Texture) [16]. In our project, the biomedical images should be classified into less than ten clusters; therefore, we used the KTH-TIPS database instead of the CUReT one. Our salivary glands can be classified according to four categories: Normal, moderate infections, mild infection, and totally infected.

14.5.1 IMAGE PREPROCESSING

Ultrasound images are mainly characterized by their low contrast and the presence of noise (speckle). The preprocessing methods aim to improve the contrast of images, in particular in the regions of interest (lesions) and reduce the noise without losing useful information for the characterization of the targeted tissue.

We can distinguish four main types of approaches to reduce the speckle:

- by Histogram Equalization (HE),
- by ultrasound image compounding
- by filtering [32],
- by wavelets, by averaging of ultrasounds acquired at different frequencies and acquisition angles.

14.5.1.1 Histogram Equalization (HE)

For a given ultrasound mode (fundamental, Tissue Harmonic Imaging (THI, *i.e.* in this mode the ultrasound image of human tissue is composed from the harmonics of reflected ultrasound signals.) or Differential-THI (DTHI, *i.e.* the ultra-sound image is obtained based on the harmonic difference of two signals with different frequencies.)), the receiver gain becomes a major control parameter for image acquisition of different patients. A high gain amplifies the received signal and allows the visualization of structures with low echogenicity. However, high echogenic structures can be saturated. To standardize as well as possible the obtained ultrasound images with different gains, we preprocessed the images in our database by equalizing the histogram of the considered images. This main idea is to distribute the gray levels of the image uniformly over the entire possible range (from 0 to 255), such as illustrated in Figure 14.10. We can notice that the histogram equalization improved the overall contrast by brightening the image, while most echogenic areas have been saturated (walls of the vein).

To improve local contrast and avoid saturation phenomena, the authors of Ref. [57] performed a local histogram equalization. Instead of the global histogram of the whole image, their method consists of equalizing the histogram on a small sliding window over the targeted image. A threshold is also set to avoid a high amplification of the noise pixels (especially in homogeneous regions) and therefore distort the image. This method is called the Contrast Limited Adaptive Histogram Equalization (CLAHE). According to CLAHE, the distribution can be uniform (as in the case of a classical histogram equalization) but it can also follow Rayleigh's law or the exponential law. Figure 14.10 shows that CLAHE performed better than the classic method. On the other hand, it is more complicated to determine the impact of the distribution choice.

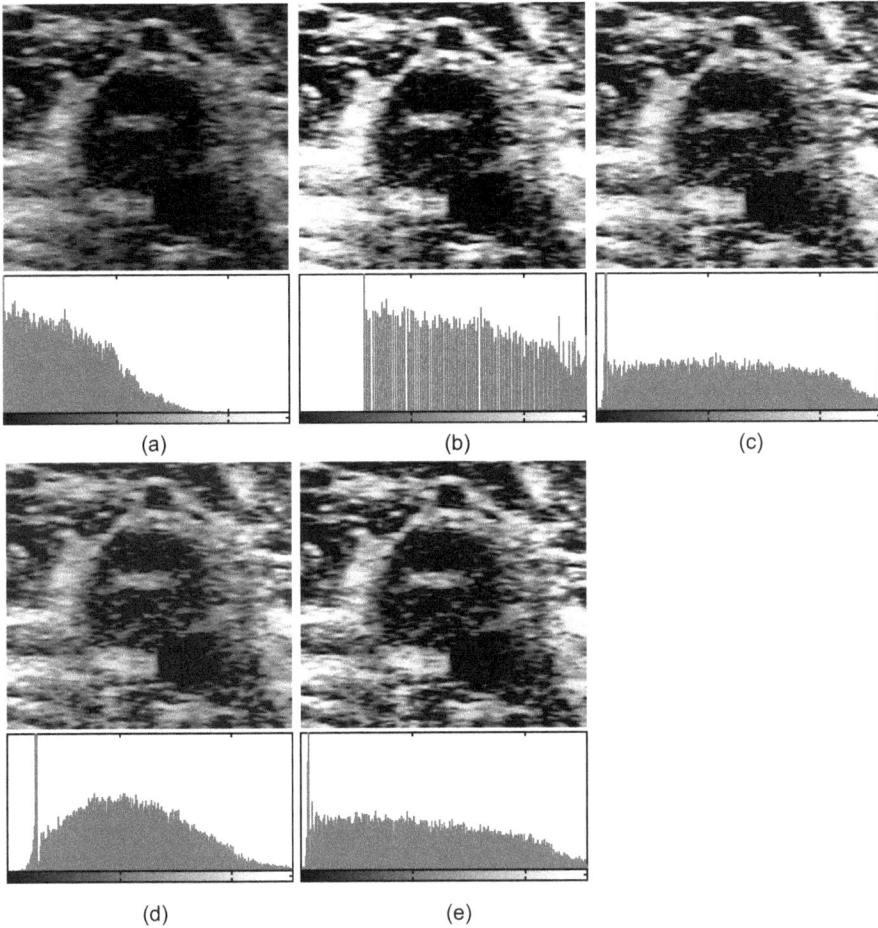

FIGURE 14.10 Histogram equalization of ultrasound images using: a classical equalizer and the CLAHE method with three different distributions (uniform, Rayleigh and exponential). (a) Original, (b) classic, (c) uniform, (d) Rayleigh, and (e) exponential.

14.5.1.2 Image Compounding

In the case of the Canon Aplio system, the latter technique named "image compounding" is implemented. It is a method that obtains sonographic information from several different angles of insonation and combines them to produce a single image. The benefits of this technique include reduced speckle, reduction in artifacts (clutter, shadowing, and echo drop-out), increased contrast resolution, and increased visibility of lesion margins (hence increased lesion conspicuity).

14.5.1.3 Image Filtering

With regard to filtering, we can distinguish two filter families:

- **Linear filters**: averaging, adaptive averaging (*i.e.* averaging in homogeneous regions)
- **Nonlinear filters**: second-order filtering (*e.g.* adaptive median filtering), Maximum A Posteriori (*e.g.* MAP Pearlman Gauss), nonlinear scattering (*e.g.* anisotropy), geometric filters, *etc.*

We did not apply any filtering methods before the feature extraction stage as the acquisition system already applied preprocessing techniques and we could not have access to the raw data.

14.5.1.4 Wavelet Transform

As for wavelets, the noise reduction procedure is generally done in three steps: calculation of the wavelet transform, modification of the wavelet coefficients to remove the noise and an inverse transform to obtain the denoised image. Techniques for performing the second stage can be grouped into three groups:

- **Wavelet shrinkage**: thresholding of the coefficients (fixed threshold or adapted to the image) [11];
- **Bayesian approach**: use of statistics of wavelet coefficients [1];
- **Filtering or diffusion approach**: use of the filters described previously in the wavelet domain [47]. As for the filtering, no wavelet transforms were applied to denoise the images.

14.5.2 TUNING OF SO PARAMETERS

Using various images of the KTH-TIPS, we tuned the various parameters of the SO, such as:

- The size of the small images N : according to our experimental study, we found that N can be selected as: 32×32, 64×64 and 128×128 pixels;
- Concerning the scale number J, we observed good results with several values of $2 \leq J \leq 6$;
- The orientation number L was selected to be in $\{1, 2, 4, 6, 8\}$
- The decomposition order M was limited to the first and the second order.

To tune the different parameters, we used the v-fold cross-validation [54]. This method consists of dividing each class into v subsets and then selecting one subset of each class to form the test base (also called the confirmation base). The other $v - 1$ subsets constitute the learning base. The images of the test base are then classified from the models generated with the learning base and the classification error is then calculated. The operation is then repeated $v - 1$ times until each subset is used as a test set. The average of the v errors is finally calculated to estimate the total classification error. These v operations must be reiterated for each set of parameters to determine the set that minimizes the classification error. In practice, we selected around 81 images from each of the ten classes, the database is partitioned into nine subsets ($v = 9$).

The cross-validation procedure, related to the v-fold cross-correlation method, requires the classification of each class with the $v - 1$ subsets of the learning base. To do this, we use a simple linear classifier: Selection of Affine Models (SAM) which uses the Principal Component Analysis (PCA) to reduce the dimension Data [30], see Figure 14.11. The objective of SAM is to obtain the most relevant features obtained from the scattering coefficients for each of the ten classes of the learning base. At first, PCA is used to analyze the correlation between the features to obtain a small number d of new variables not correlated with high variance (*i.e.* with high informative power). To perform SAM, we split the database into two subsets: one for learning and the other for testing. When learning, PCA (or another algorithm such as Linear discriminant analysis (LDA), normal discriminant analysis (NDA) or kernel PCA) is used to select an affine model for the different classes: each class is modeled by a center, the average of its elements, and by a certain number of the new uncorrelated variables with high variance. The choice of the number d of these new variables must be defined. Once each class is modeled, the test base images are labeled with the class of the most similar model. From a practical point of view, the parameter d (dimension of the space affine of a class), which must be optimized, has been selected in our simulations as one of the following values: 4, 8, 12, 14, 16, 18, 20, 24, 30, 36, 42, 50, 64, 96 or 128. Depending on the parameters of the SO, the values of d greater than the number of features are not taken into account (the dimension of the projected space cannot be greater than that of the original space).

Once all classes have been labeled, we can move on to the feature classification step of the test base. Each vector is projected into reduced spaces for each of the classes (*i.e.* the affine models) and is associated with the closest class (*i.e.* the one that has the smallest distance between the projected vector and the center). From a

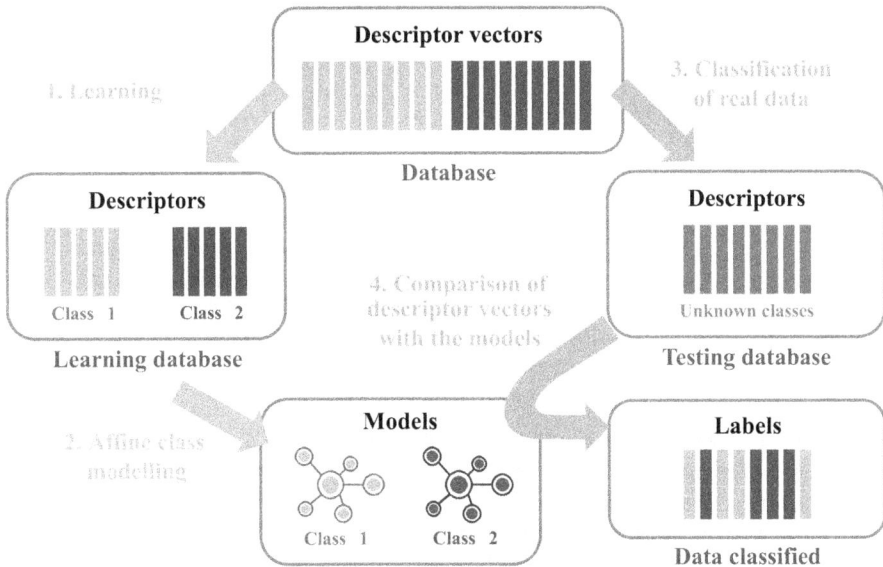

FIGURE 14.11   The concept of the Selection of Affine Models.

mathematical point [8], if we denote S_I the feature vector of the I data of the test base, then I is associated with the class k according to the following criterion:

$$k(I) = \arg\min_{k \le K} d_E\left(S_I - P_{A_{d,k}}(S_I)\right) \tag{14.17}$$

where k is the class number, d_E is the chosen distance (which can be the Euclidean distance), $P_{A_{d,k}}(S_I)$ is the projection of the S_I into the space of the kth class with d dimension.

According to our simulations and for a small learning base, the addition of rotational invariance, logarithm, scale invariance and then multiscale learning can improve gradually the classification rate. In addition, we can note that the scattering operators are efficient (with 40 images per class in the learning base) compared to the other three techniques described and applied in Ref. [53]:

- **Log Gaussian Cox (COX) [43]:** It characterizes the texture by key points which are determined from a Gaussian statistical model in different circular regions of the images;
- **Basic Image Features (BIF) [15]:** The BIF technique represents the textures using descriptors estimated from the responses of a bank of Gaussian derivative filters;
- **Sorted Random Projection (SRP) [31]:** SRP describes the texture by applying oriented filter banks, and then it uses an algorithm to sort and project the responses of the different filters.

14.5.3 FEATURE EXTRACTION, REDUCTION AND CLUSTERING

Once the SO parameters are tuned as described in the previous subsections, one should select the best classification algorithm or strategy. To reach our goal, several clustering approaches have been implemented, optimized and tested. However, our best results have been obtained using a combination of various approaches, such as k-means, DBSCAN, spectral classification and manifold untanling, t-SNE and some reduction methods.

14.5.3.1 k-means

This popular clustering algorithm, k-means, starts by dividing randomly the points in k subsets (clusters) [24]. Then it evaluates the centroids (*i.e.* the centers of the clusters) of each cluster. As long as the centroids are not stabilized (*i.e.* do not converge towards a local minimum) (Figure 14.12):

- Calculate the distance between each point and each centroid;
- Assign each point to the nearest cluster (*i.e.* whose distance from its center is the smallest);
- Update the centroid of each cluster.

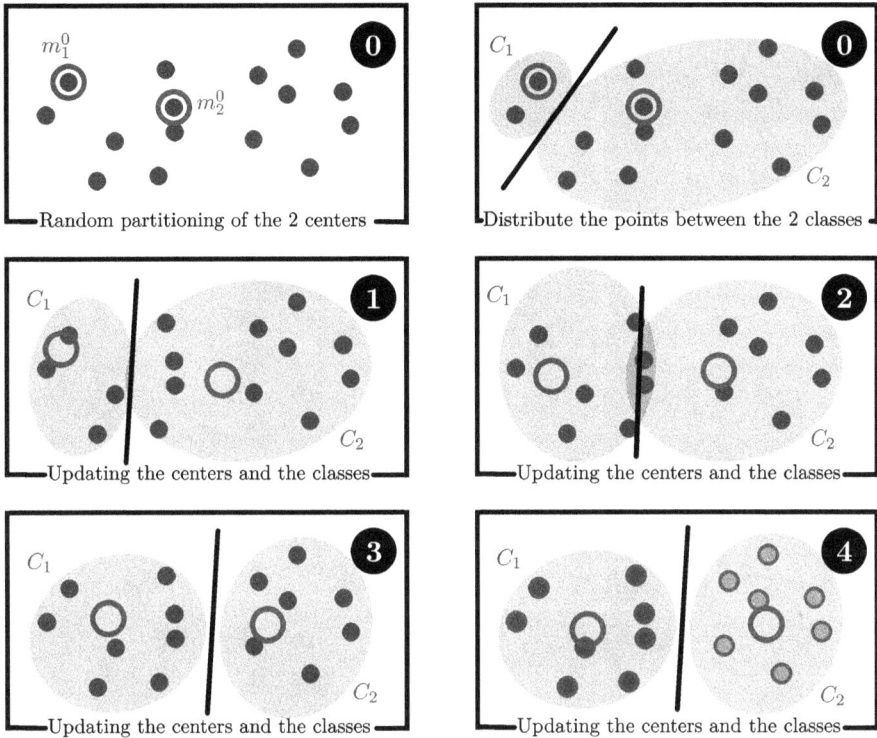

FIGURE 14.12 Convergence of k-means.

14.5.3.2 Density-Based Spatial Clustering of Applications with Noise (DBSCAN)

This algorithm [20] identifies, in the representation space, the high-density areas surrounded by low-density areas, which will form the clusters. Its principle is illustrated in Figure 14.13. Considering a set of given points in any space, the points closest to each other (*i.e.* dense) will be grouped together. In practice, the algorithm begins by randomly drawing a point A and searches for all points that are in its sufficiently close vicinity within a predefined maximum radial distance r. Every cluster should contain a minimum number N_m of points. Depending on r and N_m, three different scenarios can be obtained:

- Point A may have no neighbor and is therefore considered to be an outlier;
- A core point A should have enough points in its neighborhood;
- A reachable Point A does not have enough neighbors to be a core point, so it is assigned to the cluster of its neighbors (if its neighbors are outliers, it will also be classified as an outlier).

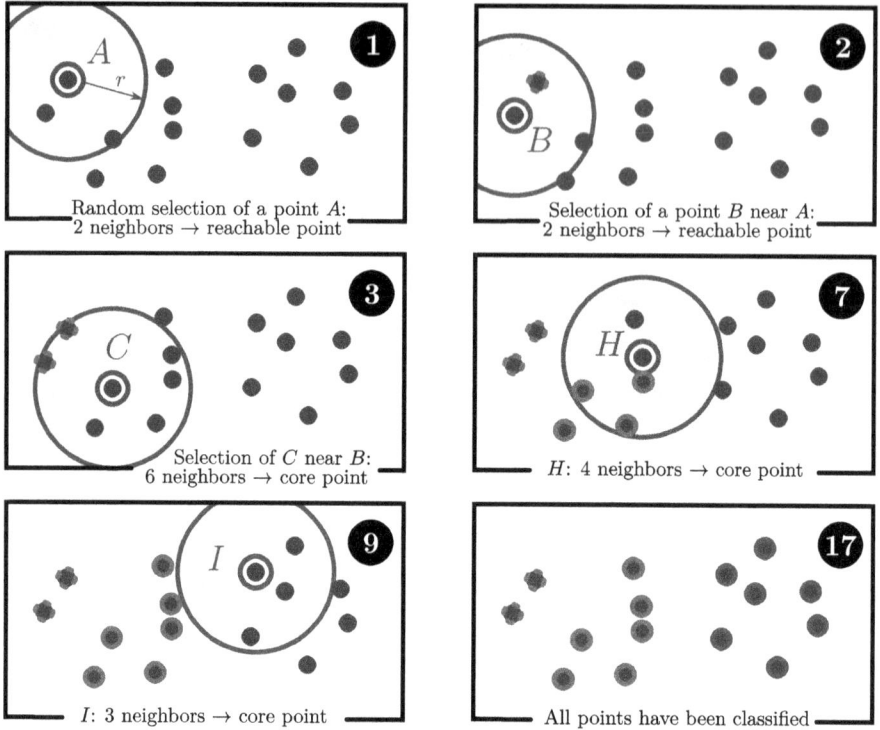

FIGURE 14.13   The concept of DBSCAN.

14.5.3.3   Spectral Classification

The main idea of Spectral Classification (CS) is to represent all the data as a weighted graph and then cut this graph at the weakest links. A graph is defined by the trio (V, E, W): V corresponds to all the nodes of the graph (each node represents a data), E corresponds to the set of pairs (v_i, v_j) standing for the link between two nodes and W corresponds to the symmetric neighborhood matrix. The coefficient W_{ij} takes as value the weight linking the node v_i to the node v_j, see Figure 14.14. To process our graph, we used the Jordan and Weiss algorithm (NJW) [42].

14.5.3.4   Manifold Untangling

The objective of this step is to describe the data preprocessed and segmented by features. The information extracted must be as relevant as possible to the type of desired classification and as small as possible to save the machine resources [19]. In practice, the raw data belong to very twisted surfaces. Feature extraction plays a key role since it must make it possible to "untwist" these surfaces (manifold untangling) so that the classes are more easily separable (see Figure 14.15). In order to achieve this goal, various transformations can be applied to the data or on the extracted features. Most of the cases, these transformations are irreversible and can be obtained from simple treatments (such as downsampling an image) or more complex, such as: Statistical

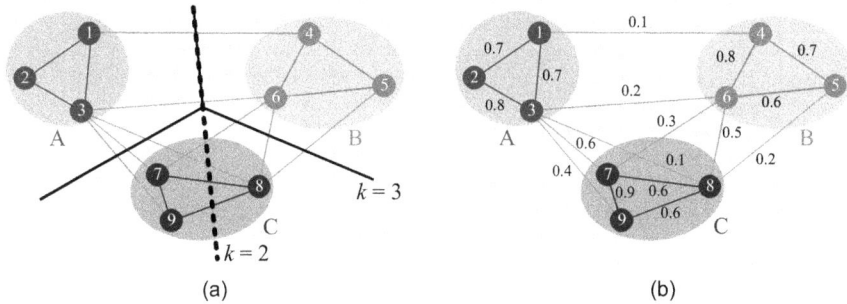

(a)   (b)

FIGURE 14.14   (a) First figure shows three circles (Blue-A, Green-B, Orange-C) representing a potential three classes of data. Each circle contains three dots (i.e elements or members).
Some dots are linked to others with lines having a weight coefficient. So we can see how the
desired class number can affect the distribution of the members in the obtained classes (two,
K=2, or three, K=3). In other words, the figure shows a dot-line (representing k=2) separating
points in two classes and two other segments (k=3) to separate the 3-circles (classes). (b) the
second figure shows the coefficients of the neighborhood matrix W (W_{ij}). We can clearly see
that the members of each class are related to strong relationships among themselves; however,
some members may still have potential links to the members of other classes.

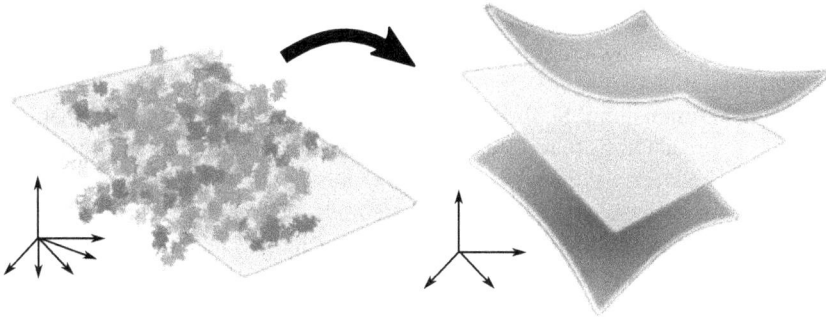

FIGURE 14.15   The concept of manifold untangling.

Feature Matrix (SFM) [56], Block Difference of Inverse Probabilities (BDIP) [25],
Power-Law Shot Noise model (PLSN) [33], *etc.*

14.5.3.5   t-Distributed Stochastic Neighbor Embedding (t-SNE)

In practice, the dimension of the representation space of raw data is very large.
Indeed, the number of extracted features, although small in size, is often greater than
three which makes the visualization of the data surface complicated: To reduce the
dimension of the representation space, one can use a popular method, t-distributed
Stochastic Neighbor Embedding (t-SNE), which has been developed in [35]. t-SNE
defines a low-dimensional space in which all distances are preserved, especially for
near points in the original space, see Figure 14.16.

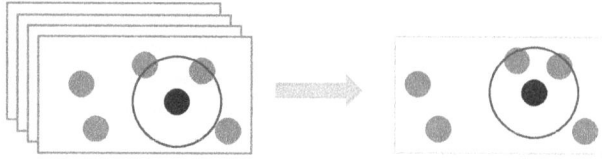

FIGURE 14.16   The concept of t-SNE.

14.5.3.6   Dimension Reduction Methods

Other dimension reduction techniques [22] can be found in the literature, such as:

- **Kernel PCA [51]:** using a kernel, it is possible to reduce the dimension by performing complex nonlinear projections. Indeed by projecting data into a space of a larger dimension and performing linear classification tasks therein, linear decision boundaries obtained in this space correspond to complex non-linear boundaries in the original space.
- **Linear discriminant analysis**: Similar to PCA, this algorithm determines the most discriminating axes to define a hyperplane on which one can project the data.
- **Locally Linear Embedding (LLE) [48]:** this technique consists in measuring the linear relations between each of the data and its closest neighbors, then finding a representation of smaller dimension which preserves its local relations.
- **Multi Dimensional Scaling (MDS) [28,55]:** Similar to the t-SNE method, this approach aims to reduce the dimension while preserving better the distances between the data.
- **Isomap**: unlike the t-SNE and MDS methods, this one represents the data in a graph to reduce the dimension while ensuring that the distances between the nodes are maintained.
- **Compression in the wavelet domain**: with a process similar to Discrete Cosine Transform, this method allows data to be compressed and it is particularly used in the compressing of JPEG 2000 images.

14.6   SIMULATIONS

Once, we validated the implementation and configured the characterization approaches (multiresolution and statistics) and classification on a reference base composed of synthetic images. The objective of our next simulations is then to classify the glands into two categories, normal or pathological. Ultrasound images of the salivary glands visually allow a medical expert to identify the Gougerot-Sjörden's syndrome. Indeed, the behavior of shear waves in the glands, which are tissue, is more consistent with the theory used to generate elastography. The second subsection presents a brief state of the art on the diagnosis of Gougerot-Sjörden's syndrome by ultrasound and elastography. Then the third subsection describes the experimental protocol. The last subsection analyzes the simulation results on clinical data.

14.6.1   SYNTHETIC IMAGES

It was previously mentioned that our simulations were first carried out on the KTH-TIPS database to better select some parameters for the scattering operators (SO). Regarding the setting, the cross-validation procedure made it possible to select the best sets of parameters (two orders of decomposition, at least four orientations and four scales) and showed that a trade-off should be made on the size of the selected small images. Different simulations have also indicated that adding scale and/or rotation invariance can help to achieve better results. We also observed that the logarithm function reinforces the information of high frequencies (*i.e.* textures) contained in higher orders (order two in particular) and considerably increases the SO performances. This can be illustrated with the t-distributed Stochastic Neighbor Embedding (t-SNE) method (see Figure 14.17): the classes become more easily dissociable with the SO features (even more using the logarithm) than with the images directly.

14.6.2   EXPERIMENTAL PROTOCOL

Our experimental protocol is summarized in Figure 14.18. Our database consists of three image types: ultrasounds with or without histogram equalization, and elastography. To limit the influence of the acquisition parameters and consider the neighboring structures of the gland, the small images of each gland can be standardized with respect to all the images extracted from the same acquisition (glands, mixed or reference). This standardization can also take place on the descriptors. A huge number of simulations have been conducted and various good results have been obtained. Therefore, a small amount of obtained results are presented in this section.

14.6.3   PAROTID AND SUBMANDIBULAR GLANDS

By mixing parotid and submandibular glands, good results were obtained by SAM with 50% of the training images then we reduced gradually the proportion of the learning base until a classification failing was reached. In unsupervised, we analyzed

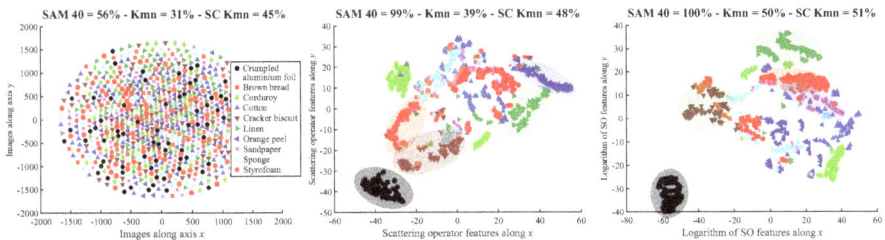

FIGURE 14.17   Manifold untangling of the scattering operators using a t-distributed Stochastic Neighbor Embedding (t-SNE) method: the multiresolution features are projected into a two dimensions space by respecting the original distances (each point represents an image of the KTH-TIPS database). The image subtitles highlight the classification rate of SAM (with 50% of images were considered in the learning base), par k-means (Kmn) and the spectral classification (SC Kmn).

Types of images	Image standadization	Features	Feature standardization	Feature fusion
Ultrasonographies (N = 64×64)	**Without**	**SO:** log: without/with $J = \log_2(N)-1$ $M = 2$ $L = 8$	**Without** (per data)	**Without**
OR	OR	AND/OR	OR	OR
Ultrasonographies with histogram equalization	**With (N):** for each image, each extracted small images is standardized according the full set of extracted images	**SOS:** $C_2(\tau_7)$ $C_3(\tau_7)$ $C_4(\tau_7)$ $C_{1,2}(\tau_7)$ $C_{2,4}(\tau_7)$ $C_{1,4}(\tau_7)$	**With (N):** per data and per order	**With (N):** Mean of all the features of a same gland
OR			AND Standardization according the full dataset	
Elastographies (N = 32×32)			OR Standardization per order according the full dataset (**npo**)	

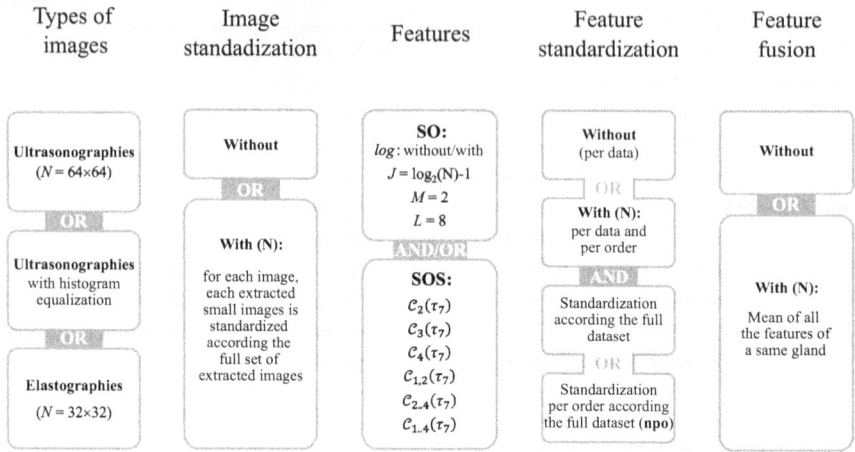

FIGURE 14.18 Experimental protocol.

the performance of two clustering algorithms (k-means and DBSCAN) and Spectral Classification (CS)).

Table 14.1 shows the classification results of SAM with 50% of glands dedicated to learning. Overall, we can observe that the classification rates obtained from ultrasound scans (around 70% on average) are superior to the ones obtained from elastographies (around 60%). In terms of ultrasounds, the simulations clearly show the

TABLE 14.1

Good Classification Rate by SAM with 50% of the Glands Devoted for Learning Task: the Classification Percentages Are Calculated for Each Combination of Features (Columns) and Each Set of Simulation Parameters (Lines)

T	Img	Features	No.	Images	SO	SO$_{log}$	C_2	C_3	C_4	$C_{1,2}$	$C_{2,4}$	$C_{1,4}$	SO+ C_2	SO+ $C_{1,2}$	SO+ $C_{2,4}$	SO+ $C_{1,4}$	SO$_{log}$+ C_2	SO$_{log}$+ $C_{1,2}$	SO$_{log}$+ $C_{2,4}$	SO$_{log}$+ $C_{1,4}$
Ultrasonographie			S01	46	73	72	60	58	56	62	75	75	74	74	73	78	81	82	78	78
		F	S02	58	71	72	63	63	64	63	71	71	73	73	74	74	75	76	77	77
	N		S03	46	71	69	58	57	57	59	73	74	72	73	75	78	77	77	78	76
	N	F	S04	58	70	68	63	64	63	64	69	68	71	70	67	67	71	72	71	71
	N		S05	47	72	72	62	58	56	63	74	75	73	73	76	78	74	74	77	77
	N	F	S06	58	69	70	63	63	64	64	70	71	72	72	72	72	74	74	75	75
	N	N	S07	47	71	66	58	57	57	59	73	74	72	73	75	79	77	76	77	77
	N	N F	S08	58	70	68	64	64	63	64	68	68	70	70	67	67	71	72	71	71
Ultrason. HE			S09	44	74	71	65	61	61	64	69	70	70	70	74	73	72	77	74	77
		F	S10	58	69	69	64	63	67	65	70	70	73	73	73	72	75	76	75	74
	N		S11	44	71	69	66	58	63	65	69	71	69	69	74	74	75	73	73	74
	N	F	S12	58	70	69	67	63	68	67	69	68	71	70	68	68	75	75	70	70
	N		S13	44	73	71	65	61	61	65	69	70	69	70	73	73	77	77	77	78
	N	F	S14	51	70	68	65	63	67	65	69	70	73	73	73	72	76	75	75	74
	N	N	S15	44	71	69	66	58	63	65	69	71	69	69	74	74	75	75	73	73
	N	N F	S16	58	70	69	66	63	68	67	69	69	71	71	68	68	75	75	70	70
Elastographie			S17	63	61	63	55	59	56	56	62	63	63	63	62	62	67	68	64	64
		F	S18	60	62	63	59	58	59	59	61	61	63	63	62	62	65	65	63	64
	N		S19	62	61	62	56	55	59	57	59	59	62	62	60	60	62	62	62	62
	N	F	S20	60	62	63	58	58	60	60	57	57	63	62	59	59	62	60	58	58
	N		S21	55	62	63	54	59	56	56	60	60	66	66	60	61	65	66	62	62
	N	F	S22	59	63	63	58	58	59	59	54	59	62	62	60	60	61	61	60	60
	N	N	S23	59	61	62	56	59	57	57	59	59	62	61	60	60	62	61	60	60
	N	N F	S24	59	62	63	58	58	60	60	57	57	63	62	59	59	62	61	58	58

importance of implementing feature extraction techniques before the classification steps (red boxes around 40% classification). For elastographies, our two description approaches seem as inefficient as vectorized images. By considering a decision rule (the closest image to one of two models) and by performing a feature standardization by order, we obtained better results (we reached in some cases the 80% good classification rate) than the ones considering the standardization by type (SO and HOS). We notice that the best performances are obtained by combining the multiresolution feature (SOlog) and statistics (C_2 or $C_{1,2}$) with a percentage up to 80% good classification. SO alone gives rates around 70% (in ultrasound).

Other simulations showed that by decreasing the size of the learning images to 25% (resp. 12.5%) on ultrasounds (without equalization) the classification rates will be affected by around 5% (resp. 10%). However, we noticed that the classification rates of rows S02 and S06 display almost the same values regardless of the learning proportion (up to 77% good ranking). This may indicate that the feature fusion by acquisition allows to increase the classification performance for small size database, by offering better stability of the models, but it can lead to an over-learning phenomenon.

We also conduct several other simulations using DBSCAN and k-means. Based on our simulations, we notice that the DBSCAN classification rates are constant around 60% good classification. This means that the configuration of this algorithm, which worked for the KTH-TIPS database, is not suitable for the base of the gland and therefore further investigation is necessary, or to develop this algorithm of partitioning by density towards a more efficient algorithm like OPTICS (*i.e.* Ordering Points to Identify the Clustering Structure) [3].

With the k-means algorithm, we obtain more satisfactory results than the ones obtained by the SAM classification. We can notice that, as in supervised, SO are more efficient in describing the ultrasound images without preprocessing while multicorrelations are more discriminating after histogram equalization. Orders one and two of statistical features seem to be the most significant since we reached 78% good classification. Unlike the supervised mode, combining the two types of descriptors did not necessarily improve the results. However, it would be interesting to combine the descriptors of the SO without equalization with the ones of the HOS after equalization. As for the supervised mode, we notice that the simulations without standardization or fusion give globally better results even if, in certain special cases (SO, C_2 and $C_{1,2}$), the fusion improves the performance. Finally, we can note a clear increase in classification rates on elastographies since we reach 71% without standardization or fusion.

Other simulations showed us that the performance of the spectral classification (with k-means) is similar to those of the k-means. The log-free SO and the second-order multicorrelations (C_2 or $C_{1,2}$) give the best results with 76% (S02) and 78% (S14) respectively. The combination of the two approaches does not improve the classification performance, but, as with k-means, the association of SO features without histogram equalization with those of HOS with equalization would surely increase the classification results.

By projecting the features in a three-dimensional plan (method t-SNE) and by analyzing their confusion matrices, sensitivities and specificities, good results have been obtained. Indeed, the supervised classification by SAM achieved 81% good classification by combining the SOlog and C_2 features without acquisition standardization or fusion, with a global standardization by order and on non-equalized ultrasound images (S01). Figure 14.19 shows the projection in a three-dimensional space of the mixed features. We observe that the pathological glands (black circles) are located on the left part of the cloud and the normal glands are on the right part. The border is not obvious, however, which may explain the lower rate of k-means (68.5%) compared to SAM (80.6%). Table 14.2 shows that the classification by SAM allows to obtain a sensitivity of 81.2% and a specificity of 78.9%, which is close to the values mentioned in the introduction obtained with experts: sensitivity average of 77.9% or 71.3%, and specificity of 86.7% or 91.9% (depending on the article considered).

In unsupervised classification, multiresolution and statistical features proved to be more discriminants by being used separately. By k-means and by CS, we approach 78% good classification. Comparing the 3D representation of the vectorized images (Figure 14.19), we find that the pathological glands and normal, as the classification rates show, become more easily separable, see the confusion matrices of Table 14.2. However, we can see that, in terms of sensitivity and specificity, unsupervised classification seems to be less efficient than SAM.

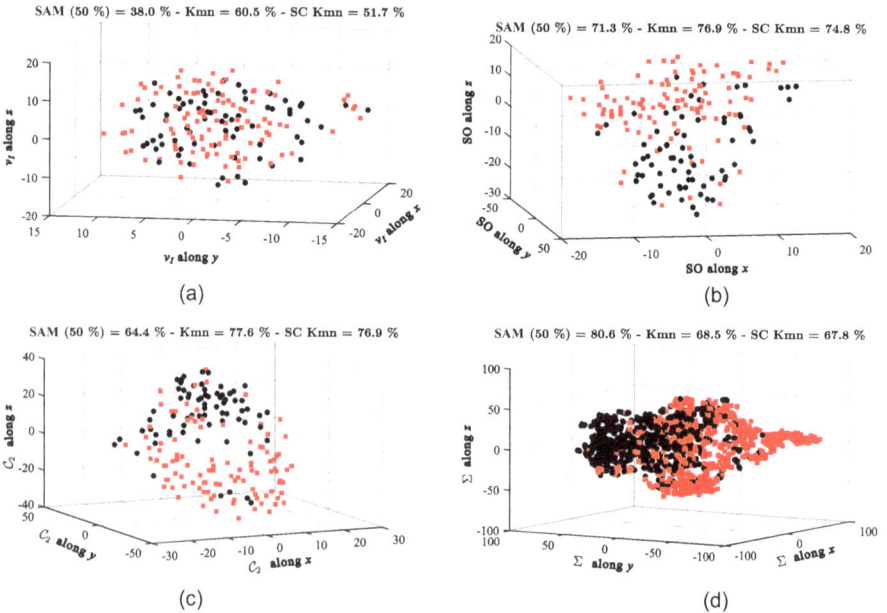

FIGURE 14.19 Manifold untangling by the t-SNE method for four sets of parameters and features: the latter are projected in a three-dimensional space preserving distances where each point represents a feature of a small image or an acquisition (because the features of each acquisition have been merged). The points represented by black circles (respectively red squares) correspond to the Gougerot class (respectively normal). The rates of good classification by SAM (50% of images are dedicated for learning task), by k-means (Kmn) and by CS (SC Kmn) are indicated above each figure. (a) S14 - Vectorised images (v_l), (b) S02 - SO features, (c) S10 - HOS features at order 2 (C_2), and (d) S01 - SO$_{\log}$ and C_2 (Σ).

TABLE 14.2
Confusion Matrices with Four Features and Classification Algorithms

C_{img} (%)	Gougerot	Normal
Gougerot	18.0	82.0
Normal	13.0	87.0
Sensibility (%)	58.1	
Specificity (%)	48.5	

(a) S14 - Vectorized image (v_I) using k-means

C_{kms} (%)	Gougerot	Normal
Gougerot	63.6	36.4
Normal	14.8	85.2
Sensibility (%)	81.1	
Specificity (%)	70.1	

(b) S02 - SO features using k-means

C_{klm} (%)	Gougerot	Normal
Gougerot	80.0	20.0
Normal	39.7	60.2
Sensibility (%)	66.8	
Specificity (%)	74.8	

(c) S10 - HOS at order 2 (C_2) using spectral clustering

C_{sm} (%)	Gougerot	Normal
Gougerot	78.1	21.9
Normal	18.0	82.0
Sensibility (%)	81.2	
Specificity (%)	78.9	

(d) S01 - SO$_{log}$ and C_2 (Σ) with SAM

14.6.4 ONLY PAROTID GLANDS

By only considering parroted glands, see Table 14.3, we reached 78% good classification, which corresponds to a sensitivity of 82.3% and a specificity of 70.3%. With unsupervised classification, we obtain very interesting rates of 80% good classification. The k-means algorithm and the CS give equivalent results with percentages reaching 85%. The confusion matrix shows that this represents a sensitivity of 84.4% and a specificity of 85.9% (for both algorithms), which is better than mixing the two types of glands.

In elastography, the supervised classification is slightly improved (from 68% to 72% at most). On the other hand, in unsupervised classification, we obtain very interesting rates up to 76% good classification with k-means. By observing the 3D projections of the elastographies in Figure 14.20, it is difficult to establish a clear border between the two classes of points but we can see that the normal parotids are rather grouped together. The best classification rate obtained by elastography is about 76% (with the k-means) and displays a sensitivity of almost 85% and a specificity of nearly 65%. We should highlight that our simulation results are very close (or slightly better) to values obtained by experts as presented in Ref. [27] (81% and 67%).

TABLE 14.3
Classification of Parotide Glands Based on k-Means

T	Img	Features	No.	Images	SO	SO$_{log}$	C_2	C_3	C_4	$C_{1,2}$	$C_{2,4}$	$C_{1,4}$	SO+C_2	SO+$C_{1,2}$	SO+$C_{2,4}$	SO+$C_{1,4}$	SO$_{log}$+C_2	SO$_{log}$+$C_{1,2}$	SO$_{log}$+$C_{2,4}$	SO$_{log}$+$C_{1,4}$
Ultrasonography			S01	42	75	79	52	66	55	53	56	55	75	75	71	71	82	82	84	84
		F	S02	54	78	81	59	63	64	59	67	69	74	76	63	63	82	81	78	74
	N		S03	42	66	78	51	47	49	52	46	46	56	56	47	47	81	81	48	48
	N	F	S04	54	73	79	55	58	50	56	52	52	61	60	51	51	79	80	52	51
			S05	56	75	79	57	66	42	53	57	56	75	76	71	71	82	82	84	84
	N		S06	61	78	81	59	64	64	59	68	68	77	76	70	70	82	81	79	78
	N		S07	56	66	78	52	47	49	51	46	46	56	56	47	47	81	81	49	48
	N	N F	S08	60	73	78	56	56	49	55	51	52	59	60	52	51	79	79	51	51
Ultrason. HE			S09	66	82	74	70	55	57	70	71	71	84	84	75	75	82	82	74	74
		F	S10	62	81	78	73	66	67	77	71	71	80	85	75	75	79	81	75	78
	N		S11	68	56	64	83	52	56	63	54	54	61	62	59	59	75	75	58	56
	N	F	S12	63	56	58	62	58	58	62	54	58	55	55	58	58	65	64	58	58
			S13	69	82	74	70	65	57	70	71	71	84	84	75	75	81	81	74	74
	N	F	S14	61	81	78	79	67	67	71	70	72	75	80	75	73	79	79	75	78
	N		S15	60	26	64	63	52	56	65	54	54	62	62	59	59	75	75	58	58
	N	N F	S16	61	56	58	63	58	59	62	58	58	57	57	59	59	62	62	58	58
Elastography			S17	74	76	74	69	51	56	70	56	57	76	76	71	71	75	76	75	75
		F	S18	60	74	74	71	73	63	73	63	63	73	73	73	73	71	70	71	71
	N		S19	74	64	64	57	56	50	59	53	53	57	57	53	53	63	63	53	53
	N	F	S20	70	53	54	50	54	49	50	51	51	50	50	51	51	56	59	51	51
			S21	61	76	74	68	51	56	70	56	60	76	76	71	71	75	76	75	75
	N	F	S22	61	55	74	71	73	55	71	56	56	73	73	73	73	70	71	71	72
	N		S23	61	64	65	57	57	50	59	53	55	57	57	55	53	63	63	53	53
	N	N F	S24	53	53	54	51	53	49	51	51	51	50	50	51	51	56	56	51	51

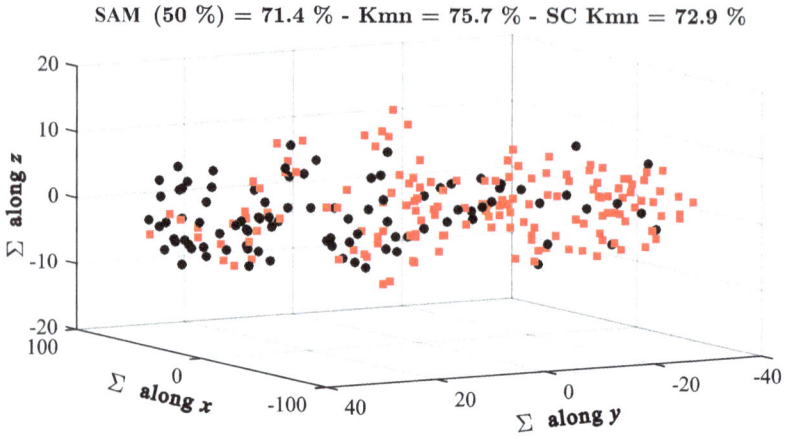

(a)

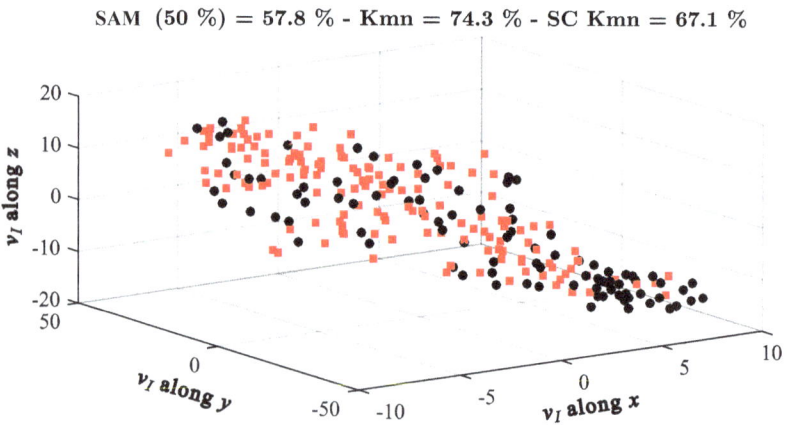

(b)

FIGURE 14.20 Manifold untangling of elastographies of parotid glands using the t-SNE method and two kinds of features. (a) S17 - Vectorized images (v_I) and (b) S17 - SO_{log} and $C_{1,2}$ (Σ).

14.6.5 ONLY SUBMANDIBULAR GLANDS

For the submandibular, the best rates are obtained for simulations with S01 and S05, *i.e.* without equalization histogram, no fusion. In general, the feature fusion leads to a decrease in the results, while we observed the opposite effect for parotids. This difference may be because the number of images extracted from the parotid glands, by their larger size, is much larger than for submandibulars (approximately 1,300 against 700 thumbnails). The maximum rate is 82% which is slightly higher than the one observed with mixed glands (81%) and parotids alone (78%). This maximum rate, acquired with multicorrelations of order two, three and four ($C_{2,3,4}$), shows a good sensitivity of 86.4% and a 74% lower specificity (for S01).

In unsupervised classification, we observed that the k-means surpass the CS with percentages reaching 86% of good ranking (against 83% with CS). The statistical features, especially order two, seem more discriminating than the features based on multiresolutions. The combined use of the two types of features achieves very interesting results (above 80%) but lower than 86% of the cross-correlation (C_2). This allows to establish a sensitivity of 87.7% and a specificity of 74.9%. We also notice that the feature standardization degrades the classification.

Concerning the elastography, we reached a similar conclusion to the ones presented in previous studies presented in Refs. [27,50] and concluded that the elastography does not seem to be helpful for the diagnosis of the submandibular glands (Table 14.4).

14.7 CONCLUSION

To analyze medical data, we proposed and implemented several processing approaches, we also implemented and combined various feature extraction and clustering methods. In our study, we considered ultrasound and elastography images. We carefully handled every step in the image processing chain with particular attention to the feature extraction step. Our approach was first validated on synthetic textures and then on medical data. The classification results obtained are promising. Therefore, these works innovative (application of signal processing tools for the characterization of venous thrombus) open up many perspectives, both medical, applicative and in image processing.

In this study, we presented a new approach to examine salivary glands and diagnose the presence of the Gougerot- Sjögren's syndrome. Various simulations have been carried out on the salivary glands with different approaches of feature extraction and classification. Overall, the results obtained are quite promising as they are

TABLE 14.4
Classification of Submandibular Glands Based on k-Means

Simulations / Features:			No.	Images	SO	SO_{log}	C_2	C_3	C_4	$C_{1,2}$	$C_{2,4}$	$C_{1,4}$	$SO+C_2$	$SO+C_{1,2}$	$SO+C_{2,4}$	$SO+C_{1,4}$	$SO_{log}+C_2$	$SO_{log}+C_{1,2}$	$SO_{log}+C_{2,4}$	$SO_{log}+C_{1,4}$
T	Img	Features																		
Ultrasonography		F	S01	51	76	67	81	59	54	81	79	79	83	83	81	81	80	80	80	80
			S02	54	73	71	86	57	50	86	49	52	84	84	64	62	80	80	74	74
	N		S03	51	64	63	70	63	64	71	66	66	65	62	64	64	67	67	64	64
	N	F	S04	54	62	57	81	58	55	59	63	62	59	60	63	61	63	60	61	61
N			S05	59	76	67	81	58	55	82	79	79	83	83	81	81	80	80	80	80
N		F	S06	61	73	66	86	57	50	66	49	52	84	84	66	62	80	80	74	74
N	N		S07	59	64	63	70	62	64	71	66	66	65	62	64	64	67	67	64	64
N	N	F	S08	61	60	57	63	63	54	59	62	63	59	59	63	63	62	61	61	61
Ultrasono. HE			S09	64	43	71	80	72	67	80	65	65	77	77	64	68	70	70	67	67
		F	S10	64	52	71	80	70	61	80	67	71	75	25	73	67	78	78	71	73
	N		S11	60	49	59	76	54	67	76	67	67	66	66	66	66	67	67	66	66
	N	F	S12	63	46	59	77	51	64	75	63	63	62	63	63	61	80	80	65	65
N			S13	61	43	71	80	69	67	80	67	66	77	77	66	68	70	70	67	67
N		F	S14	65	54	71	80	69	61	80	67	69	75	75	72	67	78	78	73	75
N	N		S15	61	49	59	76	54	67	76	67	67	66	66	66	66	67	67	66	67
N	N	F	S16	65	46	60	77	51	64	75	63	63	62	62	63	61	59	60	63	63
Elastography			S17	52	64	65	66	54	64	64	56	55	66	65	55	55	63	63	65	65
		F	S18	54	61	67	63	53	64	62	52	52	63	62	52	49	66	66	64	65
	N		S19	57	48	49	48	45	57	48	52	52	48	48	52	52	48	48	52	52
	N	F	S20	54	55	51	51	45	59	51	55	55	52	52	55	55	52	52	55	55
N			S21	52	64	65	65	54	64	68	57	58	66	66	57	56	65	65	65	65
N		F	S22	59	61	66	64	51	64	63	52	52	63	62	52	49	66	66	65	64
N	N		S23	52	48	49	49	45	57	48	52	52	49	46	52	52	48	48	52	52
N	N	F	S24	56	55	51	51	45	56	51	55	55	52	52	55	55	52	52	55	55

very similar to the ones obtained by experts. Indeed, our simulations on ultrasound images show that we can reach a sensitivity of 80% (supervised or unsupervised), mixed glands. On the other hand, our results are less good in terms of specificity (almost 80% in supervised, 75% unsupervised and around 90% for the experts). By separating the parotid glands from submandibular, we increased the performance of unsupervised classifiers (k-means and CS) which rises to 86%. On the parotids, we notably achieved a sensitivity of 84.4% and a specificity of 85.9%. The results on submandibulars show a better sensitivity (87.7%) but lower specificity (74.9%). With more data, we may reach a better modeling of the classes, and therefore a better classification.

Concerning elastography, our simulations confirmed the results of the previous studies showing that, this type of imaging gives rather satisfactory results on the parotid glands (but below ultrasound) but that, currently, it does not discriminate between the pathological submandibular glands and normal ones. Advances in elastography brought about by future acquisition systems may give better results. In terms of approaches, we notice that features based on scattering operators and higher order statistics give similar results, slightly better for the statistical features.

NOTES

1 This syndrome is an autoimmune disease characterized by decreasing the secretion of tears and saliva producing the sicca syndrome [17].
2 Which is a common disease for people above the age of 50, especially women [14]

REFERENCES

1. A. Achim, A. Bezerianos, and P. Tsakalides. Novel Bayesian multiscale method for speckle removal in medical ultrasound images. *IEEE Transactions on Medical Imaging*, 20(8): 772–783, August 2001.
2. A. Alunno, E. Bartoloni, and R. Gerli. *Sjogren's Syndrome: Novel Insights in Pathogenic, Clinical and Therapeutic Aspects*. Elsevier Science, Amsterdam, The Netherlands, June 2016.
3. M. Ankerst, M.M. Breunig, H.P. Kriegel, and J. Sander. OPTICS: Ordering points to identify the clustering structure. In *ACM SIGMOD International Conference on Management of Data*, pp. 49–60, Philadelphia, PA, June 1999.
4. A.F. Badea, M. Lupsor Platon, M. Crisan, C. Cattani, I. Badea, G. Pierro, G. Sannino, and G. Baciut. Fractal analysis of elastographic images for automatic detection of diffuse diseases of salivary glands: Preliminary results. *Computational and Mathematical Methods in Medicine*, 2013: 347238, April 2013.
5. K. Bensafia, A. Mansour, A. Boudra, S. Haddab, P. Ariès, and B. Clement. Blind separation of ECG signals from noisy signals affected by electrosurgical artifacts. *Analog Integrated Circuits and Signal Processing*, 104: 191–204, 2020.
6. K. Bensafia, A. Mansour, and S. Haddab. Blind source subspace separation and classification of ECG signals. In *Conférence Internationale en Automatique & Traitement de Signal*, Sousse, Tunisia, March 2017.
7. C. Brooker, I. Langlois-Wils, and E. Lepresle. *Le corps humain: étude, structure et fonction*. Infirmier(e)s: étudiants (Hors collection). De Boeck Supérieur, November 2000.

8. J. Bruna and S. Mallat. Invariant scattering convolution networks. *IEEE Transactions on Pattern Analysis and Machine Intelligence*, 35(8): 1872–1886, August 2013.

9. J.F. Cardoso. Eigen-structure of the fourth-order cumulant tensor with application to the blind source separation problem. In *International Conference on Acoustics, Speech, and Signal Processing*, pp. 2655–2658, Albuquerque, NM, April 1990.

10. J.F. Cardoso and A. Souloumiac. Blind beamforming for non-Gaussian signals. *IEE Proceedings F: Radar and Signal Processing*, 140(6): 362–370, December 1993.

11. S.G. Chang, B. Yu, and M. Vetterli. Spatially adaptive wavelet thresholding with context modeling for image denoising. *IEEE Transactions on Image Processing*, 9(9): 1522–1531, September 2000.

12. P. Comon. Tensor diagonalization, a useful tool in signal processing. *IFAC Proceedings Volumes*, 27(8): 77–82, July 1994.

13. P. Comon and C. Jutten. *Handbook of Blind Source Separation: Independent Component Analysis and Applications*. Elsevier Science, Amsterdam, The Netherlands, February 2010.

14. D. Cornec, V. Devauchelle-Pensec, A. Saraux, and S. Jousse-Joulin. Clinical usefulness of salivary gland ultrasonography in Sjögren's syndrome: Where are we now? *La Revue de Médecine Interne*, 37(3): 186–194, November 2016.

15. M. Crosier and L.D. Griffin. Texture classification with a dictionary of basic image features. In *Conference on Computer Vision and Pattern Recognition*, pp. 1–7, Anchorage, AK, June 2008.

16. K.J. Dana, B. Van Ginneken, S.K. Nayar, and J.J. Koenderink. Reflectance and texture of real-world surfaces. *ACM Transactions on Graphics*, 18(1): 1–34, January 1999.

17. B. Daniel, D. Françoise, and A. Isabelle. *Rhumatologie (Le livre de l'interne)*. Lavoisier, Paris, December 2014.

18. L. De Lathauwer, B. De Moor, and J. Vandewalle. Independent component analysis and (simultaneous) third-order tensor diagonalization. *IEEE Transactions on Signal Processing*, 49(10): 2262–2271, October 2001.

19. J. Di Carlo and D. Cox. Untangling invariant object recognition. *Trends in Cognitive Sciences*, 11(8): 333–341, August 2007.

20. M. Ester, H.P. Kriegel, J. Sander, and X. Xu. Density-based spatial clustering of applications with noise. *In International Conference on Knowledge Discovery and Data Mining*, pp. 226–231, Portland, OR, August 1996.

21. M. Fritz, E. Hayman, B. Caputo, and J. Eklundh. The kth-tips database. www.nada.kth. se/cvap/databases/kth-tips/, May 2004 [Online: accessed on August 23, 2018].

22. A. Géron. *Machine Learning avec Scikit-Learn: Mise en oeuvre et cas concrets*. Dunod, Paris, August 2017.

23. G.B. Giannakis and M.K. Tsatsanis. Signal detection and classification using matched filtering and higher order statistics. *IEEE Transactions on Acoustics, Speech, and Signal Processing*, 38(7): 1284–1296, July 1990.

24. J.A. Hartigan and M.A. Wong. Algorithm AS 136: A k-means clustering algorithm. *Journal of the Royal Statistical Society*, 28(1): 100–108, January 1979.

25. Y.L. Huang, K.L. Wang, and D.R. Chen. Diagnosis of breast tumors with ultrasonic texture analysis using support vector machines. *Neural Computing & Applications*, 15(2): 164–169, April 2006.

26. S. Jousse-Joulin, V. Milic, M.V. Jonsson, A. Plagou, E. Theander, N. Luciano, P. Rachele, C. Baldini, H. Bootsma, A. Vissink, A. Hocevar, S. De Vita, Z. Tzioufas, A.G. Alavi, S.J. Bowman, and V. Devauchelle-Pensec. Is salivary gland ultrasonography a useful tool in Sjögren's syndrome? A systematic review. *Rheumatology*, 55(5): 789–800, May 2016.

27. A. Knopf, B. Hofauer, K. Thürmel, R. Meier, K. Stock, M. Bas, and N. Manour. Diagnostic utility of Acoustic Radiation Force Impulse (ARFI) imaging in primary Sjögren's syndrome. *European Radiology*, 25(10): 3027–3034, October 2015.

28. J.B. Kruskal. Multidimensional scaling by optimizing goodness of fit to a nonmetric hypothesis. *Psychometrika*, 29(1): 1–27, March 1964.

29. J.L. Lacoume, P.O. Amblard, and P. Comon. *Statistiques d'ordre supérieur pour le traitement du signal*. Elsevier Science, , Amsterdam, The Netherlands, October 1997.

30. P. Lemberger, M. Batty, M. Morel, and J.L. Raffaëlli. *Big Data et Machine Learning: Les concepts et les outils de la data science*. Dunod, , Paris, October 2016.

31. L. Liu, P. Fieguth, G. Kuang, and H. Zha. Sorted random projections for robust texture classification. In *International Conference on Computer Vision*, pp. 391–398, Barcelona, Spain, November 2011.

32. C.P. Loizou, C.S. Pattichis, C.I. Christodoulou, R.S.H. Istepanian, M. Pantziaris, and A. Nicolaides. Comparative evaluation of despeckle filtering in ultrasound imaging of the carotid artery. *IEEE Transactions on Ultrasonics, Ferroelectrics and Frequency Control*, 52(10): 1653–1669, December 2005.

33. S.B. Lowen and M.C. Teich. Power-law shot noise. *IEEE Transactions on Information Theory*, 36(6): 1302–1318, November 1990.

34. J. Lu, G. Wang, and P. Moulin. Image set classification using holistic multiple order statistics features and localized multi-kernel metric learning. In *International Conference on Computer Vision*, pp. 329–336, Sydney, Australia, December 2013.

35. L. Maaten and G. Hinton. Visualizing data using t-SNE. *Journal of Machine Learning Research*, 9: 2579–2605, November 2008.

36. S. Mallat. *A Wavelet Tour of Signal Processing*. Academic Press, New York and London, 1999.

37. A. Mansour, A. Kardec Barros, and N. Ohnishi. Blind separation of sources: Methods, assumptions and applications. *IEICE Transactions on Fundamentals of Electronics, Communications and Computer Sciences*, E83-A(8):1498–1512, August 2000.

38. A. Mansour and M. Kawamoto. ICA papers classified according to their applications & performances. *IEICE Transactions on Fundamentals of Electronics, Communications and Computer Sciences*, E86-A(3):620–633, March 2003.

39. A Martin and A. Mansour. Comparative study of high order statistics estimators. In *International Conference on Software, Telecommunications and Computer Networks*, pp. 511–515, Split (Croatia), Dubrovnik (Croatia), Venice (Italy), October 10-13 2004.

40. A. Martin and A. Mansour. High order statistic estimators for speech processing. In *ITST 2005 - 5th International Conference on ITS Telecommunications*, Brest, France, June 27-29 2005.

41. M. Önder, C. Evli, E. Türk, O. Kazan, İ.Ş. Bayrakdar, Ö. Çelik, A.L.F. Costa, J.P.P. Gomes, C.M. Ogawa, R. Jagtap, and K. Orhan. Deep-learning-based automatic segmentation of parotid gland on computed tomography images. *Diagnostics*, 13(4), 2023.

42. A.Y. Ng, M.I. Jordan, and Y. Weiss. On spectral clustering: Analysis and an algorithm. In *Advances in Neural Information Processing Systems*, pp. 849–856, Cambridge, MA, December 2002.

43. H.G. Nguyen, R. Fablet, and J.M. Boucher. Visual textures as realizations of multivariate log-Gaussian Cox processes. In *Conference on Computer Vision and Pattern Recognition*, pp. 2945–2952, Colorado Springs, CO, June 2011.

44. H. Nijhuis, W. van Rooij, V. Gregoire, J. Overgaard, B.J. Slotman, W.F. Verbakel, and M. Dahele. Investigating the potential of deep learning for patient-specific quality assurance of salivary gland contours using EORTC-1219-DAHANCA-29 clinical trial data. *Acta Oncologica*, 60(5):575–581, 2021.

45. E. Prezioso, S. Izzo, F. Giampaolo, F. Piccialli, G. Dell'Aversana Orabona, R. Cuocolo, V. Abbate, L. Ugga, and L. Califano. Predictive medicine for salivary gland tumours identification through deep learning. *IEEE Journal of Biomedical and Health Informatics*, 26(10):4869–4879, 2022.

46. C.G. Puntonet, A. Mansour, C. Bauer, and E. Lang. Separation of sources using simulated annealing and competitive learning. *Neurocomputing*, 49(1-4): 39–60, December 2002.

47. Y. Rangsanseri and W. Prasongsook. Speckle reduction using wiener filtering in wavelet domain. In *International Conference on Neural Information Processing*, pp. 792–795, Singapore, November 2002.

48. S.T. Roweis and L.K. Saul. Nonlinear dimensionality reduction by locally linear embedding. *Science*, 290(5500): 2323–2326, December 2000.

49. M.S. Runge, A.M. Greganti, P.L. Masson, and J.S. Co. *Médecine interne de Netter*. Elsevier Health Sciences, Amsterdam, The Netherlands, April 2011.

50. A. Samier-Guérin, A. Saraux, S. Gestin, D. Cornec, T. Marhadour, V. Devauchelle-Pensec, L. Bressollette, M. Nonent, and S. Jousse-Joulin. Can arfi elastometry of the salivary glands contribute to the diagnosis of Sjögren's syndrome? *Joint Bone Spine*, 83(3): 301–306, 2016.

51. B. Schölkopf, A. Smola, and K.R. Müller. Kernel principal component analysis. In *International Conference on Artificial Neural Networks*, pp. 583–588, Lausanne, Switzerland, October 1997.

52. G. Sharma, S. Hussain, and F. Jurie. Local higher-order statistics (LHS) for texture categorization and facial analysis. In *European Conference on Computer Vision*, pp. 1–12, Florence, Italy, October 2012.

53. L. Sifre and S. Mallat. Rotation, scaling and deformation invariant scattering for texture discrimination. In *Conference on Computer Vision and Pattern Recognition*, pp. 1233–1240, Portland, OR, June 2013.

54. M. Stone. Cross-validatory choice and assessment of statistical predictions. *Journal of the Royal Statistical Society: Series B (Methodological)*, 36(2): 111–147, December 1974.

55. J.B. Tenenbaum, V. De Silva, and J.C. Langford. A global geometric framework for nonlinear dimensionality reduction. *Science*, 290(5500): 2319–2323, December 2000.

56. C.M. Wu and Y.C. Chen. Statistical feature matrix for texture analysis. *CVGIP: Graphical Models and Image Processing*, 54(5): 407–419, September 1992.

57. K. Zuiderveld. Contrast limited adaptive histogram equalization. In P. Heckbert (ed.), *Graphics Gems IV*, pp. 474–485. Academic Press, San Diego, CA, May 1994.

15 Intelligent Feature Selection Algorithm Using SA-SVM Classification for Skin Cancer Diagnosis

Azadeh Noori Hoshyar
Federation University Australia

Adel Al-Jumaily
University of Technology Brunei
ENSTA Bretagne
University of Western Australia
Edith Cowan University

15.1 INTRODUCTION

In the last few decades, malignant melanoma has emerged as a lethal skin cancer, and its occurrence is increasing. According to Sung et al. (2021), in 2020, more than 1,198,073 million new incidents of skin cancer were globally diagnosed, and 125,000 people died from it. Skin cancer has become a common disease in various countries but especially in Australia for people aged between 15 and 44. The occurrence is much higher than in the US, UK, and Canada, with more than 10,000 diagnoses recorded and an annual mortality of 1,250 people (Celebi et al., 2015). Fortunately, early diagnosis can significantly increase the patient's survival rate. The rapid rise, high medical costs, and death rates have prioritized the early diagnosis of this cancer (Dorrell & Strowd, 2019). Studies have confirmed that detection systems can improve the diagnosis rate of melanoma from 60.9% up to 85.4%, compared with the naked eye (Sung et al., 2021).

Melanoma is presumed to metastasize and spread to other organs of the body in comparison with other skin cancers, and its anticipation is strictly relevant to its thickness. Melanoma can be cured successfully in the early stages when the thickness is less than 1 mm. However, it is not very easy to detect melanoma in its early stages, even by experienced dermatologists (Bray et al., 2018; Celebi et al., 2015; Dorrell & Strowd, 2019). A diagnosis of melanoma is subjective and depends on dermatologists' visual perception and clinical experience. Also, this is too complex and time-consuming.

DOI: 10.1201/9781003346678-15

Dealing with numerous cases leads to losing efficiency and diagnostic accuracy. Therefore, having a second opinion is required for diagnosis with less error and better accuracy (Bray et al., 2018; Celebi et al., 2015; Dorrell & Strowd, 2019).

Machine learning (ML) algorithms have already demonstrated their effectiveness in solving many problems based on classification and are increasingly used in diagnosing diseases. These systems include different steps of pre-processing, feature extraction, feature selection, and classification.

Many researchers investigated algorithms for the steps to improve diagnosis accuracy in automatic skin cancer detection (Alyami et al., 2022; Janda et al., 2022; Thurnhofer-Hemsi & Domínguez, 2021). However, the extraction and selection of best features and using an effective ML technique are still the main concerns in this area. This is because, firstly, extracting such distinguishable attributes owing to the fine-grained variability in the appearance of lesions. Therefore, identifying features and removing excess ones is required to achieve an optimum solution for detection. Secondly, an effective learning method can lead to the best classification result after the feature selection step.

This research proposes an innovative algorithm to classify skin cancer tumors as either inoffensive or dangerous. The chapter: (i) propose a fully automated feature selection algorithm, using smart inertia-based particle swarm optimization (IPSO)-based SVM called smart IPSO-SVM, which optimizes the feature selection stage; (ii) compare our proposed feature selection algorithm with other algorithms; (iii) suggest a self-advising SVM (SA-SVM) in the area of skin cancer detection system; and (iv) undertake an analytical comparison between SA-SVM and support vector machine (SVM) along with our proposed algorithm.

The rest of the chapter is organized as follows. The literature on this topic is reviewed in Section 15.2. Then, the technical reviews of Sequential Forward Selection (SFS), IPSO, SVM, and SA-SVM are provided in Section 15.3. In Section 15.4, our proposed algorithm is explained in detail, and Section 15.5 reports the results. The comparative study is carried out in Section 15.6 to determine the performance of the proposed algorithm when compared with other state-of-the-art algorithms. Finally, this chapter concludes in Section 15.7 with a discussion and suggestions for future research.

15.2   OVERVIEW ON AUTOMATIC SKIN CANCER DETECTION

In this section, we review the literature on the methods used for different steps of skin cancer detection systems to achieve a reliable framework.

15.2.1   MELANOMA FEATURES-BASED METHODS

In skin cancer detection systems, feature extraction is to extract the parameters of a skin image to characterize the dermatological features of melanoma and perform the diagnosis based on these parameters. Clinicians rely on the features of melanoma and the method of diagnosis is essential for selecting the features. For instance, asymmetry and pigmented network are, respectively, the features extracted through ABCD-rule "Asymmetrical, Border, Color, Diameter, Evolving" and pattern analysis. However, an evaluation of the features of melanoma diagnosis is visually challenging because the content of information in dermatoscopic images is very complex and requires experienced physicians. The diagnosis methods to determine melanoma lesions in the screening process by non-dermatologists have been developed as ABCD rule, ABCD-E criteria "asymmetry, border, color, diameter and evolving", ABC-point list [A(A) BCDE] "asymmetry, abrupt cutoff of the network at the border in at least one-quarter of the circumference, three or more colours, three or more differential structures,

noticed a change (evolution) in the last 3 months", seven-point checklist, seven features for melanoma, three-point checklist, Pattern analysis, and Menzies' method (Abbasi et al., 2004; Jain et al., 2015; Malvehy et al., 2007). For instance, the ABCD rule of dermoscopy consists of four criteria: Asymmetry, Border sharpness, Color variegation, and Differential structures. The ABCDE comprises Border irregularity, Color variegation, Diameter, Evolving, and other features. The seven-point checklist contains seven criteria: A typical pigment network, a Blue-whitish veil, an Atypical vascular pattern, Irregular streaks, Irregular dots/globules, Irregular blotches, and Regression structures. The Pattern analysis consists of Global patterns and Local feature (Malvehy et al., 2007).

The automatic computing of these methods-based rules has been done in many ways. According to (Nachbar et al., 1994), symmetry has achieved the highest weight in the ABCD rule of dermatoscopy. Eedy (2003) indicated that 96% of asymmetry in melanoma cases had a score of two (both axes represent asymmetry), while in benign images, it was just about 24.2%. Many researchers have considered the asymmetry according to the axis of symmetry in the tumor. In such studies, the axis of symmetry may be identified using Fourier transform (Clawson et al., 2007), best-fit ellipse, diameter length, and principal axis (Majumder & Ullah, 2019). After that, both areas created by the axis are differentiated. In many studies, a lesion's roundness, compactness, and thinness have been considered appropriate properties of the skin cancer images (Sáez et al., 2015), emerging as an accurate geometry variable. In Sáez et al. (2015), the symmetry distance (SD) has been introduced as another measure in images. Seidenari et al. (2005) presented a method to estimate the distribution of skin lesions. Their rationale was to determine the effectiveness of distribution parameters to differentiate melanoma from normal ones. They discovered aspects concerning the non-homogeneity of the lesion region. Following their analysis, they computed the mathematical parameters such as mean, variance, and Euclidean distance. Another research (Manousaki et al., 2006) proposed estimating the distribution irregularity using the fractal dimension on the lesion's surface. Also, they computed the standard deviation to measure the sharpness of borders. M. Ramezani et al. (2014) presented an algorithm to detect malignant melanoma using computerized procedures. They extracted 187 features representing texture, border irregularity, diameter, asymmetry, and color variation. They applied principal component analysis (PCA) to reduce the number of features. They achieved an accuracy of 82.2% using their proposed algorithm. In Tan et al. (2018), the shape, color, and texture features were extracted from the lesion regions, and an algorithm was proposed. Statistical results revealed the superiority of the proposed algorithm over other methods. In another approach (Cheng et al., 2008), the different statistical properties of standard deviation, energy, mean, and entropy were computed as extracted features. An accuracy of 79% was achieved using Neural Network. Other statistical features may include pixel intensity values and a local pixel neighboring information. Other statistical approaches may incorporate pixel intensity values and a local pixel neighboring information (Akram et al., 2015).

The Gray Level Co-occurrence Matrix (GLCM) represents the popular texture features and has been employed in various applications (Ashfaq et al., 2019; Murugan et al., 2019). Other researchers have reported on feature extraction of skin cancer in the literature (Divya & Ganeshbabu, 2020; Lee & Kwon, 2020). In this study, hybrid features are extracted for the classification stage.

15.2.2 Feature Selection

Feature selection is an important process, and it is performed prior to lesion classification. Its purpose is to reduce the computational cost of classification by reducing the number of extracted feature descriptors. This diminishment is not trivial due to eliminating redundancy, which may negatively affect discriminatory power. The feature selection process may be explained as follows: the search procedure as a subset generation is performed to provide various subset candidates containing features. An evaluation criterion is considered to evaluate the subset candidates. This is compared and replaced by the prior best subset's estimated performance in a preference case. This process is repetitious until the stopping criterion is met. In the final stage, the validation and testing of the best-selected subset are performed (Yahya, 2011). Figure 15.1 shows the common feature selection process.

Different studies have been conducted to develop the feature selection process. In 2019, a review on feature descriptors was published in (Javed et al., 2019), which compared the features selection for skin lesion detection. (Maryam Ramezani et al., 2014) applied Principal component analysis to achieve the best possible subset from a set of 187 features. In another research (Lu et al., 2016), sequential forward selection (SFS) and sequential backward selection (SBS) were applied to reduce the dimensions of 71 features. The dominant features optimized the dimensionality of features within the 13–71 range and increased classification accuracy by 4%.

PSO is extensively applied in feature selection problems to search for the best feature subset of a large database of possible candidates (Moradi & Gholampour, 2016; Teck Yan Tan et al., 2018, 2019, 2020). Binary Particle Swarm Optimization is another extension of PSO where particles are considered by a point in a binary multidimensional space. This PSO type is also widely applied in the feature selection (Tan et al., 2020). In T. Y. Tan et al. (2019), the authors represented an algorithm for feature subset selection by employing PSO along with the Deep Ensemble Network

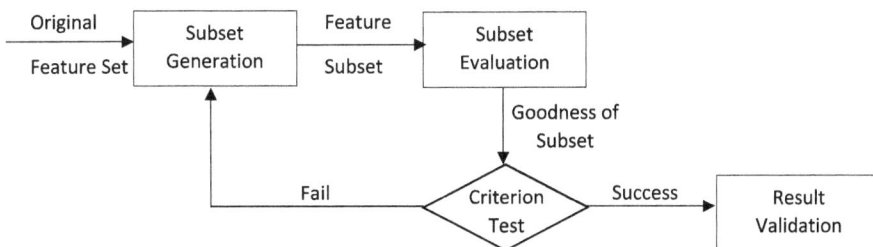

FIGURE 15.1 Common feature selection process (Yahya, 2011).

and hybrid clustering models. The learning hyper-parameters are optimized in this algorithm by a cascading PSO algorithm. The majority voting strategy combines the prediction results and generates the classification outcome.

Y. Maali and Al-Jumaily (2012) proposed a PSO-SVM feature selection algorithm in their study on sleep apnea. They could effectively reduce the number of features and select the best subset for their research. PSO computationally is less expensive than other methods and can quickly do convergence. Thus, PSO is an effective feature selection technique in many fields (Izakian & Pedrycz, 2012; Ling & Nguyen, 2012; Maali & Al-Jumaily, 2012; Tan et al., 2019).

In this study, a feature selection algorithm SFS-PSO-SVM is proposed in which the optimal subset will effectively feed the classification stage.

15.2.3 CLASSIFICATION

Lesion classification is the final step in the computerized analysis, and it estimates whether a lesion is malignant or benign. In this paper, we focus only on ML techniques for classification, as feature selection is one of our contributions. Deep learning methods, which automatically do feature extraction, have not been considered in this research. The existing systems utilize different classification methods to feature descriptors extracted in prior stages to carry out the classification task. These methods' efficiency pertains to extracted descriptors and selected classifiers (Mohemmed et al., 2009). There are many different classifiers, such as Discriminant Analysis, Artificial Neural Networks (ANN), K-Nearest Neighborhood (KNN), Support Vector Machine (SVM), Decision Trees, and Self-Advising SVM (SA-SVM). Several researchers (Cavalcanti et al., 2013; Liang et al., 2018; Maglogiannis et al., 2005) applied Discriminant analysis as a classifier to generate predefined classes from a set of observations. It works according to the values of determining measurements which are called predictors. ANN, as another tool, has been employed by Hajabdollahi et al. (2020), Huang et al. (2020), Zhang et al. (2020). This approach connects the inputs and outputs for detecting the patterns in data; it is usually employed in classification problems.

In Ghalejoogh et al. (2020), Korotkov and Garcia (2012), KNN algorithm has been considered for classifying the lesions into malignant or benign melanomas. This classifier employs the distance measure Euclidean distance to evaluate the distribution of data and classify the objects according to how close they are to the training set. SVM is another classification algorithm that highlights its powerful ability to solve nonlinear classification problems in many applications, even in high dimensions. Furthermore, SVMs prevent over-fitting by selecting a particular hyperplane among many, which can separate the data in feature space (Stoecker et al., 1992). SVM has been used as a popular technique in Afifi et al. (2019), Alfed and Khelifi (2017), Alkolifi Alenezi (2019), Singh et al. (2020), and Tan et al. (2018) for classifying skin cancer melanoma. In some other research (Chang & Chen, 2009; Chau et al., 2014), Decision trees separate the data set into different groups according to their disparity, which helps establish the classification schema. Dreiseitl et al. (2001) compared the different classification techniques of the artificial neural network, k-nearest neighborhood, support vector machine, logistic regression, and decision tree in skin cancer detection systems. The results found that SVM, ANN, and logistic regression performed effectively. In Maglogiannis and Doukas' research (Maglogiannis & Doukas, 2009), they employed and compared the performance of SVM, multinomial logistic regression, ANN, CART, Bayes networks, etc., to classify lesions on the skin. Their experimental results emphasized the superiority of SVM, followed by Bayes networks and regression. Y. Maali and A. Al-Jumaily (2013) developed a new SVM algorithm known as a Self-Advising Support vector machine (SA-SVM) which has been applied as a classifier in sleep apnea. The experimental results showed that it performed better than SVM. Ammara et al. (Masood et al., 2014) proposed a skin cancer detection system using the SA-SVM as a classifier. They used an adaptive median filter to preprocess images and proposed histogram analysis based fuzzy C mean algorithm for Level Set initialization (H-FCM-LS) as a segmentation algorithm. After texture analysis, they used SA-SVM to classify images. The experimental results show that it performed better than SVM. They indicated improved classification results when compared with standard SVM.

Therefore, this study applies SA-SVM as a classifier in our proposed algorithm. To the author's knowledge, SA-SVM has been employed in only very few studies (Masood & Al-Jumaily, 2014; Masood et al, 2015) [55,56] looking at skin cancer detection systems (Masood et al., 2015).

15.3   BRIEF TECHNICAL REVIEW OF SFS, SVM, SA-SVM AND PSO

This section briefly reviews the SVM, SA- SVM and PSO systems.

15.3.1   SEQUENTIAL FEATURE SELECTION

The sequential feature selection (SFS) (Pudil et al., 1994) is an ascendant search method of the set of superlative discriminative parameters among an initial set of parameters E_i with:

$$E_i = \{P_j, j = 1, 2, \ldots, N\} \quad E_{\text{SFSn}} \in E_i, \quad n \le NE_{\text{SFS0}} = \varphi \tag{15.1}$$

where P_j is a parameter (attribute), N is the number of parameters and E_{SFSn} is a null subset.

In this method, one parameter P_j is appended at a time to E_{SFSn} subset. Since the assessment criterion is SVM in each step, the remaining parameters P_j in E_i are added into E_{SFS} and the corresponding classification error (Err) is calculated by Equation (15.2):

$$\text{Err} = \frac{1}{q} \sum_{i=1}^{q} (d_i - a_i)^2 \tag{15.2}$$

where q is total image number of the training database, d_i is desired output. a_i is real output. In every step, the selected parameter P_j is the one in which the new E_{SFS} subset makes it possible to minimize the classification error.

15.3.2 Particle Swarm Optimization (PSO)

Particle Swarm Optimization (PSO) was initially devised by Kennedy and Eberhart in 1995 (Eberhart & Kennedy, 1995). This technique simulates social behaviors such as fish schooling or birds flocking to a promising position to achieve accurate objectives (Eberhart & Kennedy, 1995). This population-based stochastic technique is based on swarm intelligence. The PSO technique includes a set of particles or agents, known as a swarm, in that it "flies" over the solution space attempting to locate promising regions. These particles are represented as possible solutions to the optimization problem and are indicated as points in the n-dimensional search space. In standard PSO, every particle X_i includes its own velocity V_i bounded with a maximum value V_{max}, the memory of its best-obtained position is G, and the best-found neighborhood solution is considered as P_i. In this search process, PSO updates its population of particles concerning their internal velocity and position that are achieved by the experiment of all particles. Each particle adjusts its position based on the following Equations (15.3) and (15.4), respectively:

$$V_i^{(k+1)} = V_i^{(k)} + c_1 r_1 \left(P_i - X_i^{(k)} \right) + c_2 r_2 \left(G - X_i^{(k)} \right) \tag{15.3}$$

$$X_i^{(k+1)} = X_i^{(k)} + V_i^{(k+1)} \tag{15.4}$$

where k is the current generation, c_1 and c_2 are positive constants, r_1 and r_2 are random numbers on the interval [0, 1].

In this study, we use IPSO. Shi and Eberhart (1998) modified the standard PSO as "Inertia-based Particle Swarm Optimization (IPSO)" by adding inertia weight (ϖ) to it as a parameter. This inertia weight is multiplied by the prior velocity in the standard velocity equation. According to Equation (15.5), a nonzero inertia weight moves the particle in the same direction as in the prior iteration.

$$V_{id}^{(k+1)} = W * V_{id}^{(k)} + c_1 \ \text{rand()} * \left(P_{id}^{(k)} - x_{id}^{(k)} \right) + c_2 \ \text{rand()} * \left(P_{gd}^{(k)} - x_{id}^{(k)} \right) \tag{15.5}$$

where $V_{id}^{(k)}$ is the velocity of particle i at iteration (k), rand() is a random number between 0 and 1, $P_{id}^{(k)}$ is the best position of particle i at iteration (k), $x_{id}^{(k)}$ is the current position of particle i at iteration (k), $P_{gd}^{(k)}$ is the best position of the global best particle at iteration (k) and w is the nonzero inertia weight which is set based on Equation (15.6):

$$w = \frac{w_{max} - w_{min}}{\text{iter}_{max}} * \text{iter} \tag{15.6}$$

where w_{max} is the maximum value of w, w_{min} is the minimum value of w, iter_{max} is the maximum number of iterations, and iter is the current iteration number. This parameter's interpolation ensures a balance between the capacities of global and local search. The wrong value selection of this parameter will affect algorithm convergence speed. For this reason, the initial selection is critical (Pranava & Prasad, 2013).

15.3.3 Support Vector Machine (SVM)

Recently, the solid and robust theoretical foundation and superb practical performance of Support Vector Machines have attracted much interest in the ML community (Stoecker et al., 1992). SVM applies a linear separating hyperplane for creating a classifier to maximize the margin. The width of the margin between the classes is considered the optimization criterion. Margin is defined as the distance between the optimal hyperplane and a class's nearest training data points. In cases of non-linear separation of original input space, in 1992, Guyon, Boser, and Vapnik (Boser et al., 1992) presented an approach to generate nonlinear classifiers using kernel functions. SVM firstly transforms the original feature to a higher dimensional feature space. The transformation may be obtained using different nonlinear mappings. The kernel function $K(x, y)$ may be chosen to suit the problem. The common kernels include:

$$\text{Linear Kernel:} \quad K\left(x, x_j\right) = \left(x \cdot x^T\right) \tag{15.7}$$

where x is a vector of inputs and x_j is the vector of inputs at position j.

$$\text{Radial Basis Kernel (RBF):} \quad K\left(x, x_j\right) = e^{-\gamma \|x - x_j\|^2} \quad \gamma > 0 \tag{15.8}$$

where x is a feature vector, x_j is a feature vector of the jth sample, and γ is a parameter of the kernel function, must be greater than 0.

$$\text{Polynomial Kernel:} \quad K\left(x, x_j\right) = \left(1 + x \cdot x_i^T\right)^d \tag{15.9}$$

where x is a vector of feature values for an observation, x_j is a vector of feature values for a second observation, and d is a degree of the polynomial kernel (positive integer)

$$\text{Gaussian Kernel:} \quad K\left(x, x_j\right) = \exp\left(\frac{-\|x - x_j\|^2}{2\sigma^2}\right), \sigma \neq 0 \tag{15.10}$$

where x is a data point, x_j is a data point from a training dataset, $K(x, x_j)$ is the similarity between data points x and x_j, and σ is the standard deviation of the Gaussian kernel.

$$\text{Sigmoid:} \quad K\left(x, x_j\right) = \tanh\left(\gamma\left(x \cdot x_j\right) + \beta\right) \tag{15.11}$$

where x is a training example, x_j is a training example, γ is a coefficient, and β is bias.

After this transformation, the optimal hyperplane may easily be found. The achieved hyperplane is the best-case scenario concerning a maximal margin (Kecman, 2001).

In this study, SVM is selected as the method of choice using specific RBF kernels. The RBF kernel has been adopted for the following reasons (Kecman, 2001). Firstly, the Linear kernel cannot handle the problems of nonlinearly separable classification; actually, it is a particular form of RBF kernel. Secondly, the computation of the RBF kernel is much more stable than the Polynomial kernel, which may return values of zero or infinity in various cases. Thirdly, the Sigmoid kernel may only be valid (i.e., satisfies Mercer's conditions) in specific parameters. Fourthly, the RBF kernel requires fewer parameters to be calculated than Sigmoid and Polynomial kernels. We have adopted this technique as one required component for our feature selection algorithm.

15.3.4 SELF-ADVISING SUPPORT VECTOR MACHINE (SA-SVM)

The SA-SVM is a developed version of SVM, and it was introduced by Y. Maali and A. Al-Jumaily (2013). It is a non-iterative method that may extract pursuant knowledge from the training phase. In SA-SVM, pursuant knowledge may be in the form of any additional information derived about the first or second type of support vectors, e.g., distribution. More specifically, SA-SVM intends to obtain subsequent knowledge through misclassified data in SVM's training phase. These misclassified data may have two potential sources: outliers or data that could not be separated linearly by applying any kernel. This classifier advises weights according to the usage of misclassified data in the training phase to employ it along with the decision values in the SVM's test phase. The weights help the algorithm omit the outlier data. To find the misclassified data (MD), it is defined as Equation (5.12):

$$MD = \bigcup_{i=1}^{n} X_i \mid y_j \neq \text{sign}\left(\sum_{a_j > 0} y_j a_j k\left(x_i, x_j\right) + b\right) \tag{5.12}$$

It may employ any SVM decision function and kernel on the right-hand side of the equation. Although the misclassified data is prevalent during the training phase, it can also be null. The neighborhood length (NL) of each x_i in MD is described as Equation (15.13):

$$NL(x_i) = \text{minimun}_{x_j}\left(\|x_i - x_j\| \cdot y_i \neq y_j\right) \tag{15.13}$$

where $x_j, j = 1,\ldots, n$ are the training data.

The distance of x_i and x_j may be calculated according to the following equation when the training data is mapped to a higher dimension:

$$\|\theta(x_i) - \theta(x_j)\| = \left(k(x_i, x_i) + k(x_j, x_j) - 2k(x_i, x_j)\right)^{0.5} \tag{15.14}$$

Finally, the advised weight (AW) is calculated using Equation (15.15):

$$AW = \begin{cases} 0, \forall x_i \in MD, \|x_k - x_i\| > NL(x_i) \text{ or } MD = NUL \\[2mm] 1 - \dfrac{\sum x_i^{x_k - l_i}}{\sum_{x_i} NL(x_i)}, \quad x_i \in MD, \|x_k - x_i\| > NL(x_i) \end{cases} \tag{15.15}$$

The advised weights are in the range of [0, 1]. This scenario determines the closeness of test data to the misclassified data.

15.4 THE PROPOSED ALGORITHM

According to the literature, 39 features, including GLCM, texture, and shape features, are selected [8–25]. In this chapter, we create our feature set by combining feature sets in Tables 15.1–15.3.

Then, the new feature selection approach is proposed to reduce the dimensions of extracted features. This approach is called Smart IPSO-SVM and is achieved using SFC, IPSO, and Support Vector Machine. MATLAB is used to implement this algorithm, and the proposed approach's details are explained in the following section.

TABLE 15.1
Feature Set 1

GLCM Features

Autocorrelation	Contrast	Correlation	Cluster prominence	Cluster shade
Dissimilarity	Energy	Entropy	Homogeneity	Maximum probability
Sum of squares	Sum average	Sum variance	Sum entropy	Difference variance
Difference entropy	Information measure of correlation	Variance		

TABLE 15.2
Feature Set 2

Volumetric Zone Length Type Texture Features

High Grey-Level Zone Emphasis (HGZE)	Short Zone Low Gray-Level Emphasis (SZLGE)	Short Zone High Gray-Level Emphasis (SZHGE)	Long Zone Low Gray-Level Emphasis (LZLGE)
Long Zone High Gray-Level Emphasis (LZHGE)	Gray Level Non-Uniformity (GLNU)	Zone Percentage (ZP)	

TABLE 15.3
Feature Set 3

Extracted Shape Properties

Area	BoundingBox	ConvexHull	ConvexArea
PixelValues	MajorAxisLength	MaxIntensity	MeanIntensity
MinIntensity	MinorAxisLength	Orientation	Perimeter
PixelIdxList	PixelList		

In this study, a "Smart IPSO-SVM" approach is proposed for feature selection. The main task of the "Smart IPSO-SVM" algorithm is to choose the best subset of features and make the classifier learn by tuning the parameters of SA-SVM in the classification stage.

15.4.1 Initialization

Each particle is defined as an array with two parts in this step. The first part of the particle is an array including two cells with gamma and cost parameters of SVM, which can get a value between 2^{-5} and 2^5. The second part includes an array of 53 cells referring to 53 features containing weights and numbers in the 0 and 1 range. These weights indicate the significance of corresponding features.

In the first population, the first particle of IPSO is initialized by the sequential feature extraction algorithm. In other words, the SFC would propose its best-selected subset of features and feed the first particle of IPSO. The other particles in the first population of IPSO are initialized by generating a random vector with the same dimensionality.

15.4.2 Selection of Features in Particles

In each iteration, features having weight more than a specified threshold are chosen. In this study, IPSO with 40 particles is chosen with c_1 and c_2 which are set to 2.0. The threshold value for selecting a feature is set to 0.5. All the particles are sent to SVM so that their accuracy based on corresponding features can be calculated.

15.4.3 Fitness Function

To compute the fitness of each particle, the SVM algorithm is trained by the training set and the selected features from the corresponding particle. Meanwhile, the accuracy of the SVM's performance on the test set with the corresponding feature set is considered the fitness of that particle. For implementing this algorithm, cross-validation is used to generate the training and testing set.

After achieving the accuracy of particles, the best accuracy in a population would be considered as "gbest", and the best accuracy in the history of that particle would be known as "pbest". The particles in the subsequent populations are generated according to the detailed formulas described in Section 15.4.2. Apparently, in the first population, "pbest" and "gbest" are the same. After 100 iterations, the process stops, and the obtained "gbest" with its corresponding features is considered our best solution candidate.

By finishing this algorithm, the best subset of features and settings for the SVM and SA-SVM in the next step is used in the classification stage. The pseudocode for our proposed "Smart IPSO-SVM" algorithm is as follows:

Pseudocode 15.1: "Smart IPSO-SVM" Algorithm

1. Specify data sets of features to feed SFC
2. Initialize the IPSO
   i. The first particle using the proposed subset of SFC
   ii. The other particles using a random vector

3. Do until the maximum number of iterations is reached

{

4. For each particle

   i. Specify the corresponding features subset, and SVM parameters according to the particle.

   ii. Use SVM to calculate accuracy as the fitness of the particle.

5. Update particles

6. Go to 3.

7. End Do

}

8. Report the best features subset and SVM parameters

In this study, Self-advising SVM (SA-SVM) is applied in the classification stage of our proposed algorithm. The details of SA-SVM are explained in Section 15.3.4.

15.5 PERFORMANCE INDICATORS

To estimate the quality of algorithms in correctly identifying melanoma, this chapter used two performance measures of accuracy and F-score, computed using the confusion matrix, Table 15.4 [60].

The accuracy is found by measuring the percentage of correct detections and is formulated as,

$$\text{Accuracy} = \frac{t_p + t_n}{t_p + f_p + f_n + t_n} \tag{15.16}$$

The higher accuracy shows the higher predictive power of the model (Sokolova et al., 2006). The F-score as another measurement index can be calculated as follows (Sokolova et al., 2006):

$$\text{Precision} = \frac{t_p}{t_p + f_p} \tag{15.17}$$

$$\text{Recall} = \frac{t_p}{t_p + f_n} \tag{15.18}$$

$$\text{F-score} = \frac{\left(\beta^2 + 1\right) \times \text{precision} \times \text{recall}}{\beta^2 \times \left(\text{precision} + \text{recall}\right)} \tag{15.19}$$

TABLE 15.4
Confusion Matrix

	Predicted Positive	Predicted Negative
Label positive	t_p: true positive	f_n: false negative
Label negative	f_p: false positive	t_n: true negative

Precision is a function of true positives and the samples which are misclassified as positives (false positives). Recall is a function of samples that are classified correctly (true positives) and its misclassified samples (false negatives). The F-score is equally balanced in a case $\beta = 1$ because it considers precision in a case $\beta > 1$ and recall otherwise.

15.6 EXPERIMENTAL RESULTS

In this chapter, the data are collected from Sydney Melanoma Diagnostic Centre, Royal Prince Alfred Hospital, and some online resources. The database includes 338 dermoscopic and clinical view lesion images with the dimensions $100 \times 100 \times 3$. The numbers of images of each category for model training and testing are 270 and 68, respectively, with the ratio of training to testing samples of 4:1. Figure 15.2 shows image samples of melanoma and benign tumors.

(a)   (b)

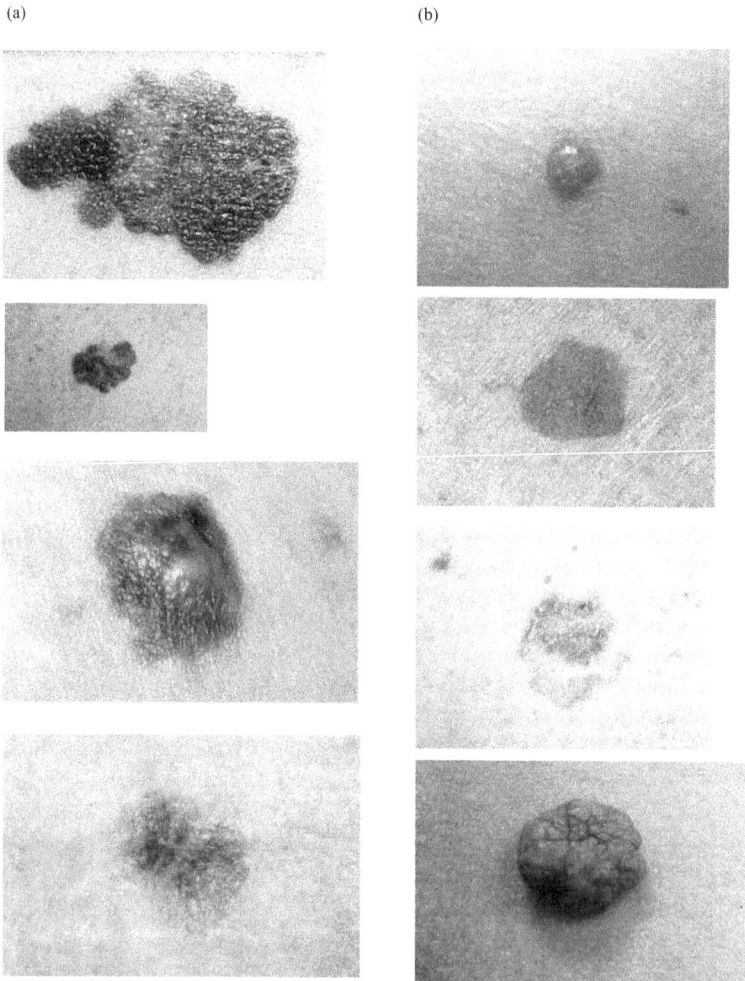

FIGURE 15.2   Image samples of skin tumor. (a) Melanoma samples and (b) benign samples.

15.6.1 COMPARISON OF OUR PROPOSED ALGORITHM WITH DIFFERENT ALGORITHMS

To evaluate the proposed algorithm, different implementations and comparisons have been conducted. The average performance in each selected set of features is used to detect melanoma and will be compared with the proposed algorithm. It is noted that the results are achieved with 5-fold cross-validations in 40 observations. Additionally, for further data analysis, the T-test is carried out between accuracies and F-scores of algorithms with the significance level set to $\alpha = 0.05$.

15.6.1.1 "SVM Classification without Feature Selection" & "SVM Classification with Sequential Feature Selection"

This section intends to determine whether the feature selection algorithm in this case study would increase the algorithm's performance on the learning classifier. For this purpose, firstly, we trained the SVM with "original features" and compared the results with the SVM, which has been trained by the subset of features obtained by the "Sequential feature selection" algorithm. Tables 15.5 and 15.6 show the results of these two algorithms.

Based on the results, the average accuracy and F-score of "SVM Classification without Feature Selection" are 81.399 and 0.878, respectively, and the "SVM Classification with Sequential Feature Selection" approach are 83.929 and 0.892, respectively. Also, it is observed that in this case study, the SFS approach could improve the results when compared with employing the original data. The results would firmly confirm the hypothesis. The statistical analysis (see Table 15.7) is carried out between the accuracies and F-scores of these two approaches. It is done by computing the p-value of the T-test and was equal to 0.005 and 0.035, respectively (Variable 1: SVM Classification without Feature Selection Algorithm, Variable 2: SVM Classification with Sequential Feature Selection).

This analysis demonstrated that the difference between accuracies and F-scores of simple SVM and the algorithm with SFS were statistically significant.

TABLE 15.5

Average Performance of "SVM Classification without Feature Selection" Algorithm

Observation No./Accuracy (%)/F-Score

1/81.304/0.875	2/82.546/0.887	3/81.385/0.88	4/80.276/0.865	5/82.5/0.884
6/79.667/0.865	7/81.223/0.878	8/81.548/0.88	9/83.225/0.891	10/79.167/0.863
11/82.522/0.887	12/79.167/0.864	13/81.385/0.88	14/82.549/0.885	15/82.549/0.885
16/81.375/0.88	17/81.291/0.875	18/81.264/0.871	19/82.427/0.881	20/80.852/0.873
21/82.63/0.887	22/80.211/0.869	23/82.549/0.885	24/82.484/0.882	25/82.511/0.885
26/82.63/0.884	27/81.364/0.875	28/81.439/0.877	29/81.954/0.883	30/81.439/0.877
31/80.322/0.87	32/81.467/0.88	33/82.511/0.884	34/82.657/0.888	35/79.113/0.864
36/75.595/0.844	37/83.685/0.892	38/81.494/0.878	39/81.439/0.877	40/80.227/0.868

TABLE 15.6
Average Performance of "SVM Classification with Sequential Feature Selection" Algorithm

Observation No./Accuracy (%)/F-Score				
1/87.159/0.92	2/85/0.907	3/83.685/0.896	4/87.175/0.915	5/86.115/0.911
6/87.094/0.912	7/83.766/0.896	8/84.849/0.902	9/83.902/0.9	10/81.494/0.879
11/77.922/0.857	12/86.147/0.911	13/87.251/0.917	14/88.366/0.926	15/88.366/0.926
16/68.479/0.756	17/86.093/0.908	18/89.61/0.934	19/86.12/0.911	20/87.121/0.914
21/87.419/0.919	22/87.284/0.918	23/84.784/0.902	24/82.522/0.888	25/79.167/0.867
26/87.175/0.916	27/82.522/0.886	28/88.231/0.926	29/68.534/0.756	30/88.366/0.924
31/88.415/0.925	32/72.024/0.777	33/83.631/0.892	34/70.925/0.769	35/79.113/0.91
36/87.284/0.92	37/82.468/0.883	38/83.869/0.899	39/84.821/0.903	40/85.92/0.904

TABLE 15.7
The T-Test Result between Accuracies and F-Scores of "SVM Classification without FS" Algorithm and "SVM Classification with SFS" Algorithm

	T-Test: Paired Two Sample for Means			
	Accuracy		F-Score	
	Variable 1	Variable 2	Variable 1	Variable 2
Mean	81.399	83.929	0.877	0.892
Observations	40	40	40	40
t Stat	−2.743		−1.863	
$P(T \leq t)$ one-tail	0.005		0.035	
$P(T \leq t)$ two-tail	0.009		0.07	

15.6.1.2   SA-SVM Classification with SFS

In this section, the "SA-SVM" technique is used as a classification approach rather than SVM, which is then optimized through the SFS. SFS is deemed to be a feature selection technique. Table 15.8 summarizes the results of this algorithm in terms of accuracy and F-score.

According to the results, the average accuracy and F-score of the "SA-SVM Classification with Sequential Feature Selection" approach are computed as 85.748 and 0.911, respectively. It demonstrates the better performance of SA-SVM versus SVM, which had been discussed in Section 15.6.1.1, with average accuracy and F-score of 83.929 and 0.892, respectively. This finding confirms the results achieved in the refereed paper in the new applied research by Y. Maali et al. (2012) [53] on "Self-Advising SVM for Sleep Apnea Classification". The reference paper and our results highlight the same direction regarding the performance of SVM and SA-SVM.

The p-value of the T-test is also computed in Table 15.9, between these two approaches and is equal to 0.004 and 0.002, respectively (Variable 1: SVM

TABLE 15.8

Average Performance of SA-SVM Classification with Sequential Feature Selection

Observation No./Accuracy (%)/F-Score				
1/87.202/0.92	2/85.985/0.913	3/83.766/0.901	4/88.366/0.926	5/88.22/0.926
6/88.312/0.925	7/82.5/0.894	8/82.614/0.891	9/83.864/0.9	10/82.522/0.885
11/82.522/0.892	12/87.744/0.923	13/84.849/0.907	14/89.524/0.934	15/89.524/0.934
16/81.364/0.888	17/87.229/0.919	18/87.284/0.92	19/87.278/0.92	20/88.312/0.925
21/87.284/0.921	22/87.284/0.92	23/82.522/0.894	24/86.648/0.918	25/81.385/0.886
26/88.277/0.926	27/87.175/0.92	28/87.167/0.92	29/81.385/0.888	30/87.175/0.92
31/87.175/0.92	32/82.549/0.889	33/84.886/0.906	34/81.385/0.888	35/87.273/0.92
36/86.226/0.916	37/87.175/0.92	38/86.039/0.913	39/84.821/0.904	40/87.121/0.917

TABLE 15.9

The T-Test Result between Accuracies and F-Scores of "SVM Classification with Sequential Feature Selection" and "SA-SVM Classification with Sequential Feature Selection"

	T-Test: Paired Two Sample for Means			
	Accuracy		F-Score	
	Variable 1	Variable 2	Variable 1	Variable 2
Mean	83.929	85.748	0.892	0.911
Observations	40	40	40	40
t Stat	−3.028		−3.247	
$P(T \leq t)$ one-tail	0.002		0.001	
$P(T \leq t)$ two-tail	0.004		0.002	

Classification with Sequential Feature Selection, Variable 2: SA-SVM Classification with Sequential Feature Selection).

These statistical test analyses show that the results achieved by these approaches are statistically different. Based on this outcome, we use SA-SVM as our algorithm's classifier because it helps improve the performance.

15.6.1.3 SVM Classification with Smart IPSO-SVM Feature Selection

In this algorithm, the Smart IPSO-SVM feature selection algorithm is used to optimize the SVM classifier. Table 15.10 summarizes the results of this algorithm.

Based on the results, the average accuracy and F-score of the "SVM Classification with Smart IPSO-SVM Feature Selection algorithm" approach are 86.526 and 0.912, respectively. In contrast, they are 85.748 and 0.911 for "SVM Classification with Sequential Feature Selection algorithm", Section 15.6.1.1, Table 15.6. The purpose of this comparison, Table 15.6 and 15.10, is to identify the performance of Smart IPSO-SVM feature selection rather than SFS. The results confirm the better performance of Smart IPSO-SVM compared to the SFS when using SVM. The p-values of the

TABLE 15.10

Average Performance of SVM Classification with Smart IPSO-SVM Feature Selection

Observation No./Accuracy (%)/F-Score

1/86.093/0.91	2/86.066/0.908	3/87.229/0.917	4/86.931/0.911	5/87.175/0.916
6/84.957/0.903	7/88.366/0.926	8/87.175/0.917	9/87.229/0.915	10/85.931/0.907
11/84.867/0.899	12/84.794/0.896	13/85.958/0.904	14/87.229/0.917	15/87.229/0.917
16/86.115/0.91	17/85.033/0.902	18/87.33/0.92	19/82.511/0.889	20/87.202/0.914
21/87.094/0.914	22/85.92/0.906	23/87.229/0.918	24/84.957/0.904	25/87.362/0.918
26/87.175/0.916	27/86.039/0.912	28/88.934/0.931	29/84.849/0.9	30/86.066/0.908
31/89.448/0.929	32/88.523/0.929	33/84.867/0.9	34/88.366/0.927	35/86.147/0.911
36/86.023/0.907	37/85.005/0.902	38/88.339/0.926	39/87.175/0.914	40/86.093/0.91

T-test are computed in Table 15.11 as 0.054 and 0.397, respectively (Variable 1: SVM Classification with Sequential Feature Selection, Variable 2: SVM Classification with Smart IPSO-SVM Feature Selection).

However, the statistical tests reveal that the results achieved by the "SVM Classification with Sequential Feature Selection algorithm" and "SVM Classification with Smart IPSO-SVM Feature Selection algorithm" are not statistically different.

Therefore, in the next section, we intend to use the findings of Section 15.6.1.2 (Better performance of SA-SVM rather than SVM) and Section 15.6.1.3 (Better performance of smart IPSO-SVM rather than SFS) to propose and investigate our algorithm.

15.6.1.4 Proposed Algorithm

In this section, we examine the performance of our proposed algorithm, "Smart IPSO-SVM Feature Selection algorithm followed by SA-SVM Classification" and compare its performance with the "SA-SVM Classification with Sequential Feature Selection" in Section 15.6.1.2. As mentioned earlier in Section 15.4.1, our algorithm

TABLE 15.11

The T-Test Result between Accuracies and F-Scores of "SVM Classification with Sequential Feature Selection" and "SVM Classification with Smart IPSO-SVM Feature Selection"

T-Test: Paired Two Sample for Means

	Accuracy		F-Score	
	Variable 1	Variable 2	Variable 1	Variable 2
Mean	85.748	86.526	0.911	0.912
Observations	40	40	40	40
t Stat	−1.646		−0.264	
$P(T \leq t)$ one-tail	0.054		0.397	
$P(T \leq t)$ two-tail	0.108		0.793	

is a hybrid technique that uses SFC, IPSO, and SVM in its process. Table 15.12 tabulates the obtained results of this algorithm.

The proposed approach, with average accuracy and F-score of 87.061 and 0.917, performed better than the other algorithm. Table 15.13 shows the T-test results, where the p-values are 0.001 and 0.011, respectively (Variable 1: SA-SVM Classification with Sequential Feature Selection, Variable 2: SA-SVM Classification with Smart IPSO-SVM Feature Selection).

These statistical tests demonstrate two things: firstly, that the results achieved by these approaches are statistically different, and secondly, our proposed algorithm could outperform the other algorithm.

15.6.2 ACCURACY AND F-SCORE VARIATIONS

To evaluate the variations in results throughout the iterations of different algorithms, Figure 15.3 depicts the propagation of accuracies and F-scores on 40 observations. The box plot is used to measure these variations.

TABLE 15.12
Average Performance of Proposed Algorithm

Observation No./Accuracy (%)/F-Score				
1/82.63/0.884	2/87.046/0.913	3/84.794/0.903	4/87.229/0.921	5/87.229/0.919
6/88.474/0.927	7/87.175/0.917	8/87.121/0.914	9/86.115/0.909	10/84.93/0.904
11/86.039/0.91	12/88.304/0.927	13/86.039/0.909	14/88.528/0.929	15/88.528/0.929
16/84.849/0.899	17/88.312/0.926	18/86.093/0.91	19/87.311/0.92	20/87.229/0.918
21/88.42/0.926	22/86.093/0.91	23/88.182/0.926	24/86.093/0.908	25/86.039/0.91
26/87.175/0.914	27/88.42/0.926	28/85.433/0.908	29/88.42/0.923	30/87.229/0.92
31/88.366/0.927	32/88.409/0.923	33/87.229/0.918	34/87.094/0.916	35/88.258/0.925
36/88.934/0.929	37/88.312/0.925	38/87.175/0.918	39/85.958/0.91	40/87.229/0.917

TABLE 15.13
The T-Test Result between Accuracies and F-Scores of "SA-SVM Classification with Sequential Feature Selection" and Proposed Algorithm

	t-Test: Paired Two Sample for Means			
	Accuracy		F-Score	
	Variable 1	Variable 2	Variable 1	Variable 2
Mean	85.748	87.061	0.911	0.917
Observations	40	40	40	40
t Stat	−3.35		−2.379	
$P(T \leq t)$ one-tail	0.001		0.011	
$P(T \leq t)$ two-tail	0.002		0.022	

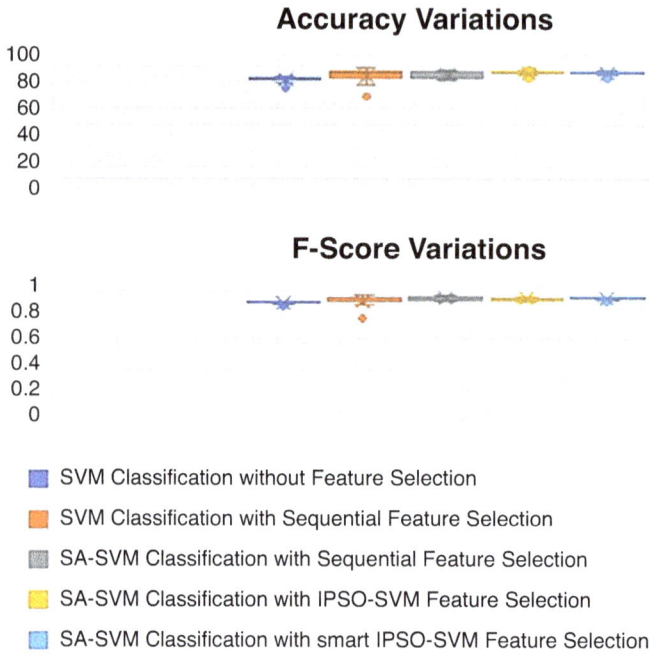

Accuracy Variations

F-Score Variations

SVM Classification without Feature Selection

SVM Classification with Sequential Feature Selection

SA-SVM Classification with Sequential Feature Selection

SA-SVM Classification with IPSO-SVM Feature Selection

SA-SVM Classification with smart IPSO-SVM Feature Selection

FIGURE 15.3   Accuracy and F-score variations for all algorithms.

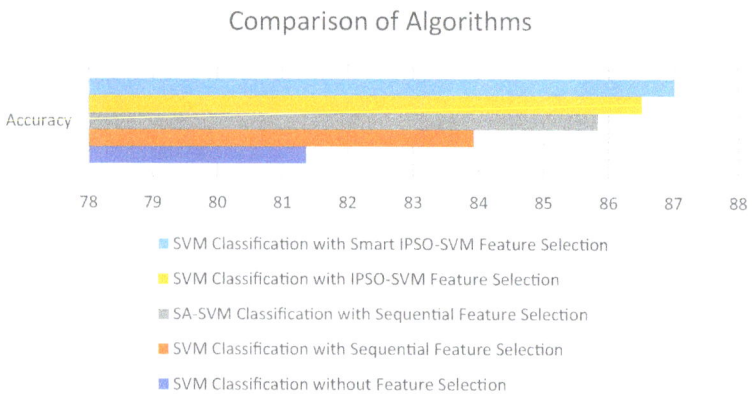

Comparison of Algorithms

Accuracy

78   79   80   81   82   83   84   85   86   87   88

SVM Classification with Smart IPSO-SVM Feature Selection

SVM Classification with IPSO-SVM Feature Selection

SA-SVM Classification with Sequential Feature Selection

SVM Classification with Sequential Feature Selection

SVM Classification without Feature Selection

FIGURE 15.4   Comparison between the mean of accuracies for all algorithms.

The above figure shows that the SA-SVM classification with Smart IPSO-SVM feature selection demonstrates the best accuracies and F-scores with fewer variations.

Figure 15.4 below compares all algorithms in terms of average accuracies. Our proposed algorithm could enhance the performance of previous algorithms, particularly with a significant increment from a range of (81–82) to over (87) when compared with SVM classification without feature selection.

Comparison of Algoritms F-Score

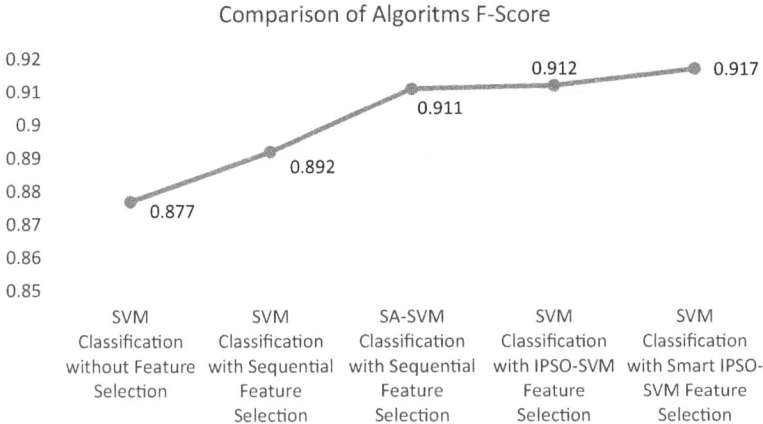

FIGURE 15.5   Comparison between the mean of F-scores for all algorithms.

Figure 15.5 follows the same interpretation as Figure 15.2, except for the considered average of F-scores instead of the accuracy average.

15.7   CONCLUSION

This chapter proposed an innovative algorithm for skin cancer detection. After extracting the different features of images, an algorithm is developed for the feature selection stage to choose the best subset of features for feeding the classification stage. The algorithm is based on IPSO and SA-SVM. In the classification stage, our skin cancer detection algorithm uses the SA-SVM as a classifier. This proposed algorithm is compared with three other algorithms and the algorithm with no feature selection. The experimental results highlight the superior performance of our proposed algorithm, "Smart Feature Selection algorithm with SA-SVM Classification", over the others. The average accuracy and F-score are estimated as 87.061 and 0.917, respectively. The statistical evaluation using a T-test shows the superiority of our algorithm when compared with other algorithms employed in this chapter.

Future research directions are built upon what has been reported in this study. The use of a large dataset with different images of the same lesion is a beneficial way to improve the accuracy of detection. These similar images can be taken from different imaging modalities, such as ultrasound, dermoscopy, etc., so that various aspects of a lesion can be considered. This can lead to more relevant information about the same tumor, such as the depth and surface of the lesion and other criteria. Thus, the acquired information would help to estimate and predict what will happen more accurately. In some cases, the sequential images taken over a period would be a good option for detection purposes.

Different segmentation methods can also be developed to perform diagnosis. Moreover, the SA-SVM as a newly developed classification technique should be further explored and improved. It can be fed with other advanced techniques in feature extraction to improve its accuracy. Finally, another contribution to this area is investigating pathology-related images with the same or improved algorithms.

REFERENCES

Abbasi, N. R., Shaw, H. M., Rigel, D. S., Friedman, R. J., McCarthy, W. H., Osman, I., Kopf, A. W., & Polsky, D. (2004). Early diagnosis of cutaneous melanoma: Revisiting the ABCD criteria. *JAMA*, *292*(22), 2771–2776. https://doi.org/10.1001/jama.292.22.2771.

Afifi, S., GholamHosseini, H., & Sinha, R. (2019). A system on chip for melanoma detection using FPGA-based SVM classifier. *Microprocessors and Microsystems*, *65*, 57–68. https://doi.org/10.1016/j.micpro.2018.12.005.

Akram, M. U., Tariq, A., Khalid, S., Javed, M. Y., Abbas, S., & Yasin, U. U. (2015). Glaucoma detection using novel optic disc localization, hybrid feature set and classification techniques. *Australasian Physical and Engineering Sciences in Medicine*, *38*(4), 643–655. https://doi.org/10.1007/s13246-015-0377-y.

Alfed, N., & Khelifi, F. (2017). Bagged textural and color features for melanoma skin cancer detection in dermoscopic and standard images. *Expert Systems with Applications*, *90*, 101–110. https://doi.org/10.1016/j.eswa.2017.08.010.

Alkolifi Alenezi, N. S. (2019). A method of skin disease detection using image processing and machine learning. *Procedia Computer Science*, *163*, 85–92. https://doi.org/10.1016/j.procs.2019.12.090.

Alyami, J., Rehman, A., Sadad, T., Alruwaythi, M., Saba, T., & Bahaj, S. A. (2022). Automatic skin lesions detection from images through microscopic hybrid features set and machine learning classifiers. *Microscopy Research and Technique*, *85*(11), 3600–3607. https://doi.org/10.1002/jemt.24211.

Ashfaq, M., Minallah, N., Ullah, Z., Ahmad, A. M., Saeed, A., & Hafeez, A. (2019). Performance analysis of low-level and high-level intuitive features for melanoma detection. *Electronics*, *8*(6), 672.

Boser, B. E., Guyon, I., & Vapnik, V. N. (1992). A training algorithm for optimal margin classifiers. *Annual Conference Computational Learning Theory (OLT 1992)* (pp. 144–152), Pittsburgh, PA, July 27–29.

Bray, F., Ferlay, J., Soerjomataram, I., Siegel, R. L., Torre, L. A., & Jemal, A. (2018). Global cancer statistics 2018: GLOBOCAN estimates of incidence and mortality worldwide for 36 cancers in 185 countries. *CA: A Cancer Journal for Clinicians*, *68*(6), 394–424. https://doi.org/10.3322/caac.21492.

Cavalcanti, P. G., Scharcanski, J., & Baranoski, G. V. G. (2013). A two-stage approach for discriminating melanocytic skin lesions using standard cameras. *Expert Systems with Applications*, *40*(10), 4054–4064. https://doi.org/10.1016/j.eswa.2013.01.002.

Celebi, M. E., Mendonca, T., & Marques, J. S. (2015). *Dermoscopy Image Analysis* (Vol. 10). CRC Press: Boca Raton, FL. https://books.google.com.au/books?id=c9uYCgAAQBAJ.

Chang, C.-L., & Chen, C.-H. (2009). Applying decision tree and neural network to increase quality of dermatologic diagnosis. *Expert Systems with Applications*, *36*(2, Part 2), 4035–4041. https://doi.org/10.1016/j.eswa.2008.03.007.

Chau, A. L., Li, X., & Yu, W. (2014). Support vector machine classification for large datasets using decision tree and Fisher linear discriminant. *Future Generation Computer Systems*, *36*, 57–65. https://doi.org/10.1016/j.future.2013.06.021.

Cheng, Y., Swamisai, R., Umbaugh, S. E., Moss, R. H., Stoecker, W. V., Teegala, S., & Srinivasan, S. K. (2008). Skin lesion classification using relative color features. *Skin Research and Technology*, *14*(1), 53–64. https://doi.org/10.1111/j.1600-0846.2007.00261.x.

Clawson, K. M., Morrow, P. J., Scotney, B. W., McKenna, D. J., & Dolan, O. M. (16 September-19 October 2007). Determination of optimal axes for skin lesion asymmetry quantification. *2007 IEEE International Conference on Image Processing, San Antonio, TX, USA*, 2007, pp. II - 453-II - 456, doi: 10.1109/ICIP.2007.4379190.

Divya, D., & Ganeshbabu, T. R. (2020). Fitness adaptive deer hunting-based region growing and recurrent neural network for melanoma skin cancer detection. *International Journal of Imaging Systems and Technology*, *30*(3), 731–752. https://doi.org/10.1002/ima.22414.

Dorrell, D. N., & Strowd, L. C. (2019). Skin cancer detection technology. *Dermatologic Clinics, 37*(4), 527–536. https://doi.org/10.1016/j.det.2019.05.010.

Dreiseitl, S., Ohno-Machado, L., Kittler, H., Vinterbo, S., Billhardt, H., & Binder, M. (2001). A comparison of machine learning methods for the diagnosis of pigmented skin lesions. *Journal of Biomedical Informatics, 34*(1), 28–36. https://doi.org/10.1006/jbin.2001.1004.

Eberhart, R., & Kennedy, J. (4-6 October 1995). A new optimizer using particle swarm theory. *Proceedings of the Sixth International Symposium on Micro Machine and Human Science (MHS'95)* (pp. 39–43), Nagoya. doi: 10.1109/MHS.1995.494215.

Eedy, D. J. (2003). Colour Atlas of Dermoscopy, 2nd enlarged and completely revised Edition (2002). *British Journal of Dermatology, 149*(3), 680–680. https://doi.org/10.1046/j.1365-2133.2003.54541.x.

Ghalejoogh, G. S., Kordy, H. M., & Ebrahimi, F. (2020). A hierarchical structure based on stacking approach for skin lesion classification. *Expert Systems with Applications, 145*, 113127. https://doi.org/10.1016/j.eswa.2019.113127.

Hajabdollahi, M., Esfandiarpoor, R., Khadivi, P., Soroushmehr, S. M. R., Karimi, N., & Samavi, S. (2020). Simplification of neural networks for skin lesion image segmentation using color channel pruning. *Computerized Medical Imaging and Graphics, 82*, 101729. https://doi.org/10.1016/j.compmedimag.2020.101729.

Huang, S., Yang, J., Fong, S., & Zhao, Q. (2020). Artificial intelligence in cancer diagnosis and prognosis: Opportunities and challenges. *Cancer Letters, 471*, 61–71. https://doi.org/10.1016/j.canlet.2019.12.007.

Izakian, H., & Pedrycz, W. (2012). A new PSO-optimized geometry of spatial and spatio-temporal scan statistics for disease outbreak detection. *Swarm and Evolutionary Computation, 4*, 1–11. https://doi.org/10.1016/j.swevo.2012.02.001.

Jain, S., Jagtap, V., & Pise, N. (2015). Computer aided melanoma skin cancer detection using image processing. *Procedia Computer Science, 48*, 736–741. https://doi.org/10.1016/j.procs.2015.04.209.

Janda, M., Olsen, C. M., Mar, V., & Cust, A. E. (2022). Early detection of skin cancer in Australia: Current approaches and new opportunities. *Public Health Research and Practice, 32*(1). https://doi.org/10.17061/phrp3212204.

Javed, R., Rahim, M. S. M., Saba, T., & Rehman, A. (2019). A comparative study of features selection for skin lesion detection from dermoscopic images. *Network Modeling Analysis in Health Informatics and Bioinformatics, 9*(1), 4. https://doi.org/10.1007/s13721-019-0209-1.

Kecman, V. (2001). *Learning and Soft Computing*. The MIT Press: Combridge, MA.

Korotkov, K., & Garcia, R. (2012). Computerized analysis of pigmented skin lesions: A review. *Artificial Intelligence in Medicine, 56*(2), 69–90. https://doi.org/10.1016/j.artmed.2012.08.002.

Lee, H., & Kwon, K. (2020). Diagnostic techniques for improved segmentation, feature extraction, and classification of malignant melanoma. *Biomedical Engineering Letters, 10*(1), 171–179. https://doi.org/10.1007/s13534-019-00142-8.

Liang, Y., Sun, L., Ser, W., Lin, F., Thng, S. T. G., Chen, Q., & Lin, Z. (2018). Classification of non-tumorous skin pigmentation disorders using voting based probabilistic linear discriminant analysis. *Computers in Biology and Medicine, 99*, 123–132. https://doi.org/10.1016/j.compbiomed.2018.05.026.

Ling, S. H., & Nguyen, H. T. (2012). Natural occurrence of nocturnal hypoglycemia detection using hybrid particle swarm optimized fuzzy reasoning model. *Artificial Intelligence in Medicine, 55*(3), 177–184. https://doi.org/10.1016/j.artmed.2012.04.003.

Lu, L., Yan, J., & de Silva, C. W. (2016). Feature selection for ECG signal processing using improved genetic algorithm and empirical mode decomposition. *Measurement, 94*, 372–381. https://doi.org/10.1016/j.measurement.2016.07.043.

Maali, Y., & Al-Jumaily, A. (2012). Hierarchical parallel PSO-SVM based subject-independent sleep apnea classification. *International Conference on Neural Information Processing (ICONIP 2012)*, Doha, Qatar, https://doi.org/10.1007/978-3-642-34478-7_61.

Maali, Y., & Al-Jumaily, A. (2013). Self-advising support vector machine. *Knowledge-Based System*, *52*, 214–222. https://doi.org/10.1016/j.knosys.2013.08.009.

Maali, Y., Al-Jumaily, A., & Laks, L. (2012). Self-advising SVM for sleep apnea classification. *CEUR Workshop Proceedings*, *944*, 24–33.

Maglogiannis, I., & Doukas, C. N. (2009). Overview of advanced computer vision systems for skin lesions characterization. *IEEE Transactions on Information Technology in Biomedicine*, *13*(5), 721–733. https://doi.org/10.1109/TITB.2009.2017529.

Maglogiannis, I., Pavlopoulos, S., & Koutsouris, D. (2005). An integrated computer supported acquisition, handling, and characterization system for pigmented skin lesions in dermatological images. *IEEE Transactions on Information Technology in Biomedicine*, *9*(1), 86–98. https://doi.org/10.1109/titb.2004.837859.

Majumder, S., & Ullah, M. A. (2019). A computational approach to pertinent feature extraction for diagnosis of melanoma skin lesion. *Pattern Recognition and Image Analysis*, *29*(3), 503–514. https://doi.org/10.1134/S1054661819030131.

Malvehy, J., Puig, S., Argenziano, G., Marghoob, A. A., & Soyer, H. P. (2007). Dermoscopy report: Proposal for standardization. Results of a consensus meeting of the International Dermoscopy Society. *Journal of the American Academy of Dermatology*, *57*(1), 84–95. https://doi.org/10.1016/j.jaad.2006.02.051.

Manousaki, A. G., Manios, A. G., Tsompanaki, E. I., Panayiotides, J. G., Tsiftsis, D. D., Kostaki, A. K., & Tosca, A. D. (2006). A simple digital image processing system to aid in melanoma diagnosis in an everyday melanocytic skin lesion unit: A preliminary report. *International Journal of Dermatology*, *45*(4), 402–410. https://doi.org/10.1111/j.1365-4632.2006.02726.x.

Masood, A., & Al-Jumaily, A. (2014). SA-SVM based automated diagnostic system for skin cancer. *Proceedings Volume 9443, Sixth International Conference on Graphic and Image Processing (ICGIP 2014)*, Beijing, 94432L (2015). https://doi.org/10.1117/12.2179094

Masood, A., Al-Jumaily, A., & Anam, K. (2014). Texture analysis based automated decision support system for classification of skin cancer using SA-SVM. *International Conference on Neural Information Processing (ICONIP 2014)*, https://doi.org/10.1007/978-3-319-12640-1_13.

Masood, A., Al- Jumaily, A. & Anam, K. (2015). Self-supervised learning model for skin cancer diagnosis. *2015 7th International IEEE/EMBS Conference on Neural Engineering (NER)* (pp. 1012–1015), Montpellier. doi: 10.1109/NER.2015.7146798.

Mohemmed, A. W., Zhang, M., & Johnston, M. (18-21 May 2009). Particle swarm optimization based adaboost for face detection. *2009 IEEE Congress on Evolutionary Computation* (pp. 2494–2501), Trondheim. doi: 10.1109/CEC.2009.4983254.

Moradi, P., & Gholampour, M. (2016). A hybrid particle swarm optimization for feature subset selection by integrating a novel local search strategy. *Applied Soft Computing*, *43*, 117–130. https://doi.org/10.1016/j.asoc.2016.01.044.

Murugan, A., Nair, S. A. H., & Kumar, K. P. S. (2019). Detection of skin cancer using SVM, random forest and kNN classifiers. *Journal of Medical Systems*, *43*(8), 269. https://doi.org/10.1007/s10916-019-1400-8.

Nachbar, F., Stolz, W., Merkle, T., Cognetta, A. B., Vogt, T., Landthaler, M., Bilek, P., Braun-Falco, O., & Plewig, G. (1994). The ABCD rule of dermatoscopy. High prospective value in the diagnosis of doubtful melanocytic skin lesions. *Journal of the American Academy of Dermatology*, *30*(4), 551–559. https://doi.org/10.1016/s0190-9622(94)70061-3.

Pranava, G., & Prasad, P. V. (6-8 February 2013). Constriction coefficient particle swarm optimization for economic load dispatch with valve point loading effects. *2013 International Conference on Power, Energy and Control (ICPEC)* (pp. 350–354), Dindigul. doi: 10.1109/ICPEC.2013.6527680.

Pudil, P., Novovičová, J., & Kittler, J. (1994). Floating search methods in feature selection. *Pattern Recognition Letters*, *15*(11), 1119–1125. https://doi.org/10.1016/0167-8655(94)90127-9.

Ramezani, M., Karimian, A., & Moallem, P. (2014). Automatic detection of malignant melanoma using macroscopic images. *Journal of Medical Signals and Sensors*, *4*(4), 281–290. https://pubmed.ncbi.nlm.nih.gov/25426432; https://www.ncbi.nlm.nih.gov/pmc/articles/PMC4236807/.

Sáez, A., Serrano, C., & Acha, B. (2015). Global pattern classification in dermoscopic images. In M. E. Celebi, T. Mendonca, J. S. Marques (eds.), *Dermoscopy Image Analysis* (pp. 183–209). CRC Press: Boca Raton, FL. https://doi.org/10.1201/b19107-7.

Seidenari, S., Pellacani, G., & Grana, C. (2005). Pigment distribution in melanocytic lesion images: a digital parameter to be employed for computer-aided diagnosis. *Skin Research and Technology*, *11*(4), 236–241. https://doi.org/10.1111/j.0909-725X.2005.00123.x.

Shi, Y., & Eberhart, R. (1998). A modified particle swarm optimizer. *1998 IEEE International Conference on Evolutionary Computation Proceedings. IEEE World Congress on Computational Intelligence (Cat. No.98TH8360)* (pp. 69–73), Anchorage, AK. doi: 10.1109/ICEC.1998.699146.

Singh, L., Janghel, R. R., & Sahu, S. P. (2020). Designing a retrieval-based diagnostic aid using effective features to classify skin lesion in dermoscopic images. *Procedia Computer Science*, *167*, 2172–2180. https://doi.org/10.1016/j.procs.2020.03.267.

Sokolova, M., Japkowicz, N., & Szpakowicz, S. (2006). Beyond accuracy, F-score and ROC: A family of discriminant measures for performance evaluation. In A. Sattar, & B. H. Kang (eds.), *AI 2006: Advances in Artificial Intelligence*. Springer: Berlin, Heidelberg. https://doi.org/10.1007/11941439_114.

Stoecker, W. V., Li, W. W., & Moss, R. H. (1992). Automatic detection of asymmetry in skin tumors. *Computerized Medical Imaging and Graphics*, *16*(3), 191–197. https://doi.org/10.1016/0895-6111(92)90073-I.

Sung, H., Ferlay, J., Siegel, R. L., Laversanne, M., Soerjomataram, I., Jemal, A., & Bray, F. (2021). Global cancer statistics 2020: GLOBOCAN estimates of incidence and mortality Worldwide for 36 cancers in 185 countries. *CA: A Cancer Journal for Clinicians*, *71*(3), 209–249. https://doi.org/10.3322/caac.21660.

Tan, T. Y., Zhang, L., & Lim, C. P. (2019). Intelligent skin cancer diagnosis using improved particle swarm optimization and deep learning models. *Applied Soft Computing*, *84*, 105725. https://doi.org/10.1016/j.asoc.2019.105725.

Tan, T. Y., Zhang, L., & Lim, C. P. (2020). Adaptive melanoma diagnosis using evolving clustering, ensemble and deep neural networks. *Knowledge-Based Systems*, *187*, 104807. https://doi.org/10.1016/j.knosys.2019.06.015.

Tan, T. Y., Zhang, L., Lim, C. P., Fielding, B., Yu, Y., & Anderson, E. (2019). Evolving ensemble models for image segmentation using enhanced particle swarm optimization. *IEEE Access*, *7*, 34004–34019. https://doi.org/10.1109/ACCESS.2019.2903015.

Tan, T. Y., Zhang, L., Neoh, S. C., & Lim, C. P. (2018). Intelligent skin cancer detection using enhanced particle swarm optimization. *Knowledge-Based Systems*, *158*, 118–135. https://doi.org/10.1016/j.knosys.2018.05.042.

Thurnhofer-Hemsi, K., & Domínguez, E. (2021). A convolutional neural network framework for accurate skin cancer detection. *Neural Processing Letters*, *53*(5), 3073–3093. https://doi.org/10.1007/s11063-020-10364-y.

Yahya, A. (2011). Feature selection for high dimensional data: An evolutionary filter approach. *Journal of Computer Science*, *7*, 800–820. https://doi.org/10.3844/jcssp.2011.800.820.

Zhang, N., Cai, Y.-X., Wang, Y.-Y., Tian, Y.-T., Wang, X.-L., & Badami, B. (2020). Skin cancer diagnosis based on optimized convolutional neural network. *Artificial Intelligence in Medicine*, *102*, 101756. https://doi.org/10.1016/j.artmed.2019.101756.

16 A Review on Advanced CNN Architecture in Diagnosing Alzheimer's Disease

Nur Amirah Abd Hamid
University Technology Malaysia

Daphne Teck Ching Lai
University Brunei Darussalam

Mohd Ibrahim Shapiai@Abd Razak
University Technology Malaysia

16.1 INTRODUCTION

Dementia and Alzheimer's Disease (AD) are often known to be related and inter-changeable, yet both terms have different definitions. Simply, dementia is a term used to describe people with cognitive impairments such as memory loss, deterioration in thinking and reasoning function, and changes in behavioral abilities. These may affect one's daily routine and social autonomy. In the year 2021, Alzheimer's Disease International (ADI) stated the number of people living with dementia globally will escalate to 78 million by the year 2030 (Gauthier et al. 2021). The rise in dementia cases is caused by various factors and is most highly related to brain-related diseases such as AD, vascular dementia, dementia with Lewy bodies, frontotemporal dementia, and young-onset dementia. Among them, AD is the most common cause of dementia, contributing 60%–80% of the cases due to the rising world population ageing (Nations 2020). AD, on the other hand, is an irreversible neurodegenerative disease that progressively changes the brain morphologies due to the degeneration of the brain cells. Hence, AD is distinctly defined as a disease related to dementia in the way of the cause of dementia. Yet, the term dementia is not solely defined as AD. It may be possible to describe other brain-related diseases.

Normally, older people over 65 years old are the most at risk of suffering from AD due to the ageing factor, and the progress of the disease is slow on-set and oblivious. Moreover, the symptoms exhibited during the mild to late stage of the disease. Thus, it requires a very attentive medical evaluation by the experts to diagnose AD earlier

DOI: 10.1201/9781003346678-16

and accurately. Hence, a better and more appropriate treatment can be facilitated. Even so, there is still no treatment that can fully cure AD, appropriate treatment to slow the progression of AD such as drug treatment (Dunn, Stein, and Cavazzoni 2021) and physical and mental support treatment to sustain their life can be provided once the disease is detected earlier. The evaluation process involves several assessments, including medical history and physical examination, cognitive assessment, blood test, and neuroimaging scan, conventionally. These processes require follow-up sessions from primary care to experts, which take longer time due to the nature of the disease which is slow progress on-set and depending on the prodromal stage (Visser et al. 2006). The observation of the medical history and behavioral patterns of the patient is the initial stage of assessment to monitor the progression of the symptoms. On top of that, the cognitive assessment is conducted to screen and evaluate the cognitive abilities of the patients based on notable tests such as the mini-mental state exam (MMSE) (Vertesi et al. 2001) and the Montreal Cognitive Assessment (MoCA) (Davis et al. 2015). In some cases, blood and urine tests are recommended to identify the presence of other causes of cognitive decline (Hort et al. 2010) or comorbidities although there is no evidence-based data explaining the significance of performing specific routine tests (Nitrini et al. 2005, Knopman et al. 2001). To support and assist experts in assessing and diagnosing AD progression and severity, neuroimaging scan using medical imaging modalities such as computed tomography (CT), positron emission tomography (PET), magnetic resonance imaging (MRI), single photon emission computed tomography (SPECT) are commonly performed (Yagis et al. 2020, Simon, Baskar, and Jayanthi 2019). Among the modalities, MRI is the most likely and safest modality to be used since it produces no radiation that can be harmful to the patients. Thus, with that in mind, the experts manually visually analyse the changes in the structure brain with the assistance of MRI modality. However, these manual evaluations are prone to human error, costly and disruptive, and require manual and thorough intervention from professional experts.

Computer-aided diagnosis (CAD) system is a powerful tool and has been widely used to assist experts in diagnosing AD in the past few decades (Ebrahimighahnavieh, Luo, and Chiong 2019, Rathore et al. 2017). The emergence of the CAD system in this domain, due to its automated analysis system, employs a machine learning approach and deep learning approach (Afzal et al. 2019). The machine learning approach has been proven to perform well in classification, regression, and decision-making tasks in various domain applications, including medical image analysis. In AD image analysis domain, the surge of research in utilizing machine learning approach to diagnose the disease has shown the successfulness of its framework to classify, localize and detect the changes in the structure of the brain. Most of the research investigate and compare the performance of different types of machine learning classifiers in classifying and detecting the AD pattern. For instance, the performance of a support vector machine (SVM) has been investigated by several researchers, as stated in their works (Plant et al. 2010, Aguilar et al. 2013, Nir et al. 2015, Altaf et al. 2017). Another example, artificial neural network (ANN), has been employed in (Aguilar et al. 2013, Escudero, Zajicek, and Ifeachor 2011) as a classifier in the machine learning framework. It can be conceded the machine learning framework, which consists of pre-processing techniques, hand-crafted feature extraction, and different types of classifiers, is powerful and robust enough to assist experts in diagnosing AD at its time. Yet, the tedious framework of machine learning algorithms, requiring manual feature extraction, has led to the exploration and utilization of recent cutting-edge technology, deep learning approach.

Convolutional neural network (CNN) is the most well-known architecture in the deep learning domain, specifically for image classification, object detection, and image segmentation task due to its outstanding performance in extracting the features automatically without prior manual feature selection. Hence, some prominent CNN architectures have been acknowledged and recognized in ILSVRC, an ImageNet challenge competition, such as AlexNet (Krizhevsky, Sutskever, and Hinton 2012), VGG (Simonyan and Zisserman 2014) and ResNet (He et al. 2016). In medical images domain, particularly AD, CNN architecture has been adopted in most studies (Ebrahimighahnavieh, Luo, and Chiong 2019), replacing conventional machine learning approaches to assist and support experts in diagnosing AD, effectively (Liu et al. 2021). The effectiveness and robustness of CNN in segmenting, localizing, and detecting the pattern changes of the brain through MRI scan images can facilitate early treatment interventions for patients. However, the outstanding performance and robustness of CNN are subjected to data-driven, requiring a large-scale dataset to achieve a better classification performance. The small-scale dataset of AD may result in architecture struggles in a common issue of overfitting, i.e., underperformance in detecting and classifying the pattern changes of AD. Besides, the insensitive to local position information (spatial invariance) is the key behavior to the outstanding performance of CNN. Even so, in medical image analysis, the local position information is significant (Schlemper et al. 2018), especially in MRI images; the local position of the hippocampus is an important part that distinguishes the pattern changes of the brain for AD. Therefore, recent studies have shown the emergence of an advanced deep learning approach involving advanced modules and CNN architectures employed for AD pattern recognition on magnetic resonance imaging (MRI) data to tackle the aforementioned issues, specifically the overfitting issue and preserving the local information (spatial invariance) issue when dealing with medical images. In dealing with the limited amount of the data, most of conventional and advanced deep learning approaches commonly work with several enhancement modules such as data pre-processing, data augmentation and transfer learning strategy. This chapter aims to review several recent literature studies that employed advanced modules on CNN architectures to solve the specific issues of AD pattern-related recognition. In other words, the contribution of this chapter can be summarized as follows:

1. Able to find out different databases that provide the dataset and the most widely used databases in this domain.
2. This chapter provides insight into recent advanced CNN approaches, mainly focusing on attention mechanisms to further improve the architecture in handling medical images.
3. The direction of future study is based on the recent approaches, considering the challenges and limitations.

Section 16.1 overviews the problem background of the study in the AD domain. Section 16.2 provides information on publicly available datasets and the current CAD system in this domain utilizing CNN architecture. In Section 16.3, the limitation of the current CAD system, the underlying challenges of the domain, and the possible future work has been highlighted. Finally, Section 16.4 conceded the study.

16.2 LITERATURE REVIEW

16.2.1 CAD System for Alzheimer's Disease Diagnosis

The emergence of automated analysis system, particularly CAD system in assisting experts to diagnose AD has proven that conventional methods are prone to human error, costly and disruptive, and require manual and thorough intervention from professional experts. Hence, the aiding of the CAD system can intervene the human error and lead to a timely diagnosis. CAD system is commonly related to the utilization of machine learning approach and deep learning approach to enhance the classification and detection task of AD pattern based on MRI images. To date, the most powerful CAD system is the deep learning-based system, to be specific, CNN since it offers the best solution in decision-making task for image domain. This section briefly reviews the literature studies on a CNN-based system for AD pattern recognition in order to answer the related research questions regarding the dataset and the recent advanced CNN-based system for AD pattern recognition to further improve the architecture in handling MRI images.

16.2.2 Dataset

Experts often diagnose AD through visual analysis of brain scan images obtained from neuroimaging modalities such as MRI, CT, PET, and SPECT. In recent years, to support and assist experts, the CNN-based CAD system has been widely used to analyze brain scan images particularly single modality structural MRI brain images as stated by Ebrahimighahnavieh, Luo, and Chiong (2019) in their works. MRI itself is non-invasive with the absence of radioactive pharmaceutical injection such as gamma rays and positron and provides good spatial resolution and tissue contrast (Beheshti et al. 2016). Since MRI can provide a high-quality spatial resolution of anatomical structure images of the brain, thus, it is useful to accurately analyse activities and changes in the brain for diagnosing AD. In addition, it has shown most of the studies employed T1-weighted structural MRI images due to the high-resolution T1-weighted MRI sequences are able to distinguish the anatomical boundaries and detect the structural changes in the brain (Frisoni et al. 2010). In the range between the years 2020 and 2021, most of the CNN-based studies in this domain employed MRI data provided by several databases such as ADNI, OASIS, MIRIAD, GARD and PND.

From Table 16.1, the most frequently used MRI data is provided by Alzheimer's Disease Neuroimaging Initiative (ADNI), followed by Open Access Series of Imaging Studies (OASIS) since both are easy to access publicly for their scientific research. The following sub-sections will briefly explain in detail on ADNI dataset and OASIS dataset, respectively.

TABLE 16.1

List of the Dataset Used in year 2020–2021

Dataset	Database	Literature References
ADNI	Alzheimer's Disease Neuroimaging Initiative (https://adni.loni.usc.edu/)	Ebrahimi, Luo, and Chiong (2020), Jiang et al. (2020), Nanni et al. (2020), Qiu et al. (2020), Tufail, Ma, and Zhang (2020b), Wang et al. (2020), AbdulAzeem, Bahgat, and Badawy (2021), Ebrahimi et al. (2021), Guo et al. (2021), Hedayati, Khedmati, and Taghipour-Gorjikolaie (2021), Kumar and Nandhini (2021), Sathiyamoorthi et al. (2021), and Zhang et al. (2021)
OASIS	Open Access Series of Imaging Studies (http://www.oasis-brains.org/)	Nawaz et al. (2020), Tufail, Ma, and Zhang (2020a), Yagis et al. (2020), Zaabi et al. (2020), Alinsaif, Lang, and Alzheimer's Disease Neuroimaging (2021), Liu et al. (2021), Wang et al. (2021)
MIRIAD	Minimal Interval Resonance Imaging in Alzheimer's Disease (https://www.nitrc.org/projects/miriad/)	Renjith and Wagaj (2020) and Goenka and Tiwari (2021)
GARD	Gwangju Alzheimer's and Related Dementia Cohort	Basher et al. (2021) and Khagi and Kwon (2021)
PND	Parelsnoer Neurodegenerative Diseases Biobank	Bron et al. (2021)

16.2.2.1 ADNI

ADNI was established in 2004 by Dr. Michael W. Weiner. Primarily, it is funded as a private-public partnership with the National Institutes of Health (NIH), National Institutes on Aging (NIA) and 20 companies, contributing in total $67 million for an initial 5-year study (ADNI-1) (Weiner et al. 2010). The study was then extended by 2 years in 2009, supported by Grand Opportunities Grant, released ADNI-GO cohort. In 2011 and 2016, ADNI-2 (Weiner et al. 2015) and ADNI-3 (Weiner et al. 2017) were released respectively by renewing the ADNI-1 Grant. Generally, ADNI is an ongoing study focusing on collecting, validating and utilizing multi-modalities data i.e., MRI and PET images, genetics, cognitive tests, CSF, and blood biomarkers for early detection and tracking of AD. The resources of the study are obtained from North American participants and can be accessed via http://adni.loni.usc.edu/.

16.2.2.2 OASIS

OASIS is an ongoing series of studies that took place in the Washington University Knight Alzheimer Disease Research Center over 15 years, compiling the data generated from multi-modalities. It provides brain neuroimaging data freely to facilitate scientific research, aiming for future discoveries and improvement in basic and clinical neuroscience. Till now, there are three cohorts of brain neuroimaging data: OASIS-1 (Marcus et al. 2007), OASIS-2 (Marcus et al. 2010) and OASI-3 (LaMontagne et al. 2019). Both OASIS-1 and OASIS-2 provide cross-sectional MRI data and longitudinal MRI data respectively and have been adopted for hypothesis-driven data analyses, development of neuroanatomical atlases, and development of segmentation algorithms. OASIS-3 is the latest released neuroimaging data, comprising data from multi-modalities i.e. longitudinal neuroimaging data, clinical, cognitive and biomarker dataset for normal aging and AD. These data are available at https://www.oasis-brains.org/ and https://central.xnat.org/.

16.2.3   CNN-Based System

The recent approach in this domain is utilizing CNN architecture within the CAD system. CNN has been widely used in this domain for the whole classification process due to its advancement in decision-making tasks and its ability to automatically extract the features without hand-crafted feature extraction and selection. Most researchers presented standalone CNN models, while in some cases, CNN is used for feature extraction and combined with other machine learning algorithms to accomplish the task. It has shown promising results in this domain with the aiding of enhancement modules such as pre-processing module, data augmentation modules and transfer learning strategies. Commonly, these enhancement modules are integrated with CNN architecture to boost the performance of CNN in detecting and classifying the pattern of AD. The main focus of these integrations is to enhance the quality and quantity of the dataset since CNN architecture requires a large-scale dataset to be trained effectively without overfitting. Several literature studies that employed the enhancement modules and were published between 2020 and 2021 have been listed in Table 16.2. From the literature studies in Table 16.2, the enhancement modules are selectively chosen according to the issue encountered. Also, it can be conceded the enhancement modules are significant and play big roles in improving and optimizing the performance of CNN architecture in detecting and classifying the pattern of AD.

The lack of data issue in ADe domain may be the major hindrance to the performance of CNN architecture, yet, has been encountered by the aforementioned enhancement modules. Even so, it is good to highlight the behaviour of CNN architecture might affect the performance of medical image analysis, particularly MRI images in this domain. According to LeCun et al. (1998), CNN is one of the types of feedforward neural networks built with the specialty of spatial invariance and are biologically inspired by the visual cortex (Basaia et al. 2019). Spatial invariance is defined as insensitive to the local position information of the images which for medical images, the local information is significant to distinguish and analyse the disease. Therefore, in recent years, several studies have investigated advancing the CNN architecture by fine-tuning the architecture to improve the performance. The following section explains in detail the advanced CNN-based system.

16.2.4   Advanced CNN-Based System

As mentioned in the previous section, this section will discuss the advanced strategy in enhancing the CNN architecture performance for effective AD classification and

TABLE 16.2

Literature Studies on CNN-Based System in Year 2020–2021

Architecture	Data	Enhancement Modules			References
		Data Pre-Processing	Data Augmentation	Transfer Learning	
AlexNet	ROI-based	×	×	√	Zaabi et al. (2020)
Inception, Xception, custom 2D CNN	Slice-based	√	√	√	Tufail, Ma, and Zhang (2020a)
DenseNet & Inception-V4	Slice-based	√	×	√	Qiu et al. (2020)
Custom-Made CNN	Slice-based	√	√	×	AbdulAzeem, Bahgat, and Badawy (2021)
5-layered CNN	ROI-based	√	×	√	Sathiyamoorthi et al. (2021)
VGG-16	Slice-based	√	√	√	Kumar and Nandhini (2021)
5-layered CNN	Slice-based	√	×	√	Guo et al. (2021)

detection. Since two issues have been the hindrance in this domain which are: (i) Lack of MRI data samples and, (ii) CNN's behavior, insensitive to the local position information, several strategies have been adopted by the researchers to tackle the issues. Conventionally, enhancement modules have been employed within the CNN architecture, mainly to improve the dataset. Still, in recent years, researchers have begun to advance the CNN architecture by modifying the neural network architecture. In terms of neural network architecture, there are three ways (i.e. depth, width, and cardinality of the neural networks) commonly considered to further improve the performance of conventional CNNs architecture. The robustness of conventional CNN architectures (from AlexNet to DenseNet) to further extract useful but intricate information leads to the emergence of advanced CNN architectures. Simply, advanced CNN architectures are defined as CNN architectures that have been integrated with advanced neural network-based modules (advanced enhancement modules) to improve its performance in classification task. Several advanced enhancement modules based on neural networks (i.e. deep separable convolution, attention mechanism, temporal convolution network) listed in Table 16.3 have been demonstrated to sharpen and alleviate the CNN architecture issues.

Convolution layer in CNN is replaced with depthwise separable convolution (DSC) to reduce the number of parameters and computational cost of training and to achieve high accuracy performance. With the network depth reduced, the idea of utilizing DSC is to produce a more compact model, allowing use on mobile-embedded devices with limited computing resources. DSC was proposed to reduce the dependency on

TABLE 16.3
Literature Studies on Advanced CNN-Based System in Year 2021

Problem Solved	Dataset	Type of Input Data	Data Pre-Processing	Data Augmentation	CNN Architecture	Transfer Learning	Year	References
Binary and three-way classification	ADNI, OASIS	Slice-based	√	×	CNN with DSC module	√ AlexNet, GoogLeNet	2021	Liu et al. (2021)
Binary classification	ADNI	Sequence-based	√	√	TCN	√ ResNet-18	2021	Ebrahimi et al. (2021)
Binary classification	ADNI	Voxel-based	√	×	3D DenseNet with connection-wise attention mechanism	×	2021	Zhang et al. (2021)
Binary classification	OASIS	Slice-based	√	√	VGG-16 with CBAM (attention mechanism)	√ VGG-16	2021	Wang et al. (2021)
Binary classification	OASIS	Slice-based	×	×	VGG-16 with attention mechanism and global average pooling (GAP) layer	√ VGG-16	2021	Abd Hamid et al. (2021)

high computational resources. Furthermore, the graphic processor unit (GPU) for running deep learning algorithms has become affordable in recent years. Together with the emergence of 5G technology, this can facilitate the deep learning processes on edge devices in the near future, by transferring the processes into cloud computing. In this manner, the concern for high computational resources requirement can be satisfied.

As mentioned before, the lack of the data sample is the major hindrance to this domain. Aside from lack of data, other forms of limitations regarding the data have also arisen such as: (i) the extraction from 3D MRI scan to 2D slices MRI images (i.e. slice-based data) might result in the features related to the ROI lost due to the brain regions spanned over the 2D slices (Ebrahimi et al. 2021). (ii) 3D CNN has been adopted on voxel-based data (3D MRI scan), but the architecture is complex and the large number of training parameters required may cause overfitting. Hence, it has motivated the combination of image-based and sequence-based data. Conventionally, CNN and RNN architecture are combined together to perform feature extraction and classification task respectively, but these two architectures are uncorrelated. Therefore, temporal convolutional networks (TCN) have been proposed since they can perform feature extraction and classification simultaneously. TCN is a CNN architecture variation with a dilated causal convolutional module that can prevent information leakage from the future to the past. Thus, it has great control over the receptive field size and improved parallelism in computational processing which in turn, reduces the need for computational resources.

Among recent advanced CNN-based system, the attention mechanism has been demonstrated to be a powerful and essential cutting-edge module in AD pattern recognition. The concept of attention mechanism is based on visual attention in human vision system, which depicts the human vision always focusing on selective parts of the whole visual screen (Zhang et al. 2021). The following sub-sections highlighted the insight of recent advanced CNN approaches, mainly utilize attention mechanisms to further improve the architecture in handling MRI images.

16.2.4.1 Attention Mechanism as Advanced Module

The attention mechanism is incorporated with the CNN architecture to improve the architecture itself by selectively capturing the most important features of MRI brain images for classification. Various type of attention mechanisms has been utilized in a large number of computer vision applications (Schlemper et al. 2018, Pesce et al. 2019, Schlemper et al. 2019, Sitaula and Hossain 2020). In the AD domain, attention mechanism is still considered as a new approach. This can be proved by the statistical number of research in 2020–2021, in which the research utilizing attention mechanism emerged in 2021. In addition, the research mostly originated from mainland China and one from Malaysia as illustrated in Figure 16.1.

The three studies have used and derived the fundamental idea of visual attention (Xu et al. 2015) in AD pattern recognition. Visual attention has been derived and employed in various ways in this domain such as: (i) connection-wise attention mechanism in Zhang et al. (2021); (ii) combination of spatial attention and channel attention mechanism in Wang et al. (2021), but with similar target, to further improve

FIGURE 16.1 Distribution of research on utilization attention mechanism in the AD domain in 2021.

the CNN architecture performance; (iii) combination channel attention with GAP layer in Abd Hamid et al. (2021).

The utilization of the attention mechanism in this domain is to tackle the CNN's behavior issue regarding the insensitivity to local position information. It has been stated in Schlemper et al. (2018), local position information is prominent for medical images analysis, particularly MRI since it provides salient features for discriminating the pattern of AD. Hence, these studies focused on preserving the local position of the MRI images for better performance of classification and detection of AD pattern as well as overcome the overfitting issue. Yet, the difference between the three studies is either the dataset, enhancement modules or the way of attention mechanism has been implemented. Even though the attention mechanism has shown a positive result in enhancing classification performance, the concept remains unclear and further research is needed.

16.3   LIMITATIONS AND FUTURE WORK

As mentioned before, deep learning-based CAD technology particularly CNN-based is the cutting-edge technology in this domain, to assist the experts in decision-making task during diagnosis. Therefore, to enhance the performance of CNN architecture in the classification task of the disease based on the changes of structural brain pattern, most of the studies aimed to encounter the two major limitations which are the insufficient and imbalanced data for research and the vanishing of salient local and spatial information of the images as listed in Table 16.4. Over time, many researchers aimed to solve the insufficient and imbalanced data by introducing solutions such as pre-processing module, data augmentation module and transfer learning strategy. In addition, some advanced CNN architectures also have been introduced to encounter the dataset issue. These solutions have proven to support the data-driven issue, however, the performance of CNN architecture is also subject to the architecture itself which leads to the second limitation, vanishing of salient local and spatial information of the images during the learning process. The second challenge has been addressed in several researches, mainly focusing on insight into CNN architecture by developing and modifying the architecture to such an extent of emerging the advanced CNN architecture such as TCN and attention mechanism in this domain. Although the advanced CNN architecture may resolve the second issue, yet, CNN architecture itself is known as 'black box', the process is difficult to understand and requires more research. Overall, the limitations are still being solved by the researchers up till now, which opens the opportunities for other researchers to discover and explore within this domain. It is good to overcome the limitations and, hence, can enhance the performance of CNN-based CAD technology in assisting the experts to diagnose the disease.

Based on the findings of this review, the following directions demand further contribution to the field:

1. Producing synthetic data similar to real data. The artificial MRI images should reflect more realistic trends of class sizes to develop more effective models to handle the real data.

TABLE 16.4
Limitations

No	Limitations	Current Solutions
1	Insufficient data for research and development/imbalanced data	**Data Augmentation** Use augmented and synthetically generated data to generate more data for CNN model training purpose. **Transfer Learning Strategy** Adopt transfer learning strategy within CNN model to supply pre-trained weight for CNN without training from scratch, requiring huge amount of data
2	Vanishing of salient local and spatial information	**Advanced CNN Module** Use combination of two or more AI algorithms to fully utilized the spatial relations between the 2D MRI sequence images for enhancing the classification task. Integrate advanced modules within the CNN architecture to sustain the salient local information, which may lost due to the dimensionality reduction during training CNN model, or geometric transformation during augmenting data.

2. More attention should be given to fine-tuning the network parameters: number of layers, activation and loss functions, kernel and stride size, and regularization methods.
3. More attention on the idea of incorporation of other advanced enhancement modules within the CNN architecture to overcome the spatial invariance issue of the CNN.
4. Bring the idea of attention mechanism into the segmentation task of salient features of the MRI images by incorporating attention mechanism with other variations of CNN, which is fully convolution network (FCN).

16.4 CONCLUSION

CNN-based CAD system has become the most interesting topic in the AD domain. The surge of studies has shown the rapid growth of the research in this domain in utilizing automated analysis system to assist experts in diagnosing AD. This chapter provides a brief review on existing CNN-based approaches between 2020 and 2021. Two major limitations have been addressed: lack of the data and vanishing salient features information. A range of enhancement modules have been used to tackle the lack of data issue, to improve the performance, including pre-processing, augmentation, transfer learning strategy, and advanced modules. On top of that, recent cutting-edge technology, attention mechanism has proved to solve the basis limitation of CNN's behavior regarding the spatial invariance. The choice of the modules has been subjected to the research objectives. In conclusion, this article will pave the way for further research on developing more effective solutions for the AD pattern recognition problem. There is a need for more accurate frameworks to be applied in actual setups. For future work, one may explore the insight of advanced CNN architecture in-depth or other advanced deep learning architecture such as FCN to encounter the common issue highlighted in this domain, which might involve the segmentation or classification task.

REFERENCES

Abd Hamid, Nur Amirah, Mohd Ibrahim Shapiai, Uzma Batool, Ranjit Singh Sarban Singh, Muhamad Kamal Mohammed Amin, and Khairil Ashraf Elias. 2021. "Incorporating attention mechanism in enhancing classification of Alzheimer's disease." In H. Fujita and H. Perez-Meana (eds.), *New Trends in Intelligent Software Methodologies, Tools and Techniques*, pp. 496–509. IOS Press: Amsterdam, The Netherlands.

AbdulAzeem, Yousry, Waleed M. Bahgat, and Mahmoud Badawy. 2021. "A CNN based framework for classification of Alzheimer's disease." *Neural Computing and Applications* 33(16):10415–10428. doi: 10.1007/s00521-021-05799-w.

Afzal, Sitara, Muazzam Maqsood, Faria Nazir, Umair Khan, Farhan Aadil, Khalid M. Awan, Irfan Mehmood, and Oh-Young Song. 2019. "A data augmentation-based framework to handle class imbalance problem for Alzheimer's stage detection." *IEEE Access* 7:115528–115539. doi: 10.1109/access.2019.2932786.

Aguilar, Carlos, Eric Westman, J-Sebastian Muehlboeck, Patrizia Mecocci, Bruno Vellas, Magda Tsolaki, Iwona Kloszewska, Hilkka Soininen, Simon Lovestone, Christian Spenger, Andrew Simmons, and Lars-Olof Wahlund. 2013. "Different multivariate techniques for automated classification of MRI data in Alzheimer's disease and mild cognitive impairment." *Psychiatry Reserarch* 212(2):89–98. doi: 10.1016/j.pscychresns.2012.11.005.

Alinsaif, Sadiq and Jochen Lang. Initiative Alzheimer's Disease Neuroimaging. 2021. "3D shearlet-based descriptors combined with deep features for the classification of Alzheimer's disease based on MRI data." *Computers in Biology and Medicine* 138:104879. doi: 10.1016/j.compbiomed.2021.104879.

Altaf, Tooba, Syed Muhammad Anwar, Nadia Gul, N. Majeed, and M Majid. 2017. "Multiclass Alzheimer disease classification using hybrid features." *Proceedings of the Future Technologies Conference (FTC)*, Vancouver, Canada.

Basaia, Silvia, Federica Agosta, Luca Wagner, Elisa Canu, Giuseppe Magnani, Roberto Santangelo, Massimo Filippi, Initiative Alzheimer's Disease Neuroimaging. 2019. "Automated classification of Alzheimer's disease and mild cognitive impairment using a single MRI and deep neural networks." *Neuroimage Clinical* 21:101645. doi: 10.1016/j.nicl.2018.101645.

Basher, Abol, Byeong C. Kim, Kun Ho Lee, and Ho Yub Jung. 2021. "Volumetric feature-based Alzheimer's disease diagnosis from sMRI data using a convolutional neural network and a deep neural network." *IEEE Access* 9:29870–29882. doi: 10.1109/access.2021.3059658.

Beheshti, Iman, Hasan Demirel, Farnaz Farokhian, Chunlan Yang, and Hiroshi Matsuda. 2016. "Structural MRI-based detection of Alzheimer's disease using feature ranking and classification error." *Computer Methods and Programs in Biomedicine* 137:177–193. doi: 10.1016/j.cmpb.2016.09.019.

Bron, Esther E., Stefan Klein, Janne M. Papma, Lize C. Jiskoot, Vikram Venkatraghavan, Jara Linders, Pauline Aalten, Peter Paul De Deyn, Geert Jan Biessels, Jurgen A.H.R. Claassen, Huub A.M. Middelkoop, Marion Smits, Wiro J. Niessen, John C. van Swieten, Wiesje M. van der Flier, Inez H.G.B. Ramakers, and Aad van der Lugt. Initiative Alzheimer's Disease Neuroimaging, and Group Parelsnoer Neurodegenerative Diseases Study. 2021. "Cross-cohort generalizability of deep and conventional machine learning for MRI-based diagnosis and prediction of Alzheimer's disease." *Neuroimage Clinical* 31:102712. doi: 10.1016/j.nicl.2021.102712.

Davis, Daniel H. J., Sam T. Creavin, Jennifer L. Y. Yip, Anna H. Noel-Storr, Carol Brayne, and Sarah Cullum. 2015. "Montreal cognitive assessment for the diagnosis of Alzheimer's disease and other dementias." *The Cochrane Database of Systematic Reviews* 2015(10):CD010775–CD010775. doi: 10.1002/14651858.CD010775.pub2.

Dunn, Billy, Peter Stein, and Patrizia Cavazzoni. 2021. "Approval of aducanumab for Alzheimer disease: The FDA's perspective." *JAMA Internal Medicine* 181. doi: 10.1001/jamainternmed.2021.4607.

Ebrahimi, Amir, Suhuai Luo, and Raymond Chiong. 2020. "Introducing transfer leaming to 3D ResNet-18 for Alzheimer's disease detection on MRI images." 2020 *35th International Conference on Image and Vision Computing New Zealand (IVCNZ)*, Wellington, New Zealand.

Ebrahimi, Amir, Suhuai Luo, Raymond Chiong, and Initiative Alzheimer's Disease Neuroimaging. 2021. "Deep sequence modelling for Alzheimer's disease detection using MRI." *Computers in Biology and Medicine* 134:104537. doi: 10.1016/j.compbiomed.2021.104537.

Ebrahimighahnavieh, Mr Amir, Suhuai Luo, and Raymond Chiong. 2019. "Deep learning to detect Alzheimer's disease from neuroimaging: A systematic literature review." *Comput Methods Programs Biomed* 187:105242. doi: 10.1016/j.cmpb.2019.105242.

Escudero, Javier, John P. Zajicek, and Emmanuel Ifeachor. 2011. "Machine learning classification of MRI features of Alzheimer's disease and mild cognitive impairment subjects to reduce the sample size in clinical trials." *2011 Annual International Conference of the IEEE Engineering in Medicine and Biology Society*, Boston, MA.

Frisoni, Giovanni B., Nick C. Fox, Clifford R. Jack Jr, Philip Scheltens, and Paul M. Thompson. 2010. "The clinical use of structural MRI in Alzheimer disease." *Nature Reviews Neurology* 6(2):67–77.

Gauthier, Serge, Pedro Rosa-Neto, José A. Morais, and Claire Webster. 2021. World Alzheimer report 2021: Journey through the diagnosis of dementia. Alzheimer's Disease International: London, England.

Goenka, Nitika, and Shamik Tiwari. 2021. "Volumetric convolutional neural network for Alzheimer detection." *2021 5th International Conference on Trends in Electronics and Informatics (ICOEI)*, Tirunelveli, India.

Guo, Huiru, Longqiang Xing, Caixia Luo, and Hui Chen. 2021. "AD diagnosis assistant system based on convolutional network." *2021 IEEE 2nd International Conference on Big Data, Artificial Intelligence and Internet of Things Engineering (ICBAIE)*, Nanchang, China.

He, Kaiming, Xiangyu Zhang, Shaoqing Ren, and Jian Sun. 2016. "Deep residual learning for image recognition." *Proceedings of the IEEE Conference on Computer Vision and Pattern Recognition*, Las Vegas, NV.

Hedayati, Rohollah, Mohammad Khedmati, and Mehran Taghipour-Gorjikolaie. 2021. "Deep feature extraction method based on ensemble of convolutional auto encoders: Application to Alzheimer's disease diagnosis." *Biomedical Signal Processing and Control* 66. doi: 10.1016/j.bspc.2020.102397.

Hort, Jackub, John T. O'Brien, Guido Gainotti, Tuula A. Pirttila, Bogdan Ovidiu Popescu, Irena Rektorová, Sandro Sorbi, and Philip Scheltens, EFNS Scientist Panel on Dementia. 2010. "EFNS guidelines for the diagnosis and management of Alzheimer's disease." *European Journal of Neurology* 17(10):1236–1248.

Jiang, Jingwan, Li Kang, Jianjun Huang, and Tijiang Zhang . 2020. "Deep learning based mild cognitive impairment diagnosis using structure MR images." *Neuroscience Letters* 730:134971. doi: 10.1016/j.neulet.2020.134971.

Khagi, Bijen, and Goo-Rak Kwon. 2021. "3D CNN based Alzheimer's diseases classification using segmented Grey matter extracted from whole-brain MRI." *JOIV: International Journal on Informatics Visualization* 5(2):200–205.

Knopman, David S., Steven T. DeKosky, Jeffrey L. Cummings, Helena Chang Chui, Jody P. Corey-Bloom, Norman R. Relkin, Gary W. Small, Bruce L. Miller, and James Clarke Stevens . 2001. "Practice parameter: Diagnosis of dementia (an evidence-based review): Report of the quality standards subcommittee of the American Academy of Neurology." *Neurology* 56(9):1143–1153.

Krizhevsky, Alex, Ilya Sutskever, and Geoffrey E. Hinton. 2012. "Imagenet classification with deep convolutional neural networks." *Advances in Neural Information Processing Systems* 25:1097–1105.

Kumar, S. Sambath, and Malaiyappan Nandhini. 2021. "Entropy slicing extraction and transfer learning classification for early diagnosis of Alzheimer diseases with sMRI." *ACM Transactions on Multimedia Computing, Communications, and Applications* 17(2):1–22. doi: 10.1145/3383749.

LaMontagne, Pamela J., Tammie L. S. Benzinger, John C. Morris, Sarah Keefe, Russ Hornbeck, Chengjie Xiong, Elizabeth Grant, Jason Hassenstab, Krista Moulder, Andrei G. Vlassenko, Marcus E. Raichle, Carlos Cruchaga, and Daniel Marcus. 2019. "OASIS-3: Longitudinal neuroimaging, clinical, and cognitive dataset for normal aging and Alzheimer disease." medRxiv:2019.12.13.19014902. doi: 10.1101/2019.12.13.19014902.

LeCun, Yann, Léon Bottou, Yoshua Bengio, and Patrick Haffner. 1998. "Gradient-based learning applied to document recognition." *Proceedings of the IEEE* 86(11):2278–2324.

Liu, Junxiu, Mingxing Li, Yuling Luo, Su Yang, Wei Li, and Yifei Bi . 2021. "Alzheimer's disease detection using depthwise separable convolutional neural networks." *Comput Methods Programs Biomed* 203:106032. doi: 10.1016/j.cmpb.2021.106032.

Marcus, Daniel S., Anthony F. Fotenos, John G. Csernansky, John C. Morris, and Randy L. Buckner. 2010. "Open access series of imaging studies: Longitudinal MRI data in nondemented and demented older adults." *Journal of Cognitive Neuroscience* 22(12):2677–2684. doi: 10.1162/jocn.2009.21407.

Marcus, Daniel S., Tracy H. Wang, Jamie Parker, John G. Csernansky, John C. Morris, and Randy L. Buckner. 2007. "Open Access Series of Imaging Studies (OASIS): Cross-sectional MRI data in young, middle aged, nondemented, and demented older adults." *Journal of Cognitive Neuroscience* 19(9):1498–1507. doi: 10.1162/jocn.2007.19.9.1498.

Nanni, Loris, Matteo Interlenghi , Sheryl Brahnam, Christian Salvatore, Sergio Papa, Raffaello Nemni, and Isabella Castiglioni, Initiative Alzheimer's Disease Neuroimaging. 2020. "Comparison of transfer learning and conventional machine learning applied to structural brain MRI for the early diagnosis and prognosis of Alzheimer's disease." *Frontiers in Neurology* 11:576194. doi: 10.3389/fneur.2020.576194.

Nawaz, Hina, Muazzam Maqsood, Sitara Afzal, Farhan Aadil, Irfan Mehmood, and Seungmin Rho. 2020. "A deep feature-based real-time system for Alzheimer disease stage detection." *Multimedia Tools and Applications*. doi: 10.1007/s11042-020-09087-y.

Nir, Talia M., Julio E. Villalon-Reina, Gautam Prasad, Neda Jahanshad, Shantanu H. Joshi, Arthur W. Toga, Matt A. Bernstein, Clifford R. Jack, Michael W. Weiner, and Paul M. Thompson, Initiative Alzheimer's Disease Neuroimaging. 2015. "Diffusion weighted imaging-based maximum density path analysis and classification of Alzheimer's disease." *Neurobiol Aging* 36(Suppl 1):S132–S140. doi: 10.1016/j.neurobiolaging.2014.05.037.

Nitrini, Ricardo, Paulo Caramelli, Cássio Machado de Campos Bottino, Benito Pereira Damasceno, Sonia Maria Dozzi Brucki, and Renato Anghinah. 2005. "Diagnosis of Alzheimer's disease in Brazil: Diagnostic criteria and auxiliary tests. Recommendations of the scientific department of cognitive neurology and aging of the brazilian academy of neurology." *Arquivos de Neuro-Psiquiatria* 63(3A):713–719.

Pesce, Emanuele, Samuel Joseph Withey, Petros-Pavlos Ypsilantis, Robert Bakewell, Vicky Goh, and Giovanni Montana. 2019. "Learning to detect chest radiographs containing pulmonary lesions using visual attention networks." *Medical Image Analysis* 53:26–38.

Plant, Claudia, Stefan J. Teipel, Annahita Oswald, Christian Böhm, Thomas Meindl, Janaina Mourao-Miranda, Arun W. Bokde, Harald Hampel, and Michael Ewers . 2010. "Automated detection of brain atrophy patterns based on MRI for the prediction of Alzheimer's disease." *Neuroimage* 50(1):162–74. doi: 10.1016/j.neuroimage.2009.11.046.

Qiu, Jingyan, Linjian Li, Yida Liu, Yingjun Ou, and Yubei Lin. 2020. "The diagnosis of Alzheimer's disease: An ensemble approach." In A. J. Tallón-Ballesteros (ed.), *Fuzzy Systems and Data Mining VI*. IOS Press: Amsterdam, The Netherlands, 93–100.

Rathore, Saima, Mohamad Habes, Muhammad Aksam Iftikhar, Amanda Shacklett, and Christos Davatzikos. 2017. "A review on neuroimaging-based classification studies and associated feature extraction methods for Alzheimer's disease and its prodromal stages." *NeuroImage* 155:530–548. doi: 10.1016/j.neuroimage.2017.03.057.

Renjith, C. V., and Santosh C. Wagaj. 2020. "MRI brain disease detection using enhanced landmark based deep feature learning." *2020 2nd International Conference on Advances in Computing, Communication Control and Networking (ICACCCN)*, Greater Noida, India.

Sathiyamoorthi, V., A. K. Ilavarasi, K. Murugeswari, Syed Thouheed Ahmed, B. Aruna Devi, and Murali Kalipindi. 2021. "A deep convolutional neural network based computer aided diagnosis system for the prediction of Alzheimer's disease in MRI images." *Measurement* 171. doi: 10.1016/j.measurement.2020.108838.

Schlemper, Jo, Ozan Oktay, Michiel Schaap, Mattias Heinrich, Bernhard Kainz, Ben Glocker, and Daniel Rueckert 2019. "Attention gated networks: Learning to leverage salient regions in medical images." *Medical Image Analysis* 53:197–207. doi: 10.1016/j.media.2019.01.012.

Schlemper, Jo, Ozan Oktay, Liang Chen, Jacqueline Matthew, Caroline Knight, Bernhard Kainz, Ben Glocker, and Daniel Rueckert. 2018. "Attention-gated networks for improving ultrasound scan plane detection." arXiv preprint arXiv:1804.05338.

Simon, Blessy C., D. Baskar, and V. S. Jayanthi. 2019. "Alzheimer's disease classification using deep convolutional neural network." *2019 9th International Conference on Advances in Computing and Communication (ICACC)*, Kochi, India.

Simonyan, Karen, and Andrew Zisserman. 2014. "Very deep convolutional networks for large-scale image recognition." arXiv preprint arXiv:1409.1556.

Sitaula, Chiranjibi, and Mohammad Belayet Hossain. 2020. "Attention-based VGG-16 model for COVID-19 chest X-ray image classification." *Applied Intelligence*. doi: 10.1007/s10489-020-02055-x.

Tufail, Ahsan Bin, Yongkui Ma, and Qiu-Na Zhang. 2020a. "Binary classification of Alzheimer's disease using sMRI imaging modality and deep learning." *Journal of Digital Imaging* 33(5):1073–1090. doi: 10.1007/s10278-019-00265-5.

Tufail, Ahsan Bin, Yongkui Ma, and Qiu-Na Zhang. 2020b. "Classification of subjects of mild cognitive impairment and Alzheimer's disease through neuroimaging modalities and convolutional neural networks." *2020 8th International Conference on Information and Communication Technology (ICoICT)*, Yogyakarta.

United Nations. 2020. World Population Ageing 2020 Highlights: Living arrangements of older persons, New York.

Vertesi, Andrea, Judith A. Lever, D. William Molloy, Brett Sanderson, Irene Tuttle, Laura Pokoradi, and Elaine Principi . 2001. "Standardized mini-mental state examination. Use and interpretation." *Canadian family physician Medecin de famille canadien* 47:2018–2023.

Visser, Pieter Jelle, Arnold Kester, Jellemer Jolles, and Frans Verhey. 2006. "Ten-year risk of dementia in subjects with mild cognitive impairment." *Neurology* 67(7):1201–1207.

Wang, Ruyue, Li Zeng, Hanhui Li, Linfa Lu, Ji Li, and Xiaonan Luo. 2020. "Alzheimer's disease classification based on one dimensional convolutional neural network." *2020 8th International Conference on Digital Home (ICDH)*, Dalian.

Wang, Shui-Hua, Qinghua Zhou, Ming Yang, and Yu-Dong Zhang. 2021. "ADVIAN: Alzheimer's disease VGG-inspired attention network based on convolutional block attention module and multiple way data augmentation." *Front Aging Neurosci* 13:687456. doi: 10.3389/fnagi.2021.687456.

Weiner, Michael W., Dallas P. Veitch, Paul S. Aisen, Laurel A. Beckett, Nigel J. Cairns, Jesse Cedarbaum, Michael C. Donohue, Robert C. Green, Danielle Harvey, and Clifford R. Jack Jr. 2015. "Impact of the Alzheimer's disease neuroimaging initiative, 2004 to 2014." *Alzheimer's & Dementia* 11(7):865–884.

Weiner, Michael W., Dallas P. Veitch, Paul S. Aisen, Laurel A. Beckett, Nigel J. Cairns, Robert C. Green, Danielle Harvey, Clifford R. Jack Jr, William Jagust, and John C. Morris. 2017. "The Alzheimer's disease neuroimaging initiative 3: Continued innovation for clinical trial improvement." *Alzheimer's & Dementia* 13(5):561–571.

Weiner, Michael W., Paul S. Aisen, Clifford R. Jack Jr, William J. Jagust, John Q. Trojanowski, Leslie Shaw, Andrew J. Saykin, John C. Morris, Nigel Cairns, and Laurel A. Beckett. 2010. "The Alzheimer's disease neuroimaging initiative: Progress report and future plans." *Alzheimer's & Dementia* 6(3):202–211.

Xu, Kelvin, Jimmy Ba, Ryan Kiros, Kyunghyun Cho, Aaron Courville, Ruslan Salakhudinov, Rich Zemel, and Yoshua Bengio. 2015. "Show, attend and tell: Neural image caption generation with visual attention." *International Conference on Machine Learning,* Lille.

Yagis, Ekin, Luca Citi, Stefano Diciotti, Chiara Marzi, Selamawet Workalemahu Atnafu, and Alba G. Seco De Herrera. 2020. "3D convolutional neural networks for diagnosis of Alzheimer's disease via structural MRI." *2020 IEEE 33rd International Symposium on Computer-Based Medical Systems (CBMS),* Rochester, MN.

Zaabi, Marwa, Nadia Smaoui, Houda Derbel, and Walid Hariri. 2020. "Alzheimer's disease detection using convolutional neural networks and transfer learning based methods." *2020 17th International Multi-Conference on Systems, Signals & Devices (SSD),* Monastir, Tunisia.

Zhang, Jie, Bowen Zheng, Ang Gao, Xin Feng, Dong Liang, Xiaojing Long . 2021. "A 3D densely connected convolution neural network with connection-wise attention mechanism for Alzheimer's disease classification." *Magn Reson Imaging* 78:119–126. doi: 10.1016/j.mri.2021.02.001.

17 Combination of Sensors-Based Monitoring System and Internet of Things (IoT)

A Survey and Framework for Remote and Intensive Care Unit Patients

Mohammad Aminul Islam, Kazi Zehad Mostofa, Hamidreza Mohafez, and Md Jakir Hossen
University of Malaya

Foo Wah Low
Universiti Tunku Abdul Rahman

Mikhail Vasiliev
ClearVue PV technologies

Syed Mohammed Shamsul Islam and Mohammad Nur-E-Alam
Edith Cowan University

17.1 INTRODUCTION

Information and communication technologies (ICTs) are now fundamental to modern human civilization, providing access to information and connections with the outside world. The IoT expands the capabilities of the internet and has a significant impact on socio-economic development, including women's empowerment, rural health, e-commerce, and education. Medical IoT devices, with their sensors, enable remote sensing, monitoring, and data collection for healthcare applications like medical imaging, patient monitoring, and clinical information systems [1]. However, the best performance of these IoT devices is significantly dependent on superfast computing systems and high-volume data storage amenities. Therefore, to overcome

DOI: 10.1201/9781003346678-17

the common issues related to the computing and storage constraints of IoT devices, the cloud platform is one of the alternatives to make up for such limitations [2]. IoT devices can convert observed physical quantities into digital data that can be stored and processed in the cloud, which offers numerous advantages such as mass storage, self-service access, scalability, and advanced security. Combining cloud computing services with IoT devices has become the most popular and recommended architecture, which can significantly enhance patient quality of life through the provision of cloud-IoT-based healthcare services [2–4]. The integration of cloud computing services and IoT devices is a promising architectural paradigm for improving patient care and creating personalized health services, such as prescription reminders and access to online health information. In addition, it can be used to provide real-time monitoring of the daily health activities of elderly and senior citizens. A prediction has been made that the number of IoT devices connected to the internet will exceed 75 billion by the year 2025 which will bring great challenges related to scalability and the ability of the cloud to process IoT-collected data in real time. However, the scalability challenges of processing IoT-collected data in real-time will require localized processing and data storage, while the implementation of medical IoT devices in architecture with a computing layer located near IoT devices is also gaining popularity for further research and investigation [5,6]. Edge devices can process vital health information, reducing the volume of data sent to distant cloud data centers and enabling faster reaction times for vital healthcare applications such as real-time monitoring and health condition notifications for patients.

The healthcare and life sciences industries are in the midst of switching from reactive, costly episodic care models to proactive, digitally enabled care models that offer better value for patients, with the IoT providing an opportunity to connect patients, providers, and payers and improve patient-centricity, effectiveness, and efficiency [7]. The healthcare industry is experiencing disruption from IoT technology including big data, AI, mobile apps, 3D printing, advanced sensors, voice technology, and interoperable electronic health records (EHRs), which offer new opportunities for MedTech companies while also automating and creating new jobs using automation and robotics [8]. Remote patient care using IoT can solve the problem of long travel for healthcare services and help healthcare professionals consistently monitor patients with ongoing conditions. Connectivity allows healthcare professionals to assist outpatients through recommendations, medication, and biometric measurements using sensors and remote equipment. IoT devices can generate a schedule of patients' daily physical condition information, and live video and audio streams can be used to monitor patients' current condition without the need for exchange [9–12]. Figure 17.1 presents the major uses of IoTs in this era of modern communication technology and the possible future scope of IoTs in healthcare systems overall.

We aim to investigate the possibilities of continuous remote patients monitoring that can be achieved through the deployment of a non-stop patient checking and control system that utilizes biosensors, local Wi-Fi, and the Internet of Things (IoT) to store and analyze patient data. In addition, an emergency alert system can be developed to send signals or messages to the appropriate departments, including relatives, doctors, and police, reducing the need for physical contact between patients and front-line health workers and service providers.

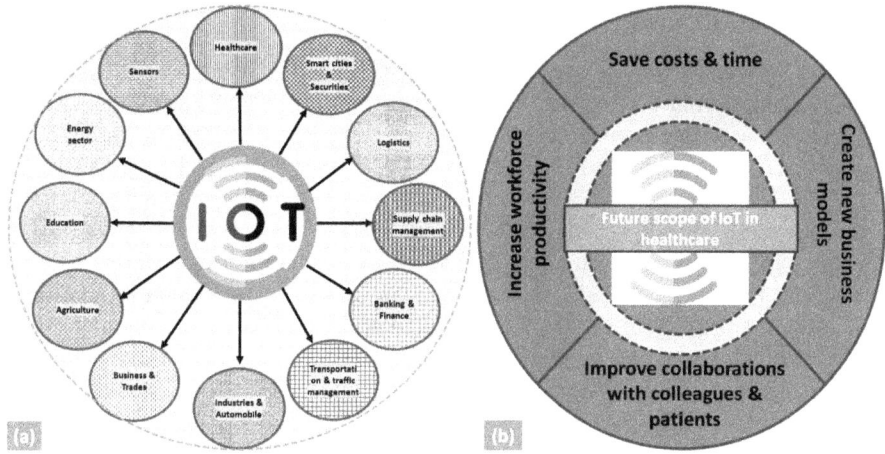

FIGURE 17.1 The common areas of frequently uses of IoTs with the combination of modern technologies (a), and the prospects of IoT sector in healthcare (b).

17.2 BACKGROUND

IoT-enabled home automation and monitoring systems have gained popularity for elderly healthcare, allowing remote monitoring of daily activities and health status, with affordable and plentiful sensors. A prior pilot study demonstrated a privacy-aware, intelligent home care system that learned user activity patterns and automatically sent alerts for unusual circumstances, providing peace of mind to family members and caretakers. In another study, patients using a home-monitoring sensor system before and after joint replacement found the technology adequate, but informal support networks greatly aided in using the technology at home [13–15]. Several studies have been conducted using glucose sensor-based mobile technology (active system) in persons with diabetes (types 1 and 2) to study the patients' attitudes and experiences of online access to their electronic medical records [16–20]. These studies have found that the use of technology can have both positive and negative effects on patients, such as empowering them, providing psychosocial support, and relieving anxiety, while also causing concerns about excessive information demands and potential socioeconomic biases. Therefore, technological solutions should be customized for each individual and used with professional support. However, concerns about security, privacy, and integrity arise with the use of modern technology. The experiences of community-dwelling older people with telecare technology for home safety have been found to be diverse, with limitations on active aging due to certain technologies being confined to the home [21,22]. Older people may face the stigma of being weak and useless, despite the importance of independence. Service providers' support is critical because lack of knowledge can hinder proper equipment usage. The studies reviewed focused on first-generation devices that involved user-activated alarm buttons. Further research is necessary to examine experiences with newer systems, such as second-generation passive alarms that detect specific risks without user input and third-generation systems that send data to an internet portal for caregiver monitoring.

17.2.1   Basic Concept of IoT Project Studies

IoT has revolutionized the way people live, leading to the emergence of high-tech lifestyles, smart homes, communities, industries, and eco-friendly solutions. Despite the significant progress made in IoT technology, there are still obstacles and challenges that must be addressed to realize its full potential. These challenges cut across various IoT domains, such as applications, enabling technologies, and social and environmental impacts, among others. To understand the recent IoT development, two aspects are critical: technical building blocks integration from three primary industries, namely electronics, communications, and software, which involve multiple subdomains, including analog and digital electronics, wireless communication modules, protocol stacks, embedded software, database management, and analytics. Another significant feature is the large number of diverse use cases, including smart cities, energy management, environmental monitoring, building automation, patient monitoring, health tracking, home automation, fleet management, tracking systems, and more. Managing an IoT project can be a challenging task, given the diverse use cases and technologies involved [15,23–25].

17.2.2   A Sensor System for Notifications of Deviant Behaviors and Routines in the Home

Wireless environmental IoT sensors have been utilized in the homes of elderly individuals to gather data on their movements, light and temperature in the room, and the usage of machines such as smart electricity plugs and lamps [26]. These sensors are connected to the Sense Smart Region platform through a gateway, allowing for the collection and analysis of data. Notifications are then sent via a mobile app to family members or healthcare staff regarding any changes in the daily behavior or routines of the elderly individuals, requiring a supervisory visit. Notifications are also sent via a mobile app to individuals living in housing with special services for those with functional disabilities, reminding them to complete various tasks such as starting daily activities, taking a shower, and doing the dishes. This is a passive monitoring system that does not require input from the occupants of the homes [27]. Cameras were also connected to sensors in both types of accommodations to monitor the care recipient, particularly at night. Figure 17.2 provides an overview of the process of remotely monitoring the physical data of patients.

17.2.3   Classification of Health Monitoring Sensors

Wearable technology and wireless communications have made it possible to remotely monitor a person's health by collecting data from medical sensors and wearable devices. These sensors and gadgets can be integrated into a variety of accessories, including clothing, wristbands, glasses, socks, hats, and shoes, as well as other devices like smartphones, headphones, and watches [28]. Medical sensors are divided into two categories: non-contact sensors and contact sensors, with subcategories of therapeutic and monitoring sensors for contact sensors and further subcategories for non-contact sensors based on their usage. Health-monitoring sensors

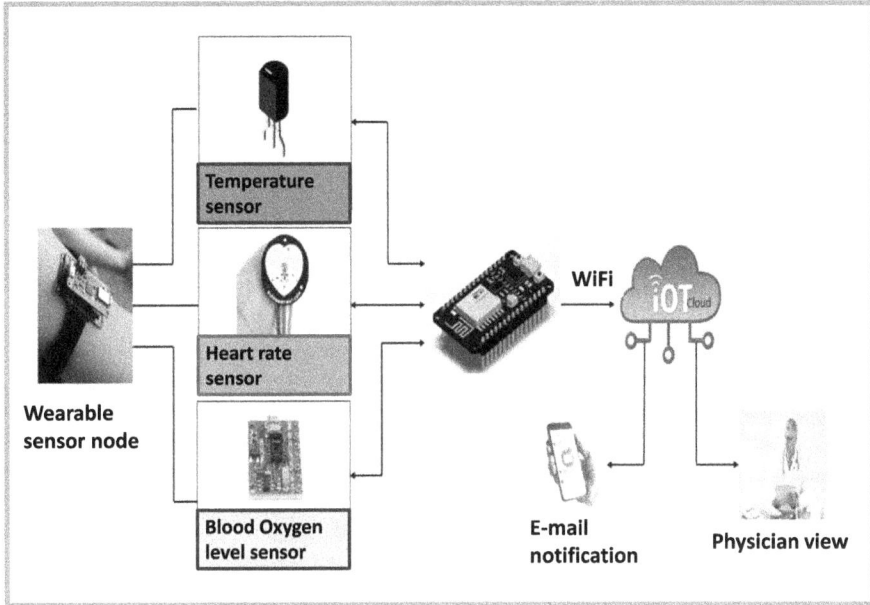

FIGURE 17.2 Overview of a remote physical data monitoring.

can be divided into contact and non-contact sensors, with contact sensors used to monitor physiological actions, chemical levels, and therapy-related activities such as medication and stimulation. Non-contact sensors are used to monitor behaviors related to fitness, wellness, and rehabilitation. Medical sensors, wearables, and the IoT have potential applications in monitoring vital signs in hospitals, aging in place and motion, assisting individuals with motor and sensory impairments, and conducting medical and behavioral research. Pressure sensors are used in a range of medical devices, including anesthesia delivery machines, oxygen concentrators, sleep apnea machines, ventilators, kidney dialysis machines, infusion and insulin pumps, blood analyzers, respiratory and blood pressure monitoring equipment, hospital beds, surgical fluid management systems, and pressure-operated dental instruments. A schematic diagram (Figure 17.3) presents the types of sensors that are commonly used for various medical applications [29].

Temperature sensors are utilized not only for digital thermometers and organ transplant system temperature monitoring and control but also in various medical equipment such as anesthesia delivery machines, sleep apnea machines, ventilators, kidney dialysis machines, blood analyzers, medical incubators, and neonatal intensive care units to monitor patient temperature. Flow sensors are incorporated in anesthesia delivery systems, oxygen concentrators, sleep apnea devices, ventilators, respiratory monitoring, gas mixing, and electro-surgery devices, which use high-frequency electric current to cut, coagulate, desiccate, or destroy tissue such as tumors. Image sensors have multiple applications including radiology, fluoroscopy, cardiology, mammography, dental imaging, endoscopy, external observation, minimally invasive surgery, laboratory equipment, ocular surgery and observation, and artificial retinas.

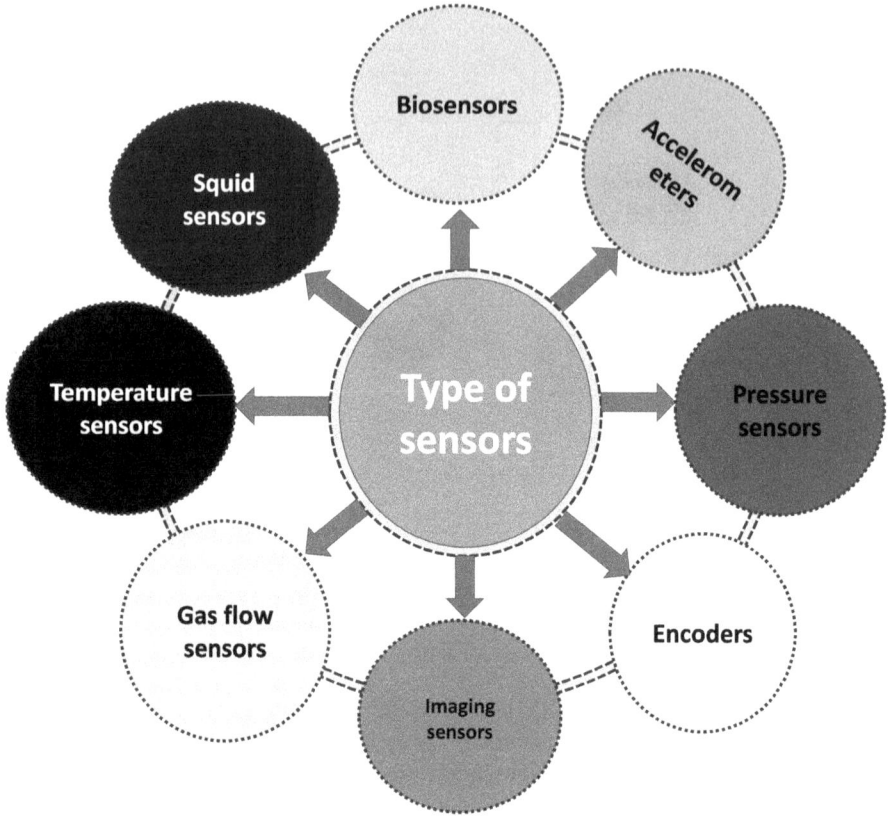

FIGURE 17.3   Various types of sensors available in market that are used in many medical applications.

Accelerometers are integrated into various health monitoring devices such as blood pressure monitors, patient monitoring devices, heart pacemakers, and defibrillators. Biosensors are utilized to screen for drug addiction, infectious diseases, pregnancy, and blood glucose and cholesterol levels. Superconducting quantum interference devices (SQUIDs) are used in magnetoencephalography (MEG) and magnetocardiography (MCG) systems to examine neuronal activity inside the brain with incredible sensitivity. Encoders are used in numerous medical devices such as X-ray machines, MRI machines, computer-assisted tomography equipment, medical imaging systems, blood analyzers, surgical robots, laboratory sample-handling equipment, sports and healthcare equipment, and other non-critical medical devices [28–30].

17.3   STATE-OF-THE-ART SENSORS-BASED MONITORING HEALTH SYSTEMS

Numerous studies have been undertaken on various subjects and technologies related to IoT health care. The use of home automation and monitoring systems for the care of elderly people has accelerated with the advent of the Internet of Things (IoT), which provides cheap and abundant sensors. For instance, remote monitoring can be used to monitor daily activities or health status in chronic illnesses. Earlier research showed how intelligent, privacy-aware home care support could automatically send out notifications if an unexpected circumstance was identified by using a sensor system and learning the user's daily activity patterns [31,32]. This was a passive system that didn't need any user interaction. The major goal was to provide a solution to the concern of whether a loved one was okay, and the case study demonstrated how to preserve the freedom of elderly individuals who live alone while providing caretakers and/or family members with peace of mind [33,34]. A healthcare monitoring system using a Raspberry Pi has been presented by S. Rajkumar et al. [35] and tracks the patients' body temperature, heart rate, and movement. However, it does not give the patient a reminder to take the necessary medications. Using a Raspberry Pi, another work has been reported related to developing a healthcare monitoring system that enables clinicians and close family members to assess the patients' health status, whereas a less expensive Arduino Uno has been used for data display on the webpage [36]. The proper data collection from electronic physical health sensors has been described by Danilo F. S. Santos et al. However, it does not offer a way to keep the patient's environment at its best [37]. Another proposal for a health monitoring system that uses an Arduino Uno to deliver information to hospital administration and patients' loved ones about their health status is evident in the Ref. [38]. However, it does not include an SMS option to inform the patient when to take their medication.

H. Cai et al. have explored the issue of acquiring, reading, and storing data in the cloud using the internet and have also offered a solution. It is challenging to read and save the data in the proper format in this situation due to the various data inputs occurring at regular intervals. As a result, this system provides a way to do it [39]. An efficient and smart healthcare system in an IoT environment that can monitor a patient's basic health signs such as patients' heart rates and temperatures as well as the room condition in real-time has been proposed in Ref. [40]. A healthcare monitoring system utilizing a microprocessor and radio frequency identified data (RFID) tags is described by S. F. Khan [41]. IoT is strengthened by using this combo. It does not, however, control the equipment in accordance with the patients' medical needs. The additional feature that is suggested in this research is the creation of messages to remind people to take their prescribed medications on time. The fact that the doctor and close family members will also receive the SMS alert is another special benefit of this approach. This will make it easier for the doctor and the patient's family to provide effective care.

Historically, medical decisions were based on a combination of a physician's expertise, a patient's physical symptoms, and lab results. However, with the ongoing technological revolution, sensors have become increasingly important in the medical field. These sensors are used in a variety of settings, including dental offices, labs, intensive care units, hospitals, and home care products. Medical devices rely on sensors for disease and injury detection, prevention, evaluation, and treatment. Hospital beds equipped with sensors that promote blood flow to expedite patient recovery and lower healthcare expenses are an example of sensor usage. Additionally, sensors play a critical role in ventilators, providing life-saving treatment in the ongoing fight against COVID-19. Sensors play a crucial role in the medical industry by converting various stimuli into electrical signals that can be processed in electronic medical equipment. They enable monitoring of vital signs and other health parameters, increase the intelligence of medical devices, and are integrated into both equipment and patient bodies. Healthcare organizations require real-time, reliable, and accurate diagnostic data from remotely monitored devices in hospitals, clinics, and homes due to the growing and aging population's demand for unique medical gadgets, including various sensors [42].

Since people increasingly desire to have a longer and healthier life, medical treatments are expected to be delivered more frequently in the comfort of their homes. This has resulted in a decrease in the cost, size, intelligence, and ease of use of medical devices. Smart medical equipment with advanced sensor technologies is now being developed for home care, enabling the collection of crucial data for rapid diagnosis and treatment. The sensors used in these devices are highly precise, repeatable, and consume less power while providing digital output. This results in simplified designs, reduced development costs and time-to-market, and the creation of more affordable and compact medical devices [43]. Commonly used sensors in healthcare equipment include those that measure force, pressure, airflow, oxygen, pulse oximetry, temperature, and barcode scanning and decoding. More information about vital sensors and frequently used products in the healthcare system can be found in Ref. [44].

17.4   SIGNIFICANCE OF INTERNET OF THINGS (IoT) ON HEALTH/MEDICARE

IoT has emerged as a game-changing technology that is transforming various aspects of modern society. It has found extensive use in diverse sectors such as business, transportation, and healthcare. Before the advent of IoT, doctors and patients had limited options to communicate, such as visits, teleconferences, and text messages. The continuous monitoring of patients' health and providing guidance was a challenge for healthcare professionals. With IoT-enabled devices, remote monitoring of patients has become a reality, enabling doctors to provide exceptional treatment and keep patients healthy and safe. The use of IoT has led to an increase in patient engagement and satisfaction while reducing hospital stays and preventing readmissions [45]. By improving patient outcomes and reducing healthcare costs, IoT has revolutionized the healthcare industry, transforming the way devices and people interact in delivering healthcare solutions. The advantages of IoT applications in healthcare benefit patients, families, physicians, hospitals, and insurance companies.

17.4.1   IoT FOR PATIENT MONITORING

The use of fitness bands and wireless-connected medical equipment, including blood pressure and heart rate monitors, glucometers, and more, can significantly enhance patients' access to personalized care. These devices can be programmed to alert users to remember things like appointments, changes in blood pressure, and a variety of other things. IoT has changed people's lives by making it feasible to continuously

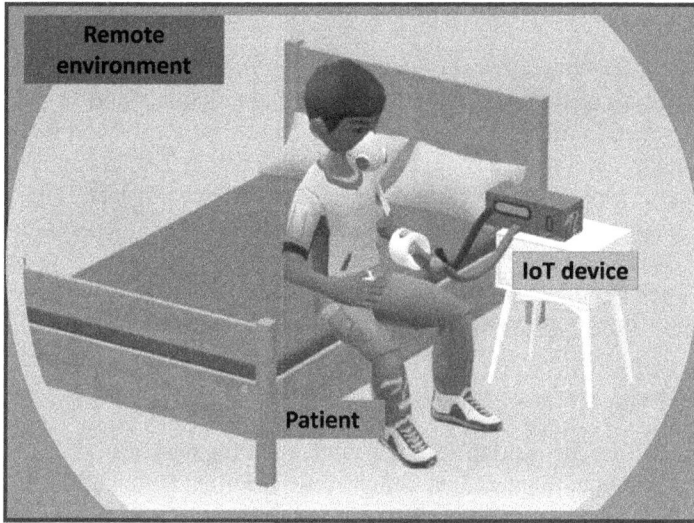

FIGURE 17.4 Outlook of IoT based remote patient monitoring. (Reproduced from Ref. [46] with the permission.)

follow medical issues, especially those of elderly patients, and thus have an influence on single-person households and families. Any interruptions or modifications to a person's routine activities can be reported to family members and concerned healthcare professionals via an alarm system. Figure 17.4 shows an IoT-based patient monitoring system.

IoT healthcare market worth was estimated at USD 73.5 billion in 2021 and is expected to reach USD 190 billion by 2028 at a CAGR of 25.9%. IoT can automate patient care workflow through next-gen healthcare facilities, mobility solutions, and emerging technologies. Remote patient monitoring devices can save healthcare businesses $300 billion annually, along with other IoT platform technological benefits [47,48].

17.4.2 IoT for Doctors and Hospitals

Healthcare IoT technology utilizes communication and sensor data to enhance patient diagnostics, monitoring, and treatment delivery. Cloud computing enables medical devices to collect and exchange data, which can be analyzed at high speeds. IoT sensors in healthcare can be worn or integrated into medical devices and send data for analysis. The use of dynamic patient data in healthcare IT and IoT leads to quicker diagnoses and countermeasures against health deterioration. Wearable devices with IoT sensors are a significant trend in the healthcare IoT market. However, collaborative developments between research institutes and healthcare IoT market players can lead to innovative and commercially viable solutions.

17.4.3 HEALTHCARE IoT: MARKET DRIVERS AND CHALLENGES

The use of IoT for remote patient monitoring is expected to steadily increase, particularly for residential end-users, as well as the adoption of smart home technologies. IoT is making it easier for patients to navigate the medical system by providing network-enabled healthcare devices for remote monitoring. Doctors and patients can use smartphone apps and software to collect and analyze health data. However, as the number of healthcare IoT solutions and technologies grows, it becomes more challenging to ensure that they are all connected and communicating with each other, leading to potential misinterpretation of data and negative impacts on patient health. This may slow the growth rate of the global healthcare IoT market [49,50].

17.4.4 IoT FOR HEALTH INSURANCE COMPANIES

IoT is transforming the insurance industry by providing new and adaptable ways of operating. A significant portion of companies have already begun utilizing IoT to create connected homes or buildings, and it is anticipated that connected devices will drive future growth in insurance revenue. IoT and Insurance Tech are used to promote transparency between insurance companies and their customers in various operational procedures, such as underwriting, pricing, claims handling, and risk assessment. Customers who use IoT devices to monitor their normal activities and adherence to treatment plans may receive rewards from insurers for sharing health data generated by IoT devices, resulting in reduced insurance claims. Insurance companies may utilize IoT devices and the data they collect to confirm claims [49–51].

17.5 INTEGRATED APPROACHES (SENSORS AND IoT-BASED) FOR HEALTHCARE MONITORING

The industrial healthcare sector has seen consistent progress in smart healthcare monitoring systems that could potentially transform the way healthcare is delivered. While these systems can automate patient monitoring and enhance procedure management, their effectiveness in clinical procedures is still a topic of debate. The efficiency, clinical acceptability, strategies, and suggestions to improve current healthcare monitoring systems are being thoroughly examined. The latest concept of IoT-based healthcare monitoring enables remote healthcare monitoring for patients in their remote locations, making it the most significant advancement in the healthcare sector [52]. The current challenges faced by healthcare practitioners have been considered while exploring significant improvements in IoT-based healthcare monitoring system design. However, detecting and evaluating these systems in comparison to other systems of a similar nature may pose challenges in the healthcare monitoring area. Typically, IoT-based systems for remote healthcare monitoring consist of three major elements: Cloud, IoT gateways, and data acquisition units [53,54]. The initial step in the IoT-based healthcare monitoring system is data acquisition, which relies primarily on end-users' smart devices and sensors. Patient information is collected from these devices and sensors and undergoes pre-processing at the IoT gateway before being transmitted to the cloud. Various data analysis techniques are employed in the cloud to collect and comprehend pertinent data which is utilized by medical professionals for examination and evaluation.

17.5.1 DATA ACQUISITION

Data collection is a critical aspect of utilizing IoT-based networks for remote healthcare monitoring, which involves gathering biological data using biosensors. Typically, biosensors generate analog signals that are susceptible to noise and low amplitude, making it essential to preprocess the signals to ensure that the information is accurate and persistent. This is crucial because inaccurate data could lead to incorrect treatment decisions. Biosensors can be classified into wearable and non-wearable smart sensors and gadgets. Wearable sensors are attached to a patient's body to monitor their health status, while non-wearable sensors are placed in the patient's surroundings to track ambient conditions. These sensors generate vast amounts of data that can be transmitted to healthcare providers to facilitate informed decision-making. Smart sensing networks are essential for monitoring patients in remote settings, and many innovative companies have developed wearable smart sensing gadgets to reduce hospital stays and clinical procedures for managing chronic diseases. Luis et al. provide an example of a real-time healthcare monitoring system that uses a mobile application connected with a biometric bracelet to monitor the health status of older persons living in geriatric facilities [55]. The biometric bracelet collects information such as blood oxygenation, pulse rate, and body temperature and transmits the data to the mobile application in real-time. The illustration of the data collection process for an IoT project is shown in Figure 17.5.

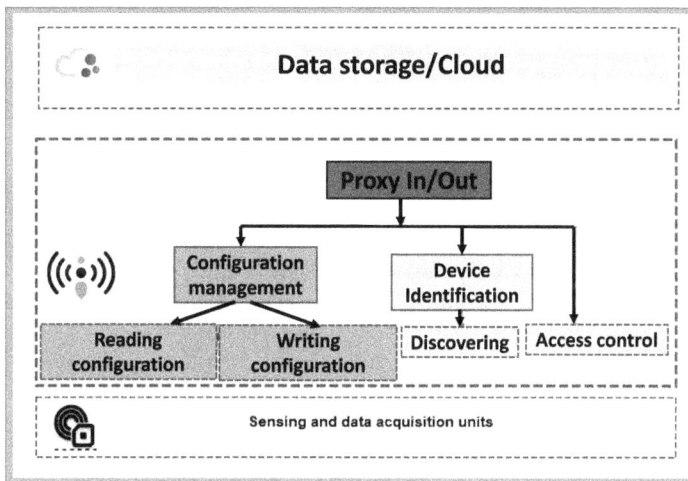

FIGURE 17.5 Data collecting process of an IoT project. (Redrawn by using the concept of Ref. [39].)

17.5.2 IoT GATEWAYS

IoT gateways play a crucial role in remote healthcare monitoring systems by serving as interfaces that connect sensors, smart devices, and the cloud. It can act as a bridge between smart items and applications since data supplied to the cloud or received from it would go through IoT gateways. Data standardization, data preprocessing, and network connectivity, however, could be functions for IoT gateways. The IoT gateways provide data normalization by taking the various data sets and converting them into standard data formats as various sensor categories gather data. Additionally, IoT gateways preprocess data before transmitting it to the cloud, reducing the amount of data that needs to be transferred, retrieved, and stored. This helps to lower transmission costs, ensure data accuracy, and improve overall system efficiency. Furthermore, IoT gateways offer network connectivity for the sensors in IoT-based remote healthcare monitoring systems, reducing their limitations in connecting large networks.

17.5.3 CLOUD

IoT-based healthcare monitoring systems rely on servers for storing and retrieving data that are crucial in identifying and evaluating anomalous healthcare interventions. The servers play a critical role in processing the signals gathered from related sensors by the IoT gateway. Once the signals are preprocessed, the IoT gateway transmits the healthcare data to the cloud for additional data mining. In the data mining process, critical healthcare data is categorized into bio-signals to enable effective analysis and interpretation. Thus, servers and data mining are essential components of IoT-based healthcare monitoring systems that enable the efficient processing, analysis, and interpretation of healthcare data for improved patient care.

17.5.4 PERFORMANCE EVALUATION OF IoT SENSORS

Sensors are the brain and heart of any healthcare monitoring system and must be trustworthy, compact, precise, have quick data transmission, consume minimal power, and provide accurate outputs. Wearable sensors face challenges in achieving both accuracy and size, and medical-grade sensors are bulky and require specialized equipment and skilled workers. IoT sensor-based applications also require authentication, security, and privacy, with various protocols available on the market. However, the security mechanisms of these apps are continually susceptible to attack [56].

17.6 STATUS AND CHALLENGES OF COMBINATORIAL HEALTHCARE MONITORING APPROACHES

Digital healthcare systems that utilize technologies such as IoT and big data are expected to enable seamless connections between patients and providers across different healthcare systems. Medical wearables are increasingly being connected to these systems over the Internet for real-time healthcare monitoring [57]. S. Zeadally

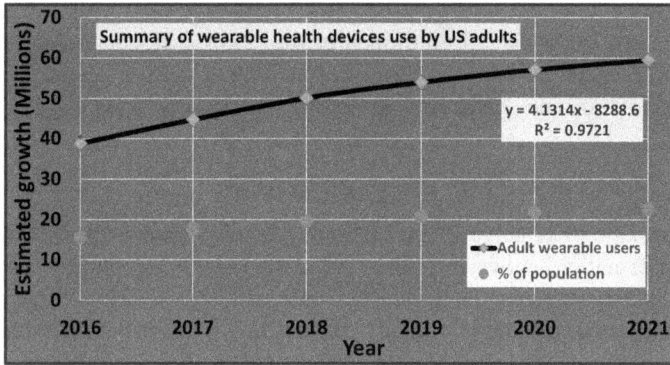

FIGURE 17.6   Estimation of growth of wearable health devices used by US adults.

and Bello have identified challenges and potential solutions for utilizing IoT and big data analytics in smart healthcare to improve healthcare access [58]. The combination of big data and IoT technologies is expected to connect patients and providers across different healthcare systems. Figure 17.6 shows the adoption rate (in millions) of medical wearables among US adults [59,60].

17.6.1   SECURITY AND PRIVACY

IoT devices may threaten users' security and privacy, especially in the case of unauthorized access. Patients' personal and health data is transmitted to the cloud through connected devices, including medical equipment and mobile devices. The device layer is vulnerable to various threats such as tag cloning, spoofing, radio frequency (RF) blocking, and cloud polling. A man-in-the-middle attack can inject commands into a device through communication diversion during cloud polling. Direct connection attacks can be launched using service discovery protocols such as Bluetooth low energy and universal plug and play. Denial of service (DoS) attacks can also jeopardize patient safety and healthcare systems. However, duplicating resources is not always feasible, particularly for implanted life-critical equipment. Detection of potential security risks is a challenge due to the increasing number and complexity of software and hardware vulnerabilities [58,61].

The proliferation of IoT devices has led to increasing security and privacy concerns, with default authentication and unsecured web-based interfaces expanding the attack surface. Wearable devices, including embedded sensors and medical implants, lack security standards and are vulnerable to attacks [62–64]. Recent wireless networking technologies like Wi-Fi, BLE, and ZigBee used in healthcare require security protection against various attacks. Centralized databases containing personal data, such as electronic health records and genomic data, must also be guarded against hackers and dangerous software to ensure security and privacy [65,66]. Privacy and confidentiality issues are a concern for both patients and healthcare professionals. Incorporating linked technology into medical information systems may compromise patient data security and raise privacy concerns. The sharing of patient data across multiple applications and linking a person's location to pharmacy purchases also raises privacy concerns [67,68]. Additionally, the use of different providers who must give law enforcement access to private information can further compromise privacy [69]. These concerns may deter patients from adopting new technology.

17.6.2 HEALTH INFORMATION EXCHANGE BARRIERS

Health Information Exchange (HIE) enables secure and reliable electronic exchange of health information among different healthcare institutions, improving the delivery of healthcare services. There are currently three approaches to achieve HIE: consumer-mediated exchange, directed exchange, and query-based exchange [70]. Consumer-mediated exchange allows patients to access their electronic records, track their health status, and amend their self-reports. Directed exchange occurs when healthcare institutions share important information with other professionals involved in the patient's care. Query-based exchange happens when healthcare organizations request prior medical records of new patients. However, the implementation of HIE systems faces security and privacy issues, which cause delays. Authorized insiders may abuse their access rights by disclosing patient medical records with unapproved individuals for personal or financial gain. Unapproved insiders, such as hospital staff who don't directly care for patients or previous employees with unrestricted access to their records, may break regulations and exploit the system's security. Additionally, unauthorized intruders may attempt to hack into the system directly or by posing as a member of the medical staff. These issues with HIE systems can compromise patient privacy and confidentiality, leading to legal and ethical implications. Addressing these challenges is crucial to ensure the effective implementation of HIE systems and improve healthcare services [70–73]. Hospital security lapses can cost them up to $7 million due to cybercrime in healthcare [74,75]. Significant breaches in organizations like Anthem, CareFirst, Primera, and UCLA Health systems resulted in 143 million patient records being compromised, which is equivalent to 45% of the American population, according to Ref. [59]. A cyber security assessment by the Healthcare Information and Management Systems Society in 2015 found that 64% of healthcare companies experienced foreign intrusions in the previous year [60]. According to Bloomer News, 90% of all healthcare organizations have experienced most attacks. In addition, as opposed to the financial, governmental, or educational sectors, the healthcare and medical industries experience most data breaches [76,77].

17.6.3 DEVICE COMMUNICATION

One of the biggest challenges in implementing connected or smart health is communication. With many devices equipped with sensors to collect data, they often use their own language to communicate with servers, and different manufacturers use unique proprietary protocols that can prevent sensors from different brands from communicating with each other. This fragmented software environment and privacy concerns can lead to valuable information being isolated on "data islands," undermining the primary purpose of the IoTs. Moreover, the use of wireless network technology to connect medical devices raises additional issues, such as collisions between Wireless Personal Area Networks (WPANs) operating in the same frequency channel, which can result in decreased performance and potentially problematic situations, particularly for healthcare delivery. As a result, it's essential to ensure that medical devices work as intended when connected using various wireless communication technologies [78,79]. The complexity of smart health systems, including their many features, can make it challenging for healthcare professionals to adopt them. Interoperability is crucial for users and service providers to collaborate effectively within specific IoT domains, but this is complicated by various regulatory organizations governing different aspects of IoT [80,81]. Achieving true interoperability in a connected health system involves data flowing through both one-to-one and one-to-many connections, enabling the exchange of information across multiple interfaces that require system collaboration. Devices must be able to communicate using various formats and protocols for authentication and encryption in healthcare settings. Device management will require directories of device functionality, protocols, terminologies, and standard compliance. Despite the standardization of "plug and play" interoperability in non-health domains, achieving this level of interoperability with medical equipment remains challenging.

17.6.4 Collection and Management of Data

Digital healthcare using IoT sensor devices poses several data management issues. Medical sensors that are implanted or worn produce a constant flow of data, which is varied and consists of various data formats. Connected components, such as user devices, networks, and systems, are present in a connected health scenario, which makes it difficult to address constantly changing cyber-physical components. To address this, appropriate data-driven learning strategies should be used during the design process, as proper data analysis can provide valuable information about patients' health conditions. The volume and speed of data created in healthcare settings, along with the lack of standardized data-gathering formats, provide several issues. It is important to ensure the integrity of the data, as erroneous decisions may result from inaccurate data. Strong authentication mechanisms are required to ensure that healthcare data is submitted from legitimate, registered clinics, hospitals, and medical institutions because healthcare data frequently comes from multiple sources. Researchers, pharmaceutical companies, and healthcare providers have developed various data analysis techniques to extract useful knowledge from patients' data [59,82]. Healthcare data can be complex and difficult to obtain accurately due to varying definitions and measures that are constantly evolving. The length of stay (LOS)metric, for example, can have different definitions and interpretations, leading to inaccurate data. Additionally, integrating big data in healthcare is challenging due to the varying frequency of updates required for different types of data and the need to maintain data integrity while updating. Proper document control is necessary to avoid risking data integrity, but maintaining up-to-date databases can be costly due to maintenance expenses and the laws of Health Insurance Portability and Accountability Act (HIPAA) [83,84].

17.6.5 STANDARDIZATION

The deployment and standardization of IoT technologies have benefited from the efforts of numerous organizations. The proposals made by the Machine-to-Machine European Telecommunications Standards Institute (ETSI) and Internet Engineering Task Force (IETF) Working Groups had the biggest impact on the standardization of IoTs. To create a comprehensive solution that contributes to the development of standards for the future Internet, all fresh and emerging ideas should be combined [85,86]. The standardization of IoTs will enable the creation of comprehensive and effective IoT-based healthcare systems.

17.6.6 INTERACTIVE REPORTING AND VISUALIZATION

When dealing with big data applications, it is important to differentiate between reporting and analysis. Simply reporting data is insufficient, as these applications should extract meaningful insights from large volumes of data. Therefore, accurate algorithms are needed to produce concise and precise insights, which can be presented in visually appealing graphs and statistics. It is essential to create effective visualizations that highlight trends and issues within healthcare to ensure that the insights are useful and informative [87].

17.7 A NEW FRAMEWORK FOR NONSTOP PATIENT MONITORING USING BIOSENSORS AND WI-FI

We introduce a new package for continuous patient monitoring and instrument control via biosensors and Wi-Fi. In response to the COVID-19 outbreak, our system monitors critical physical parameters, including oxygen levels, blood pressure, and temperature. The sensors are connected to a microcontroller that transmits data to a cloud-based database via Wi-Fi. Specifically, we employ an LM35 sensor, which is a high-precision temperature sensor, to measure temperature [88]. Temperature sensors, LM35 and TMP36, are both suitable for human body temperature measurement; however, LM35 has been chosen due to higher accuracy and wider range from its counterpart [89]. LM35, developed by Texas Instrument, is an integrated-circuit temperature device that has accuracies of $\pm 1/4°C$ at room temperature, working in wide range ($-55°C$ to $+150°C$). The output voltage is also linearly proportional to Celsius (Centigrade) temperature. The features of LM35 are low output impedance, calibrated directly in degree (Centigrade) and linear $+10.0$ mV/°C scale factor. It is also suitable for remote applications [90]. An integrated pulse oximetry monitor biosensor module (MAX30100) has been used to measure the oxygen levels in the blood of patients. Pulse oximeter monitors SpO_2 of a patient based on two wavelengths, red (660 nm) and IR (940 nm). Absorption characteristics of HbO_2 and Hb are obtained by switching the wavelength transmission alternatively through a body part such as a finger or external pinna which is reflected and captured by a photodiode sensor [91].

The system also includes internal LEDs, photodetectors, optical elements, and low-noise electronics with ambient light rejection [92]. This is a weak bio-signal which is boosted using an inverting operational amplifier. The resulting output of

OPAMP filter represents the light absorbed by the body. The peak-to-peak signal amplitudes (V_{pp}) of both the wavelengths are measured and converted to their respective root-mean-squared voltage (V_{rms}), resulting in a proportional value shown in Equation (17.1).

$$\text{Ratio} = \left(\text{Red\_AC\_}V_{rms}/\text{Red\_DC}\right)/\left(\text{IR\_AC\_}V_{rms}/\text{IR\_DC}\right) \qquad (17.1)$$

The oxygen saturation value is obtained through a look-up table and corresponding ratio values, while the heart rate is computed based on the sampling rate of the pulse oximeter's Analog-to-Digital Converter (ADC) using a pulse sensor [93]. A pulse sensor has been used to monitor heart rate (BPM) of the patient that consists of a central LED with an additional noise filter. This LED activates the heartbeat rate detection sensor. The operating principle of this sensor for the heartbeat rate is very simple. The heartbeat rate is the time ratio between two consecutive heartbeats. Blood is squeezed into capillary tissues as blood circulates in the human body. Due to this, capillary tissue volume is increased, but after each heartbeat, this volume is reduced. This change in capillary tissue volume affects the heart rate pulse sensor's LED light, which transmits light after each heartbeat. There are three pins in this sensor, namely Ground, V_{cc}, and the A0 input signal. The LED is in ON condition when it is coupled with the NodeMCU. It works with the aid of an internet link, either in 3 or 5 V [94,95]. NodeMCU is like a microcontroller, and it can be configured to connect to the Internet for the Internet of Things (IoT). The NodeMCU development board is an open-source board based on Esp8266 microcontroller with an integrated Wi-Fi transceiver. NodeMCU is a complete environment of hardware and software for IoT [96].

IoT gateways play a crucial role in remote healthcare monitoring systems by serving as physical or virtual interfaces that connect detectors, smart devices, and the cloud. They act as a bridge between smart objects and applications, enabling the transfer of data to and from the cloud. One of the primary functions of IoT gateways is to provide data normalization. With various sensor categories gathering data, the IoT gateways translate the diverse data sets into common data formats, making it easier to process and analyze the data. IoT gateways also carry out data preprocessing before transmitting the information to the cloud. This reduces the volume of data that needs to be transmitted, which helps to reduce the requirements for transfer, retrieval, and storage. Additionally, IoT gateways provide network connectivity for sensors in IoT-based remote healthcare monitoring systems, reducing the limitations in linking up extensive networks. Overall, the role of IoT gateways is crucial in ensuring the seamless and efficient transfer of data in remote healthcare monitoring systems [42].

In the proposed system there are three different types of sensors (oxygen sensor, heart rate sensor, and temperature sensor) must need to be active during the live observation. Figure 17.7 shows the circuit diagram of these sensors that are calibrated using the microcontroller. Temperature sensor LM35 and Pulse sensor working principles are similar. Both sensors sense data in analog mode shown in Figure 17.7a and c. The analog pin A0 of the microcontroller is connected to the Data pin of the sensor. The microcontroller's ADC converts that data to electrical signal and transmit to the cloud database with the help of Wi-Fi connection. To measure blood oxygen level data shown in Figure 17.7b, the digital pins of microcontrollers D1 and D2 are connected to the SCL and SDA pins of the sensor. The sensor senses the physical parameters (analogue datasets) of patient and sends these to the microcontroller where an ADC converts the data to an electrical signal waveform [97].

The data from the oxygen sensor, heart rate sensor, and temperature sensor initially sensed and stored in the cloud database using a local Wi-Fi connection. A cloud platform is used to upload and save the patient's health condition sensed by the sensors. These data can be accessed for further investigation and monitored from the cloud. The data has been divided into three different conditions (low, normal, and high) as summarized in Table 17.1.

The trial is performed separately for the three different sensors to obtain more accurate values. The condition is set to design the alert system as shown in Table 17.2.

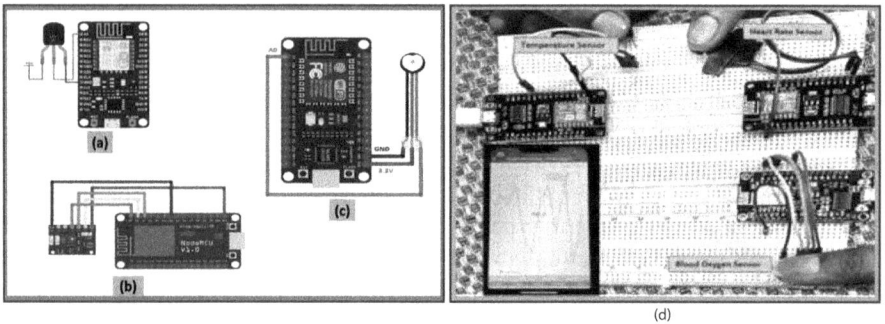

FIGURE 17.7 Circuit diagram (a)–(c) and real-time monitoring prototype of the proposed system (d).

TABLE 17.1
Patient's Physical Condition Specification

Patient's Condition	Temperature (°F)	Oxygen Level (%)	Heart Rate (BPM)
Normal	96–100	95–100	60–90
Low	<96	<94	<60
High	>100	-	>90

TABLE 17.2
Critical Condition Specification

Sensor	Condition	Remark
Temperature	>100°F	Critical condition
Blood oxygen level	<94%	Critical condition
Heart rate level	<60 BPM	Critical condition

To obtain real-time data, the experiment is performed on patient to measure health condition shown in Figure 17.7d reflects that the data is viewable from the mobile screen which is working as a display (Figure 17.9). The system output includes temperature, heart rate, blood oxygen level which are shown from the figures below. These sensors are being calibrated using microcontroller.

The Wi-Fi built into the NodeMCU will send all the data to the cloud via IoT, where the doctor can view it from anywhere using an active internet connection. The parameter output of the patient is shown in Figure 17.8a–c. This data can be accessed from the cloud by the authorized users using the IoT application platform. If any data is out of the ordinary, the Hyper Text Transfer Protocol (HTTP) function of the cloud will be activated and start sending email alert to the specific email address. The process will be continued with a fixed time interval until the patient gets back to normal condition or sensor gets removed. The temperature below 90°F is considered as room temperature and temperature between 96°F and 103°F is considered as

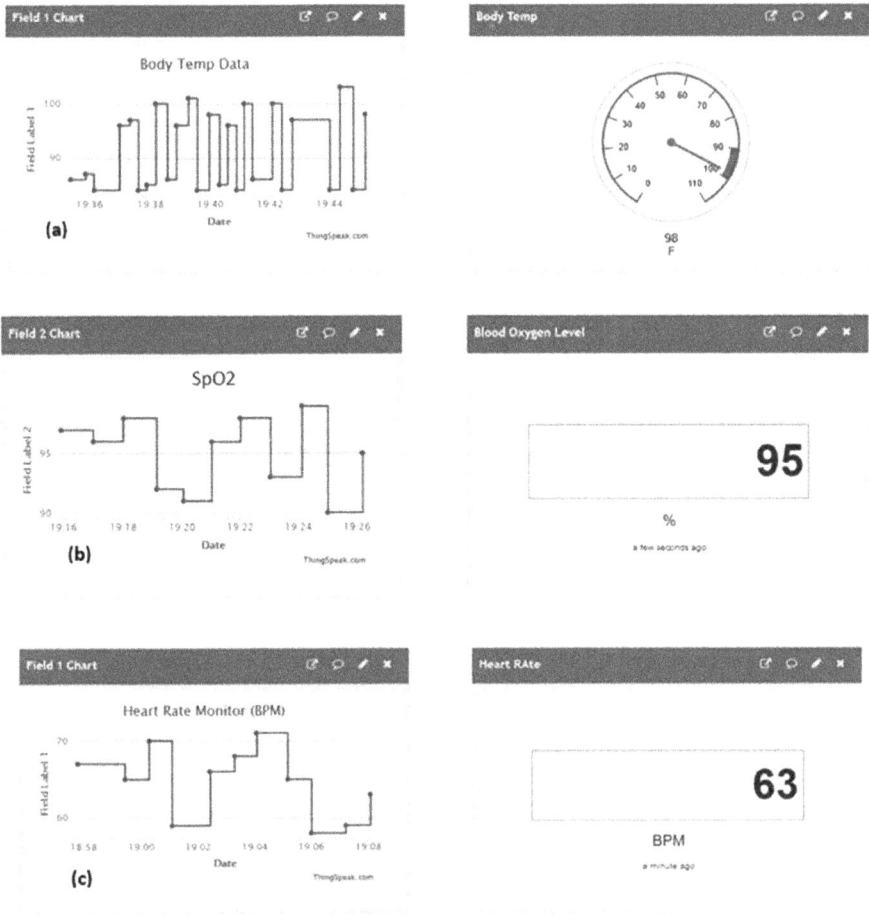

FIGURE 17.8 Real time output of patient data's visualization. Body temperature (a), blood oxygen level (b), and heart rate (c).

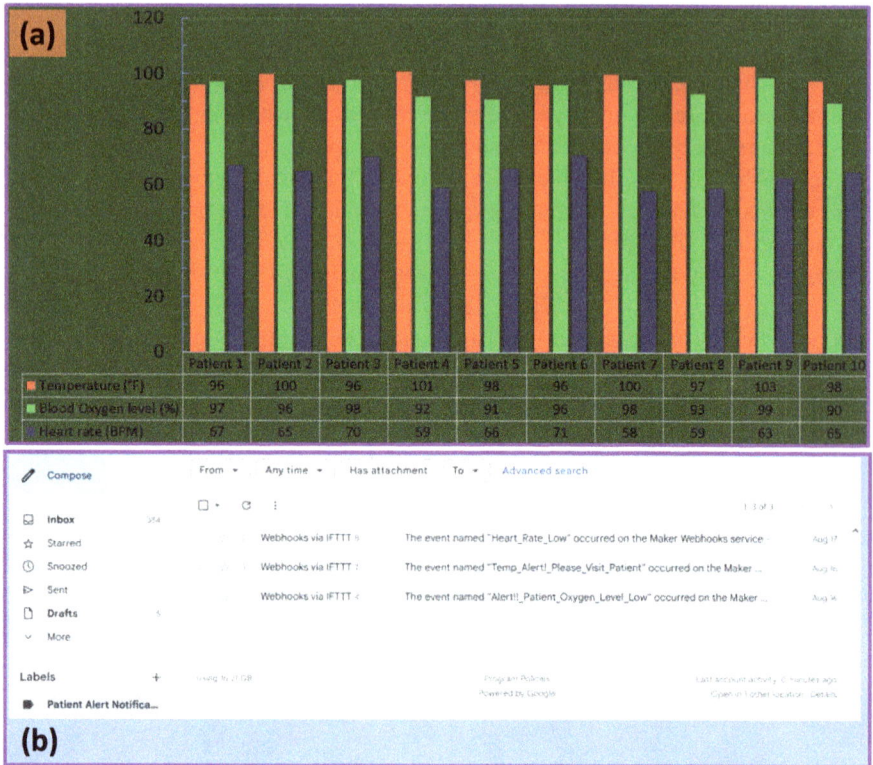

FIGURE 17.9   Monitored data of ten patients (a), and email alert sent for patients (b).

the body temperature shown in Figure 17.8a. Moreover, ThingSpeak cloud has own mobile application. Hence, it is more convenient to access the data using a mobile phone. The experiment was performed on ten individual patients aged between 59 and 78 years and acquired data that is shown in Figure 17.9. From Figure 17.9b the email alert sent only for patient 4 and 9 due to high body temperature. For patients 4, 7, and 9, heart rate alert has been sent due to abnormal condition. Blood oxygen level is in critical condition and alert has been activated for patients 4, 5, 8, and 10.

17.8   PROSPECT OF COMBINATION OF SENSORS-BASED MONITORING SYSTEM AND INTERNET OF THINGS (IoT)

The proposed combinatorial strategy has the potential to reduce the number of individuals affected by pandemics like COVID-19. However, implementing the suggested design may limit residents' freedom, privacy, and human rights. It is important to note that the design is only intended for emergency situations and will not compromise people's privacy in other circumstances. Data protection laws may allow for exceptions during these brief incursions. There have been proposals in this area where image sensors are used to instantly determine a person's mood or heart rate;

this information is then shared anonymously (without sharing the photos) to safeguard the users' privacy. Therefore, the suggested architecture will guarantee data in an anonymous form to protect user privacy. There are numerous techniques to extend the lifespan of IoT devices without sacrificing their functionality. By developing more energy-efficient technology (such as wearables, sensors, actuators, drones, and robots), designing specific radio protocols to improve energy conservation, activating hardware only when and where necessary, and enabling energy-efficient data transmission, the energy gap in communications and networks can be minimized.

One potential solution to address wireless environmental problems like fading, shadowing, and interference is to utilize IoT architecture. However, creating dependable, concise, and robust behavior in low-power IoT networks poses a significant challenge due to the limited energy, computational, and overhead budgets. Nonetheless, it is possible to implement an energy-efficient IoT network by utilizing one or a combination of the aforementioned techniques. Additionally, analyzing the benefits and drawbacks of popular and emerging message and communication protocols can facilitate the creation of more interoperable solutions that support IoT applications. However, it is crucial for the government, ministries, responsible authorities, health professionals, and the public to collaborate in adopting and trusting available technologies, including IoT and sensor-based devices and programs, to minimize the impact of pandemics like COVID-19.

17.9   CONCLUSION

We have conducted a thorough survey on successful design implementations in healthcare monitoring, coupled with a design approach that is readily applicable for the continuous monitoring of patients from remote locations. The newly proposed IoT-based patient monitoring system using temperature sensor, Pulse Oximeter device and Heart-Rate sensor configured with NodeMCU can support healthcare providers to work frequently even in those areas where the coronavirus and its variants can spread easily by the cough or nose aerosols released from the affected patients in the air. It is a highly comfortable and cheap device which is wearable on the wrist. By connecting to the Internet of Things (IoT) application the sensor reading is transmitted to the cloud database real time from where the data is visible with an active internet connection. Thus, this gives advantage to the doctors, relatives, or others to observe patient's condition from far away. The real time implementation system with the cloud database comes with login system ensuring security which can provide confidence to user and to the medical stuffs.

REFERENCES

1. J. Gubbi, R. Buyya, S. Marusic, and M. Palaniswami, "Internet of Things (IoT): A vision, architectural elements, and future directions," *Future Generation Computer Systems*, vol. 29, no. 7, pp. 1645–1660, 2013.
2. A. Botta, W. de Donato, V. Persico, and A. Pescapé, "Integration of cloud computing and Internet of Things: A survey," *Future Generation Computer Systems*, vol. 56, pp. 684–700, 2016.

3. Q. Zhang, L. Cheng, and R. Boutaba, "Cloud computing: State-of-the-art and research challenges," *Journal of Internet Services and Applications*, vol. 1, no. 1, pp. 7–18, 2010.

4. D. Gachet, M. de Buenaga, F. Aparicio, and V. Padron, "Integrating internet of things and cloud computing for health services provisioning: The virtual cloud carer project," In *2012 Sixth International Conference on Innovative Mobile and Internet Services in Ubiquitous Computing*, 2012, Palermo, 2012, pp. 918–921. doi: 10.1109/IMIS.2012.25.

5. Evans, D., "The internet of things how the next evolution of the internet is changing everything", Scirp.org., 2011 [Online]. Available: https://www.scirp.org/(S(lz5mqp453edsnp55rrgjct55))/reference/referencespapers.aspx?referenceid=1871901. [Accessed on 29-Oct-2022].

6. S. El Kafhali and K. Salah, "Efficient and dynamic scaling of fog nodes for IoT devices, *The Journal of Supercomputing* vol. 73, no. 12, pp. 5261–5284, 2017. Available online: https://www.researchgate.net/publication/317380217_Efficient_and_Dynamic_Scaling_of_Fog_Nodes_for_IoT_Devices. [Accessed on 29-Oct-2022].

7. Ordr, "10 internet of things (IoT) healthcare examples,", 01-Sep-2020. Available online: https://ordr.net/article/iot-healthcare-examples/. [Accessed on 30-Oct-2022].

8. C. Y. Koh, "The impact of IoT in healthcare," Linkedin.com, 21-Nov-2019. Available online: https://www.linkedin.com/pulse/impact-iot-healthcare-c-y-koh. [Accessed on 30-Oct-2022].

9. R. Nivetha, S. Preethi, P. Priyadharshini, B. Shunmugapriya, B. Paramasivan, and J. Naskath, "Smart health monitoring system using IoT for assisted living of senior and challenged people," *International Journal of Scientific & Technology Research*, vol. 9, no. 2, pp. 4285–4288, 2020.

10. G. J. Joyia, R. M. Liaqat, A. Farooq, and S. Rehman, "Internet of Medical Things (IOMT): Applications, benefits and future challenges in healthcare domain," *Journal of Communications,* vol. 12, no. 4, 2017. Available online: http://www.jocm.us/uploadfile/2017/0428/20170428025024260.pdf. [Accessed on 30-Oct-2022].

11. P. Rizwan, B. M. Rajasekhara, and K. Suresh, "Design and development of low investment smart hospital using internet of things through innovative approaches," *Biomedical Research (Aligarh)*, vol. 28, no. 11, , pp. 4979–4985, 2017.

12. L. Barolli, A. Poniszewska-Maranda, and T. Enokido, Eds., "Complex, intelligent and software intensive systems," *Proceedings of the 14th International Conference on Complex, Intelligent and Software Intensive Systems (CISIS-2020)*. Cham: Springer International Publishing, 2021.

13. P. A. Laplante and N. Laplante, "The internet of things in healthcare: Potential applications and challenges," *IT Professinal*, vol. 18, no. 3, pp. 2–4, 2016.

14. A. Grgurić, M. Mošmondor, and D. Huljenić, "The SmartHabits: An intelligent privacy-aware home care assistance system," *Sensors (Basel)*, vol. 19, no. 4, p. 907, 2019.

15. S. Grant, A. W. Blom, I. Craddock, M. Whitehouse, and R. Gooberman-Hill, "Home health monitoring around the time of surgery: Qualitative study of patients' experiences before and after joint replacement," *BMJ Open*, vol. 9, no. 12, p. e032205, 2019.

16. H. Rexhepi, R.-M. Åhlfeldt, Å. Cajander, and I. Huvila, "Cancer patients' attitudes and experiences of online access to their electronic medical records: A qualitative study," *Journal of Health Informatics*, vol. 24, no. 2, pp. 115–124, 2018.

17. M. D. Ritholz, O. Henn, A. A. Castillo, H. Wolpert, S. Edwards, L. Fisher, and E. Toschi, "Experiences of adults with type 1 diabetes using glucose sensor-based mobile technology for glycemic variability: Qualitative study," *JMIR Diabetes*, vol. 4, no. 3, p. e14032, 2019. doi: 10.2196/14032.

18. N. Brew-Sam, M. Chhabra, A. Parkinson, K. Hannan, E. Brown, L. Pedley, K. Brown, K. Wright, E. Pedley, C. J. Nolan, C. Phillips, H. Suominen, A. Tricoli, and J. Desborough, "Experiences of young people and their caregivers of using technology to manage type 1 diabetes mellitus: Systematic literature review and narrative synthesis," *JMIR Diabetes,* vol. 6, no. 1, p. e20973, 2021. doi: 10.2196/20973.

19. E.-Y. Lee, J.-S. Yun, S.-A. Cha, S.-Y. Lim, J.-H. Lee, Y.-B. Ahn, K.-H. Yoon, and S.-H. Ko, "Personalized Type 2 diabetes management using a mobile application integrated with electronic medical records: An ongoing randomized controlled trial," *International Journal of Environmental Research and Public Health*, vol. 18, no. 10, p. 5300, 2021. doi: 10.3390/ijerph18105300.

20. J. C. Pickup, H. M. Ford, and K. Samsi. "Real-time continuous glucose monitoring in type 1 diabetes: A qualitative framework analysis of patient narratives," *Diabetes Care*, vol. 38, no. 4, pp. 544–50, 2015. doi: 10.2337/dc14-1855.

21. R. Mieronkoski et al., "The internet of things for basic nursing care: A scoping review," *International Journal of Nursing Studies*, vol. 69, pp. 78–90, 2017.

22. C. Karlsen, M. S. Ludvigsen, C. E. Moe, K. Haraldstad, and E. Thygesen, "Experiences of commu-nity-dwelling older adults with the use of telecare in home care services: A qualitative systematic review," *JBI Database of Systematic Reviews and Implementation Reports*, vol. 15, no. 12, pp. 2913–2980, 2017.

23. S. Kumar, P. Tiwari, and M. Zymbler, "Internet of Things is a revolutionary approach for future technology enhancement: A review," *Journal of Big Data*, vol. 6, no. 1, 2019, https://doi.org/10.1186/s40537-019-0268-2.

24. "Capture calendar," Archive-it.org. Available online: https://wayback.archive-it.org/12090/*/https://ec.europa.eu/digital-single-market/en/news/ehealth-action-plan-2012-2020-innovative-healthcare-21st-century. [Accessed on 30-Oct-2022].

25. P. Bodin, "IoT project case study #1," Linkedin.com, 20-Mar-2016. Available online: https://www.linkedin.com/pulse/iot-project-case-study-1-pascal-bodin. [Accessed on 04-Nov-2022].

26. S. Saguna, C. Åhlund, and A. Larsson, "Experiences and challenges of providing IoT-based care for elderly in real-life smart home environments," In A. Y. Zomaya (Ed.), *Scalable Computing and Communications*. Cham: Springer International Publishing, 2020, pp. 255–271.

27. J. Dugstad, V. Sundling, E. R. Nilsen, and H. Eide, "Nursing staff's evaluation of facili-tators and barriers during implementation of wireless nurse call systems in residential care facilities. A cross-sectional study," *BMC Health Services Research*, vol. 20, no. 1, p. 163, 2020.

28. P. Singh, "Internet of things based health monitoring system: Opportunities and chal-lenges," *International Journal of Advanced Research in Computer Science,* vol. 9, pp. 224–228, 2018.

29. R. Thusu, "Sensors facilitate health monitoring," Available online: https://www.fier-ceelectronics.com/components/sensors-facilitate-health-monitoring. [Accessed on 25-Nov-2022].

30. N. Patel, "Internet of things in healthcare: Applications, benefits, and challenges," *Peerbits*. Available online: https://www.peerbits.com/blog/internet-of-things-health-care-applications-benefits-and-challenges.html. [Accessed: 30-Oct-2022].

31. M. Peyroteo, I. A. Ferreira, L. B. Elvas, J. C. Ferreira, and L. V. Lapão, "Remote moni-toring systems for patients with chronic diseases in primary health care: Systematic review," *JMIR MHealth UHealth*, vol. 9, no. 12, p. e28285, 2021.

32. D. Kim, H. Bian, C. K. Chang, L. Dong, and J. Margrett, "In-home monitoring technol-ogy for aging in place: Scoping review," *Interactive Journal of Medical Research*, vol. 11, no. 2, p. e39005, 2022.

33. W. Little and W. Little, "Chapter 13: Aging and the elderly," In *Introduction to Sociology: 1st Canadian Edition*. Victoria: BCcampus, 2014, https://opentextbc.ca/introductiontosociology/chapter/chapter13-aging-and-the-elderly/.

34. R. Joshi, A. Joseph, S. Mihandoust, K. C. Madathil, and S. R. Cotten, "A mobile appli-cation-based home assessment tool for patients undergoing joint replacement surgery: A qualitative feasibility study," *Applied Ergonomics*, vol. 103, no. 103796, p. 103796, 2022.

35. S. Rajkumar, M. Srikanth, and N. Ramasubramanian, "Health monitoring system using Raspberry PI," In *2017 International Conference on Big Data, IoT and Data Science (BID)*, 2017, Pune, 2017, pp. 116–119, doi: 10.1109/BID.2017.8336583. pp. 116–119.

36. N. Sawant, "Remote healthcare monitoring system with home automation," *International Journal for Research in Applied Science and Engineering Technology (IJRASET)*, vol. 7, no. 8, pp. 255–259, 2019.

37. D. F. S. Santos, H. O. Almeida, and A. Perkusich, "A personal connected health system for the Internet of Things based on the constrained application protocol," *Computers and Electrical Engineering*, vol. 44, pp. 122–136, 2015.

38. M. Akash, M. Rathod, G. U. Kharat, and R. S. Bansode, "Patient health monitoring IoT system," Ijariie.com. Available online: https://ijariie.com/AdminUploadPdf/Iot_Based_Patient_Monitoring_system_ijariie12292_converted.pdf. [Accessed on 30-Oct-2022].

39. H. Cai, B. Xu, L. Jiang, and A. V. Vasilakos, "IoT-based big data storage systems in cloud computing: Perspectives and challenges," *IEEE Internet Things Journal*, vol. 4, no. 1, pp. 75–87, 2017.

40. M. M. Islam, A. Rahaman, and M. R. Islam, "Development of smart healthcare monitoring system in IoT environment," *SN Computer Science*, vol. 1, no. 3, 2020, https://doi.org/10.1007/s42979-020-00195-y.

41. S. F. Khan, "Health care monitoring system in Internet of Things (IoT) by using RFID," In *2017 6th Int'l Conference on Industrial Technology and Management (ICITM)*, Cambridge, 2017, pp. 198–204. doi: 10.1109/ICITM.2017.7917920.

42. A. A. Mohammed, M. A. Burhanuddin, M. S. Talib, M. E. Hameed, and M. F. Ali, "A review on IoT-based healthcare monitoring systems for patient in remote environments," *European Journal of Molecular & Clinical Medicare*, vol. 7, no. 3, 2020, 2227- 2235.

43. S. Abdulmalek, et al., "IoT-based healthcare-monitoring system towards improving quality of life: A review," *Healthcare (Basel)*, vol. 10, no. 10, p. 1993, 2022.

44. "Complete guide to medical sensors," sps.honeywell.com. Available online: https://sps.honeywell.com/us/en/support/blog/siot/complete-guide-to-medical-sensors-benefits-and-applications. [Accessed on 02-Nov-2022].

45. Vantage Market Research, "Internet of things in Healthcare Market size & share to surpass $190 bn by 2028;," Vantage Market Research, 14-Oct-2022. Available online: https://www.globenewswire.com/en/news-release/2022/10/14/2534568/0/en/Internet-of-Things-in-Healthcare-Market-Size-Share-to-Surpass-190-Bn-by-2028-Vantage-Market-Research.html. [Accessed on 30-Oct-2022].

46. "IoT Covid patient health monitor in quarantine," Nevon Projects, 27-Oct-2020. Available online: https://nevonprojects.com/iot-covid-patient-health-monitor-in-quarantine/. [Accessed on 02-Nov-2022].

47. E. Altynpara, "The internet of medical things: Its role in healthcare and how to implement it," Forbes, 01-Apr-2022. Available online: https://www.forbes.com/sites/forbestechcouncil/2022/04/01/the-internet-of-medical-things-its-role-in-healthcare-and-how-to-implement-it/?sh=7a84823453f9. [Accessed on 30-Oct-2022].

48. "Healthcare IoT market," Futuremarketinsights.com. Available online: https://www.futuremarketinsights.com/reports/healthcare-iot-market. [Accessed on 30-Oct-2022].

49. "IoT in Healthcare Market," Futuremarketinsights.com. Available online: https://www.futuremarketinsights.com/reports/iot-in-healthcare-market. [Accessed on 02-Nov-2022].

50. "Internet of Things in healthcare market size report, 2019-2025," Grandviewresearch.com. [Online]. Available: https://www.grandviewresearch.com/industry-analysis/internet-of-things-iot-healthcare-market. [Accessed on 02-Nov-2022].

51. M. Mitra, "IoT in insurance sector: Home, auto and health insurance," Mantra Labs, 31-Aug-2017. Available online: https://www.mantralabsglobal.com/blog/iot-in-insurance-sector/. [Accessed on 30-Oct-2022].

52. V. Pardeshi, S. Sagar, S. Murmurwar, and P. Hage, "Health monitoring systems using IoT and raspberry Pi: A review," In *Proceedings of International Conference on Innovative Mechanisms for Industry Applications (ICIMIA 2017)*, Bengaluru, India, 21-23 Febuary 2017, pp. 134–137. Scirp.org. [Online]. Available: https://www.scirp.org/(S(czeh2tfqw2orz553k1w0r45))/reference/referencespapers.aspx?referenceid=3039824. [Accessed: 30-Oct-2022].

53. F. Al-Turjman, M. H. Nawaz, and U. D. Ulusar, "Intelligence in the Internet of Medical Things era: A systematic review of current and future trends," *Computer Communications*, vol. 150, pp. 644–660, 2020.

54. A. M. Rahmani et al., "Exploiting smart e-Health gateways at the edge of healthcare Internet-of-Things: A fog computing approach," *Future Generation Computer Systems*, vol. 78, pp. 641–658, 2018.

55. L. A. Durán-Vega et al., "An IoT system for remote health monitoring in elderly adults through a wearable device and mobile application," *Geriatrics (Basel)*, vol. 4, no. 2, p. 34, 2019.

56. J. Kim, A. S. Campbell, B. E.-F. de Ávila, and J. Wang, "Wearable biosensors for healthcare monitoring," *Nature Biotechnology*, vol. 37, no. 4, pp. 389–406, 2019.

57. M. Yang and Y. Hara-Azumi, "Implementation of lightweight eHealth applications on a low-power embedded processor," *IEEE Access*, vol. 8, pp. 121724–121732, 2020.

58. S. Zeadally and O. Bello, "Harnessing the power of Internet of Things based connectivity to improve healthcare," *Internet of Things*, vol. 14, no. 100074, p. 100074, 2021.

59. S. Zeadally, F. Siddiqui, Z. Baig, and A. Ibrahim, "Smart healthcare: Challenges and potential solutions using Internet of Things (IoT) and big data analytics", *PSU Research Review*, vol. 4, no. 2, pp. 149–168, 2020.

60. Available online: https://www.emerald.com/insight/content/doi/10.1108/PRR-08-2019-0027/full/html [Accessed on 25-Nov-2022].

61. "The internet of things (IoT) security and privacy risks," Datasilk.com. [Online]. Available: https://datasilk.com/iot-cybersecurity/. [Accessed on 02-Nov-2022].

62. "Shodan," Shodan. [Online]. Available: https://www.shodan.io/. [Accessed on 30-Oct-2022].

63. U. Umar, S. Nayab, R. Irfan, M. A. Khan, and A. Umer, "E-Cardiac care: A comprehensive systematic literature review," *Sensors (Basel)*, vol. 22, no. 20, p. 8073, 2022.

64. A. K. Das, S. Zeadally, and D. He, "Taxonomy and analysis of security protocols for Internet of Things," *Future Generation Computer Systems*, vol. 89, pp. 110–125, 2018.

65. S. Nazir, Y. Ali, N. Ullah, and I. García-Magariño, "Internet of Things for healthcare using effects of mobile computing: A systematic literature review," *Wireless Communications and Mobile Computing*, vol. 2019, pp. 1–20, 2019.

66. N. Drissi, S. Ouhbi, M. A. Janati Idrissi, L. Fernandez-Luque, and M. Ghogho, "Connected mental health: Systematic mapping study (preprint)," JMIR Preprints, 2020.

67. D. Koutras, G. Stergiopoulos, T. Dasaklis, P. Kotzanikolaou, D. Glynos, and C. Douligeris, "Security in IoMT communications: A survey," *Sensors (Basel)*, vol. 20, no. 17, p. 4828, 2020.

68. I. K. Poyner and R. S. Sherratt, "Privacy and security of consumer IoT devices for the pervasive monitoring of vulnerable people," In *Living in the Internet of Things: Cybersecurity of the IoT -2018*, 2018.

69. C. Ye, W. Cao, and S. Chen, "Security challenges of blockchain in Internet of things: Systematic literature review," *Transactions on Emerging Telecommunications Technologies*, vol. 32, no. 8, 2021. doi:10.1002/ett.4177.

70. Healthaffairs.org. [Online]. Available: https://www.healthaffairs.org/content/31/3/527.abstract. [Accessed on 31-Oct-2022].

71. J. Szerejko, "Reading between the lines of electronic health records: The health information technology for economic and clinical health act and its implications for health care and information security," 2015.

72. A. T. Strauss et al., "A user needs assessment to inform health information exchange design and im-plementation," *BMC Medical Informatics and Decision Making*, vol. 15, no. 1, p. 81, 2015.

73. "U.S. Department of Health & Human Services - Office for Civil Rights," Hhs.gov. [Online]. Available: https://ocrportal.hhs.gov/ocr/smartscreen/main.jsf [Accessed on 31-Oct-2022].

74. L. Agha, "The effects of health information technology on the costs and quality of medical care," *Journal of Health Economics*, vol. 34, pp. 19–30, 2014.

75. R. W. Anwar, T. Abdullah, and F. Pastore, "Firewall best practices for securing smart healthcare en-vironment: A review," *Applied Sciences (Basel)*, vol. 11, no. 19, p. 9183, 2021.

76. P. C. Paul, J. Loane, F. McCaffery, and G. Regan, "Towards design and development of a data security and privacy risk management framework for WBAN based healthcare applications," *Applied System Innovation*, vol. 4, no. 4, p. 76, 2021.

77. L. Delany, L. Signal, and G. Thomson, "International trade and investment law: A new framework for public health and the common good," *BMC Public Health*, vol. 18, no. 1, p. 602, 2018.

78. D. V. Dimitrov, "Medical Internet of Things and big data in healthcare," *Healthcare Informatics Research*, vol. 22, no. 3, pp. 156–163, 2016.

79. A. Gawanmeh, "Open issues in reliability, safety, and efficiency of connected health," In *2016 IEEE First International Conference on Connected Health: Applications, Systems and Engineering Technologies (CHASE)*, Washington, DC, 2016, pp. 1–6. doi: 10.1109/CHASE.2016.60..

80. C. de Grood, A. Raissi, Y. Kwon, and M. J. Santana, "Adoption of e-health technology by physicians: A scoping review," *Journal of Multidisciplinary*, vol. 9, pp. 335–344, 2016.

81. A. Ghosh and S. Dey, "'Sensing the mind': An exploratory study about sensors used in E-health and M-health applications for diagnosis of mental health condition," In *Internet of Things*. In C. Chakraborty, U. Ghosh, V. Ravi, and Y. Shelke (Eds.), Cham: Springer International Publishing, 2021, pp. 269–292.

82. D. Tse, C.-K. Chow, T.-P. Ly, C.-Y. Tong, and K.-W. Tam, "The challenges of big data governance in healthcare," In *Proceedings of 17th IEEE International Conference on Trust, Security And Privacy in Computing and Communications/12th IEEE International Conference on Big Data Science and Engineering (TrustCom/BigDataSE)*, New York, 2018, pp. 1632–1636. doi: 10.1109/TrustCom/BigDataSE.2018.00240.

83. I. Anagnostopoulos, S. Zeadally, and E. Exposito, "Handling big data: Research challenges and future directions," *The Journal of Supercomputing*, vol. 72, no. 4, pp. 1494–1516, 2016.

84. Office for Civil Rights (OCR), "HIPAA home," Hhs.gov, 09-Jun-2021. [Online]. Available: https://www.hhs.gov/hipaa/index.html. [Accessed on 31-Oct-2022].

85. K. Stuurman and I. Kamara, "IoT standardization: The approach in the field of data protection as a model for ensuring compliance of IoT applications?," In *2016 IEEE 4th International Conference on Future Internet of Things and Cloud Workshops (FiCloudW)*, 2016, pp. 336–341.

86. "IERC-European research cluster on the internet of things," Internet-of-things-research. eu. [Online]. Available: https://www.internet-of-things-research.eu/. [Accessed on 31-Oct-2022].

87. "What is Data Analytics: Understanding big data analytics," Intellipaat Blog, 13-Oct-2022. [Online]. Available: https://intellipaat.com/blog/what-is-data-analytics/. [Accessed on 31-Oct-2022].

88. B. Oyebola and O. Toluwani, "LM35 based digital room temperature meter: A simple demonstration (October 15, 2017)", *Equatorial Journal of Computational and Theoretical Science*, vol. 2, no. 1. Available at SSRN: https://ssrn.com/abstract=3053601.

89. A. H. Kioumars and L. Tang, "Wireless network for health monitoring: Heart rate and temperature sensor," *2011 Fifth International Conference on Sensing Technology*, 2011, pp. 362–369, doi: 10.1109/ICSensT.2011.6137000.

90. H. Mansor, M. H. A. Shukor, S. S. Meskam, N. Q. A. M. Rusli and N. S. Zamery, "Body temperature measurement for remote health monitoring system," *2013 IEEE International Conference on Smart Instrumentation, Measurement and Applica-tions (ICSIMA)*, 2013, pp. 1–5, doi: 10.1109/ICSIMA.2013.6717956.

91. D. Murali, D. R. Rao, S. R. Rao and M. Ananda, "Pulse oximetry and IoT based cardiac monitoring integrated alert system," In *Proceedings of International Conference on Advances in Computing, Communications and Informatics (ICACCI)*, 2018, pp. 2237–2243, doi: 10.1109/ICACCI.2018.8554425.

92. Available online: https://www.esp8266learning.com/max30102-pulse-and-heart-rate-monitor-sensor-and-esp8266.php. (Accessed on 25/07/2022).

93. Available online: https://embedded-computing.com/articles/measuring-Ievels-portable-medical-wearable-devices/# [Accessed on 27-Feb-2023].

94. C. Patil and A. Chaware, "Heart (pulse rate) monitoring using pulse rate sensor, piezo electric sensor and NodeMCU," *2021 8th International Conference on Computing for Sustainable Global Development (INDIACom)*, New Delhi, 2021, pp. 337–340.

95. "Pulse sensor," Components101. Available online: https://components101.com/sensors/pulse-sensor. [Accessed on 06-Sep-2022].

96. D. A. Aziz, "Web server-based smart monitoring system using ESP8266 node MCU module," *International Journal of Scientific & Engineering Research*, vol. 9, pp. 801–808, 2018.

97. F. N. Ngajieh and C. E. Weiber, "Arduino dynamic wireless sensor network system," Thesis, Blekinge Institute of Technology School of Engineering, Sweden, 2015.

18 End-to-End Solutions for the Remote Monitoring of Post-Operative Prehabilitation Program
IoT Solution Challenges

Khalid Al-Naime and Adnan Al-Anbuky
Auckland University of Technology

18.1 INTRODUCTION

Abdominal cancer is one of the most frequent and dangerous cancers in the world, particularly among the elderly, and is considered one of the leading causes of death in New Zealand and throughout the world [1,2]. A worldwide population of over 230 million people undergo major abdominal surgery each year, and, with easier access to healthcare, this number is likely to increase [3,4]. Major surgery is associated with a significant deterioration in quality of life, as well as a 20%–40% reduction in post-operative physical function [5]. In the absence of prehabilitation or rehabilitation pro-grammes, the results showed that only 30% of patients had returned to preoperative levels of function 8 weeks after surgery, and 50% had attained baseline functional ability after 6 months [5]. Prehabilitation is likely to be most useful in older cancer patient cohorts, who are more likely to have complex co-morbidities as well as sensory, balance or cognitive impairments and reduced functional abilities [6]. Exercise training has been shown to positively affect the maintenance of functional capacity and facilitate recovery from surgery. A literature review on the role of optimising functional exercise capacity in the surgical population during the preoperative time frame has demonstrated that this approach can have a positive impact on reducing postoperative complications, decreasing the length of stay in hospital, and improving quality of life [7]. There have been various studies conducted with the concept of prehabilitation as a primary intervention [8]. Figure 18.1 shows a schematic illustrating a comparison of the fitness levels of patients who are involved and those who are not involved in a prehabilitation programme before surgery [4,7].

18.1.1 THE ROLE OF PHYSICAL ACTIVITY

For the majority of cancer patients in an elderly population, participating in physical activity has been shown to lessen all-cause mortality, morbidity and disability and

DOI: 10.1201/9781003346678-18

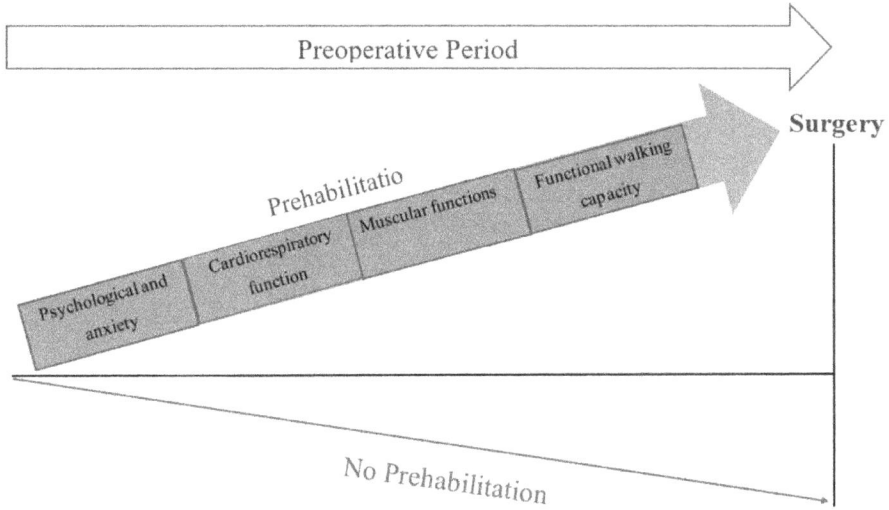

FIGURE 18.1 Schematic of the progress of patients during the preoperative period. The green line depicts the progression of patients who participate in the prehabilitation programme. The red line indicates the patients who did not participate in the prehabilitation programme [7,9,10].

to be beneficial for improvements in health outcomes in cardiovascular, metabolic and respiratory diseases and cancer [11]. The American College of Sports Medicine and the American Heart Association recommend that regular physical activity is essential for healthy aging and should include moderate-intensity aerobic activity for 30 minutes per day at least five times per week, along with strengthening, balance, and flexibility exercises [12]. Engaging in regular physical activity helps to provide a protective effect against many functional limitations [13]. An elevated level of physical activity is associated with a significantly reduced risk for the incidence of colorectal cancer, and lower physical activity is known to be a risk factor for the development of abdominal cancer [14]. Physical activity has also been associated with improved short-term surgical outcomes [15], thus facilitating early recovery. Therefore, attempts to improve quality of life should be made by enhancing the level of physical activity patients engage in prior to surgery.

As the population ages, there is a reduction in physical function [16], often resulting in frailty. It is thus of importance to preserve and/or enhance cardiopulmonary function, muscle and bone strength and mobility [17] in cancer patients due to the adverse effects of surgery, chemotherapy and radiotherapy. Not only does physical activity have a positive effect physically, but also psychologically [16]. Physical inactivity is known to be the fourth most important risk factor for overall death, according to the World Health Organization [18], and has an association with different types of cancer, including colorectal and breast [19].

18.1.2   An Overview of the Different Types of Abdominal Cancers

The term "abdominal cancer" refers to cancers that affect the digestive system and abdominal organs such as the stomach, liver, large intestine, small intestine, pancreas, gallbladder, oesophagus, and rectum [20]. It happens when damaged or old cells rapidly divide and multiply, resulting in a malignant mass tumour [20]. As an example, Figure 18.2a and b depicts the various stages of colon and stomach cancers.

The main abdominal cancer types are colorectal cancer, pancreatic cancer, and stomach cancer [21]. Depending on the type of abdominal cancer, symptoms may vary. Many people experience no symptoms in the early stages of colorectal cancer, liver cancer, stomach cancer, or pancreatic cancer. Most of the above types have some general common symptoms such as [22]:

- Abdominal pain
- Appetite loss
- Blood in the stool
- Noticeable increase in fatigue and/or weakness
- Unexplained weight loss
- Nausea
- Vomiting

However, treatment options are determined by the stage of cancer. Each has its own set of drawbacks. A combination of treatments may be recommended by doctors. One of the most common treatments, particularly in early-stage cancer, is surgery [23]. To be ready for major surgery, the patient must undergo extensive preparations in terms of psychological fitness, physical fitness, nutrition, and general health conditions. Prehabilitation is one of the common fitness programmes recommended for abdominal patients undergoing major surgery [15,17,24,25].

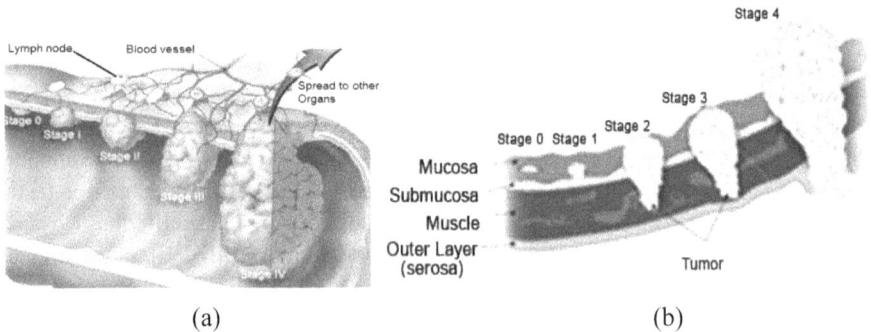

         (a)   (b)

FIGURE 18.2   (a) Stages of colon cancer. (b) Stages of stomach cancer [10].

18.1.2.1   Existing Prehabilitation Programmes

The two main prehabilitation models that are currently applied to patients undergoing major surgery are supervised and unsupervised programmes [15,17,24,25]. Patients who are involved in a supervised programme usually perform the prescribed physical exercises under the direct supervision of a healthcare professional. A supervised in-person programme provided by a healthcare professional is considered the gold standard in terms of safety and effectiveness [26,27]. Most supervised prehabilitation programmes documented in the literature occur in a hospital setting and tend to be supervised by an experienced health professional (e.g., physiotherapist, exercise physiologist, or trained nurse.). This type of prehabilitation is usually limited to those individuals who live in close proximity to a hospital [28]. Prehabilitation exercise sessions are typically performed in a gymnasium, and the type of equipment used in these settings ranges from simple (commercial equipment) [29] to sophisticated cycle ergometers with pre-set workload parameters [28,30]. The number of required visits varies between studies, but most supervised programmes require participants to attend between two to three sessions per week [1,30]. Exercise intensity/workload settings for the programmes are often based on cardiopulmonary exercise testing (CPET) findings. All these factors enable more controlled exercise prescription. However, supervised prehabilitation demands a high level of resources and is often limited to those individuals who live close to the prehabilitation facilities [28]. Figure 18.3 shows the general structure of supervised prehabilitation programmes.

The home-based prehabilitation programme (unsupervised) offers flexibility for patients to perform the prescribed activity in their home or community centre (gymnasium) [30]. Another advantage of the home-based programme is that it overcomes geographical barriers; patients living outside the region may access services when they might not have had the opportunity due to the distance and time required to

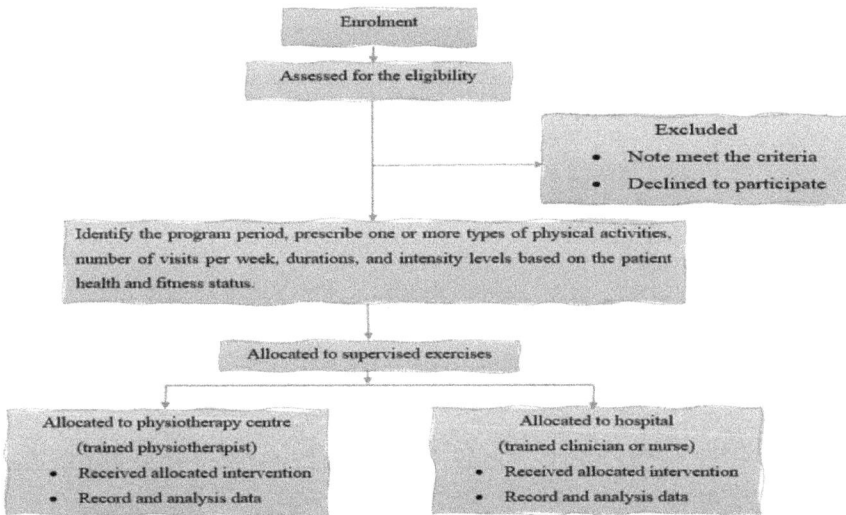

FIGURE 18.3   General structure of supervised prehabilitation programme [10].

travel to the referral centres [30,31]. Despite prescribed physical exercises being based on home-based prehabilitation programmes (unsupervised), they are similar to those in a supervised prehabilitation programme [1,15,17,25,30]. However, the lack of group sessions means that patients do not have access to peer support, which is felt to be important in addition to exercise. Furthermore, some patients believe that home-based approaches rely more on self-motivation than face-to-face appointments (supervised guidance) [31]. Phone calls and video interventions are common telehealth services used to guide patients during home-based prehabilitation programmes [27,28,32]. Based on pilot research [33,34], the feasibility of adopting a home-based prehabilitation programme has been demonstrated. According to the findings of these trials, the majority of patients in the prehabilitation group were below their baseline values 4 weeks following surgery, indicating that the home-based programme was insufficient to encourage physical activity and daily life modifications. Lack of direction and supervision by a kinesiologist, resulting in patients not following the programme as advised, could be one reason for not returning to baseline measures. Figure 18.4 below shows the general procedure for unsupervised prehabilitation programmes.

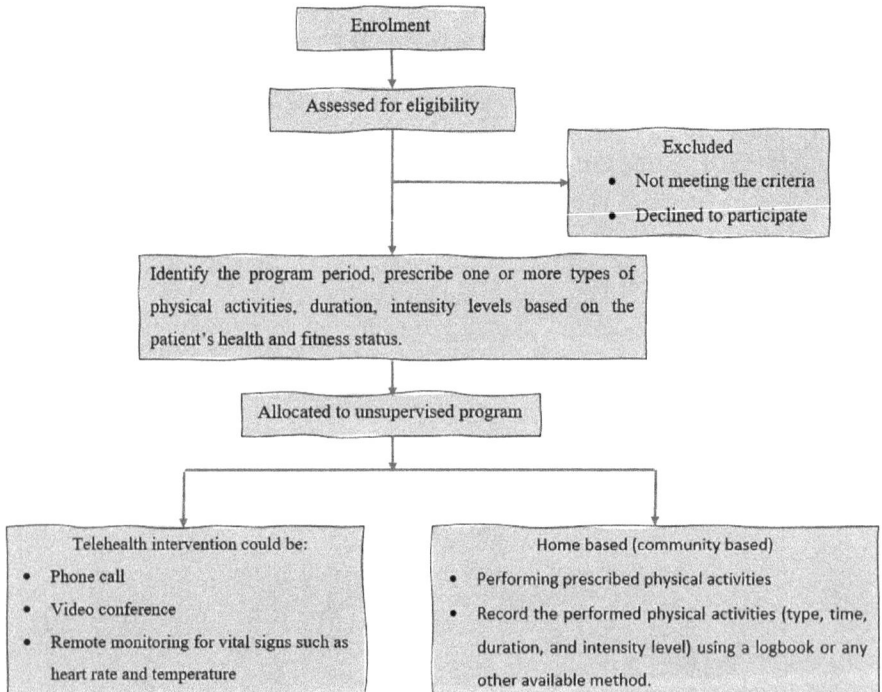

FIGURE 18.4   General structure of unsupervised prehabilitation programme [10].

18.2   BARRIERS FOR PREHABILITATION PROGRAMMES

Even though the number of supervised and home-based programmes has increased over the last decade, there are many obstacles preventing large numbers of patients from being involved in such programmes. One key problem is access to prehabilitation facilities and resources. For example, people diagnosed with cancer who live outside of large urban regions have a much lower chance of surviving than those who live in cities [7,8]. Patients in rural and regional areas are less likely to have their treatment monitored by an oncologist [9] or to have access to multidisciplinary specialist oncology services [9,10]. The long-distance travel to vital cancer services can have a significant impact on a patient's quality of life and survival [11]. There is also consistent evidence that cancer patients face barriers to attending in-person supervised programmes. These include transportation, parking and time, as well as a desire to avoid additional hospital appointments [35]. Furthermore, limited numbers of healthcare centres and staff will impact the number of patients who can participate in supervised prehabilitation programmes. In addition, transportation, healthcare centre visits, and demand on facilities are an additional burden for both the patient and the healthcare system [35].

Unsupervised prehabilitation programmes address a number of these issues but also have some problems. For example, there may be uncertainty that home-based exercises are performed at the intended frequency and intensity, and the lack of monitoring means that patient safety is an issue with unsupervised programmes. It has been shown that some cancer patients prefer flexible home-based programmes [23], but a study investigating experiences of such programmes found that some patients felt that greater involvement from healthcare professionals would increase engagement, especially if the patients were sedentary [24]. Given the inherent challenges of each approach, we contend that debating the best delivery mode (supervised versus unsupervised home-based) is a futile and counterproductive exercise. Rather, consideration should be given to how the limitations of any delivery mode can be addressed to provide programmes that are best suited to the local context. A number of the prehabilitation trials described below addressed this by combining supervised sessions with home-based elements [32,36]. Community-based programmes that provide more readily available assistance have also yielded promising preliminary results [23].

Telehealth has the potential to address this health disparity and improve health outcomes [23] by providing an alternative for those who are unable to travel due to caring or work commitments [24,25], conflicting clinical appointments, or treatment-related symptoms. The nature of telehealth interventions might range from instructive or helpful websites to computerised questionnaires and live online chats [23]. When presented to patients preparing for cancer surgery, patient education sessions involving audio-visual or multi-media treatments have been demonstrated to improve satisfaction and knowledge [23]. Evidence on telehealth in the perioperative context for cancer patients, on the other hand, is currently restricted to small cohort studies [24,25].

18.3   IoT HOSTING MIXED MODE PREHABILITATION MODEL

As discussed in our previous works [37,38], the key positive factors of existing prehabilitation programmes (supervised and unsupervised) were extracted to design a mixed mode prehabilitation programme. IoT special design techniques were developed to host this model [38]. Figure 18.5 shows that the three main IoT parts (wearable sensor device WSD, gateway, and cloud (ThingSpeak (TS)) are integrated to form the cyber physical system, which provides the framework for the mixed mode prehabilitation model. The WSD is normally attached to the participant's (or patient's) ankle, and the gateway could be either stationary (located at home, gym, or physiotherapy centre) or mobile (carried by the participant while performing the physical exercise). Processed WSD data is transmitted to the gateway for further processing and calculations, then communicated to the Cloud TS platform for long-term repository, visualisation and further analysis [13].

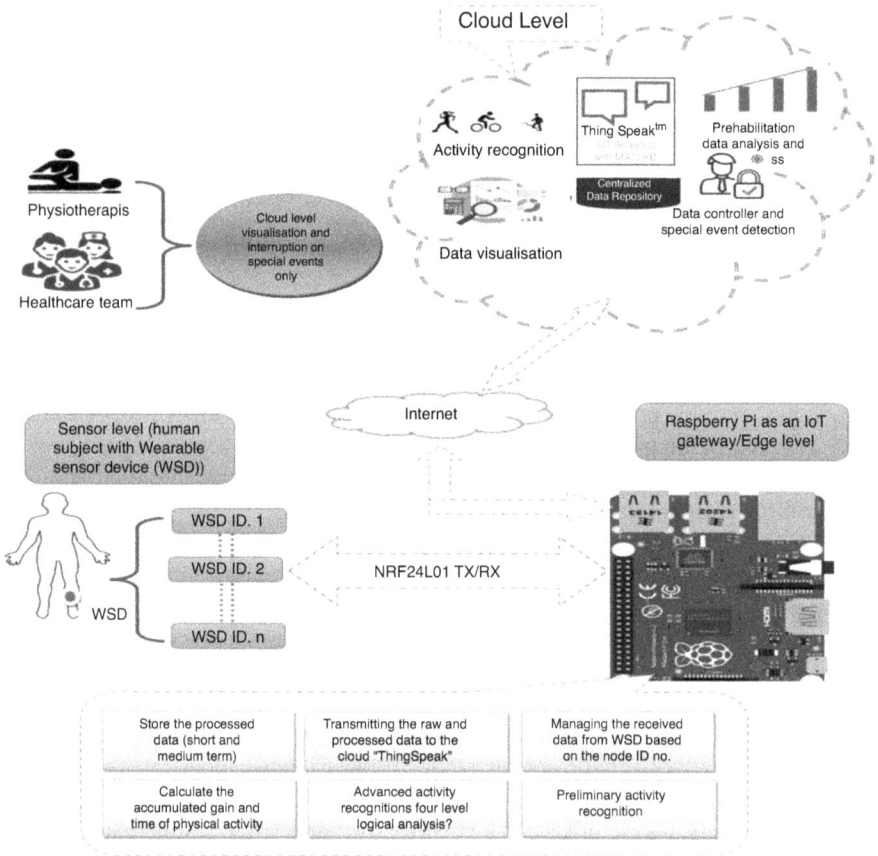

FIGURE 18.5   IoT architecture for cyber-physical system that supports the Mixed Mode Prehabilitation programme.

The architecture proposed for the IoT-compatible wearable prehabilitation exercise activity monitoring system is shown in the figure below. Architectural features utilise computational resources on three main levels. These are the wearable wireless sensing level, the IoT gateway (or internet edge) level, and the cloud level. Each of these steps plays an important role in facilitating the smooth operation of the entire remote monitoring process of the patient's prehabilitation.

The proposed wearable wireless activity tracker includes sensor modules that are responsible for detecting human movement in real time. Here, the tracker provides four main functionalities, namely, sensor data acquisition (including sensor selection, sample rate, and acquisition duration), data storage (short-term data storage for motion detection purposes and long-term backup data storage), data processing (including data calibration and FFT signal processing), and data communication (to regulate message and data communication patterns). Embedded computing plays an important role in the WSD (such as FFT techniques) that reduces the amount of transmitting data to the upper level and reduces the communication overheads. The above points have been considered in the conceptual design of the WSD.

Gateways (e.g., Raspberry Pi, laptop, or smart mobile devices) can process one or more wearable sensors that are involved in one or more types of detection. These can be multiple wireless portable devices used by one user or multiple users. The gateway has three main functions:

1. Data communication with WSD on one side and the internet on the other. In this context, the gateway acts as a protocol converter from the WSD protocol to the internet connectivity-related protocol and vice versa.
2. Multi-stages of activity recognition and prehabilitation efforts calculations.
3. Data storage for short- and medium-term repository of movement and processed data. Data can be managed with Mongo DB and MySQL databases.

At the cloud level, the platform should use both user interactions and higher-level data analytics throughout the pre-approval process. In addition, long-term process data and event monitoring take place in real time during the entire prehabilitation cycle. Key functions occurring at the cloud level include an HTTP protocol for gateway device communication for transferring data from a gateway to the cloud, and a cloud data repository that consists of a health and knowledge data repository. On the other side, the long-term data analysis, deep learning, alert messaging, and big data management are all considered in the conceptual design during the selection of the cloud services. The support of the data available in the cloud repository is intended for long-term storage and further analysis. In addition, the classification of movement activities, the detection of prehabilitation progress model events and the creation of various screens for data visualisation also take place at the cloud level.

18.3.1 Participants and Data Collection Method

Forty-three participants were involved in this study, which was approved by the Auckland University of Technology Ethics Committee (AUTEC reference number 19/212). The age of the participants ranged from 20 to 91 years. One of this study's

TABLE 18.1
Summary of Participants in This Study by Age, Gender, and Health Status [10]

Gender	Age Range	Health Condition	Participant Nos
Male	20–27	Young healthy	3
Male	48–57	Middle aged healthy	3
Female	44–51	Middle aged healthy	3
Male	65–91	Elderly healthy	11
Female	65–83	Elderly healthy	8
Male	65–81	Elderly abdominal health condition (cancer patient)	9
Female	60–78	Elderly abdominal health condition (cancer patient)	6

goals was to recruit a broader age range of participants for a variety of reasons, including research data validation, a lack of cancer patients volunteering, and enriching data collection to build a strong database that could cover a broader range of people.

However, for the purpose of this study, participants were divided into three age groups (young, middle-aged, and elderly), with most participants being above 65 years of age to represent the demographics of people undergoing major abdominal surgery. A group of male and female participants with health conditions awaiting abdominal surgery also participated in the study, as shown in Table 18.1.

Over 60% of the participants did not complete the prehabilitation programme (some dropped out of the programme and others came to the AUT physiotherapy gymnasium a few times only), which was intended to last for 4–6 weeks. The shortest participation period was two sessions of 35–60 minutes per week or over 2 weeks. Approximately 90% of participants performed the physical activities using the AUT physiotherapy gymnasium, while the other 10% of the participants were already involved in a special AUT exercise programme for people over 60, called "never2old." The prescribed physical activities during each session were performed under direct supervision. For the data processing and data transferring to the cloud, a standby base station gateway was installed at the gymnasium, as described in the previous works [37,38]. The initial session used a general-purpose database developed for activity detection for all participants. This was followed by a subsequent personal database that had been established for each participant. Finally, additional outdoor sessions as well as home-based sessions were added to the AUT gymnasium sessions for validation purposes. Twelve percentage of the total participants were involved in these additional activities.

18.3.2 ACTIVITY RECOGNITION CHALLENGES

The supervised, unsupervised, and mixed mode prehabilitation programmes were built on different types of physical activities, intensities, durations, and frequencies. However, one of the main challenges facing the remote monitoring of the prehabilitation programme is activity recognition. Accordingly, the percentage of activities recognised is considered the most important factor in evaluating the system's performance. A special technique was applied to recognise the different physical activities to validate the data collected from the participants [38]. The technique was used for the recognition of the nine common physical activities (walking, running, treadmill, cross trainer, cycling, rowing, step up and staircase ascending and descending) prescribed for the prehabilitation of patients undergoing major abdominal surgery [37,38]. However, in this study, special artificial intelligence (AI) based on knowledge base and logical approaches has been explored to enhance activity recognition and credited physical effort calculations. To reduce the impact of overlaps of power spectrum frequency and amplitude measures among the recognised activities, the extracted recognised activities are examined using four simple logical methods based on short-term history data. The first recognition method is a direct method (M1) and is based on the direct preliminary activity recognition data, which are stored in a separate table as an initial database [38]. Every 4 seconds of incoming data (frequency and amplitude (intensity) measures) are compared with a relational table comprising activity codes representing average frequency and intensity values for each exercise, as explained in our previous works [9,10,37,38]. The second (M2) and third (M3) recognition methods are similar. Both are based on estimating recognition consistency by examining a short-term history of the data recognised using the first method [38]. Each minute of incoming data is stored in a temporal buffer used to identify the majority of occurrences of specific activities detected via the first method. If the consistency of a particular activity is more than a certain percentage of the total activity, then all the data throughout that minute are credited as the dominant single activity. If this consistency is not achieved, the system shows nonspecific activity (NSA). The third method is similar to the second, with an additional moving one-minute window. This improves the recognition timing precision by covering the transition time between activities. The fourth method (M4) keeps track of the type of recognised physical movement over a predefined span of time. When the time expires, it checks on the consistency of the recognition through identifying the movement represented in the majority of the instances and removes the outliers. Following recognition, the output results of all the four methods are stored in a Comma Separated Values (CSV) file for further analysis. The system stores the full summary of incoming and processed data for the full prehabilitation period of each patient. Accordingly, there are three distinct database types that are subjected to the same techniques for movement outlier detection and correction.

The first category is the personalised database which contains training data taken from the same individual while engaging in various physical activities. Reference identifiers extracted from these training data are then used for the four methods of activity recognition [38]. The second database type is based on multiple factors (age, health condition, and fitness level). Each group has a common database. In this stage, the categorisation is based on age groups due to the lack of participants "two groups each four participants."

The third type of database is a general database (non-personalised) that covers all ages and conditions and is developed from physical activity information gathered from all participants. A common value for each physical activity was calculated from the group data and compiled in a shared (non-personalised) database and then

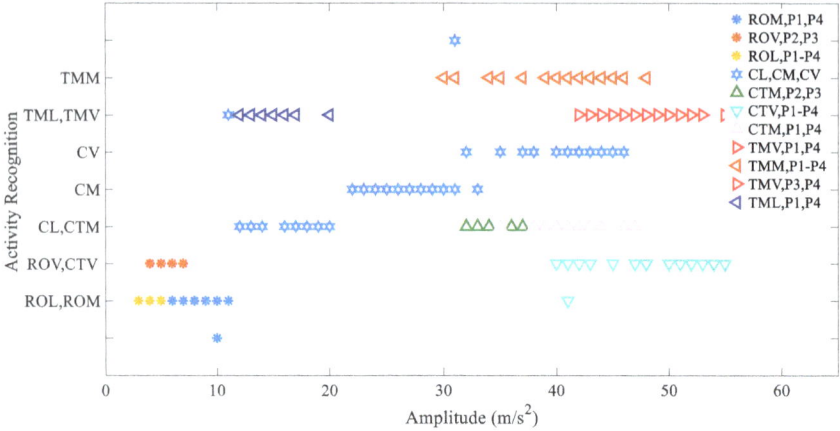

FIGURE 18.6   Personalised activity recognition from four participants performing the same activity at varying times [10].

utilised as a point of reference for new users when they began undertaking various physical activities. Analysis was performed on four participants of each age group who conducted the same physical activities on the same pieces of fitness equipment three times in the same environment.

Figure 18.6 depicts the four participants (P1, P2, P3, and P4) performing treadmill (TM), cycling (CL), rowing (RO), and cross trainer (CT) activities at varied intensities on different days. In the initial session, all participants performed the same four activities, and their performance data was stored in the system as a reference for each participant (personalised database). The measurement precision percentages of the activity recognition for the same person ranged from 70% in TM to more than 95% in CL and CT activities.

Where: CL, CM and CV are cycling low, moderate and vigrous intensity, respectively; ROL, ROM and ROV are rowing low, moderate and vigrous intensity, respectively, TML, TMM and TMV are treadmil light, moderate and vigrious intensity, respectively, and CTM is cross trainer moderate intensity.

The limited participant number (four participants across four physical activities) meant the categorised database (data collected from a group of people that share commmon features such as body mass index (BMI), age group, fitness level, and so on) was created based on age group only, rather than fitness level. Accordingly, the categorised database was selected for participants aged 65 and above in this test (elderly people only). Therefore, the recognition percentages in the categorised database are showing lower percentages of regonition in some activities such as treadmill and walking. Data was also analysed using the same methods from another four participants aged 48–73 years, who performed an additional four physical activities (walking, leg press, step up and ascending and descending a staircase) at different exercise intensities. The outcomes of both groups for the eight physical activities can be seen in Table 18.2 and Figure 18.7. The table shows the comparison between activity recognition percentages in personalised, categorised and non-personalised

TABLE 18.2

Percentage of Recognition of Each Activity for Personalised, Categorised, and Non-Personalised Data

Physical Activity	Personalised (%)	Categorised (%)	Non-Personalised (%)
Cycling (C)	94	88.5	85
Cross Trainer (CT)	90.5	80.7	76
Rowing (RO)	90	84	81
Leg Press (LP)	90	81	78
Treadmill Walking (TM)	82	69	67
Walking (W)	73.5	58	56
Staircase (STA)	30–45	28–38	25–35
Step Up (SUP)	35–45	31–41	25–35

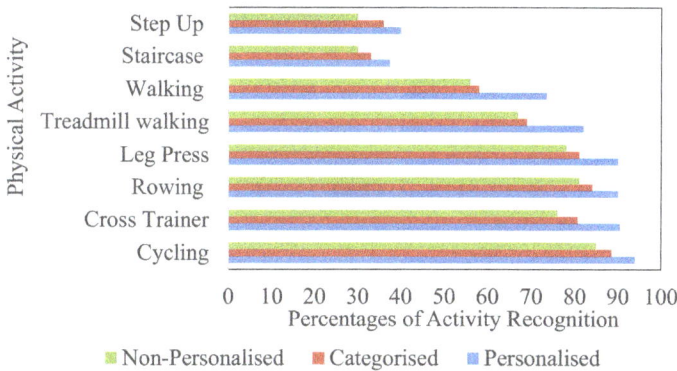

FIGURE 18.7 Histogram of eight activity recognitions based on different types of databases.

databases of eight different physical activities. Figure 18.7 shows the percentages of activity recognitions against different databases.

The physical activities illustrated in the above table can be divided into two groups. The first group represents those activities (cycling, cross trainer, rowing and leg press) where a high percentage of the activity recognition was evident when using both personalised and non-personalised databases (Table 18.2). The high level of recognition in this group of activities was because each activity had minimal overlap in amplitude and frequency measures, and the participant was in a fixed position (sitting on a bike saddle). This group of activities was further classified into two categories based on the movement patterns associated with them. Cycling and cross-trainer are classified as the first category, where the participants were asked to perform the three levels of intensity based on velocity measures of 50, 70 and 90 rpm, which represented intensities of light, moderate and vigorous, respectively. Rowing and leg press are classified as the second category, where the movement velocity was similar for each intensity level, resulting in relatively stable accelerometery parameters (frequency and amplitude) across intensity levels. All four recognition methods achieved high recognition percentages for both cycling and cross-trainer. In contrast, there is a slight difference in the analysis of the second category (rowing and leg press), as the range of movement of the lower limbs was relatively small (around 1 m), producing low amplitude and corresponding frequency measures. In addition, the repetition in the movement velocity will produce consistency in the extracted accelerometer components as well.

The second group represented the physical activities with the lowest percentage of recognition and included staircase, step up, walking and treadmill walking. The lower percentage of recognition for these four activities could be due to having more overlap with other physical activities and whole human body movements with different cadences. Land-based and treadmill walking also showed low recognition rates, particularly when using the non-personalised recognition system. This is probably due to factors such as the individual and age-related differences in gait parameters (step rate and foot position at heel strike) that can influence accelerometery readings [39]. In addition, the participants did not necessarily walk at the same speed for a given intensity during the test, which leads to variation in the output data that may have a negative effect on the recognition percentages [39]. The other two physical activities with low recognition percentages (staircase and step up) present a clear example of inconsistency and overlapping.

18.4   MIXED MODE PREHABILITATION PROGRAMME RESULTS

Forty-three participants, aged between 20 and 91 years old, participated in this research. Participants had a variety of activity levels, and 17 of those had health conditions (abdominal cancer patients). Only 17 people completed the typical period (4–6 weeks) for the prehabilitation programme. This was due to factors related to geographical location and patient health status as well as the COVID-19 pandemic restrictions. The remaining participants performed the intervention activities over a period of 1–3 weeks. Participants were categorised into two groups. The first group consisted of healthy young and elderly participants, whereas the second group consisted of elderly abdominal cancer patients.

18.4.1   Supervised Prehabilitation

Both groups performed their exercise intervention at the AUT physiotherapy gymnasium. Participants joined the programme at varying dates based on their availability, timetable, and referral date from the hospital. Participants attended exercise sessions two to four times per week. Table 18.3 illustrates the number of participants, physical exercises, and intensity of each group. Each exercise session was supervised by a researcher and a physiotherapist and lasted between 35–55 minutes.

The direct supervision ensured that the participants performed the physical activity as prescribed and enabled a comparison between the system activity recognition and performed physical activity for validation purposes. The 4-week prehabilitation programme was selected instead of 6 weeks to ensure that all participants/patients had the same time frame. The setting for the ideal target gain for every participant was 20 points for the 4-week programme. There was no substantial difference

TABLE 18.3

Participant Numbers and Group Description, and the Type and Intensity of the Physical Activities Undertaken at the Gymnasium

Health Condition	Physical Exercises	Intensity	Participants
Young and middle-aged healthy fit	Cycling, treadmill, rowing, leg press, cross trainer, and walking	M, V	3
Elderly healthy fit	Cycling, treadmill, rowing, leg press, and walking	M,V	6
Elderly healthy unfit	Cycling, treadmill, rowing, and walking	L, M, V	4
Elderly abdominal health condition	Cycling, treadmill, rowing, leg press, and walking	L, M, V	4

Where L, M, and V are the light, moderate, and vigorous intensities respectively.

among the first three recognition logical methods compared with M4, so only M1 and M4 were selected for comparisons with the idle target gain. Furthermore, for equity purposes, only 110 minutes per week (two 55-minute sessions) of physical activities performed at various intensities were considered for the analysis. Data from two sessions per week were selected because most participants attended two sessions per week and two separate sessions of data provided more clarity for graphing results. Therefore, each subject performed 440 minutes of physical activities at varying intensities for a 4-week prehabilitation programme. This equated to a total gain of 20 points throughout the 4 weeks. The reason for setting the 20 points gain credit target for all participants was because these 20 points equated to weekly exercise recommendations for improving fitness [37], additionally, to validate if the target was feasible for healthy individuals and participants who had abdominal cancer with relatively low levels of fitness.

Figure 18.8a and b depicts the accumulated time and gain of fit young and elderly healthy participants, respectively. The young and older healthy participants were able to perform physical activities at high intensities and achieve a high credit gain of 90%–115% relative to the ideal target gain. However, the second group of cancer patients were only able to achieve between 65% and 85% of the target gain. The main reason for missing the 20-point credit target was because this group of participants were not able to achieve the specific intensity of the activity due to fitness issues.

Figure 18.9a and b demonstrates an example of 4 weeks of prehabilitation exercises for a healthy elderly participant and an abdominal cancer patient who were classified as having poor aerobic fitness via cardiopulmonary exercise testing. Figure 18.9a shows a healthy elderly patient who performed a 4-week prehabilitation programme and had an accumulated gain of 105% based on M4 and slightly less than 100% of the target gain when using M1 calculations. In contrast, the elderly unfit

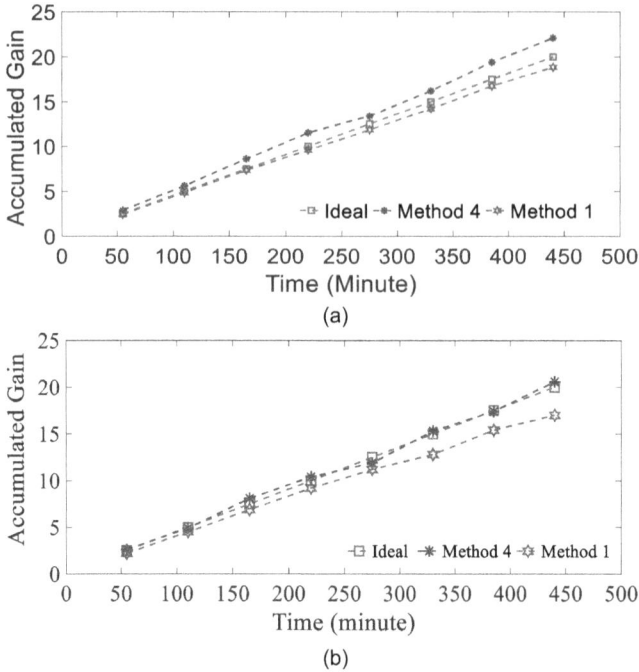

FIGURE 18.8 Accumulated gain versus time for (a) healthy young, and (b) healthy elderly participants [10].

abdominal cancer patient was unable to reach the 20-point target gain and only achieved 65%–77% of the ideal gain. This would indicate that programmes, and the associated points gain, may have to be tailored for patients who have different levels of health and fitness. It would seem that healthy participants were better conditioned to cope with higher intensity activity than the cancer patients. Therefore, some adjustments would need to be made by the physiotherapist or clinician for the patients with low fitness levels.

18.4.2 UNSUPERVISED (HOME-BASED) PREHABILITATION

Three participants completed 2 weeks of an unsupervised (home-based) prehabilitation programme: one elderly abdominal cancer patient waiting for the scheduled surgery and two middle-aged healthy people. The patients conducted three sessions per week, with each session being a period of 25–30 minutes (two sessions in the local gym and one outdoor session), and the total accumulated time was ~150 minutes, with varying intensity levels. The outdoor activity only involved walking at varying intensities. The other two participants continued for 2 weeks, doing six sessions (three sessions per week, totalling 100–130 minutes). Their activity consisted of three sessions of 45–60 minutes at the gym (cycling, treadmill, cross trainer, and rowing) at varying intensity levels, and three additional sessions in outdoor environments (walking at varied intensities) and each session lasting between 30 and 45 minutes.

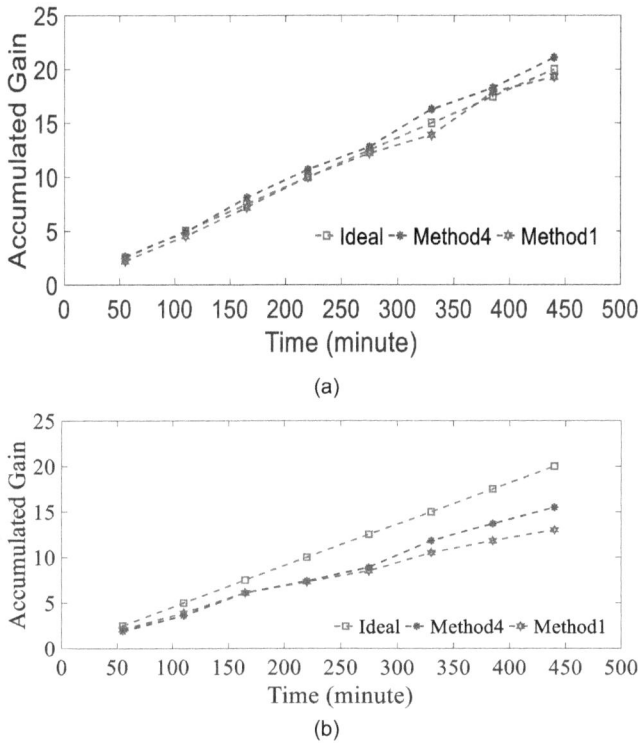

FIGURE 18.9 Accumulated time and gain of (a) a fit elderly patient, and (b) an unfit elderly patient [10].

The portable Raspberry Pi type zero (RPiZ) and a WSD shown in Figure 18.10 were given to all three participants and joined to the local internet connection for the RpiZ. A fabric strip band to fixate the WSD on the ankle was given to all participants with a USB cable for the WSD for battery charging. Specific TS channels were assigned to each participant, preventing data from being mixed up with other participants.

Real-time processed WSD data transmitting to the RpiZ for activity recognition, short-term storage data and further calculations, as shown in Figure 18.11 below, then uploaded to the cloud TS for long-term repository, visualisation, and further analysis.

During the first lab session with the AUT physiotherapist, the preliminary data for the first and second participant were available, but the data for the third participant were not. As a result, the personalised database was used for the first and second participants, while the categorised database served as a reference for the third participant for activity recognition.

Figure 18.12a and b shows how the system was able to collect data from various activities, then store and analyse the offline data in the gateway, and push the data to the cloud TS. Figure 18.13a and b depicts a snapshot at various points during physical activity (Figure 18.13).

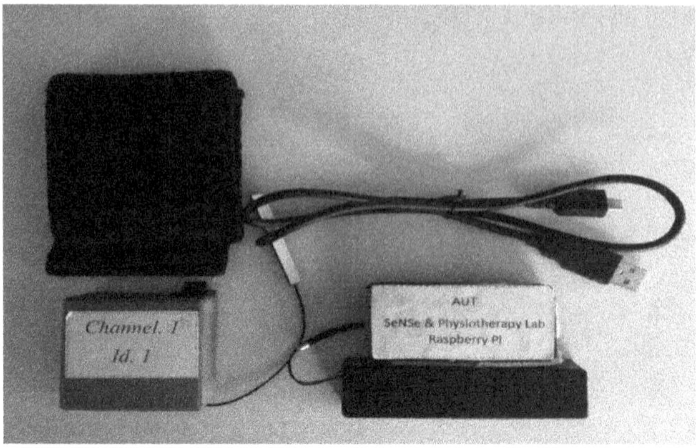

FIGURE 18.10   Portable RPiZ and WSD.

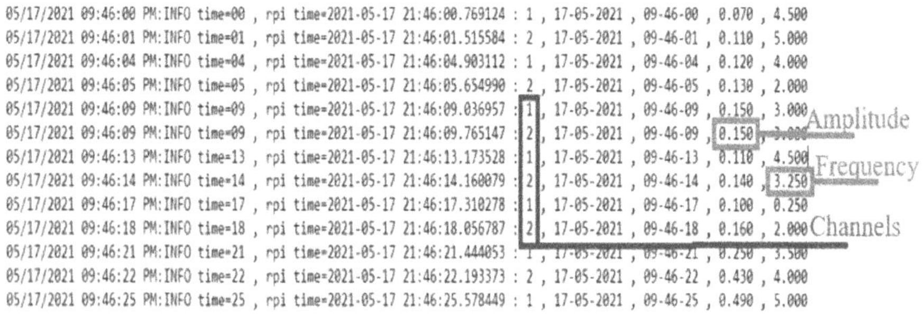

```
05/17/2021 09:46:00 PM:INFO time=00 , rpi time=2021-05-17 21:46:00.769124 : 1 , 17-05-2021 , 09-46-00 , 0.070 , 4.500
05/17/2021 09:46:01 PM:INFO time=01 , rpi time=2021-05-17 21:46:01.515584 : 2 , 17-05-2021 , 09-46-01 , 0.110 , 5.000
05/17/2021 09:46:04 PM:INFO time=04 , rpi time=2021-05-17 21:46:04.903112 : 1 , 17-05-2021 , 09-46-04 , 0.120 , 4.000
05/17/2021 09:46:05 PM:INFO time=05 , rpi time=2021-05-17 21:46:05.654990 : 2 , 17-05-2021 , 09-46-05 , 0.130 , 2.000
05/17/2021 09:46:09 PM:INFO time=09 , rpi time=2021-05-17 21:46:09.036957 : 1 , 17-05-2021 , 09-46-09 , 0.150 , 3.000
05/17/2021 09:46:09 PM:INFO time=09 , rpi time=2021-05-17 21:46:09.765147 : 2 , 17-05-2021 , 09-46-09 , 0.150 , 3.000
05/17/2021 09:46:13 PM:INFO time=13 , rpi time=2021-05-17 21:46:13.173528 : 1 , 17-05-2021 , 09-46-13 , 0.110 , 4.500
05/17/2021 09:46:14 PM:INFO time=14 , rpi time=2021-05-17 21:46:14.160079 : 2 , 17-05-2021 , 09-46-14 , 0.140 , 3.250
05/17/2021 09:46:17 PM:INFO time=17 , rpi time=2021-05-17 21:46:17.310278 : 1 , 17-05-2021 , 09-46-17 , 0.100 , 0.250
05/17/2021 09:46:18 PM:INFO time=18 , rpi time=2021-05-17 21:46:18.056787 : 2 , 17-05-2021 , 09-46-18 , 0.160 , 2.000
05/17/2021 09:46:21 PM:INFO time=21 , rpi time=2021-05-17 21:46:21.444053 : 1 , 17-05-2021 , 09-46-21 , 0.250 , 3.500
05/17/2021 09:46:22 PM:INFO time=22 , rpi time=2021-05-17 21:46:22.193373 : 2 , 17-05-2021 , 09-46-22 , 0.430 , 4.000
05/17/2021 09:46:25 PM:INFO time=25 , rpi time=2021-05-17 21:46:25.578449 : 1 , 17-05-2021 , 09-46-25 , 0.490 , 5.000
```
Amplitude
Frequency
Channels

FIGURE 18.11   Raspberry Pi logfile showing the transmitting/receiving date and time, node ID, amplitude, and frequency.

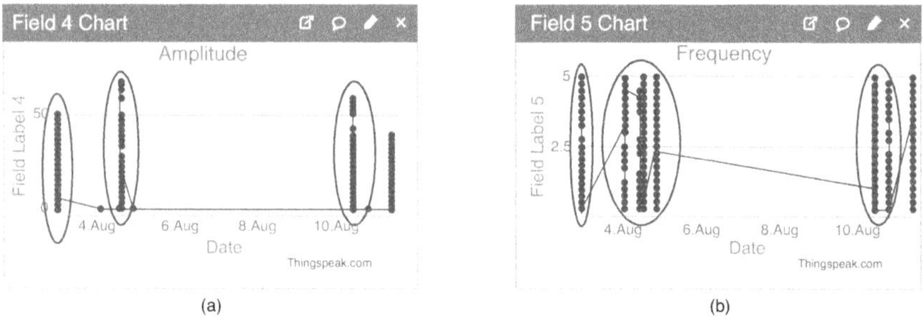

(a)   (b)

FIGURE 18.12   (a) Amplitude. (b) Frequency for three sessions at different dates and times, respectively [10].

The system recognised 82% of the actual activities that were performed in the gym, while this value was around 68% for outdoor activities (walking at varying intensities), due to the overlapping between walking and some other activities. The percentage of the ideal gain for the three participants was 78%, 72%, and 68%, respectively. These percent values show that the design and application of the IoT

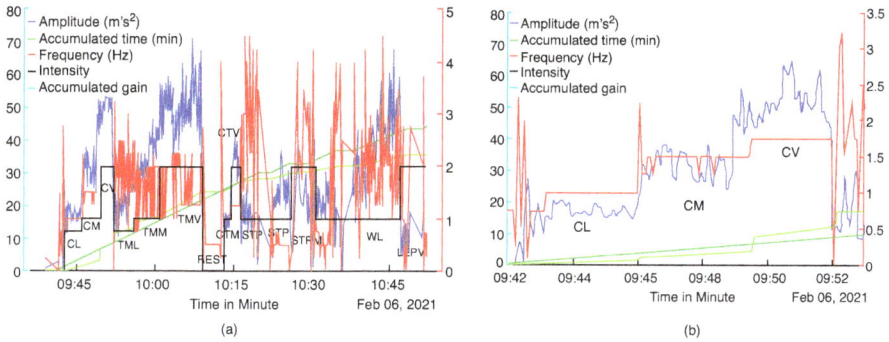

FIGURE 18.13   (a) Retracing data and applying different analysis methods. (b) Time windows for cycling activity at low, moderate, and vigorous intensity [10,38].

remote monitoring system can support a mixed mode prehabilitation programme in terms of real-time data collection, processing, visualisation, and programme follow-up. This, in effect, would reduce barriers to the prehabilitation required by practices and programme implementation.

18.5   CONCLUSION

The test conducted on the designed cyber-physical system demonstrated the ability to support a mixed mode prehabilitation programme using both supervised indoor and unsupervised outdoor implementation. This has been achieved through the use of a wearable device supported by a portable internet gateway. The approach could minimise prehabilitation barriers, in particular, for remote and limited resources areas.

Furthermore, the findings of the experiments show how personalisation on and non-personalisation logical analysis of movement dynamics encourages alignment with reality. The test carried out using seven activities shows that the system is able to recognise more than 70% of personalised data, while the percentages were considerably <55% for non-personalised data. This highlights the importance of establishing a personalised database for monitoring key prehabilitation activities.

Further improvement in movement recognition may take place through the use of deep learning and AI. AI, through the use of neural networking for deep learning, could aid in training the subject database, transitioning it from non-personalised to categorised, and then to a personalised database, without the need for human intervention. This would mean that each time a subject performs a prescribed physical activity, the system starts a comparison with the previous database. This information could be used to generate an algorithm that would then be able to predict a new database based on the training data, and the database would be updated accordingly. The integration of AI into the system could potentially reduce the need for healthcare worker or technician intervention to update the database and decrease unnecessary health centre visits. The challenge here is to efficiently utilise the resources of the portable gateway for embedding the deep learning algorithm.

18.6 ETHICAL APPROVAL

This study was approved by the Auckland University of Technology Ethics Committee (AUTEC reference number 19/212).

REFERENCES

1. C. Grimmett et al., "The role of behavioral science in personalized multimodal prehabilitation in cancer," *Frontiers in Psychology,* vol. 12, p. 261, 2021.
2. R. Costilla, M. Tobias, and T. Blakely, "The burden of cancer in New Zealand: A comparison of incidence and DALY metrics and its relevance for ethnic disparities," *Australian and New Zealand Journal of Public Health,* vol. 37, no. 3, pp. 218–225, 2013.
3. T. G. Weiser et al., "An estimation of the global volume of surgery: A modelling strategy based on available data," *The Lancet,* vol. 372, no. 9633, pp. 139–144, 2008.
4. V. Wynter-Blyth and K. Moorthy, "Prehabilitation: Preparing patients for surgery," *BMJ: British Medical Journal (Online),* vol. 358, 2017, doi: https://doi.org/10.1136/bmj.j3702.
5. V. A. Lawrence et al., "Functional independence after major abdominal surgery in the elderly," *Journal of the American College of Surgeons,* vol. 199, no. 5, pp. 762–772, 2004.
6. J. Moore et al., "Implementing a system-wide cancer prehabilitation programme: The journey of greater Manchester's 'Prehab4cancer'," *European Journal of Surgical Oncology,* vol. 47, no. 3, pp. 524–532, 2021.
7. F. Carli and G. S. Zavorsky, "Optimizing functional exercise capacity in the elderly surgical population," *Current Opinion in Clinical Nutrition & Metabolic Care,* vol. 8, no. 1, pp. 23–32, 2005.
8. E. McAuley, A. Szabo, N. Gothe, and E. A. Olson, "Self-efficacy: Implications for physical activity, function, and functional limitations in older adults," *American Journal of Lifestyle Medicine,* vol. 5, no. 4, pp. 361–369, 2011.
9. K. Al-Naime, A. Al-Anbuky, and G. Mawston, "IoT based pre-operative prehabilitation program monitoring model: Implementation and preliminary evaluation," In *2022 4th International Conference on Biomedical Engineering (IBIOMED),* Yogyakarta, 2022, pp. 24–29. doi: 10.1109/IBIOMED56408.2022.9988432
10. K. A. M. Al-Naime, "Cyber physical system for pre-operative patient prehabilitation," Auckland University of Technology, 2022.
11. T. Christensen, T. Bendix, and H. Kehlet, "Fatigue and cardiorespiratory function following abdominal surgery," *Journal of British Surgery,* vol. 69, no. 7, pp. 417–419, 1982.
12. W. L. Haskell et al., "Physical activity and public health: Updated recommendation for adults from the American College of Sports Medicine and the American Heart Association," Circulation, vol. 116, no. 9, p. 1081, 2007.
13. S. P. Mullen, E. McAuley, W. A. Satariano, M. Kealey, and T. R. Prohaska, "Physical activity and functional limitations in older adults: The influence of self-efficacy and functional performance," Journals of Gerontology Series B: Psychological Sciences and Social Sciences, vol. 67, no. 3, pp. 354–361, 2012.
14. J. Quadrilatero and L. Hoffman-Goetz, "Physical activity and colon cancer," *Journal of Sports Medicine and Physical Fitness,* vol. 43, pp. 121–38, 2003.
15. D. Santa Mina, C. Scheede-Bergdahl, C. Gillis, and F. Carli, "Optimization of surgical outcomes with prehabilitation," *Applied Physiology, Nutrition, and Metabolism,* vol. 40, no. 9, pp. 966–969, 2015.

16. L. K. Sprod et al., "Exercise and cancer treatment symptoms in 408 newly diagnosed older cancer patients," *Journal of Geriatric Oncology,* vol. 3, no. 2, pp. 90–97, 2012.

17. C. Stevinson, D. A. Lawlor, and K. R. Fox, "Exercise interventions for cancer patients: Systematic review of controlled trials," *Cancer Causes & Control,* vol. 15, no. 10, pp. 1035–1056, 2004.

18. L. Y. Blackman, "Lifestyle physical activity among older adults: The health implications," The University of Utah, 2017.

19. D. E. Warburton, C. W. Nicol, and S. S. Bredin, "Health benefits of physical activity: The evidence," *CMAJ,* vol. 174, no. 6, pp. 801–809, 2006.

20. B. C. Center, "Abdominal Cancer," (in English), 2022.

21. N. H. Jessen et al., "Cancer suspicion, referral to cancer patient pathway and primary care interval: A survey and register study exploring 10 different types of abdominal cancer," *Family Practice,* vol. 38, no. 5, pp. 589–597, 2021.

22. M. M. Koo et al., "The nature and frequency of abdominal symptoms in cancer patients and their associations with time to help-seeking: Evidence from a national audit of cancer diagnosis," *Journal of Public Health,* vol. 40, no. 3, pp. e388–e395, 2018.

23. I. Dimopoulou et al., "Pituitary-adrenal responses following major abdominal surgery," *Hormones,* vol. 7, no. 3, pp. 237–242, 2008.

24. F. Carli et al., "Randomized clinical trial of prehabilitation in colorectal surgery," *Journal of British Surgery,* vol. 97, no. 8, pp. 1187–1197, 2010.

25. T. Smith, "A long way from home: Access to cancer care for rural Australians," *Radiography,* vol. 18, no. 1, pp. 38–42, 2012.

26. R. J. Copeland et al., "Psychological support and behaviour change interventions during the perioperative period for people with a cancer diagnosis; Consensus statements for use from Macmillan, The Royal College of Anaesthetists and the National Institute for Health Research," 2020.

27. R. U. Newton, D. R. Taaffe, S. K. Chambers, N. Spry, and D. A. Galvão, "Effective exercise interventions for patients and survivors of cancer should be supervised, targeted, and prescribed with referrals from oncologists and general physicians," *Journal of Clinical Oncology,* vol. 36, no. 9, pp. 927–928, 2018.

28. M. West et al., "Effect of prehabilitation on objectively measured physical fitness after neoadjuvant treatment in preoperative rectal cancer patients: A blinded interventional pilot study," *British Journal of Anaesthesia,* vol. 114, no. 2, pp. 244–251, 2015.

29. C. J. Molenaar, N. E. Papen-Botterhuis, F. Herrle, and G. D. Slooter, "Prehabilitation, making patients fit for surgery-a new frontier in perioperative care," *Innovative Surgical Sciences,* vol. 4, no. 4, pp. 132–138, 2019.

30. D. Martin et al., "Feasibility of a prehabilitation program before major abdominal surgery: A pilot prospective study," *Journal of International Medical Research,* vol. 49, no. 11, p. 03000605211060196, 2021.

31. F. Wu, O. Rotimi, R. Laza-Cagigas, and T. Rampal, "The feasibility and effects of a telehealth-delivered home-based prehabilitation program for cancer patients during the pandemic," *Current Oncology,* vol. 28, no. 3, pp. 2248–2259, 2021.

32. P. A. T. Collaborative et al., "SupPoRtive Exercise Programmes for Accelerating REcovery after major ABdominal Cancer surgery trial (PREPARE-ABC): Pilot phase of a multicentre randomised controlled trial," *Colorectal Disease,* vol. 23, no. 11, pp. 3008–3022, 2021.

33. C. Gillis et al., "Prehabilitation versus rehabilitation: A randomized control trial in patients undergoing colorectal resection for cancer," *Anesthesiology,* vol. 121, no. 5, pp. 937–947, 2014.

34. C. Li et al., "Impact of a trimodal prehabilitation program on functional recovery after colorectal cancer surgery: A pilot study," *Surgical Endoscopy,* vol. 27, no. 4, pp. 1072–1082, 2013.

35. V. Ferreira et al., "Maximizing patient adherence to prehabilitation: What do the patients say?" *Supportive Care in Cancer,* vol. 26, no. 8, pp. 2717–2723, 2018.
36. E. Karlsson et al., "Feasibility of preoperative supervised home-based exercise in older adults undergoing colorectal cancer surgery: A randomized controlled design," *PLoS One,* vol. 14, no. 7, p. e0219158, 2019.
37. K. Al-Naime, A. Al-Anbuky, and G. Mawston, "Human movement monitoring and analysis for prehabilitation process management," *Journal of Sensor and Actuator Networks,* vol. 9, no. 1, p. 9, 2020.
38. K. Al-Naime, A. Al-Anbuky, and G. Mawston, "Remote monitoring model for the preoperative prehabilitation program of patients requiring abdominal surgery," *Future Internet,* vol. 13, no. 5, p. 104, 2021.
39. C. Price, T. Schmeltzpfenning, C. J. Nester, and T. Brauner, "Foot and footwear bio-mechanics and gait," In A. Luximon (Ed.), *Handbook of Footwear Design and Manufacture.* Elsevier: Amsterdam, Netherlands, 2021, pp. 79–103.

For Product Safety Concerns and Information please contact our EU
representative GPSR@taylorandfrancis.com
Taylor & Francis Verlag GmbH, Kaufingerstraße 24, 80331 München, Germany

www.ingramcontent.com/pod-product-compliance
Lightning Source LLC
Chambersburg PA
CBHW060425220326
41598CB00021BA/2291

* 9 7 8 1 0 3 2 3 8 7 6 9 7 *